MOLECULAR BASIS OF ION CHANNELS AND RECEPTORS INVOLVED IN NERVE EXCITATION, SYNAPTIC TRANSMISSION AND MUSCLE CONTRACTION

In Memory of Professor Shosaku Numa

ANNALS OF THE NEW YORK ACADEMY OF SCIENCES
Volume 707

MOLECULAR BASIS OF ION CHANNELS AND RECEPTORS INVOLVED IN NERVE EXCITATION, SYNAPTIC TRANSMISSION AND MUSCLE CONTRACTION

In Memory of Professor Shosaku Numa

Edited by Haruhiro Higashida, Tohru Yoshioka, and Katsuhiko Mikoshiba

The New York Academy of Sciences
New York, New York
1993

Library of Congress Cataloging-in-Publication Data

Molecular basis of ion channels and receptors involved in nerve excitation, synaptic transmission and muscle contraction: in memory of Professor Shosaku Numa / edited by Haruhiro Higashida, Tohru Yoshioka, and Katsuhiko Mikoshiba.
    p. cm. — (Annals of the New York Academy of Sciences, ISSN 0077-8923 ; v. 707)
    Contains the proceedings of a conference of the New York Academy of Sciences, held at the Ibuka Memorial Hall, Waseda University, Japan, in Jan. 1993. Cf. Pref.
    Includes bibliographical references and index.
    ISBN 0-89766-823-5 (cloth : acid-free paper). — ISBN 0-89766-824-3 (pbk. : acid-free paper)
    1. Ion channels — Congresses. 2. Cellular signal transduction — Congresses. 3. Neurotransmitter receptors — Congresses. 4. Neuromuscular transmission — Congresses. 5. Muscle contraction — Congresses. 6. Numa, Shosaku, 1929- — Congresses. I. Higashida, Haruhiro. II. Yoshioka, Tohru. III. Mikoshiba, Katsuhiko, 1945- . IV. Numa, Shosaku, 1929- . V. Series.
Q11.N5 vol. 707
[QH603.I54]
500 s — dc20
[591.87'5]
                                                                          93-44950
                                                                               CIP

CCP
Printed in the United States of America
ISBN 0-89766-823-5 (cloth)
ISBN 0-89766-824-3 (paper)
ISSN 077-8923

ANNALS OF THE NEW YORK ACADEMY OF SCIENCES

Volume 707
December 20, 1993

MOLECULAR BASIS OF ION CHANNELS AND RECEPTORS INVOLVED IN NERVE EXCITATION, SYNAPTIC TRANSMISSION AND MUSCLE CONTRACTION (IN MEMORY OF PROFESSOR SHOSAKU NUMA)[a]

Editors
HARUHIRO HIGASHIDA, TOHRU YOSHIOKA, and KATSUHIKO MIKOSHIBA

CONTENTS

---

[a] This volume is the result of a conference entitled **Molecular Basis of Ion Channels and Receptors Involved in Nerve Excitation, Synaptic Transmission and Muscle Contraction (In Memory of Professor Shosaku Numa)** held by the New York Academy of Sciences on January 12–15, 1993 in Tokyo

Financial assistance was received from:

Major funders
- Ministry of Education, Science, and Culture of Japan—Monbusho Science Fund/Grant-in-Aid "Nerve Impulse Signaling"
- The Pharmaceutical Manufacturer's Associations of Tokyo and Osaka

Supporters
- Inoue Science Foundation
- Kashima Foundation
- Kowa Life Science Foundation
- Mitsubishi Science Foundation
- Uehara Life Science Foundation
- U.S. Army Medical Research and Development Command

Contributors
- Asahi Glass Ware Foundation
- Axon Instruments, Inc.
- Chubu Electric Power Company/Basic Technology Institute
- The Green Cross Corp.
- Merck, Sharp & Dohme Research Laboratories—Japan
- Monsanto Company
- Nihon Kohden Corporation
- Nissan Science Foundation
- Parke-Davis Pharmaceutical Research
- Pfizer Central Research
- Saibikai Foundation

Foreword

On the First New York Academy of Sciences Conference in Japan

It is said that the New York Academy of Sciences has such a long history and authority that a conference published as an *Annals* of the Academy is referred to as a "bible" by researchers. That such a conference could be held in Japan serves as proof that the academic level of the life sciences in Japan has progressed and that the number of Japanese researchers in this field has increased. As an official in charge of the promotion of science in Japan, I am quite pleased by such developments.

Needless to say, science has no national boundaries. Scientific truth is common property to all mankind and should be shared by everybody. It is also true that for stimulating academic creativity international conferences are extremely important, as researchers of different cultural backgrounds discuss relevant matters and challenge one another intellectually.

Molecular research of ion channels and receptors is one of the key topics in understanding the functions of living organs including the nervous system. This field has grown in importance and many Japanese researchers are now actively involved in related studies. The late Professor Shosaku Numa of Kyoto University contributed much to this field. Unfortunately, he passed away last February. Here, I would like to pay my respects to Professor Numa, and express my hope that many young Japanese researchers will follow in his footsteps and continue research in this field.

To our participants from abroad, I welcome you to Japan. Please make the most of this occasion to understand even a fragment of our culture and society. I heartily hope your stay in Japan will be comfortable and fruitful.

Last but not least, I would like to express my gratitude for the strenuous efforts of those people who have made this conference possible, and I sincerely hope the conference bears much fruit and contributes to scientific development.

YOSHIKAZU HASEGAWA
Director-General
Science and International Affairs Bureau
Ministry of Education, Science and
Culture of Japan

Introduction

HARUHIRO HIGASHIDA

Department of Biophysics
Kanazawa University School of Medicine
Kanazawa, Japan

TOHRU YOSHIOKA

Department of Molecular Neurobiology
School of Human Sciences
Waseda University
Tokorozawa, Saitama, Japan

KATSUHIKO MIKOSHIBA

Department of Molecular Neurobiology
Institute of Medical Science
University of Tokyo
Tokyo, Japan

This volume contains the proceedings of the first conference of the New York Academy of Sciences to be held in Japan. This conference took place at the Ibuka Memorial Hall of Waseda University, one of the oldest universities in Tokyo, and was organized in part by a group of researchers who were supported by the Grant-in-Aid (No. 03225102) for Scientific Research in the Priority Area of "Impulse signaling," from the Japanese Ministry of Education, Science and Culture.

The aim of the conference was to summarize recent advances in our understanding of signal transduction, from the receptors and effectors, in this case the various ion channels of neurons, smooth or striated muscle cells, endocrine cells and other tissues. The 250 participants exposed to the latest work on structure-function relationships of voltage-dependent Na^+, Ca^{2+}, K^+ and Cl^- channels, ligand-gated channels, cGMP-gated channels, intracellular Ca^{2+} channels such as inositol-1,4,5-trisphosphate and ryanodine receptors, and muscarinic and glutamate receptors. The presentations also encompassed diseases causally linked to altered ion channels and receptors, for example muscular dystrophy (linked to the nifedipine-binding Ca^{2+} channel), cystic fibrosis (Cl^- channel), malignant hyperthermia (ryanodine-sensitive Ca^{2+}-release channel), and hyperkalemic periodic paralysis (Na^+ channel). In addition, the characteristics of ion channels and transmitter receptors that underlie their roles in processes as diverse as membrane excitability, memory process, mind, muscle contraction, secretion, and gene expression were also discussed. It is our hope that this volume, which presents the golden fruits of the application of molecular biological techniques to the study of membrane proteins and of signal transduction pathways will provide a valuable resource in which readers may find clues to problems as lofty as the decoding of higher brain function.

The rapid developments in biomedical research during the past 15 years have brought major advances in ligand binding techniques for monitoring receptor sites, and in biophysical techniques and gene engineering for evaluating structure/function relationships of receptors of ion channels. The invention of the patch-clamp technique by Profs. E. Neher and B. Sakmann, for which they were awarded the 1991 Nobel Prize, was revolutionary not merely in the narrow field of classical electrophysiology but also in practically every branch of medical and biological science. Their technique made possible detailed observations of the behavior of single molecules, in the form of ion currents in their natural environment and in real time, a remarkable achievement. The subsequent cloning and sequencing of the DNAs and hence determination of the primary sequence of amino acids in ion channel molecules has likewise had an enormous impact on the way we view single ion channels.

The molecular constituents of ion channels and of receptor/ion channel complexes were first identified by the protein purification techniques of Prof. Jean Pierre-Changeux of the Pasteur Institute and Prof. William Catterall of Washington University. Changeux began the characterization of the nicotinic acetylcholine receptor a long time ago in connection with his and J. Monod's examination of allosteric interactions in protein molecules. Catterall's interest in Na^+ ion channels dates back to his study of neuroblastoma cell lines when he was in Marshall Nirenberg's laboratory at NIH. Subsequently, based on the information gathered from these purified proteins, Prof. Shosaku Numa of Kyoto University and his colleagues succeeded in isolating and expressing complementary DNAs coding for nicotinic ACh receptors and for Na^+ ion channels.

Numa's crystal clear studies have since then been extended to many other channels and receptors. Numa was successful in isolating cDNA clones for nifedipine-binding Ca^{2+} ion channels, ryanodine-binding Ca^{2+}-release channels, cyclic nucleotide-gated cation channels, and muscarinic acetylcholine receptors. FIGURE 1 is taken from his Harvey Lecture at the Rockefeller University in 1989[1] and his Rita Levi-Montalcini Lecture at Georgetown University in 1991.[2] FIGURE 1 shows schematically the two-dimensional topology of ion channel molecules and of GTP binding protein-coupled receptors in lipid bilayers. This volume covers mainly the molecules illustrated in FIGURE 1. Additional topics covered at the conference are various Cl^- ion channels and the ion channel–related diseases.

Although we initially planned this symposium to honor Numa's seminal contributions towards our understanding of muscle and brain at the molecular biological level and to celebrate his retirement at the age of 63 from his position as the chairman of the Departments of Medical Chemistry and Molecular Genetics in the Kyoto University Faculty of Medicine, sadly he passed away on February 15, 1992 (one month before his planned retirement) after battling colon cancer for nearly three years.

Therefore, this symposium and the resulting *Annals* provide a fitting tribute to his memory. Further information on the late Shosaku Numa can be found in his obituary.[3]

FIGURE 2 shows Shosaku Numa relaxing at Cold Spring Harbor, New York, in

I   Voltage-gated channels

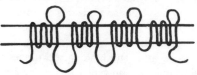

II   Neurotransmitter-gated channels

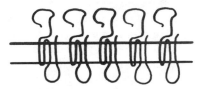

III   Intracellular channels

IV   Cyclic nucleotide-gated channels

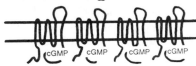

V   G-protein-coupled receptors

FIGURE 1. Schematic representations of members of five different families of neurotransmitter receptors and ion channels. Modified from Numa.[1]

the summer of 1988 where he, Prof. David A. Brown of University College London, and Haruhiro Higashida all attended a symposium on signal transduction.[4,5] Probably, this was the last time that he was in good health because at the end of 1988 his health began to deteriorate.

We would like to acknowledge the advice and encouragement of Professor Emeritus Setsuro Ebashi of the University of Tokyo, Prof. David Gadsby of The Rockefeller University, and Prof. Olaf Andersen of Cornell University Medical College. We greatly appreciate the efficient organizational and editorial work by staff of the New York

FIGURE 2. A snapshot taken at Cold Spring Harbor, New York in 1988. From left, H. Higashida, S. Numa, and D. A. Brown.

Academy of Sciences, especially Ms. Geraldine Busacco, Renée Wilkerson, and Mary K. Brennan. We thank the more than 20 sponsors, including the Tokyo and Osaka Pharmaceutical Manufacturers Associations and the Japanese Ministry of Education, Science and Culture, for their financial support. We thank Ms. Yuhki and Ms. Nakajima of Waseda University and Ms. Yamashita and Ms. Ozasa of Kanazawa University for their secretarial assistance. Finally, we are grateful to all of the domestic and overseas speakers and poster presenters for their scientific contributions to conference.

REFERENCES

1. NUMA, S. 1989. A molecular view of neurotransmitter receptors and ion channels. Harvey Lect. **83:** 121–165.
2. NUMA, S. 1991. Neurotransmitter receptors and ionic channels. From structure to function. Fidia Res. Found. Neurosci. Award Lect. **5:** 23–44.
3. HAYAISHI, O. 1992. Obituary of Shosaku Numa. Trends Biochem. Sci. **17:** 327–328.
4. NUMA, S., K. FUKUDA, T. KUBO, A. MAEDA, I. AKIBA, H. BUJO, J. NAKAI, M. MISHINA & H. HIGASHIDA. 1988. Molecular basis of the functional heterogeneity of the muscarine acetylcholine receptor. Cold Spring Harbor Symp. Quant. Biol. **53:** 295–301.
5. BROWN, D. A., H. HIGASHIDA, P. R. ADAMS, N. V. MARRION, & T. G. SMART. 1988. Role of G-protein-coupled phosphatidylinositide system in signal transduction in vertebrate neurons: Experiments on neuroblastoma hybrid cells and ganglion cells. Cold Spring Harbor Symp. Quant. Biol. **53:** 375–384.

Preface

SETSURO EBASHI

National Institute for Physiological Sciences
Okazaki, Aichi 444 Japan

This symposium is dedicated to the memory of the late Professor Shosaku Numa, who not only pioneered the field expressed by the title of this symposium, but whose accomplishments represent the most important part of that field.

I became acquainted with him when he was working with the late Professor F. Lynen in Munich early in 1960. At that time, I was in a state of excitement because of the success I had achieved in the exploration of the role of Ca^{2+} in muscle contraction. This helped me to overcome my initial hesitation in making a new friend.

Thereafter I always found him gentle, polite, and cooperative—a man who spoke and acted with sincerity and honesty. To me he was a person very different from the "iron man" that some perceived in his last years. I do not deny that his incredible effort to advance scientific knowledge was motivated by his strong competitive mind and his zeal for victory. But he was a devoted missionary of a religion called "science."

I am an old-timer and am ignorant about the recent findings concerning channels and receptors that underlie the signal transduction across the membrane. Therefore I am not really qualified to offer introductory remarks at the beginning of this memorable symposium. What I can do is give a bit of a historical overview of chemical transmitter research. Since this is the first time that the New York Academy of Sciences has held a meeting in Japan, it may not be unreasonable if my talk places emphasis on the accomplishments of Japanese scientists.

The most well-known work accomplished in early days by a Japanese was the chemical identification of adrenaline by Jokichi Takamine in 1901. Takamine's finding was made at a particularly good time. Despite the fact that he conducted his research in the United States, English researchers immediately took up adrenaline as the core of the chemical transmission concept. Prior to Takamine's work, J. J. Abel, a distinguished American chemical physiologist, had isolated a substance he called epinephrine in 1898. Unfortunately, it was not epinephrine but still conjugated with benzoic acid.

J. N. Langley was the key person at the dawn of chemical transmitter theory. He became aware of similarities between the effect of the extract of adrenal gland, now identified as adrenaline, and sympathetic stimulation, thus approaching the concept of chemical transmission. Subsequently, J. R. Elliott in 1904 proposed that adrenaline was the transmitter of sympathetic nerves. In 1921, Langley wrote "Autonomic Nervous System," in which he set forth the magnificent view of this system—essentially the same view we embrace today.

As is well known, the role of acetylcholine in the neuromuscular junction of skeletal muscle was finally established by Dale and his colleagues in 1936, but prior to that a farsighted view was presented by W. R. Hess in Zurich in 1923, and by Kenmatsu Shimizu, working in Hess' laboratory in 1926. A large part of the work subsequently reported by Dale *et al.* in 1936 was described in Shimizu's earlier paper, though the latter was somewhat more preliminary. However, Shimizu became a clinician and never returned to the laboratory after returning to Japan.

There are a few scientific anedcotes related to acetylcholine worthy of mention. Even after Dale received the Nobel Prize in 1937, there were quite a few physiologists who did not accept the role of acetylcholine at the neuromuscular junction, but insisted on the electrical transmission theory. This situation continued until the elegant electrophysiological work of P. Fatt and B. Katz in 1951. It seems to me some symbolism could be found in this story in the difference between a chemical and a physical approach, or between a biochemical and a physiological mind set. In this particular case the former won, but this might not be a general principle.

Another instance was the work of E. M. Keil and E. J. M. Sichel in 1936. They showed that a vigorous contraction of a frog skeletal muscle fiber could be induced by application of a minute amount of acetylcholine on its outer surface, but no response at all occurred when even an enormous amount of acetylcholine was injected into the interior of the muscle cell by a micropipette. This was the first indication of the position of a receptor in the process of signal transduction across the membrane. Sichel, a unique biologist with an original mind, described his remarkable findings only in brief abstracts and did not publish any regular papers.

The works of the Japanese referred to above were carried out in laboratories outside Japan, but work on glutamate and GABA were entirely domestic efforts.

The excitatory effect of glutamate and the inhibitory effect of GABA on the brain were first noted by Takashi Hayashi in 1954 and 1956, respectively. His observations with glutamate preceded those of other researchers and his findings with GABA coincided with observations by K. Florey's group.

Frankly speaking, however, his experiments were not refined from the viewpoint of modern science, so some people hesitated to accept him as a true pioneer. He was also a famous mystery writer, and this did not add to his reputation, but detracted from it. Nevertheless, his foresight in sensing two important neurotransmitters is really deserving of higher evaluation — perhaps we have been unfair to him.

Since 1960, there have been many Japanese who have greatly contributed to the neurotransmitter research, but since they carried out their research in the international arena, I shall not enumerate them here.

As we shall see in this symposium, the revolutionary progress of cell and molecular biology has revealed fundamental molecular processes common to various kinds of cells, irrespective of their origin. Our new task is to apply this fundamental knowledge to our understanding of organs and tissues.

Structure and Modulation of Na$^+$ and Ca^{2+} Channels

WILLIAM A. CATTERALL

Department of Pharmacology, SJ-30
School of Medicine
University of Washington
Seattle, Washington 98195

INTRODUCTION

The voltage-sensitive sodium channel is responsible for the increase in sodium permeability during the initial rapidly rising phase of the action potential in neurons. Upon depolarization, sodium permeability first increases dramatically and then after approximately one millisecond decreases to the baseline level. This biphasic behavior is described in terms of two experimentally separable processes that control sodium channel function: activation, which controls the rate and voltage dependence of sodium permeability increase after depolarization, and inactivation, which controls the rate and voltage dependence of the subsequent return of sodium permeability to the resting level during a maintained depolarization. The sodium channel can therefore exist in three functionally distinct states or groups of states: resting, active, and inactivated. Both resting and inactivated states are nonconducting, but channels that have been inactivated by prolonged depolarization are refractory unless the cell is repolarized to allow them to return to the resting state.

Depolarization of neurons by the sodium current activates voltage-gated calcium channels. Calcium ions moving into the cell cause a plateau depolarization prolonging the action potential and serve as an intracellular second messenger to initiate exocytosis of neurotransmitters and intracellular biochemical events including protein phosphorylation and gene expression. This chapter reviews some of the basic structural features of sodium and calcium channels using the sodium channel from rat brain and the L-type calcium channel from rabbit skeletal muscle as primary examples and focuses on recent advances on two aspects of the cell and molecular biology of the sodium channels that may be critical determinants of neuronal excitability: the molecular mechanism of sodium channel inactivation and mechanisms of modulation of sodium channel function by protein phosphorylation.

ION CHANNEL SUBUNITS

Structures of Sodium Channel Subunits

The purified sodium channel from rat brain consists of three polypeptides: α of 260 kD, β1 of 36 kD, and β2 of 33 kD (FIG. 1).[1] The β2 subunit is covalently

1

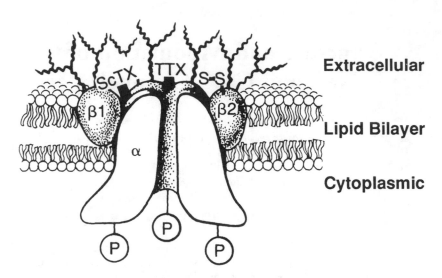

FIGURE 1. Subunit structure of the brain sodium channel. A view of a cross section of a hypothetical sodium channel consisting of a single transmembrane α subunit of 260 kD in association with a β1 subunit of 36 kD and a β2 subunit of 33 kD. The β1 subunit is associated noncovalently while the β2 subunit is linked through disulfide bonds. All three subunits are heavily glycosylated on their extracellular surfaces and the α subunit has receptor sites for α-scorpion toxins (ScTx) and tetrodotoxin (TTX). The intracellular surface of the α subunit is phosphorylated by multiple protein kinases (P).

attached to the α subunit by disulfide bonds while the β1 subunit is associated noncovalently. The subunits are present in a 1:1:1 stoichiometry and the sum of their molecular weights (329,000) agrees closely with the oligomeric molecular weight of the solubilized sodium channel. Antibodies against either the β1 or β2 subunits immunoprecipitate nearly all brain sodium channels indicating that they all have a heterotrimeric structure.[2,3] Sodium channels from eel electroplax contain only a single α subunit while sodium channels from rat skeletal muscle have α and β1 subunits.[1] Reagents derived from studies of purified sodium channels led directly to cloning of their genes.[4-9]

Identification of the genes encoding the sodium channel subunits and the determination of their primary structures by Professor Shosaku Numa and his colleagues in 1984, was a major advance in studies of the molecular properties of the voltage-gated ion channels. Oligonucleotides encoding short segments of the electric eel electroplax sodium channel and the antibodies directed against it were used to isolate cDNAs encoding the entire polypeptide from expression libraries of electroplax mRNA.[4] The deduced amino acid sequence revealed a protein with four internally homologous domains, each containing multiple potential alpha-helical transmembrane segments (FIG. 2). The wealth of information contained in this deduced primary structure has revolutionized research on voltage-gated ion channels.

The cDNAs encoding the electroplax sodium channel were used to isolate cDNAs encoding three distinct, but highly homologous, rat brain sodium channels (Types I, II, and III[5,6]). cDNAs encoding the alternatively spliced Type IIA sodium channel

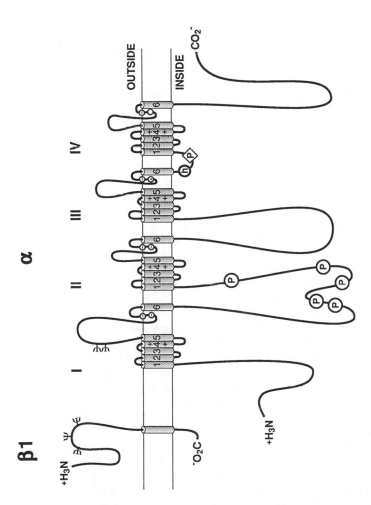

FIGURE 2. Primary structures of the α and β1 subunits of the sodium channel illustrated as transmembrane folding diagrams. The bold line represents the polypeptide chains of the α and β1 subunits with the length of each segment approximately proportional to its true length in the rat brain sodium channel. Cylinders represent probable transmembrane alpha helices. Other probable membrane-associated segments are drawn as loops in extended conformation like the remainder of the sequence. Sites of experimentally demonstrated glycosylation (Ψ), cAMP-dependent phosphorylation (P in a circle), protein kinase C phosphorylation (P in a diamond), amino acid residues required for tetrodotoxin binding (small circles with +, −, or open fields depict positively charged, negatively charged, or neutral residues, respectively), and amino acid residues that form the inactivation particle (h in a circle).

were isolated independently by screening expression libraries with antibodies against the rat brain sodium channel α subunit.[7–9] These sodium channels have a close structural relationship. In general, the similarity in amino acid sequence is greatest in the homologous domains from transmembrane segment S1 through S6 while the intracellular connecting loops are not highly conserved.

The primary structures of sodium channel β1 subunits have been determined only recently.[10] The β1 subunit cloned from rat brain is a small protein of 218 amino acids (22,821 daltons) with a substantial extracellular domain having four potential sites of N-linked glycosylation, a single alpha-helical membrane-spanning segment, and a very small intracellular domain (FIG. 2). Distinct β1 subunits may form specific associations with different α subunits and contribute to the diversity of sodium channel structure and function. Beta-2 subunits have not yet been cloned, but it may be anticipated that they also add to the potential diversity of sodium channel structure and function.

Functional Expression of Sodium Channel Subunits

Alpha subunit mRNAs isolated from rat brain by specific hybrid selection with Type IIA cDNAs[7] and RNAs transcribed from cloned cDNAs encoding α subunits of rat brain sodium channels[7–9,11,12] are sufficient to direct the synthesis of functional sodium channels when injected into *Xenopus* oocytes. These results establish that the protein structures necessary for voltage-dependent gating and ion conductance are contained within the α subunit itself.

Although α subunits alone are sufficient to encode functional sodium channels, their properties are not normal. Inactivation is slow relative to that observed in intact neurons and its voltage dependence is shifted to more positive membrane potentials. Co-expression of low molecular weight RNA from brain can accelerate inactivation, shift its voltage dependence to more negative membrane potentials, and increase the level of expressed sodium current.[7,8] These results suggested that the low molecular weight β1 and β2 subunits may modulate functional expression of the α subunit. Co-expression of RNA transcribed from cloned β1 subunits directly demonstrates this modulation.[10] Co-expression of β1 subunits in *Xenopus* oocytes accelerates the decay of the sodium current fivefold, shifts the voltage-dependence of sodium channel inactivation 20 mV in the negative direction, and increases the level of sodium current 2.5-fold. Evidently, β1 subunits are essential for normal functional expression of rat brain sodium channels.

Sodium channel α subunits can also be functionally expressed in mammalian cells in culture. Stable lines of Chinese hamster ovary (CHO) cells expressing the Type IIA sodium channel generate sodium currents with normal time course and voltage dependence, even though there is no evidence that these cells express an endogenous β1 subunit to form a complex with the transfected α subunit.[13,14] Evidently, β1 subunits do not have as important a functional impact when the α subunit is expressed in the genetic background of a mammalian somatic cell.

The α subunits expressed in CHO cells have normal pharmacological properties as well. They have high affinity receptor sites for saxitoxin and tetrodotoxin and are inhibited by low concentrations of tetrodotoxin.[11,12] The voltage dependence of their activation is shifted in the negative direction and they are persistently activated by veratridine in a stimulus-dependent manner.[13] Their inactivation is slowed by α-scorpion toxins.[14] In addition, they are inhibited in a strongly frequency- and voltage-dependent manner by local anesthetic, antiarrhythmic, and anticonvulsant drugs.[15] Thus, the receptor sites for all of these diverse pharmacological agents are located on the α subunits.

L-type Calcium Channels

Four physiological classes of voltage-gated calcium channels have been defined based on electrophysiological and pharmacological properties.[16] Members of all four physiological classes of calcium channels are expressed in neurons. L-type, dihydropyridine-sensitive calcium channels mediate long-lasting calcium currents and are the most abundant calcium channels in muscle tissues where they initiate excitation-contraction coupling. The major form of the skeletal muscle L-type calcium channel is a complex of five subunits[17-19]: α1 (165–190 kD), α2 (143 kD), β (55 kD), γ (30 kD), and δ (24–27 kD). The α1 subunit alone can function as a voltage-gated ion channel when expressed in *Xenopus* oocytes[20] or mammalian cells.[21] The cDNA sequence of the α1 subunit predicts a protein of 1,873 amino acids, whose structure is similar to the sodium channel α subunit.[22] Two size forms of this subunit are present in skeletal muscle: a major form of about 1,700 amino acids (190 kD) and a full-length form of 212 kD.[23,24]

The dihydropyridine-sensitive L-type calcium channels from brain are multisubunit complexes that include α1, α2δ, and β subunits analogous to the skeletal muscle L-type Ca channel.[25] Monoclonal antibody MANC1 against the α2δ subunits immunoprecipitates up to 82% of dihydropyridine-sensitive L-type calcium channels from different brain regions.[25,26] Molecular cloning experiments show that rat brain expresses at least four major classes of Ca channel α1 subunit (designated rbA, rbB, rbC, and rbD[27]). The rbC and rbD isoforms are most closely related to DHP-sensitive Ca channels from various tissues, are specifically labeled by dihydropyridines,[28] and direct the synthesis of functional dihydropyridine-sensitive calcium channels when expressed in heterologous cells,[29] indicating that the brain expresses two distinct forms of L-type Ca channel.

N-type Calcium Channels

N-type Ca channels are distinct in that they have only been described in neurons and are blocked by the peptide neurotoxin, ω-conotoxin GVIA (ω-CgTx[30,31]). The presence of ω-CgTx–sensitive calcium channels at the neuromuscular junction and the correlation between ω-CgTx–sensitive N-type Ca channels and the sensitivity of neurotransmitter release to ω-CgTx[32-34] has led to the proposal that N-type Ca channels are concentrated at presynaptic nerve termini and that they play a major role in mediating chemical synaptic transmission. Omega-CgTx–sensitive N-type Ca channels are a target for modulation by neurotransmitters and neuropeptides.[35] The ω-CgTx–sensitive calcium channels purified from brain contain subunits analogous to those of neuronal L-type calcium channels: α1, α2δ, and β.[36-38] They are substrates for phosphorylation by cAMP-dependent protein kinase and protein kinase C.[37]

Recent work has led to molecular cloning of a gene encoding a ω-CgTx–sensitive calcium channel and determination of the primary structure of this protein.[28,40] In contrast to the rbC and rbD calcium channels, the rbA and rbB isoforms are only moderately related to L-type Ca channels and are, therefore, of particular interest with respect to identifying Ca channels that mediate specific functions in neurons. The rbB gene encodes a calcium channel α1 subunit of 2,336 amino acid residues. Anti-peptide antibodies directed against a unique sequence in this α1 subunit immuno-precipitate ω-CgTx–labeled N-type calcium channels specifically, identifying the protein product of this gene as an N-type calcium channel.[28] Its ion conductance activity is blocked by ω-Cg-Tx when expressed in mammalian cells.[40] CNB1 antibodies immunoprecipitated up to 50% of the total N-type calcium channels labeled with $[^{125}I]$ω-CgTx. These results suggest that there may be multiple subtypes of ω-conotoxin-sensitive N-type calcium channels that are differentially recognized by this antibody. Immunoblotting with CNB1 revealed proteins of 210 and 240 kD that precisely co-migrated with ω-conotoxin–binding activity on sucrose gradient sedimentation.[39] These two size forms of the α1 subunit of an N-type calcium channel may reflect differential posttranscriptional or posttranslational processing of the carboxyl terminal region as has been described for the α1 subunit of skeletal muscle calcium channels.

STRUCTURE AND FUNCTION OF ION CHANNELS: THE SODIUM CHANNEL AS AN EXAMPLE

A Functional Map of the Sodium Channel Alpha Subunits

A major goal of current research on the voltage-gated ion channels is to define the structural components responsible for specific aspects of channel function. Two main experimental approaches have proven valuable in these studies. Antibodies against short, approximately 20-residue peptide segments of the principal α subunits of the sodium channels have been used to probe domains that are required for specific channel functions or that can be covalently labeled by neurotoxins or protein phosphorylation. Mutations have been introduced into cDNAs encoding the principal α subunits by oligonucleotide-directed mutagenesis, expressed in recipient cells, and analyzed by electrophysiological recording. Work applying these methods to sodium channels is described below.

Voltage-dependent Activation

The steep voltage dependence of activation of the voltage-sensitive ion channels is their unique characteristic. It requires that they have charged amino acid residues or strongly oriented dipoles within the membrane electric field of the phospholipid bilayer.[41] The movement of these gating charges or voltage sensors under the force of the electric field is believed to initiate a conformational change in the channel protein resulting in activation. The requirement for transmembrane movement of multiple charges during Na^+ channel activation has focused attention on the S4 segments of the voltage-sensitive ion channels, which are both positively charged and hydrophobic. These unique structures, consisting of repeated motifs of a positively charged amino acid residue, usually arginine, followed by two hydrophobic residues, were first noted in the amino acid sequence of the electroplax Na$^+$ channel.[4] Conservation of this amino acid sequence among different voltage-sensitive ion channels, first noted for Na$^+$ channels from electroplax and brain, is striking across this broad range of ion channels from diverse species.

Several authors have independently proposed that these S4 segments have a transmembrane orientation and are the gating charges or voltage sensors of the Na$^+$ channel.[1,5,42-44] Direct experimental support for designation of the S4 segments as the voltage sensors for activation of the voltage-gated ion channels has been provided by site-directed mutagenesis experiments on both Na$^+$ channels[9,45] and K$^+$ channels.[46] Neutralization of one to three positively charged amino acid residues in the S4 segments causes a progressive reduction in the steepness of the voltage-dependent activation of sodium channels as expected if these positively charged amino acid residues serve as gating charges. The effect of neutralization of different charged residues is not equivalent, indicating that they do not all move through a comparable fraction of the membrane electric field. Since the electric field is not expected to be strictly uniform through the membrane, the relative distance moved by the gating charges cannot be directly inferred from the fraction of the field through which they move so this value cannot be used to define a detailed molecular mechanism.

If the S4 helices must move through the protein structure of the sodium channel as the channel activates, the size and shape of the amino acid side chains might affect the voltage dependence of gating making it easier or more difficult for the gating segments to move. In fact, mutation of a hydrophobic residue in an S4 segment from leucine to phenylalanine causes a 20 mV shift in the voltage dependence of gating to more positive membrane potentials.[9] Moreover, mutation of positively charged amino acid residues in S4 helices from arginine to lysine, which retains positive charge, can cause a large shift in the voltage dependence of channel activation, and the shifts in voltage dependence of activation caused by mutation of arginine residues to uncharged glutamine residues are not exactly correlated with the number of charges neutralized, suggesting that size and shape of the residues may also be important.[45] Overall, these mutational analyses provide strong evidence that the S4 segments are indeed the voltage sensors of the voltage-gated ion channels.

The Transmembrane Pore

Tetrodotoxin inhibits sodium channels by binding to a receptor site on the α subunit. This receptor site is widely considered to be located near the extracellular end of the transmembrane pore of the sodium channel such that binding of the cationic toxins at that site impedes access of transported monovalent cations to the pore.[47-49] Neutralization of glutamine-387 by site-directed mutagenesis and expression of the modified channels in *Xenopus* oocytes cause a complete loss of tetrodotoxin inhibition of the expressed sodium channels.[50] This residue is located just outside transmembrane segment S6 in domain I of the sodium channel (FIG. 2). The corresponding residues in the other domains are also negatively charged and neutralization of them by site-directed mutagenesis also dramatically reduces tetrodotoxin binding.[51] In addition, tyrosine-374 in skeletal muscle sodium channels is also required for the high affinity binding of tetrodotoxin.[52] It is located two residues from the required negatively charged residues in the first domain. This residue is changed to cysteine in cardiac (h1) sodium

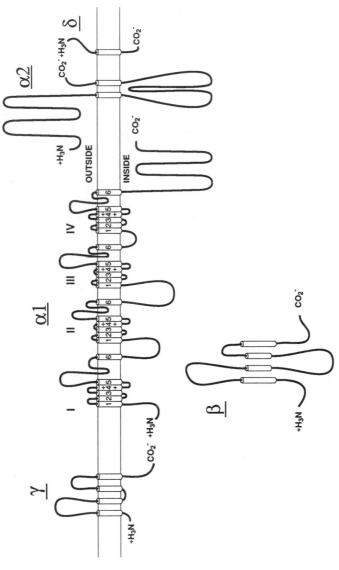

FIGURE 3. Subunit structure of skeletal muscle calcium channels. Transmembrane folding models of the calcium channel subunits derived from primary structure determination and analysis. Cylinders represent predicted α-helical segments in the transmembrane regions of the α1, α2δ, and γ subunits and in the peripherally associated β subunit. The transmembrane folding patterns are derived only from hydropathy analysis for α2δ and γ and from a combination of hydropathy analysis and analogy with the current models for the structures of sodium and potassium channels for α1. The transmembrane arrangement of α2δ is not well-defined by hydropathy analysis and the indicated structure should be taken as tentative.

channels causing them to have 200-fold reduced affinity for tetrodotoxin compared to muscle sodium channels.[52] Evidently, these residues in analogous positions in each domain form a single binding site for tetrodotoxin in or near the extracellular end of the transmembrane pore of the sodium channel. This region may contribute to formation of both the tetrodotoxin receptor site and the extracellular opening of the transmembrane pore.

Essentially all models for the structure of the voltage-gated ion channels include a transmembrane pore in the center of a square array of homologous transmembrane domains. Each domain would contribute one-fourth of the wall of the pore. Short segments (designated SS1 and SS2) between proposed transmembrane alpha helices S5 and S6 have been suggested to be membrane-associated and contribute to pore formation.[43,44] Recent studies support proposals in which the S5 and S6 alpha-helical segments and the short segments SS1 and SS2 are intimately involved in pore formation. As described above, the receptor site for tetrodotoxin, an extracellular pore blocker of sodium channels, includes acidic amino acid residues immediately on the extracellular side of transmembrane segment S6 in each domain.[50,51] The receptor site for verapamil, a probable intracellular pore blocker of calcium channels, involves residues immediately on the intracellular side of transmembrane segment IVS6.[53] These results argue that the ends of S6 segments form part of the intracellular and extracellular openings of the transmembrane pore.

Recent results on potassium channels provide direct evidence that regions analogous to the SS1 and SS2 segments form the lining of the transmembrane pore. Site-directed mutagenesis and formation of chimeric potassium channels between isoforms with different ion conductance and pharmacological properties have revealed that several amino acid residues located near the extracellular ends of the S5 and S6 segments are required for blocking the channel from the extracellular side by the polypeptide charybdotoxin and by tetraethylammonium ion.[54] Moreover, residues in the center of the segment containing SS1 and SS2 are important for blocking the channel from the *intracellular* side by tetraethylammonium ion. These results argue that the short segments SS1 and SS2 may traverse the membrane in an extended conformation placing the residues between them on the intracellular side of the channel. These short segments may therefore form the inner walls of the transmembrane pore and the residues between them may form an intracellular binding site for tetraethylammonium ion. Consistent with this idea, minor changes in the amino acids in this segment have dramatic effects on ion selectivity.[55,56]

A key role for residues at the extracellular mouth of this putative pore region in determination of ion selectivity of sodium channels is indicated by recent mutagenesis results.[57] Mutation of lysine-1422 and alanine-1714 to negatively charged glutamate residues caused a dramatic change in the ion selectivity of the sodium channel from sodium-selective to calcium-selective (FIG. 2). In addition, these changes created a high affinity site for calcium binding and blocking of monovalent ion conductance through the sodium channel, as has been previously described for calcium channels.[58,59]

Inactivation

Depolarization of the membrane of excitable cells results in a transient inward Na$^+$ current that is terminated within a few milliseconds by the process of inactivation.

Perfusion of the intracellular surface of the sodium channel with proteolytic enzymes prevents inactivation, implicating intracellular structures in the inactivation process.[41] The α subunit of sodium channels consists of four homologous domains connected by cytoplasmic linker sequences.[4–6] Antibodies directed against the intracellular linker between homologous domains III and IV ($L_{III/IV}$, FIG. 2) completely block fast inactivation of affected single sodium channels.[60,61] Expression of the sodium channel as two polypeptides with a cut between domains III and IV slows inactivation approximately 20-fold[45] and small insertions in this loop also slow inactivation.[62] Phosphorylation of a single serine residue in $L_{III/IV}$ by protein kinase C slows inactivation.[63] The amino acid sequence of $L_{III/IV}$ contains several clustered positively and negatively charged residues. Surprisingly, these highly conserved residues are not essential for fast sodium channel inactivation.[64] However, deletions of 10-amino acid segments within $L_{III/IV}$ can completely block fast sodium channel inactivation supporting an essential role for this segment in the inactivation process.[64] To assess the role of hydrophobic amino acids within $L_{III/IV}$ in inactivation, site-directed mutants were constructed in which conserved hydrophobic residues were altered, expressed in *Xenopus* oocytes or transfected Chinese hamster cells, and analyzed by whole cell voltage clamp and single channel recording.

A mutation was constructed in which the contiguous hydrophobic residues isoleucines-1488, phenylalanine-1489, and methionine-1490 (IFMQ3) were substituted with glutamine. RNA encoding mutant Na^+ channel α subunits was co-injected into *Xenopus* oocytes with RNA encoding the β1 subunit, and the expressed channels were analyzed by two-microelectrode voltage clamp recording.[65] Na^+ currents recorded in oocytes injected with RNA encoding Na^+ channel mutant IFMQ3 show a dramatic removal of fast inactivation (FIG. 4). The half-time for the decay of the sodium current is slowed about 4,000-fold. Only a minor shift (about 6 mV) was observed in the voltage-dependence of peak Na^+ conductance. Thus, the mutation IFMQ3 results in a specific and potent inhibition of the fast inactivation process.

The role of each amino acid in the cluster IFM in Na^+ channel inactivation was examined by substitution of each individually with glutamine (I1488Q, F1489Q, M1490Q).[65] While mutations I1488Q and M1490Q showed only mild effects, mutant F1489Q displayed greatly slowed, biphasic inactivation. For strong depolarizations, a small fraction of the current inactivated quickly, but most of the current failed to inactivate by the end of a 50-msec pulse. A mean of 86% of the Na^+ current remained at the end of a 50-msec pulse. The time course of decay of the Na^+ current was almost as slow as for mutant IFMQ3 during long test pulses to −10 mV. These results identify phe1489 as the critical amino acid residue within the hydrophobic cluster IFM. Single F1489Q channels open early in the pulse but continue to reopen for the duration of the pulse instead of inactivating. The increased probability of reopening of single channels evidently causes the noninactivating component of Na^+ current observed at the macroscopic level. The cluster of hydrophobic amino acid residues at positions 1488–1490 may form an essential part of the fast inactivation gate of the Na^+ channel.

The structure and function of the inactivation gate of the Na^+ channel resemble the "hinged lids" of allosteric enzymes.[66] Hinged lids have been defined structurally by X-ray crystallography and molecular modeling and therefore provide a valuable model for the unknown structure of the Na^+ channel inactivation gate. They consist

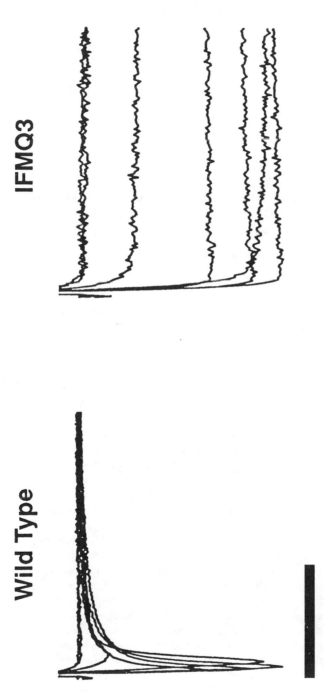

FIGURE 4. Inactivation of wild-type and mutant sodium channels. RNA encoding wild-type or IFMQ3 sodium channel α subunits was transcribed *in vitro* and injected into *Xenopus* oocytes together with RNA encoding β1 subunits. Sodium currents expressed in the oocytes we're recorded by whole cell voltage clamp using two-microelectrode voltage clamp procedure. Sodium currents were elicited by voltage steps from a holding potential of −100 mV to test potentials of −50 to 0 mV in 10 mV increments. Calibration bar is 20 msec.

of structured loops of 10 to 20 residues between two hinge points and serve as rigid lids that fold over the active sites of allosteric enzymes to control substrate access. Binding of allosteric ligands causes a conformational change of the lid to open or close the active site. By analogy, $L_{III/IV}$ may function as a rigid lid to control Na^+ entry to and exit from the intracellular mouth of the pore of the Na^+ channel (FIG. 5).[65] This hinged lid may be held in the closed position during inactivation by a hydrophobic latch formed by the hydrophobic cluster IFM. Glycine and proline residues on either side of the IFM domain that are conserved in all five cloned rat Na^+ channels that have been functionally expressed may function as hinge points[67] allowing the inactivation gate region of $L_{III/IV}$ to move in and out of the channel pore.

MODULATON OF SODIUM CHANNEL FUNCTION BY PROTEIN PHOSPHORYLATION

cAMP-dependent Protein Kinase

The possibility of modulation of sodium channel function by cAMP-dependent phosphorylation was first suggested by biochemical experiments showing that the α subunit of the sodium channel purified from rat brain was rapidly phosphorylated by cAMP-dependent protein kinase on at least three sites.[68] Sodium channel α subunits in intact synaptosomes are rapidly phosphorylated in response to agents that increase cAMP, and neurotoxin-activated ion flux through sodium channels is reduced concomitantly.[69] Stimulation of rat brain neurons in primary cell culture with agents that increase cAMP also causes rapid phosphorylation of sodium channel α subunits.[70] Substantial phosphorylation is observed at the basal level of cAMP in the cultured neurons and a twofold increase is observed upon stimulation.

The physiological effect of phosphorylation of sodium channels is revealed most clearly by analysis of the effect of direct phosphorylation of sodium channels in excised membrane patches by purified cAMP-dependent protein kinase.[71] Phosphorylation of the inside-out membrane patches from rat brain neurons or transfected CHO cells reduces peak sodium currents approximately 50% with no change in the time course or the voltage dependence of activation or inactivation of the sodium current (FIG. 6,B).

The sites of phosphorylation of the sodium channel by cAMP-dependent protein kinase have been identified by a combination of two-dimensional phosphopeptide mapping, immunoprecipitation of phosphopeptides with site-directed anti-peptide antibodies, and microsequence determination.[72,73] Four sites of in vitro phosphorylation are clustered in the large intracellular loop connecting homologous domains I and II (FIG. 2). These sites are all phosphorylated in intact neurons, but their different rates of phosphorylation in vitro suggest that a subset of the sites may play a predominant role in channel regulation. Mutation of individual serine residues followed by expression and functional analysis will be required to define the role of each site in regulation by cAMP-dependent phosphorylation.

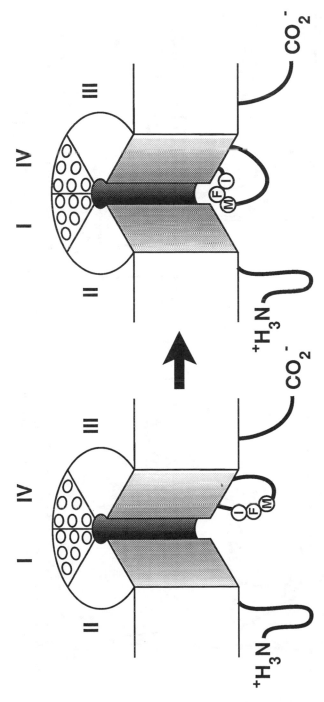

FIGURE 5. A hinged-lid model for Na^+ channel inactivation. Segment L_{III-IV} is depicted as a hinged lid that occludes the transmembrane pore of the Na^+ channel during inactivation. Phe1489 is illustrated in a pore-blocking position in the inactivated state.

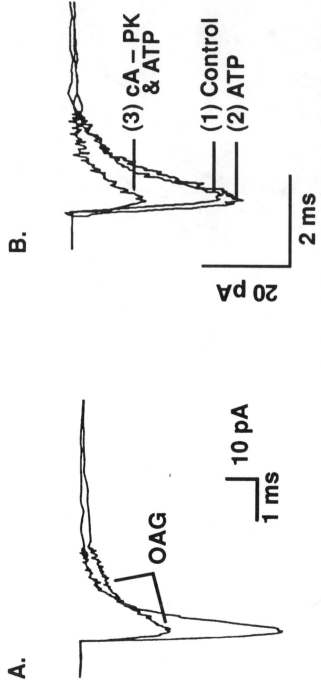

FIGURE 6. Differential modulation of sodium currents by protein phosphorylation. (A) Sodium currents were recorded in the cell-attached patch configuration in Chinese hamster ovary cells expressing Type IIA sodium channel α subunits during depolarizations from a holding potential of −110 mV to a test potential of 0 mV. Ensemble average currents were calculated from macropatches containing up to 30 active sodium channels. Current traces are illustrated under control conditions and after activation of protein kinase C with the synthetic diacylglycerol oleylacetylglycerol (OAG). (B) Sodium currents were recorded from the same cells in the excised patch clamp configuration during depolarization from a holding potential of −130 mV to a test potential of −20 mV. Current traces are illustrated under control conditions, after addition of 1 mM ATP, and after addition of 1 mM ATP and 2 μM cAMP-dependent protein kinase. Ensemble average currents were recorded from macropatches containing approximately 15 or more active sodium channels.

Protein Kinase C

Alpha subunits of purified sodium channels from rat brain are also phosphorylated by protein kinase C[74,75] suggesting that they may be modulated by the calcium/diacylglycerol signaling pathway. In agreement with this suggestion, sodium currents in neuroblastoma cells are reduced by treatment with fatty acids that can activate protein kinase C,[76] and sodium currents in *Xenopus* oocytes injected with rat brain mRNA are reduced by treatment with phorbol esters that activate protein kinase C.[77,78] Activation of protein kinase C in rat brain neurons or in Chinese hamster ovary (CHO) cells transfected with cDNA encoding the type IIA sodium channel α subunit by treatment with diacylglycerols causes two functional effects: slowing of inactivation and reduction of peak current (FIG. 6,A[79]). Both of these actions are prevented by prior injection of the pseudosubstrate inhibitory domain of protein kinase C into the cells indicating that they reflect phosphorylation by protein kinase C. Moreover, both effects can be observed in excised, inside-out membrane patches by phosphorylating sodium channels directly with purified protein kinase C. These results support the conclusion that protein kinase C can modulate sodium channel function by phosphorylation of the α subunit of the sodium channel protein itself as observed with purified sodium channels.

The intracellular loop connecting domains III and IV has been implicated in sodium channel inactivation as described above. This segment has a consensus sequence for phosphorylation by protein kinase C centered at serine1506. Mutagenesis of this serine residue to alanine blocks both of the modulatory effects of protein kinase C.[63] Evidently, phosphorylation of this site is required for both slowing of sodium channel inactivation and reduction of peak sodium current by protein kinase C. It likely slows inactivation by a direct effect on the structure and/or closure of the inactivation gate itself.

PERSPECTIVE

The structural basis for the functional properties of the voltage-gated ion channels is rapidly being elucidated. Development of a molecular map of these channels has evolved through contributions from many laboratories. However, the molecular template for this map originated in the determination of the primary structures of the principal subunits of the sodium and calcium channels by Professor Shosaku Numa and his colleagues. Their work opened the way for molecular analysis of these channels and many other proteins considered in this volume. Their determination of the primary structures of important signaling molecules is unparalleled. Professor Numa's untimely death at the apex of his scientific powers is a loss of substantial proportions for this field. His contributions will be sorely missed.

REFERENCES

1. CATTERALL, W. A. 1986. Molecular properties of voltage-sensitive sodium channels. Ann. Rev. Biochem. 55: 953–985.
2. WOLLNER, D. A., D. J. MESSNER & W. A. CATTERALL. 1987. Beta 2 subunits of sodium channels

from vertebrate brain. Studies with subunit-specific antibodies. J. Biol. Chem. 262: 14709–14715.

3. McHugh-Sutkowski, E. & W. A. Catterall. 1990. β1 subunits of sodium channels. Studies with subunit-specific antibodies. J. Biol. Chem. 265: 12393–12399.

4. Noda, M., S. Shimizu, T. Tanabe, T. Takai, T. Kayano, T. Ikeda, H. Takahashi, H. Nakayama, Y. Kanaoka, N. Minamino, K. Kangawa, H. Matsuo, M. Raftery, T. Hirose, S. Inayama, H. Hayashida, T. Miyata & S. Numa. 1984. Primary structure of *Electrophorus electricus* sodium channel deduced from cDNA sequence. Nature 312: 121–127.

5. Noda, M., T. Ikeda, T. Kayano, H. Suzuki, H. Takeshima, M. Kurasaki, H. Takahashi & S. Numa. 1986. Existence of distinct sodium channel messenger RNAs in rat brain. Nature 320: 188–192.

6. Kayano, T., M. Noda, V. Flockerzi, H. Takahashi & S. Numa. 1988. Primary structure of rat brain sodium channel III deduced from the cDNA sequence. FEBS Lett. 228: 187–194.

7. Goldin, A. L., T. P. Snutch, H. Lubbert, A. Dowsett, J. Marshall, V. Auld, W. Downey, L. C. Fritz, H. A. Lester, R. Dunn, W. A. Catterall & N. Davidson. 1986. Messenger RNA coding for only the α subunit of the rat brain Na channel is sufficient for expression of functional channels in *Xenopus* oocytes. Proc. Natl. Acad. Sci. USA 83: 7503–7507.

8. Auld, V. J., A. L. Goldin, D. S. Krafte, J. Marshall, J. M. Dunn, W. A. Catterall, H. A. Lester, N. Davidson & R. J. Dunn. 1988. A rat brain Na+ channel alpha subunit with novel gating properties. Neuron 1: 449–461.

9. Auld, V. J., A. L. Goldin, D. S. Krafte, W. A. Catterall, H. A. Lester, N. Davidson & R. J. Dunn. 1990. A neutral amino acid change in segment IIS4 dramatically alters the gating properties of the voltage-dependent sodium channel. Proc. Natl. Acad. Sci. USA 87: 323–327.

10 Isom, I. I., K. S. De Jongh, B. F. X. Reber, J. Offord, H. Charbonneau, K. Walsh, A. L. Goldin & W. A. Catterall. 1992. Primary structure and functional expression of the β1 subunit of the rat brain sodium channel. Science 256: 839–842.

11. Noda, M., T. Ikeda, T. Suzuki, H. Takeshima, T. Takahashi, M. Kuno & S. Numa. 1986. Expression of functional sodium channels from cloned cDNA. Nature 322: 826–828.

12. Suzuki, H., S. Beckh, H. Kubo, N. Yahagi, H. Ishida, T. Kayano, M. Noda & S. Numa. 1988. Functional expression of cloned cDNA encoding sodium channel III. FEBS Lett. 228: 195–200.

13. Scheuer, T., V. J. Auld, S. Boyd, J. Offord, R. Dunn & W. A. Catterall. 1990. Functional properties of rat brain sodium channels expressed in a somatic cell line. Science 247: 854–858.

14. West, J. W., D. E. Patton, A. L. Goldin & W. A. Catterall. 1992. A cluster of hydrophobic amino acid residues required for fast Na+-channel inactivation. Proc. Natl. Acad. Sci. USA 89: 10910–10914.

15. Ragsdale, D. S., T. Scheuer & W. A. Catterall. 1991. Frequency and voltage-dependent inhibition of type IIA Na+ channels, expressed in a mammalian cell line, by local anesthetic, antiarrhythmic, and anticonvulsant drugs. Mol. Pharmacol. 40: 756–765.

16. Tsien, R. W., P. T. Elinor & W. A. Horne. 1991. Molecular diversity of voltage-dependent calcium channels. Trends Neurosci. 12: 349–354.

17. Campbell, K. P., A. T. Leung & A. H. Sharp. 1988. The biochemistry and molecular biology of the dihydropyridine-sensitive calcium channel. Trends Neurosci. 11: 425–430.

18. Catterall, W. A., M. J. Seagar & M. Takahashi. 1988. Molecular properties of dihydropyridine-sensitive calcium channels in skeletal muscle. J. Biol. Chem. 263: 3535–3538.

19. Catterall, W. A. 1991. Functional subunit composition of voltage-gated calcium channels. Science 253: 1499–1500.

20. Mikami, A., K. Imoto, T. Tanabe, T. Niidome, Y. Mori, H. Takeshima, S. Narumiya & S. Numa. 1989. Primary structure and functional expression of the cardiac dihydropyridine-sensitive calcium channel. Nature 340: 230–233.

21. Perez-Reyes, E., H. S. Kim, A. E. Lacerda, W. Horne, X. Y. Wei, D. Rampe, K. P. Campbell,

A. M. Brown & L. Birnbaumer. 1989. Induction of calcium currents by the expression of the alpha 1-subunit of the dihydropyridine receptor from skeletal muscle. Nature 340: 233–236.

22. Tanabe, T., H. Takeshima, A. Mikami, V. Flockerzi, H. Takahashi, K. Kangawa, M. Kojima, H. Matsuo, T. Hirose & S. Numa. 1987. Primary structure of the receptor for calcium channel blockers from skeletal muscle. Nature 328: 313–318.

23. DeJongh, K. S., D. K. Merrick & W. A. Catterall. 1989. Subunits of purified calcium channels: a 212-kDa form of α1 and partial amino acid sequence of a phosphorylation site of an independent β subunit. Proc. Natl. Acad. Sci. USA 86: 8585–8589.

24. DeJongh, K. S., C. Warner, A. A. Colvin & W. A. Catterall. 1991. Characterization of the two size forms of the α1 subunit of skeletal muscle L-type calcium channels. Proc. Natl. Acad. Sci. USA 88: 10778–10782.

25. Ahlijanian, M. K., R. E. Westenbroek & W. A. Catterall. 1990. Subunit structure and localization of dihydropyridine-sensitive calcium channels in mammalian brain, spinal cord, and retina. Neuron 4: 819–832.

26. Westenbroek, R. E., M. K. Ahlijanian & W. A. Catterall. 1990. Clustering of L-type Ca^{2+} channels at the base of major dendrites in hippocampal pyramidal neurons. Nature 347: 281–284.

27. Snutch, T. P., J. P. Leonard, M. M. Gilbert, H. A. Lester & N. Davidson. 1990. Rat brain expresses a heterogeneous family of calcium channels. Proc. Natl. Acad. Sci. USA 87: 3391–3395.

28. Dubel, S. J., T. V. B. Starr, J. Hell, M. K. Ahlijanian, J. J. Enyeart, W. A. Catterall & T. P. Snutch. 1992. Molecular cloning of the α-1 subunit of an ω-conotoxin-sensitive calcium channel. Proc. Natl. Acad. Sci. USA 89: 5058–5062.

29. Williams, M. E., D. H. Feldman, A. F. McCue, R. Brenner, G. Velicelebi, S. B. Ellis & M. M. Harpold. 1992. Structure and functional expression of $α_1$, $α_2$, and β subunits of a novel human neuronal calcium channel subtype. Neuron 8: 71–84.

30. Bean, B. P. 1989. Classes of calcium channels in vertebrate cells. Annu. Rev. Physiol. 51: 367–384.

31. Hess, P. 1990. Calcium channels in vertebrate cells. Annu. Rev. Neurosci. 13: 337–356.

32. Hirning, L. D., A. P. Fox, E. W. McCleskey, B. M. Olivera, S. A. Thayer, R. J. Miller & R. W. Tsien. 1988. Dominant role of N-type Ca^{2+} channels in evoked release of norepinephrine from sympathetic neurons. Science 239: 57–61.

33. Stanley, E. F. & G. Goping. 1991. Characterization of a calcium current in a vertebrate cholinergic presynaptic nerve terminal. J. Neurosci. 11: 985–993.

34. Robitaille, R., E. M. Adler & M. P. Charlton. 1990. Strategic location of calcium channels at transmitter release sites of frog neuromuscular synapses. Neuron 5: 773–779.

35. Hille, B. 1992. G protein coupled mechanisms and nervous signaling. Neuron 9: 187–195.

36. McEnery, M. W., A. M. Snowman, A. H. Sharp, M. E. Adams & S. Synder. 1991. Purified ω-conotoxin GVIA receptor of rat brain resembles a dihydropyridine-sensitive L-type calcium channel. Proc. Natl. Acad. Sci. USA 88: 11095–11099.

37. Ahlijanian, M. K., J. Striessnig & W. A. Catterall. 1991. Phosphorylation of an α1-like subunit of an ω-conotoxin-sensitive brain calcium channel by cAMP-dependent protein kinase and protein kinase C. J. Biol. Chem. 266: 20192–20197.

38. Sakamoto, J. & K. P. Campbell. 1991. A monoclonal antibody to the β subunit of the skeletal muscle dihydropyridine receptor immunoprecipitates the brain ω-conotoxin GVIA receptor. J. Biol. Chem. 266: 18914–18919.

39. Westenbroek, R. E., J. Hell, S. Dubel, C. Warner, T. Snutch & W. A. Catterall. 1992. Biochemical properties and subcellular localization of an N-type calcium channel α1 subunit. Neuron 9: 1099–1115.

40. Williams, M. E., P. F. Brust, D. H. Feldman, S. Patthi, S. Simerson, A. Maroufi, A. F. McCue, G. Velicelebi, S. B. Ellis & M. M. Harpold. 1992. Structure and functional expression of an ω-conotoxin-sensitive human N-type calcium channel. Science 257: 389–395.

41. Armstrong, C. M. 1981. Sodium channels and gating currents. Physiol. Rev. 61: 644–682.
42. Greenblatt, R. E., Y. Blatt & M. Montal. 1985. The structure of the voltage-sensitive sodium channel: inferences derived from computer-aided analysis of the *Electrophorus electricus* channel primary structure. FEBS Lett. 193: 125–134.
43. Guy, H. R. & P. Seetharamulu. 1986. Molecular model of the action potential sodium channel. Proc. Natl. Acad. Sci. USA 83: 508–512.
44. Guy, H. R. & F. Conti. 1990. Pursuing the structure and function of voltage-gated channels. Trends Neurosci. 13: 201–206.
45. Stühmer, W., F. Conti, H. Suzuki, X. Wang, M. Noda, N. Yahadi, H. Kubo & S. Numa. 1989. Structural parts involved in activation and inactivation of the sodium channel. Nature 339: 597–603.
46. Jan, L. Y. & Y. N. Jan. 1992. Structural elements involved in specific K+ channel functions. Ann. Rev. Physiol. 54: 537–555.
47. Narahashi, T. 1974. Chemicals as tools in the study of excitable membranes. Physiol. Rev. 54: 813–889.
48. Hille, B. 1975. The receptor for tetrodotoxin and saxitoxin: a structural hypothesis. Biophys. J. 15: 615–619.
49. Ritchie, J. M. & R. B. Rogart. 1977. The binding of saxitoxin and tetrodotoxin to excitable tissue. Rev. Physiol. Biochem. Pharmacol. 79: 1–49.
50. Noda, M., H. Suzuki, S. Numa & W. Stühmer. 1989. A single point mutation confers tetrodotoxin and saxitoxin insensitivity on the sodium channel II. FEBS Lett. 259: 213–216.
51. Terlau, H., S. H. Heinemann, W. Stühmer, M. Pusch, F. Conti, K. Imoto & S. Numa. Mapping the site of block by tetrodotoxin and saxitoxin of sodium channel II. FEBS Lett. 293: 93–96.
52. Satin, J., J. W. Kyle, M. Chen, P. Bell, L. L. Cribbs, H. A. Fozzard & R. B. Rogart. 1992. A point mutation of TTX-resistant cardiac Na channels confers three properties of TTX-sensitive Na channels. Science 256: 1202–1205.
53. Striessnig, J., H. Glossmann & W. A. Catterall. 1990. Identification of a phenylalkylamine binding region within the α1 subunit skeletal muscle Ca2+ channels. Proc. Natl. Acad. Sci. USA 87: 9108–9112.
54. Miller, C. 1991. 1990: Annus mirabilis for potassium channels. Science 252: 1092–1096.
55. Yool, A. J. & T. L. Schwarz. 1991. Alternation of ionic selectivity of a K+ channel by mutation of the H5 region. Nature 349: 702–704.
56. Kirsche, G. E., J. A. Drewe, H. A. Hartmann, M. Taglialatela, M. DeDiasi, A. M. Brown & R. H. Joho. 1992. Differences between the deep pores of K+ channels determined by an interacting pair of nonpolar amino acids. Neuron 8: 499–505.
57. Heinemann, S. H., H. Terlau, W. Stühmer, K. Imoto & S. Numa. 1992. Calcium channel characteristics conferred on the sodium channel by single mutations. Nature 356: 441–443.
58. Almers, W., E. W. McCleskey & P. T. Palade. 1984. A nonselective cation conductance in frog muscle membrane blocked by micromolar external Ca++. J. Physiol. (Lond.) 353: 565–583.
59. Hess, P. & R. W. Tsien. 1984. Mechanism of ion permeation through calcium channels. Nature 309: 453–456.
60. Vassilev, P. M., T. Scheuer & W. A. Catterall. 1988. Identification of an intracellular peptide segment involved in sodium channel inactivation. Science 241: 1658–1661.
61. Vassilev, P. M., T. Scheuer & W. A. Catterall. 1989. Inhibition of inactivation of single sodium channels by a site-directed antibody. Proc. Natl. Acad. Sci. USA 86: 8147–8151.
62. Patton, D. E. & A. L. Goldin. 1991. A voltage-dependent gating transition induces use-dependent block by tetrodotoxin of rat IIA sodium channels expressed in *Xenopus* oocytes. Neuron 7: 637–647.
63. West, J. W., R. Numann, B. J. Murphy, T. Scheuer & W. A. Catterall. 1991. A phosphorylation site in a conserved intracellular loop that is required for modulation of sodium channels by protein kinase C. Science 254: 866–868.

64. Patton, D. E., J. W. West, W. A. Catterall & A. L. Goldin. 1992. Amino acid residues required for fast Na⁺-channel inactivation: Charge neutralizations and deletions in the III-IV linker. Proc. Natl. Acad. Sci. USA **89**: 10905–10909.

65. West, J. W. Personal communication.

66. Joseph, D., G. A. Petsko & M. Karplus. 1990. Anatomy of a conformational change: hinged lid motion of the triosephosphate isomerase loop. Science **249**: 1425–1428.

67. Vermersch, P. S., J. J. G. Tesmer, D. D. Lemon & F. A. Quiocho. 1990. A pro to gly mutation in the hinge of the arabinose-binding protein enhances binding and alters specificity. J. Biol. Chem. **265**: 16592–16603.

68. Costa, M. R., J. E. Casnellie & W. A. Catterall. 1982. Selective phosphorylation of the alpha subunit of the sodium channel by cAMP-dependent protein kinase. J. Biol. Chem. **257**: 7918–7921.

69. Costa, M. R. & W. A. Catterall. 1984. Cyclic AMP-dependent phosphorylation of the alpha subunit of the sodium channel in synaptic nerve ending particles. J. Biol. Chem. **259**: 8210–8218.

70. Rossie, S. & W. A. Catterall. 1987. Cyclic AMP-dependent phosphorylation of voltage-sensitive sodium channels in primary cultures of rat brain neurons. J. Biol. Chem. **262**: 12735–12744.

71. Li, M., J. W. West, Y. Lai, T. Scheuer & W. A. Catterall. 1992. Functional modulation of brain sodium channels by cAMP-dependent phosphorylation. Neuron **197**: 1151–1159.

72. Rossie, S., D. Gordon & W. A. Catterall. 1987. Identification of an intracellular domain of a sodium channel having multiple cyclic AMP-dependent phosphorylation sites. J. Biol. Chem. **262**: 17530–17535.

73. Rossie, S. & W. A. Catterall. 1989. Phosphorylation of the alpha subunit of rat brain sodium channels by cAMP-dependent protein kinase at a new site containing Ser686 and Ser687. J. Biol. Chem. **264**: 14220–14224.

74. Costa, M. R. & W. A. Catterall. 1984. Phosphorylation of the alpha subunit of the sodium channel by protein kinase C. Cell Mol. Neurobiol. **4**: 291–297.

75. Murphy, B. J. & W. A. Catterall. 1992. Phosphorylation of purified rat brain Na⁺ channel reconstituted into phospholipid vesicles by protein kinase C. J. Biol. Chem. **267**:16129–16134.

76. Linden, D. J. & A. Routtenberg. 1989. Cis-fatty acids, which activate protein kinase C, attenuate Na⁺ and Ca²⁺ currents in mouse neuroblastoma cells. J. Physiol. **419**: 95–119.

77. Sigel, E. & R. Baur. 1988. Activation of protein kinase C differentially modulates neuronal Na⁺, Ca²⁺, and γ-aminobutyrate type A channels. Proc. Natl. Acad. Sci. USA **85**: 6192–6196.

78. Dascal, N. & I. Lotan. 1991. Activation of protein kinase C alters voltage dependence of a Na⁺ channel. Neuron **6**: 165–175.

79. Numann, R., W. A. Catterall & T. Scheuer. 1991. Functional modulation of brain sodium channels by protein kinase C phosphorylation. Science **254**: 115–118.

Structure and Function of Sodium Channels

MASAHARU NODA[a]

Division of Molecular Neurobiology
National Institute for Basic Biology
Okazaki 444, Japan

INTRODUCTION

The sodium channel is a transmembrane protein responsible for the voltage-dependent modulation of the sodium ion permeability of excitable membranes and thus plays an essential role in generating action potentials. A new approach to studying the structure and function of this membrane protein has been provided by recombinant DNA technology. Cloning and nucleotide sequence analysis of cDNAs have allowed the elucidation of the primary structures of the sodium channel proteins and have afforded insight into the evolution of these proteins. Furthermore, expression of the cloned cDNAs and their mutants produced by site-directed mutagenesis has made it possible to investigate the structural basis for the function of this ionic channel.

PRIMARY STRUCTURE

The primary structures of the sodium channel from the electric organ of the eel *Electrophorus electricus*[1] and the three distinct sodium channels (designated as sodium channels I, II, and III) from rat brain[2,3] were elucidated by cloning and sequence analysis of the cDNAs. FIGURE 1 shows the alignment of the amino acid sequences of the four sodium channels. The *Electrophorus* sodium channel and rat sodium channels I, II, and III consist of 1,820, 1,998 [or 2,009 (see the legend to FIG. 1)], 2,005, and 1,951 amino acid residues (including the initiating methionine), respectively. Homology matrix comparison of the amino acid sequences revealed the presence of four internal repeats (I–IV) that exhibit sequence homology.[1-4] The regions corresponding to these repeats are highly conserved among the four sodium channels, whereas the remaining regions, all of which are assigned to the cytoplasmic side of the membrane, are less well conserved, except for the short segment between repeats III and IV. A large insertion of 135–194 amino acids occurs in the region between repeats I and II of the rat sodium channels, compared with the *Electrophorus* counterpart. The inserted segments and their carboxy-terminal–neighboring regions contain several potential

---

[a] Address correspondence to: Masaharu Noda, Division of Molecular Neurobiology, National Institute for Basic Biology, 38 Nishigonaka, Myodaiji-cho, Okazaki 444, Japan.

sites of phosphorylation by cyclic AMP–dependent protein kinase, which are conserved in the three rat sodium channels.

PROPOSED TRANSMEMBRANE TOPOLOGY AND SECONDARY STRUCTURE

The *Electrophorus* and rat sodium channels show similar hydropathy profiles.[1-3] Each of the four internal repeats has five hydrophobic segments (S1, S2, S3, S5, and S6) and one positively charged segment (S4), all of which exhibit predicted secondary structure.[1-3] It seems reasonable to assume that the four repeated units of homology are oriented in a pseudosymmetric fashion across the membrane. This suggests the presence of an even number of transmembrane segments in each repeat, because no additional hydrophobic segments are predicted outside the repeats. Furthermore, the sodium channels have no hydrophobic prepeptide and may, like some transmembrane proteins devoid of a cleavable prepeptide, have their amino terminus on the cytoplasmic side of the membrane. Thus the transmembrane topology of the sodium channel molecule has been assigned in such a way that each internal repeat contains six presumably α-helical membrane-spanning segments (S1–S6) and that the amino- and carboxy-terminals are on the cytoplasmic side of the membrane (FIG. 2).[2] In addition to these transmembrane segments, the loop between S5 and S6 was also postulated to form a hairpin into the channel, the so-called SS1 and SS2 segments after Guy *et al.*,[5] thus forming the channel pore lining (FIG. 2). The proposed transmembrane topology is consistent with five of the six potential N-glycosylation sites that are conserved in all four sodium channels as well as with all eight potential cyclic AMP–dependent phosphorylation sites that are conserved in the three rat sodium channels.

The voltage-dependent gating of the sodium channel implies the presence of a voltage sensor, which is thought to be a collection of charges or equivalent dipoles moving under the influence of the membrane electric field.[6,7] In fact, this movement can be measured as a gating current.[8,9] The finding that the equivalent of four to six charges must move fully across the membrane to open one sodium channel[6] suggests the intramembranous location of many dipoles that move by smaller distances. The unique structure of the positively charged segment S4 is strikingly well conserved among the four sodium channels (FIG. 1). Segment S4 in repeats I, II, III, and IV contains four, five, six, and eight arginine or lysine residues, respectively, at every third position (except for the arginine residue closest to the carboxyl end of segment S4 in repeat III residing at the fourth position), with mostly nonpolar residues intervening between the basic residues. We proposed that the positive charges in this segment, many of which presumably form dipoles, represent the voltage sensor.[1,2] They would move outward in response to depolarization, causing conformational changes and possible rearrangement of ion pairs. The presence of four homologous repeats in a single sodium channel molecule is consistent with the sigmoid activation kinetics characteristic of this channel.[6]

Furthermore, clustered positively charged residues (predominantly lysine) are conserved in the region between segment S6 of repeat III and segment S1 of repeat IV, and clustered negatively charged residues in the region following segment S6 of repeat

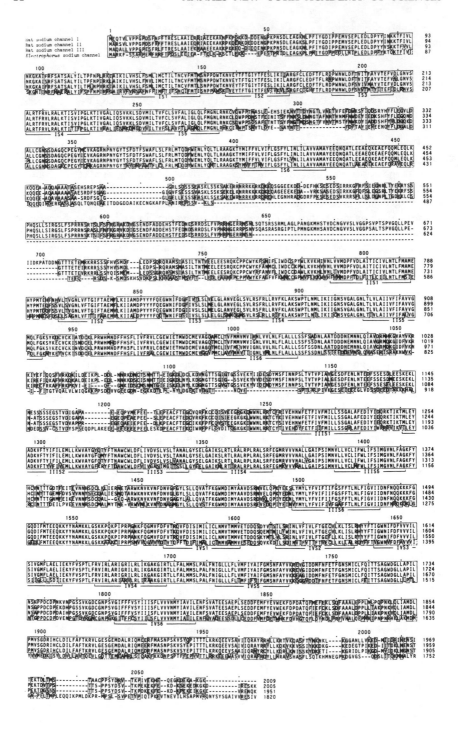

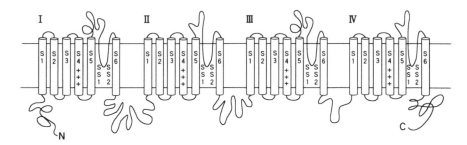

FIGURE 2. Proposed transmembrane topology of the sodium channels. Segments S1–S6 in each repeat (I–IV) are shown by cylinders. In addition, two shorter putative membrane-associated regions are shown between each S5 and S6 segments, labeled SS1 and SS2, after notation of Guy.[5] (Based on Noda et al.[2])

IV.[2] We speculated that these regions, which are assigned to the cytoplasmic side of the membrane, are involved in the inactivation of the sodium channel.[2,10]

These tentative assignments of the functional regions based on the analysis of the primary structure have been confirmed one by one by the investigation of the effects of side-directed mutations of rat sodium channel II on its functional properties.

EXPRESSION OF THE SODIUM CHANNEL mRNAS

The three sodium channel mRNAs exhibit different temporal and regional expression patterns in the rat central nervous system (CNS).[11,12] FIGURE 3(A) shows the blot hybridization analysis of total RNA from adult rat brain, using probes specific for the sodium channels I, II, and III (lanes 1, 2, and 3, respectively) or the common probe (lane 4). The estimated sizes of the major RNA species specific for sodium channels I, II, and III were ~9,000, ~9,500, and ~9,000 nucleotides, respectively.[11]

FIGURE 1. Alignment of the amino acid sequences of rat sodium channels I (top), II (second row), III (third row), and the *Electrophorus electricus* sodium channel (bottom). The one-letter amino acid notation is used. Sets of three or four identical residues at one position are enclosed with solid lines, and the fourth residue regarded as conservative substitutions at the same position is enclosed with broken lines. Conservative substitutions are defined as pairs of residues belonging to one of the following groups: S, T, A, and G; N, D, E, and Q; H, R, and K; M, I, L, and V; F, Y, and W. Gaps (–) have been inserted to achieve maximum homology. Amino acid residues are numbered beginning with the initiating methionine, and numbers of the residues at the right-hand end of individual lines are given. Positions in the aligned sequences including gaps are numbered beginning with that of the initiating methionine, and position numbers are given above the sequences. The putative transmembrane segments S1–S6 in each of repeats I–IV are indicated; the termini of these segments have been tentatively assigned. The amino acid differences resulting from the nucleotide differences found among the individual clones are as follows (position numbers in the aligned sequences are given in parentheses). Sodium channel I: deletion (694–704); Asn (503); Lys (617); Asn (849). Sodium channel II: Lys (522); His (1,057); Met (1,144); Gly (1,274). Sodium Channel III: Ile (271); Leu (279); Thr (356); Lys (533); Arg (1,149). (Based on Noda et al.[1,2] and Kayano et al.[3])

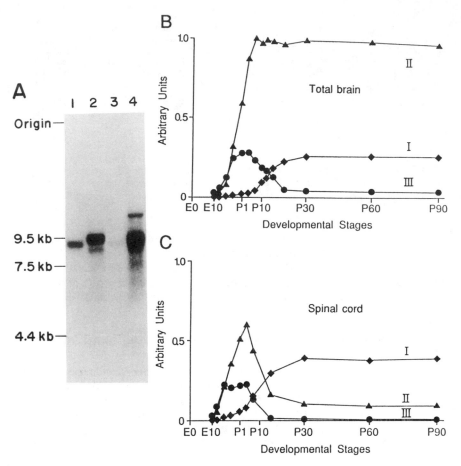

FIGURE 3. RNA blot hybridization analysis. (**A**) Analysis of total RNA from brain using the RNA probes specific for sodium channels I (Lane 1), II (Lane 2), and III (Lane 3) and the RNA probe common to them (Lane 4). RNA samples (25 μg each) from adult rat brain were electrophoresed on 1.0% agarose gel. The specific radioactivities of the RNA probes were $5.6–7.6 \times 10^8$ dpm/μg. Autoradiography was performed at $-70°C$ for 21 h with an intensifying screen. (**B** and **C**) Temporal expression patterns of sodium channel I, II, and III mRNAs in total brain and spinal cord. Diamonds, triangles, and circles represent the relative abundance of sodium channel I, II, and III mRNAs, respectively; similar hybridization efficiencies for the three specific probes are assumed. The densitometric values have been corrected for differences in autoradiographic exposure times and in specific activities of the probes. Averaged data comprising 2–5 independent RNA samples (deviation from the means being within ±10 %) are shown for total brain at E12, P4, P7, P15, P30, P60, and P90 and for spinal cord at E12, P7, P30, P60, and P90. (**A** from Suzuki *et al*. Reprinted with permission.[11] **B** and **C** from Beckh *et al*.[12] Reprinted with permission.)

FIGURE 3 (B and C) shows the temporal expression patterns of the three sodium channel mRNAs in brain and spinal cord, respectively.[12] The initial increase of all three sodium channel mRNAs occurs several days earlier in the spinal cord than in the brain. The temporal expression patterns reveal two time-dependent switches in sodium channel mRNA expression in the CNS: (1) a switch in expression of sodium channel III and I mRNAs is observed in all the regions studied at a time when both mRNA levels are about half-maximal and (2) a second switch in expression of sodium channel II and I mRNAs is observed in the spinal cord and medulla-pons.[12] These findings suggest that sodium channel III is expressed predominantly at fetal and early postnatal stages, whereas sodium channel I is expressed predominantly at late postnatal stages. The expression of sodium channel II is suggested to occur throughout the developmental stages studied, being subject to greater regional variability.

EXPRESSION OF FUNCTIONAL SODIUM CHANNEL FROM cDNA

mRNAs specific for rat sodium channels I, II, and III were synthesized by transcription *in vitro* of the respective cDNAs using the bacteriophage SP6 promoter. *Xenopus* oocytes injected with the sodium channel II–specific or the sodium channel III–specific mRNA show a transient inward current (up to 19–26 µA in Ringer's solution) when the holding membrane potential is shifted from − 100 mV to − 10 mV under voltage clamp.[11,13] The inward current is blocked by tetrodotoxin (TTX) as well as by saxitoxin (STX). On the other hand, oocytes injected with the sodium channel I–specific mRNA exhibit only a small TTX-sensitive response.[13] FIGURE 4(A) shows an example of the dose-response curves for TTX obtained from oocytes injected with the sodium channel II–specific mRNA. The apparent dissociation constant for TTX (K_{TTX}) ranges from 10 nM to 14 nM. When the external Na$^+$ concentration is lowered by replacement with tetraethylammonium, tetramethylammonium, or sucrose, the TTX-sensitive inward current is reduced in a dose-dependent manner, being virtually abolished at ∼ 3 mM Na$^+$ (FIG. 4,B). No significant difference in functional properties is observed between sodium channels II and III expressed in oocytes. FIGURE 4(C) shows the peak inward current-voltage (I-V) relation for sodium channel II expressed in an oocyte. The maximum current occurs at a potential of − 13.8 ± 4.5 mV.

The properties of sodium channel II expressed in oocytes were analyzed[14] according to the model of Hodgkin and Huxley,[6] assuming three activation gates. The voltage dependence of steady-state activation of sodium channel II is in good agreement with that of sodium channels described for the rat peripheral nerve [15] (FIG. 5,A), and its single-channel properties closely resemble those reported for cultured rat muscle cells[16] (see also FIG. 10). However, the steady-state inactivation of sodium channel II occurs at less negative potentials, compared with the data obtained for rat peripheral nerve[15] and muscle[17] (FIG. 5,B). A consequence of this shift is that the activation and inactivation curves overlap over a wider potential range than in peripheral nerve or muscle. This wider overlap in the potential range between − 60 mV and − 40 mV is expected to produce a slowly inactivating inward Na$^+$ current at potentials near the threshold of action potential firing. A persistent inward Na$^+$ current in this potential range has been reported for hippocampal neurons,[18] where it is thought to aid the repetitive firing of action potentials.

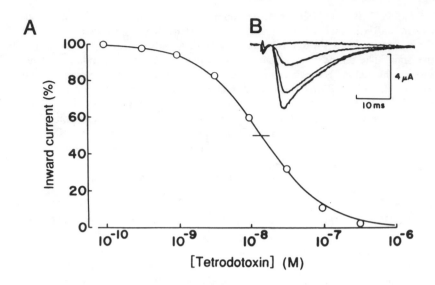

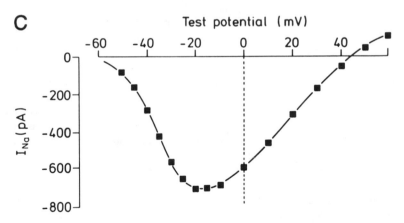

FIGURE 4. Expression of functional sodium channels from cDNA. (**A**) Effect of TTX on depolarization-activated whole-cell inward currents in *Xenopus* oocytes injected with the sodium channel II–specific mRNA. The curve represents the dose-response relation expected from a K_{TTX} of 14 nM (indicated by a horizontal bar) according to the equation $y = (1 + T/K_{TTX})^{-1}$, where T is the TTX concentration. (**B**) Effects of changes in external Na^+ concentration replaced by tetraethylammonium ions. The four records (from bottom to top) were obtained at external Na^+ concentrations of 118 mM, 80 mM, 41 mM, and 2.75 mM, respectively. (**C**) Peak inward current versus voltage relation. The current records were obtained from a membrane patch of an oocyte injected with the sodium channel II–specific mRNA. By interpolation, the reversal potential is $V_{rev} = 45$ mV (**A** and **B** from Noda *et al.*[13] Reprinted from *Nature*, Copyright 1986 Macmillan Journals Limited. **C** from Stühmer *et al.*[14] Reprinted with permission.)

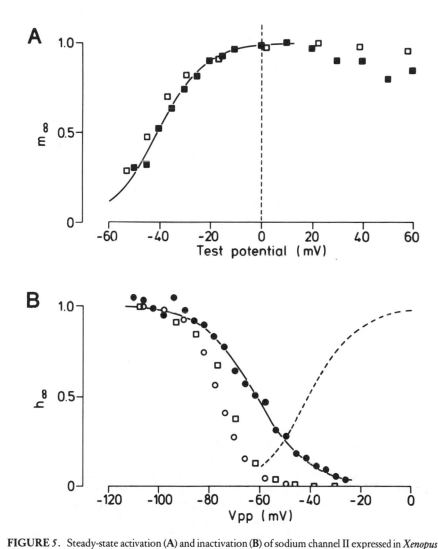

FIGURE 5. Steady-state activation (**A**) and inactivation (**B**) of sodium channel II expressed in *Xenopus* oocytes. (**A**) Steady state activation parameter m_∞ as a function of test potential (filled symbols). The smooth line represents the best fit to the equation $m_\infty = 1/[1 + \exp\left(\dfrac{V^m_{1/2} - V_t}{a_m}\right)]$ with $V^m_{1/2} = -40.5$ mV and $a_m = 9.4$ mV. The open symbols represent corresponding values from rat peripheral nerve.[15] (**B**) Steady state inactivation parameter b_∞ as a function of prepulse potential (V_{pp}) (filled symbols). Currents were elicited by a test pulse to -10 mV, following conditioning prepulses of 36 msec duration to potentials between -110 mV and -26 mV. The solid line represents a non-linear least-squares fit to the equation $b_\infty = 1/[1 + \exp\left(\dfrac{V_{pp} - V^h_{1/2}}{a_h}\right)]$ with $V^h_{1/2} = -62$ mV and $a_h = 10.6$ mV. For comparison, equivalent data from rat peripheral nerve[15] are shown as open squares, and the open circles represent values for rat twitch muscle[16] plotted using $V^h_{1/2} = -76$ mV and $a_h = 5.7$ mV. The activation relation from (**A**) (dashed line) is also shown to indicate the potential range where activation and inactivation overlap. (Reprinted with permission from Stühmer et al.[14])

There has been considerable controversy as to the subunit structure of the sodium channel. The sodium channels purified from the electric organ of *E. electricus*[19,20] and from chick cardiac muscle[21] consist of a single large polypeptide of M_r ~ 260,000, whereas those purified from rat brain[22] and from rat and rabbit skeletal muscle[23] contain, in addition to the large polypeptide (α-subunit), one or two smaller polypeptides of M_r 33,000–43,000 (β-subunits). The findings described above indicate that the mRNAs derived from the rat brain cDNAs encoding the sodium channel large polypeptide can direct the formation of functional sodium channels in *Xenopus* oocytes. The functional properties of the sodium channels expressed from the cDNAs are comparable to those of the sodium channels produced in oocytes injected with poly(A)$^+$ RNA from rat brain. Therefore, the function of the β-subunits is not clear at the moment. Nevertheless a low-molecular-weight fraction from the total-brain mRNA has been shown to modulate sodium channel inactivation.[24]

MAPPING OF FUNCTIONAL REGIONS

Voltage Sensor

The ability to respond to a change of the transmembrane voltage is one of the peculiar properties of the voltage-gated ion channels and is a basic mechanism underlying the electrical excitability of nerve and muscle membranes. To test the hypothesis that the positive charges in segment S4 serve as the voltage sensor, we introduced point mutations into segment S4 to replace positively charged amino acid residues by neutral or negatively charged residues, and analyzed the functional properties of the resultant mutant sodium channels expressed in *Xenopus* oocytes.[25] The results showed that reducing the net positive charge in segment S4 of repeat I causes a decrease in apparent gating charge, as manifested by a reduction in the steepness of the potential dependence of activation (FIG. 6,A). A roughly inverse relationship was observed between the apparent gating charge and the decrease in total net positive charge (FIG. 6,B). This finding provides experimental evidence that the positive charges in this segment are involved in the voltage-sensing device for activation of the sodium channel.

All the modifications in S4 caused a shift in the range of activation along the voltage axis [TABLE 1 and Stühmer *et al.* (figure 3)[25]] including a substitution of one unchanged residue for another.[27] Channel gating involves transition among several states. Mutations that change the stability of the closed and/or open states alter the equilibrium distribution between them and thereby shift the voltage dependence of the channel. The effect of a mutation on the shift must be expected most when the mutation resides in the voltage sensor region. Therefore, the shifts of activation and inactivation are consistent with the idea that segment S4 forms the voltage sensor. The unique structural features of segment S4 are strikingly well conserved in the calcium channel and the potassium channel, which show amino-acid sequence homology with the sodium channel. Therefore, this mechanism seems to be shared by other voltage-gated ionic channels as well.[26]

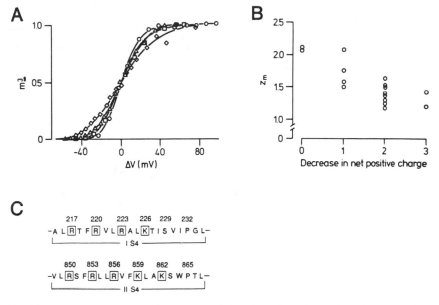

FIGURE 6. (A) Comparison of the steepness of the voltage dependence of steady-state activation ($m_\infty^{\frac{1}{3}}$) for the wild-type (○) and the mutant sodium channels K226Q (□), R217Q·K226Q (△), and R217Q·R220Q·R223Q (◇). Single representative experiments are shown. For the purpose of making changes in slope readily visible, the activation curve of each mutant has intentionally been shifted along the voltage axis to have the same voltage of half activation as the wild-type. The continuous lines correspond to the best fits of the data points according to the equation $m_\infty = 1/[1 + \exp\left(\frac{z_m e_o(V - V_{1/2}^m)}{kT}\right)]$. The valence of the apparent single-gate change, Z_m, is 2.1, 1.7, 1.5, and 1.1, respectively. These values represent the valence of the gating charge contributed by a single gate if a hypothetical channel comprising three identical gating subunits is assumed. (B) Changes in Z_m with decreases in the total number of net positive charges in segment S4 of repeat I at amino acid positions 217, 220, 223, and 226. The values plotted as Z_m are the average Z_m values given in TABLE 1, but with one more digit to avoid overlapping of similar data points. Replacement of a positively charged residue by a negatively charged one is counted as a decrease of two net positive charges. (C) Amino acid sequences[2] of segment S4 in repeat I (IS4) and repeat II (IIS4) of wild-type rat sodium channel II. The termini of the segments are tentatively assigned. Positively charged residues are boxed with solid lines and the numbers[2] of the relevant residues are given. (From Stühmer et al.[25] Reprinted from *Nature*, Copyright 1989 Macmillan Journals Limited.)

Inactivation Gate

The presence of four internal repeats suggests that the sodium channel evolved by duplications of an ancestral gene.[1] To examine whether individual repeats or their combinations can form functional sodium channels, we prepared mRNAs encoding single repeats or several contiguous repeats by transcription *in vitro* of the corresponding cDNAs (FIG. 7). Next, we investigated the effects of cleavage or deletion of putative cytoplasmic regions on sodium channel function at the same time. The results in this study[25] suggested that all four repeats are required for expression of functional channels,

TABLE 1. Properties of Wild-type and Mutant Sodium Channels Expressed in *Xenopus* Oocytes

Mutant	ΔQ	Activation $V^m_{1/2}$ (mV)	Z_m (e_0)	n	Inactivation $V^h_{1/2}$ (mV)	a_h (mV)	n
Wild type	0	-32 ± 7	2.1 ± 0.2	13	-61 ± 9	10.4 ± 1.3	9
R217Q	1	-34 ± 7	2.1 ± 0.3	6	-66 ± 8	9.6 ± 1.1	6
R220Q	1	-40 ± 10	1.5 ± 0.1	4	-67 ± 21	9.9 ± 0.8	4
R223Q	1	-20 ± 9	1.6 ± 0.2	5	-55 ± 14	11.8 ± 2.8	4
K226Q	1	-13 ± 2	1.8 ± 0.2	5	-67 ± 6	11.3 ± 1.0	5
K226E	2	-3 ± 6	1.2 ± 0.1	3	-71	9.9	2
K226D	2	-12 ± 10	1.5 ± 0.2	3	-69 ± 14	10.4 ± 2.2	3
K226R	0	-32 ± 4	2.1 ± 0.2	5	-70 ± 7	10.4 ± 1.7	4
S229R	-1 (0)	-25 ± 3	2.1 ± 0.2	3	-61 ± 5	10.4 ± 0.9	3
R217Q·R220Q	2	-51 ± 5	1.2 ± 0.2	5	-87 ± 5	8.6 ± 0.9	3
R217Q·R223Q	2	-49 ± 2	1.3 ± 0.1	4	-78 ± 2	8.6 ± 0.9	4
R217Q·K226Q	2	-14 ± 10	1.5 ± 0.1	4	-83 ± 7	9.6 ± 1.8	3
R220Q·R223Q	2	-28 ± 16	1.6 ± 0.2	4	$-$	$-$	$-$
R220Q·K226Q	2	-14 ± 5	1.4 ± 0.1	3	-61 ± 3	10.8 ± 2.8	3
R223Q·K226Q	2	-6 ± 6	1.4 ± 0.2	4	-70 ± 14	12.4 ± 1.2	3
K226R·S229R	-1 (0)	-25 ± 4	2.0 ± 0.1	4	-76 ± 10	9.2 ± 0.7	4
S229K·P232R	-2 (0)	-17 ± 7	1.9 ± 0.1	4	-63 ± 8	9.9 ± 1.2	4
R217Q·R220Q·R223Q	3	-44 ± 13	1.2 ± 0.2	3	-74	10.8	2
R217Q·R220Q·K226Q	3	-41 ± 7	1.4 ± 0.2	4	-82 ± 12	13.1 ± 3.4	3
K862Q	1 (0)	-22 ± 5	2.1 ± 0.1	5	-70 ± 7	9.6 ± 0.7	4
K859Q·K862Q	2 (0)	-5 ± 7	2.1 ± 0.2	6	-60 ± 12	13.1 ± 3.4	5
K226Q·K859Q·K862Q	3 (1)	11 ± 9	1.8 ± 0.3	5	-83 ± 8	8.9 ± 1.0	4
ΔN	0	-31 ± 3	2.0 ± 0.2	5	-68 ± 11	13.8 ± 1.5	5
cX-1	0	$-$	$-$	4	$-$	$-$	4
cY-1	0	-46 ± 7	2.0 ± 0.3	3	-75 ± 10	11.8 ± 2.3	3
cY-2	0	-33 ± 2	2.0 ± 0.1	3	-78 ± 5	13.1 ± 1.4	3
ΔY	0	-31 ± 3	2.2 ± 0.2	5	-71 ± 8	12.4 ± 1.2	5
cZ-1	0	-36 ± 6	2.2 ± 0.2	4	-60 ± 7	13.8 ± 1.5	3
cZ-2	0	-37 ± 4	1.8 ± 0.1	3	-61	11.3	2
ΔC	0	-40 ± 8	2.2 ± 0.2	5	-80 ± 9	8.3 ± 1.1	3

Data are given as means $\pm$ s.d. for the wild-type sodium channel and the sodium channels with point mutations in segment S4 of repeats I and/or II (upper part) and for those with a deletion, a cut, or a cut/addition (lower part). All the data are taken from cell-attached macro patch recordings. ΔQ is the reduction in positive charge caused by the mutation; the values in parentheses are the reduction in positive charge at amino acid positions 217, 220, 223, and 226 of segment S4 of repeat I. $V^m_{1/2}$ and $V^h_{1/2}$ are the single-gate equilibrium potentials of activation and inactivation. Z_m is the valence of the apparent single-gate charge for activation and a_h is the slope factor, which is inversely proportional to the maximal derivative of the inactivation curve. The average values for Z_m and a_h are rounded to two digits. n is the number of oocytes used, which were taken from at least two different series of successful injections; several patches were obtained from most oocytes. No inactivation data were available for R220Q·R223Q. cX-1 yielded currents too small to allow reliable measurements. *Xenopus laevis* oocytes were injected with the wild type (0.2 µg/µl) or a mutant mRNA (0.2–0.5 µg/µl; total concentration of an equimolar mixture for c-type mutants) and incubated for 4–7 days. (Based on Stühmer et al.[25])

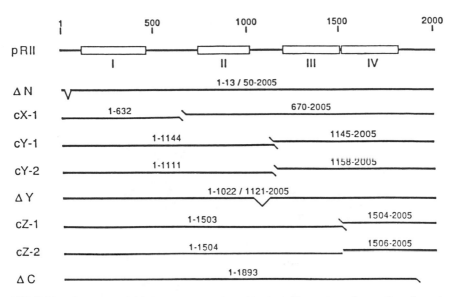

FIGURE 7. Structures of deletion mutants and combinations. The regions of rat sodium channel II carried by the individual mutants are shown by horizontal lines with the numbers of the constituent amino acid residues; oblique lines at the termini of some constructs indicate the presence of short additional sequences resulting from the strategy used.[25] V-shaped lines indicate internal deletions. On the top, the protein-coding region is shown and amino acid numbers are given above the diagram. The coding regions for the four internal repeats (I–IV) are indicated by open boxes. (Based on Stühmer et al.[25])

and therefore the repeats must assemble spontaneously when separate mRNAs coding for the various repeats are co-injected.

In FIGURE 8(A), macroscopic currents recorded from oocytes implanted with the wild type and the mutant cY-2, cZ-1, or cZ-2 are shown. Apart from their magnitude, the currents produced by the mutant cY-2, which has a cut/addition between repeats II and III, are similar to the wild-type. By contrast, the currents evoked by the mutants cZ-1 and cZ-2, which have a cut/addition or a cut, respectively, between repeats III and IV, are characterized by a dramatic decrease in the rate of inactivation. A quantitative comparison of the steady-state and kinetic properties of cY-2 and cZ-1 with those of the wild-type channel is shown in FIGURE 8 (B and C). For both mutants, the steady-state properties of activation and inactivation are similar to those of the wild-type. However, the τ_h of the mutant cZ-1 becomes nearly voltage-independent, being about 30-fold greater at strong depolarizations than that of the wild-type.

By contrast with the wild-type channel, which exhibits short openings clustered at the beginning of the depolarization,[14] openings of the mutant cZ-1 sometimes lasted for periods as long as the voltage step (80 msec)[25] (FIG. 9,A). The open-time histograms for elementary currents flowing through the wild type and mutant cZ-1 channels are shown in FIGURE 9(B). The mean open time of the mutant cZ-1 (5.8 msec) is more than one order of magnitude larger than that of the wild type channel (0.43 msec).

Our results show that cleavage of the linkage between repeats III and IV of the

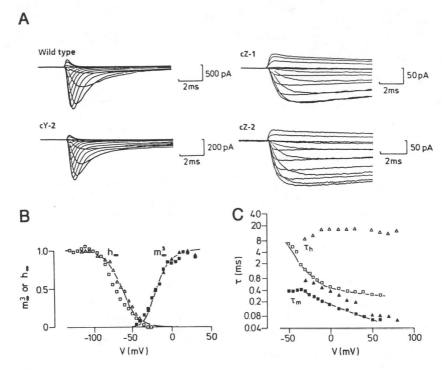

FIGURE 8. (**A**) Current responses to depolarizations in oocytes injected with the wild type (0.20 μg/μl) and the mutant sodium channel cY-2 (0.33 μg/μl), cZ-1 (0.33 μg/μl), or cZ-2 (0.5 μg/μl) mRNA. Responses were evoked by depolarizations ranging from −60 to +70 mV (+90 mV for cY-2) in 10 mV steps from a holding potential of −120 mV. Averages of 16 individual traces. Temperature: 15°C. (**B**) Voltage dependence of steady-state activation (m_∞^3, filled symbols) and inactivation (h_∞, open symbols) for cY-2 (■,□) and for cZ-1 (▲,△). The smooth lines represent the best fit of the data for the wild-type. (**C**) Voltage dependence of the time constants of activation (τ_m, filled symbols) and inactivation (τ_h, open symbols) for the same mutant channels as in **A** and **B**. The continuous lines correspond to the wild-type data. (From Stühmer et al.[25] Reprinted from *Nature*, Copyright 1989 Macmillan Journals Limited.)

sodium channel causes a strong reduction in the rate of inactivation. This finding, together with the similar effect observed for the wild-type sodium channel treated with intracellularly applied endopeptidases, supports the view that this region, located on the cytoplasmic side of the membrane, is involved in the inactivation of the sodium channel.[25]

Toxin Binding Site and Channel Pore Lining

A single point mutation (E357Q) in the region between S5 and S6 of repeat I decreases the sensitivity to TTX and STX by at least four orders of magnitude (FIG. 10, A–C).[28] This mutation substitutes the glutamic acid residue at position 387 with glutamine, effectively neutralizing a negative charge. The properties of the mutant

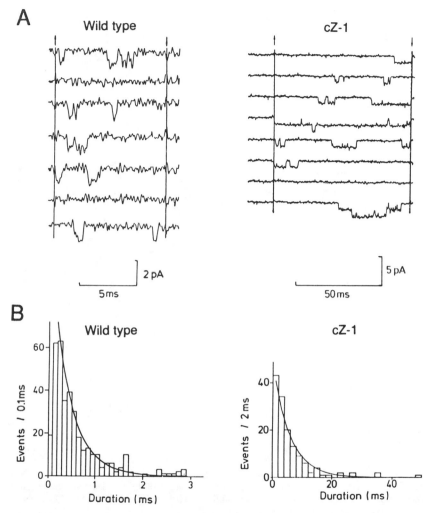

FIGURE 9. (A) Single-channel currents recorded from a patch on an oocyte implanted with the wild type or the mutant sodium channel cZ-1. For the wild type, responses to 10 msec depolarization to − 32 mV from a holding potential of − 83 mV are shown. For the cZ-1 mutant, responses to successive 80 msec depolarization to − 20 mV from − 100 mV are shown. (B) Distribution of open times of elementary current pulses of the wild-type sodium channel and the cZ-1 mutant. The respective distributions are fitted with single exponentials with decay time constants of 0.43 msec and 5.8 msec. (Reprinted with permission from Stühmer *et al*.[14,25])

E387Q resemble closely those of natural channels exposed to trimethyloxonium (TMO). TMO treatment of frog sodium channels yields a loss of TTX sensitivity, a more linear instantaneous I-V relation, and a threefold reduction of conductance with little modification of gating.[29] These effects are reproduced in the mutant E387Q in

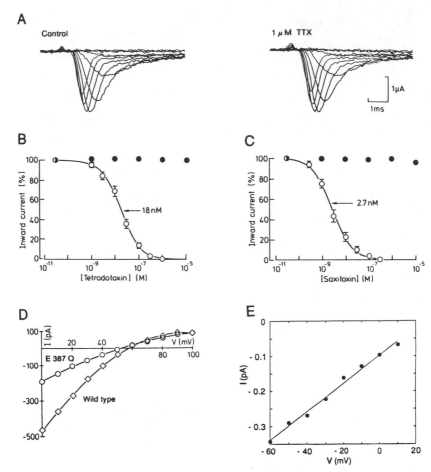

FIGURE 10. (A) Whole-cell current responses of the mutant E387Q recorded with a two-electrode voltage clamp. Current records from control condition and the perfusion with 1 μM TTX are shown. Depolarizing steps were between − 60 and + 30 mV, in steps of 10 mV from a holding potential of − 80 mV. (B and C) Dose-response curves for the wild type (open circles) and the mutant E387Q (filled circles) to TTX and STX. The smooth lines are fitted to the open circles according to the equation $y = [1 + (T/IC_{50})^n]^{-1}$, where T is the toxin concentration and n the Hill coefficient. Error bars are ± SD and indicated only when larger than the symbol size. (B) For the wild type, the IC_{50} value for TTX is 18 nM with a Hill coefficient of 1.1. Data were averaged from 7 (wild type) and 8 (mutant) experiments. (C) For the wild type, the IC_{50} value for STX is 2.7 nM with a Hill coefficient of 1.1. Data were averaged from 4 (wild type) and 7 (mutant) experiments. (D) Current-voltage relationships for tail currents after a 0.5-msec depolarizing pulse to + 30 mV. The wild-type tail current amplitudes have been scaled down by a factor of 1.2 so that the outward currents become comparable. All the data were obtained from macro-patches. (E) Noise analysis of mutant E387Q. A single-channel current-voltage relationship over a large voltage-range is shown. Single-channel currents (filled circles) obtained by noise analysis from one inside-out patch are plotted versus test voltage. Pipette solution: NFR; bath (internal) solution: 30 mM NaCl, 90 mM KCl, 10 mM HEPES, 10 mM EGTA, pH 7.2. The straight line represents a linear regression line ($r = 0.99$) with a slope of 4.0 ± 0.2 pS (wild type; 19 pS) and an extrapolated reversal potential of 24 ± 2 mV. (A–D from Noda et al.[28] E from Pusch et al.[30])

the absence of TMO (FIG. 10, D and E), making this position a probable site of action for TMO.

Another mutation (D384N) that neutralizes the aspartic acid D384 (three residues apart from position 387) to asparagine again renders the channel insensitive to TTX and STX (FIG. 11, A and B).[30] Mutant D384N has a very low permeability for any of the following ions: Cl^-, Na^+, K^+, Li^+, Rb^+, Ca^{2+}, Mg^{2+}, NH_4^+, TMA^+, TEA^+. However, asymmetric charge movements similar to the gating currents of the Na^+-selective wild-type are still observed (FIG. 11,C). The TTX- and STX-binding site is thought to reside close to the sodium channel pore since channel modifications that affect toxin binding reduce the inward current. These findings suggest that residues D384 and E387 are located in the extracellular mouth or inside the ion-conducting pore of the channel.

These two residues belong to the short segment SS2 in the region between the hydrophobic segments S5 and S6 in repeat I (FIG. 2). In each repeat the S5-S6 region is thought to contain two short segments, SS1 and SS2, that may partly span the membrane as a hairpin. SS2 segments have been postulated as forming part of the channel lining.[5] Actually, all mutations at the equivalent position in the SS2 region of the other repeats (E942 and E945 in repeat II, K1422 and M1425 in repeat III, A1714 and D1717 in repeat IV) strongly reduce toxin sensitivity.[31] This suggests that these pairs of residues of the four repeats form part of the extracellular mouth and make the determinants of TTX and STX sensitivity.

Recently, Heinemann et al.[32] reported that the single mutations K1422E and A1714E in the SS2 segment of repeats III and IV, respectively, which occur at the equivalent positions in the calcium channels, alter the ion selectivity of the sodium channel to resemble those of calcium channels. Furthermore, the channel carrying both mutations is not only permeable to Ca^{2+} and Ba^{2+}, but is also selective for Ca^{2+} over Na^+ at their physiological concentrations. These findings suggest that these sites of the sodium channel and corresponding sites of the calcium channel form part of the selective filters of these channels and supports the view that the SS1–SS2 region of voltage-gated ionic channels forms part of the channel lining.[32]

CONCLUDING REMARKS

The complete amino acid sequences of the *Electrophorus* electroplax sodium channel and the three distinct sodium channels from rat brain have been elucidated by cloning and sequencing the cDNAs. The deduced primary structure suggests functional regions involved in the operation of this voltage-gated ionic channel. Expression of the cloned cDNAs yields functional sodium channels in *Xenopus* oocytes. The functional properties, including single-channel characteristics, of the sodium channel have been studied.

Furthermore, functional regions have been mapped by the analysis of mutant channels produced by site-directed mutagenesis. The results obtained show the validity of the structural prediction based on the primary structure. Future studies will be directed to work out how the amino acids are arranged in three dimensions.

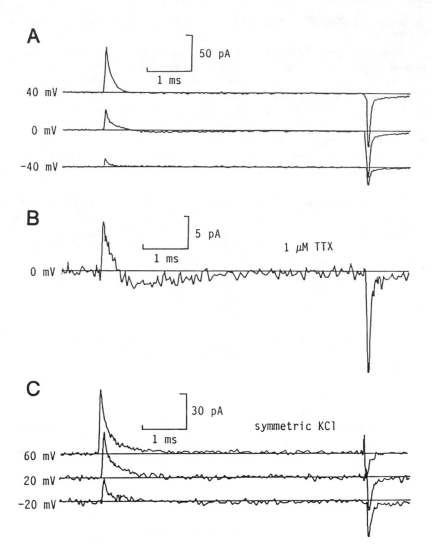

FIGURE 11. Voltage-clamp currents of mutant D384N. (**A**) Traces from a cell-attached patch with NFR in the recording pipette obtained by stepping the membrane voltage from the holding voltage of − 100 mV to the indicated values. The intracellular potential of the oocyte was monitored with a microelectrode, filled with 2 M KCl. (**B**) Trace obtained under identical conditions as the traces in **A** except that 1 μM TTX was included in the pipette filling solution. (**C**) Current traces from an inside-out patch from a different oocyte than in **A** [holding voltage in **C**: − 97 mV]. The solution in **C** was (symmetrical): 10 mM HEPES, 10 mM EGTA, 100 mM KCl, pH 7.2. In **A, B,** and **C** linear leakage and capacitive currents were subtracted by a P/4 method with a "P/4 holding" of − 140 mV. (Reprinted with permission from Pusch *et al.*[30])

ACKNOWLEDGMENT

I would like to thank the late Prof. Shosaku Numa for his continuous encouragement for a long period.

REFERENCES

1. Noda, M., S. Shimizu, T. Tanabe, T. Takai, T. Kayano, T. Ikeda, H. Takahashi, H. Nakayama, Y. Kanaoka, N. Minamino, K. Kangawa, H. Matsuo, M. S. Raftery, T. Hirose, S. Inayama, H. Hayashida, T. Miyata & S. Numa. 1984. Nature 312: 121–127.
2. Noda, M., T. Ikeda, T. Kayano, H. Suzuki, H. Takeshima, M. Kurasaki, H. Takahashi, H. Takahashi & S. Numa. 1986. Nature 320: 188–192.
3. Kayano, T., M. Noda, V. Flockerzi, H. Takahashi & S. Numa. 1988. FEBS Lett. 228: 187–194.
4. Numa, S. & M. Noda. 1986. Ann. N.Y. Acad. Sci. 479: 338–355.
5. Guy, H. R. & F. Conti. 1990. Trends Neurosci. 13: 201–206.
6. Hodgkin, A. L. & A. F. Huxley. 1952. J. Physiol. 117: 500–544.
7. Hille, B. 1992. Ionic Channels of Excitable Membranes. Sinauer Associates. Sunderland, MA.
8. Armstrong, C. M. & F. Bezanilla. 1973. Nature 242: 459–461.
9. Keynes, R. D. & E. Rojas. 1974. J. Physiol. (Lond.) 239: 393–434.
10. Armstrong, C. M. 1981. Physiol. Rev. 61: 644–683.
11. Suzuki, H., S. Beckh, H. Kubo, N. Yahagi, H. Ishida, T. Kayano, M. Noda & S. Numa. 1988. FEBS Lett. 228: 195–200.
12. Beckh, S., M. Noda, H. Lübbert & S. Numa. 1989. EMBO J. 8: 3611–3616.
13. Noda, M., T. Ikeda, H. Suzuki, H. Takeshima, T. Takahashi, M. Kuno & S. Numa. 1986. Nature 322: 826–828.
14. Stühmer, W., C. Methfessel, B. Sakmann, M. Noda & S. Numa. 1987. Eur. Biophys. J. 14: 131–138.
15. Neumcke, B. & R. Stämpfli. 1982. J. Physiol. (Lond.) 329: 163–184.
16. Sigworth, F. J. & E. Neher. 1980. Nature 287: 447–449.
17. Almers, W., W. M. Roberts & R. L. Ruff. 1984. J. Physiol. (Lond.) 347: 751–768.
18. French, C. R. & P. W. Gage. 1985. Neurosci. Lett. 56: 289–293.
19. Agnew, W. S., S. R. Levinson, J. S. Brabson & M. A. Raftery. 1978. Proc. Natl. Acad. Sci. USA 75: 2606–2610.
20. Miller, J. A., W. S. Agnew & S. R. Levinson. 1983. Biochemistry 22: 462–470.
21. Lombet, A. & M. Lazdunski. 1984. Eur. J. Biochem. 141: 651–660.
22. Hartshorne, R. P. & W. A. Catterall. 1981. Proc. Natl. Acad. Sci. USA 78: 4620–4624.
23. Barchi, R. L. 1983. J. Neurochem. 40: 1377–1385.
24. Krafte, D. S., T. P. Snutch, J. P. Leonard, N. Davidson & H. A. Lester. 1988. J. Neurosci. 8: 2859–2868.
25. Stühmer, W., F. Conti, H. Suzuki, X. Wang, M. Noda, N. Yahagi, H. Kubo & S. Numa. 1989. Nature 339: 597–603.
26. Liman, E. R., P. Hess, F. Weaver & G. Koren. 1991. Nature 353: 752–775.
27. Auld, V. J., A. L. Goldin, D. S. Krafte, W. A. Catterall, H. A. Lester, N. Davidson & R. J. Dunn. 1990. Proc. Natl. Acad. Sci. USA 87: 323–327.
28. Noda, M., H. Suzuki, S. Numa & W. Stühmer. 1989. FEBS Lett. 259: 213–216.
29. Spalding, B. C. 1980. J. Physiol. (Lond.) 305: 485–500.
30. Pusch, M., M. Noda, W. Stühmer, S. Numa & F. Conti. 1991. Eur. Biophys. J. 20: 127–133.
31. Terlau, H., S. H. Heinemann, W. Stühmer, M. Pusch, F. Conti, K. Imoto & S. Numa. 1991. FEBS Lett. 293: 93–96.
32. Heinemann, S. H., H. Terlau, W. Stühmer, K. Imoto & S. Numa. 1992. Nature 356: 441–443.

Molecular Aspects of Ion Permeation through Channels

KEIJI IMOTO

Department of Medical Chemistry
Kyoto University Faculty of Medicine
Kyoto 606-01, Japan

INTRODUCTION

The essential function of ion channels is to provide a passage for ions to permeate across the membrane. However, most of the ion channels behave not as simple pores but exhibit ion selectivity. This fascinating property has led us to many investigations of various types of ion channels, such as the nicotinic acetylcholine receptor in the end-plate of neuromuscular junctions and the sodium channel of squid giant axons. In these tissues, the channels are densely packed and relatively homogeneous. Combined with classical biophysics, electrophysiological analyses successfully constructed the framework to explain how these ion channels operate.[1] In most cells, however, channels are less dense and many species of channels are often co-localized, keeping us from studying those ion channels in detail. Furthermore, we did not have experimental methods to verify the biophysical interpretations of channel functions.

Introduction of molecular biology has revolutionized our approach to the understanding of ion channels.[2] Firstly, cloning and sequence analysis of complementary DNAs (cDNAs) has revealed the primary structure of ion channels, providing the perspective of further biochemical analyses and a rational basis for tertiary structure prediction. Secondly, channel proteins can be over-expressed from the cloned cDNAs in various cells for functional characterization and possibly for future X-ray crystallography. And thirdly, we can modify the channel proteins using recombinant DNA techniques, which allows us to test the biophysical interpretations by analyzing the site-specifically mutated ion channels. This article deals with identification of the channel-forming region and analysis of the effects of site-specific mutations on ion permeation properties of the nicotinic acetylcholine receptor channel and the voltage-gated sodium channel. Possible inferences on the molecular basis of ion permeation through ion channels are also discussed.

NICOTINIC ACETYLCHOLINE RECEPTOR CHANNEL

Identification of Pore-forming Region

The nicotinic acetylcholine receptor (AChR) is an archetypal ligand-gated channel. It is a pentameric transmembrane protein composed of four kinds of homologous subunits assembled in the molar stoichiometry $\alpha_2\beta\gamma\delta$ and arranged pseudosymmetrically around a central channel. The hydropathy analysis revealed that each subunit has four hydrophobic segments (M1–M4), which are presumably transmembrane segments.

The *Torpedo* and bovine AChR channels, expressed in *Xenopus* oocytes by injecting the respective sets of mRNA specific to α, β, γ, and δ subunits, showed different single-channel conductance at low divalent cation concentration.[3] By analyzing hybrid AChRs in which a bovine subunit was substituted for the corresponding *Torpedo* subunit, the difference in conductance between the *Torpedo* and bovine channels was found to be ascribable, at least partly, to a difference in their δ subunits. To localize that part of the δ-subunit molecule responsible for the difference, we measured single-channel conductance of the AChRs containing chimeric δ subunits in which portion of the *Torpedo* δ subunit were systematically replaced by homologous regions of the bovine δ subunit. The results suggested that the region comprising the M2 hydrophobic segment and its vicinity contains an important determinant of the rate of ion transport through the AChR channel. They also suggested that this region is responsible for the reduction in channel conductance caused by divalent cations.

In an attempt to identify those amino acid residues that interact with permeating ions, we systematically introduced mutations into the *Torpedo* AChR subunits cDNAs so that the net charge of the charged or glutamine residues around the M2 segment was altered (Fig. 1).[4] These mutant AChRs were tested for single-channel conductance. The results showed that reduction in net negative charge of glutamic acid 262 of the α subunit (αE262) and/or of the residues at the equivalent positions of the β, γ, and δ subunits resulted in decreased conductance. An approximately inverse relationship was found between channel conductance and change in total net negative charge. This observation suggested that the residues in this cluster are located equidistantly from the channel axis. Analogous experiments in the cluster at aspartic acid 238 of the α subunit (αD238) and the residues at the equivalent positions of the β, γ, and δ subunits revealed a similar inverse relationship. Reducing the net negative charge of the cluster of glutamic acid 241 of the α subunit (αE241) and the residues at the equivalent positions of the other subunits showed a much stronger reduction in conductance. These findings suggested that these three clusters of negatively charged and glutamine residues at the α subunit positions 238, 241, and 262 and the equivalent positions of the other subunits are major determinants of the rate of ion transport, the intermediate cluster being more critical. These amino acid residues are next to or near the M2 segment, probably forming three anionic ring-like structures.

Single-channel conductance of the AChR channel is affected by divalent cations. The mutation αE262K reduced sensitivity to extracellular Mg^{2+} without affecting that to cytoplasmic Mg^{2+}. On the other hand, the mutation αD238K reduced sensitivity to only cytoplasmic Mg^{2+}. These results suggested that the anionic rings at αD238 and αE262, which are involved in interactions with Mg^{2+}, constitute part of the cytoplasmic and the extracellular ring of the channel, respectively. The inward and the outward current of the mutant δE255Q were less sensitive to extracellular and cytoplasmic Mg^{2+}, respectively. The intermediate ring is therefore involved in the interaction with Mg^{2+}, and was suggested to be located between the cytoplasmic and extracellular

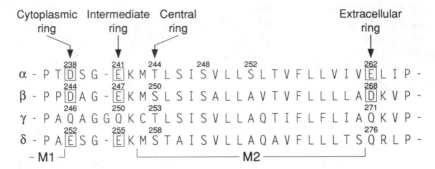

FIGURE 1. Regions surrounding the M2 segment of the α, β, γ, and δ subunits of the *Torpedo californica* AChR. The relevant amino acid sequences with one-letter code are aligned,[22] and the positions of the M1 and M2 hydrophobic segments are indicated. The four arrows indicate the cytoplasmic, intermediate, central, and extracellular rings. Negatively charged residues are boxed with solid lines. The numbers of amino acid residues subjected to site-specific mutations are given. (From Imoto *et al.*[5]. Reprinted with permission.)

rings. The one-sided effect on Mg^{2+} sensitivity of the mutations provided experimental evidence to prove the membrane topology of the three anionic rings, in which the M2 hydrophobic segment is a transmembrane segment with the M1–M2 portion on the cytoplasmic side and with the M2–M3 portion on the extracellular side. Since both the cytoplasmic and intermediate ring are located in the M1–M2 portion, this portion probably forms part of the transmembrane segment containing the M2 segment. This transmembrane segment probably constitutes at least part of the channel lining (FIG. 2).

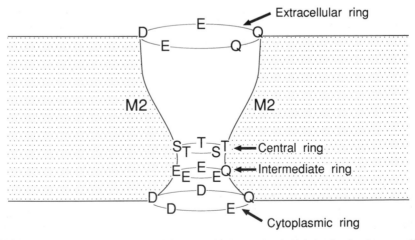

FIGURE 2. Schematic presentation of three anionic rings and one uncharged polar ring as major determinants of the rate of ion transport through the AChR channel. Amino acid residues in the four rings are represented in one-letter code.

Channel Constriction

For further characterization of the ion-conducting pore of the AChR channel, uncharged polar amino acid residues of the AChR subunits between the cytoplasmic and the extracellular ring were mutated so that the size and polarity of their side chains were altered (FIG. 1).[5] Threonine 244 of the α subunit (αT244) and the uncharged polar residues at the equivalent positions of the other subunits are predicted to reside at 0.8 α-helix turn from the residues forming the intermediate ring, facing the channel pore. When single-channel conductances of mutant channels (in which the residues αT244, βS250, and δS258 were altered) were plotted against the volume of the substituted side chain, two classes of effect became evident, one resulting from hydrophobic substitutions and the other resulting from polar substitutions. In each case, an approximately inverse relationship was found between channel conductance and size of the substituted side chain. When the substituted side chains compared were similar in size, stronger reductions in conductance were observed for hydrophobic substitutions than for polar substitutions. These results indicated that both the size and polarity of the residues αT244, βS250, and δS258 are critical for determining the rate of ion transport, and suggested that these residues come into close contact with permeating cations. Thus their side chains likely take part in forming a narrow channel constriction. These residues, together with γT253, probably form a ring-like structure. Interestingly, the extent of reduction in conductance varied depending on the subunit into which the mutations were introduced. Stronger effects were generally observed for the mutations of δS258, particularly when the subunit stoichiometry of $\alpha_2\beta\gamma\delta$ was taken into consideration. In contrast, the mutation of γT253 caused much smaller changes in conductance. The results suggested that individual residues at this position, particularly γT253, are not symmetrically arranged. Similar mutational studies of αS248 and the serine residues at the equivalent positions of the other subunits as well as αS252 suggested that the channel pore is larger in cross section at these positions.

Ion Permeation Properties of Mutated Channels

The AChR channel was regarded as a water-filled pore[6] because the permeability sequence of this channel is like that of free solution mobility of alkali metal cations ($Rb^+ \sim Cs^+ > K^+ > Na^+ > Li^+$). The single-channel conductance sequence of the AChR channel, however, was $K^+ > Rb^+ > Cs^+ > Na^+ > Li^+$, which is different from the mobility sequence. This finding showed that there must be at least one AChR channel region that selects between large alkali metal cations not according to their mobility in water. Since mutations in the three anionic rings affected the channel conductance, we examined contributions of the three anionic rings in determining the permeability and conductance sequence of the AChR channel (FIG. 1).[7] The mutant δE255Q, in which the negative charge was neutralized, showed reduced conductances for all cations tested, but most strikingly the conductance ratios of Cs^+ to K^+ (g_{Cs}/g_K) and Rb^+ to K^+ (g_{Rb}/g_K) were much smaller than those of the wild-type channel. The conductance sequence was $K^+ > Rb^+ > Na^+ > Cs^+ > Li^+$. The permeability sequence of this mutant was $K^+ > Rb^+ \sim Na^+ > Cs^+ > Li^+$. Thus both the permeability and conductance sequences for this mutant channel showed a shift in ion selectivity in the same direction in terms of conductance and permeability ratios g_{Cs}/g_K and P_{Cs}/P_K. An analogous, but less marked, change in conductance ratios was observed for the mutant βE247Q. The changes in conductance ratios caused by the mutation δE255Q were compensated by combining this mutation with an additional mutation in the intermediate ring in another subunit, γQ250E, which by itself did not appreciably affect the ion permeation properties. The results showed that the side chains of negatively charged amino acids in the intermediate ring are important in determining ion selectivity.

The mutant δE255D, in which the size of the side chain was reduced without altering its charge, showed g_{Cs}/g_K and g_{Rb}/g_K ratios larger than those of the wild-type channel. The conductance sequence of this mutant was $Rb^+ \sim K^+ > Cs^+ > Na^+ > Li^+$. A similar change in conductance sequence was observed for the mutant βE247D. These results suggested that the size of side chains in the intermediate ring is also important in alkali metal cation selection. No substantial change in conductance ratios resulted from the mutation αE262K in the extracellular ring or from the mutation αD238K in the cytoplasmic ring.

To gain more insight into the molecular basis of the ion selection mechanism, we estimated the pore size of mutant AChR channels by measuring permeability for various organic cations.[8] The sequence of permeability relative to Na^+ of the wild-type channel was ammonium > methylamine > ethylamine > ethanolamine > diethanolamine > tris(hydroxymethyl)aminomethane. From the relationship between the observed permeability ratios and the ion radius, the pore size of the wild-type channel was estimated to be 7.43 Å. These results were consistent with the previous observations.[9] The mutant γQ250N, in which the size of the side chain of the amino acid residue in the intermediate ring was reduced without altering the net charge, showed significantly increased permeability ratios of organic cations. The pore size of this mutant was estimated to be 7.98 Å. Analogous substitutions in the α, β, and δ subunits also resulted in increased permeability ratios, although different in degree among the subunits. Because changes in the size of the side chain are expected to result in alterations of the pore size, our observation provided experimental evidence that the physical dimension of the pore at the intermediate ring is a determinant of the permeability properties of the AChR channel.

The mutant δE255Q, in which the negative charge in the intermediate ring was neutralized, markedly reduced the permeability ratios of organic cations (except for ammonium ion). The pore size of this mutant was estimated to be 6.17 Å. A similar but less marked reduction in permeability ratio was caused by analogous mutations αE241Q and βE247Q. The mutation γQ250K resulted in a larger reduction of the permeability ratios (except for ammonium ion) than did the mutation δE255Q. The changes in the permeability ratio caused by the mutation δE255Q were compensated for by combining this mutation with an additional mutation in the intermediate ring in another subunit, γQ250E, being analogous to that observed for the conductance ratios of alkali metal cations. These results suggested that the negative charge at the intermediate ring is another determinant of the permeability of organic cations. We further observed that the changes caused by δE255Q were also compensated for by combining an additional mutation γQ250N, which reduced the size of the side chain. The permeability ratios of organic cations were not affected by changing the size (αD238E and αE262D) or the net negative charge (αD238K and αE262K) of the side chains in the cytoplasmic and extracellular rings.

The effect on the permeability ratios of changing net negative charge in the intermediate ring may suggest that the channel would be less permeable to bulky compounds because reduction in net negative charge would weaken electrostatic attraction of organic cation into the narrow channel constriction. Alternatively, however, the results suggested that decreasing the net negative charge in the intermediate ring reduces the pore size. Our finding suggested that negative charges in the intermediate ring are also structurally important in sustaining the channel pore by electrostatic repulsion between the charged side chains.

Predicted Architecture of AChR Channel

This series of mutagenesis work on the AChR channel allowed prediction of the architecture of the open channel. The inner wall of the channel is, at least partly, composed of the M2-containing transmembrane segments of each subunits. There are three anionic rings and one uncharged polar ring, which are major determinants of the rate of ion transport through the AChR channel (FIG. 2). Among those, the residues in the central ring and those in the intermediate ring form a narrow channel constriction, while the channel pore is larger in cross section on both sides of this constriction. Since the uncharged polar residues of the central ring and the anionic and uncharged polar residues of the intermediate ring are adjacent to each other on the assumed α-helices of the M2-containing transmembrane segments of the individual subunits, the channel constriction is suggested to be confined to a small region comprising these two rings, which is close to the cytoplasmic side of the membrane. A simplistic estimate of its length would be one α-helix turn (5.4 Å), which agrees with the value obtained from streaming potential measurements (3–6 Å).[10] The negatively charged residues in the cytoplasmic and extracellular rings may be important for concentrating permeant cations at the mouths of the channel.

SODIUM CHANNEL

Identification of Pore-forming Region

The voltage-gated sodium channel is a membrane protein that is essential for the generation of action potentials in excitable cells. When the primary structures of rat sodium channels I and II were elucidated by cloning and sequencing of the cDNAs, the S2 hydrophobic segment in each of the four internal repeats of homology was hypothesized to contribute to forming the inner wall of the channel,[11] partly on the analogy of the postulated structure of the AChR. Mutations in the S2 segment, however, did not significantly affect the single-channel conductance of the sodium channel (Imoto, unpublished observation). Meanwhile, a single point mutation E387Q of rat sodium channel II was found to abolish sensitivity of the sodium channel to tetrodotoxin and saxitoxin.[12] In addition, another mutation D384N almost completely eliminated ionic currents without preventing gating function as judged by gating currents.[13] These results suggested that D384 and E387 are located at the extracellular mouth or inside the ion-conducting pore of the channel. D384 and E387 belong to the short segment SS2 in the region between the S5 and S6 hydrophobic segments in repeat I. In each repeat the S5–S6 region was thought to contain two short segments, SS1 and SS2, that may partly span the membrane as a hairpin and the SS2 segments was postulated to form part of the channel lining.[14,15] We systematically introduced mutations in the region encompassing the SS2 segment of each of the four repeats, focussing mainly on charged residues, and tested the mutated channels for sensitivity to tetrodotoxin and saxitoxin and single-channel conductance.[16]

In all four repeats, mutations were found that made the IC_{50} (half-inhibitory concentration of the peak current) for tetrodotoxin and/or saxitoxin more than 100 times larger than the wild-type values. These mutations involved changes in the charge of the residues D384, E387, E942, E945, K1422, M1425, A1714, and D1717 (FIG. 3). Charge mutations at other positions produced only minor or insignificant changes in toxin sensitivity. Mutations without a change in the net charge (D384E and M1425Q) and at other positions of the SS2 segment (W368Y and W943Y) also had minor or insignificant effects. All of the mutations involving a decrease in net negative charge that strongly reduced toxin sensitivity also caused a marked decrease in single-channel conductance. These results showed that the sensitivity to tetrodotoxin and saxitoxin of the sodium channel is strongly reduced by mutations of specific amino acid residues in the SS2 segment of each of the four internal repeats. These residues were located in two clusters, the residues in each cluster being equivalently positioned in the aligned sequences. It was suggested, therefore, that these two clusters of predominantly negatively charged residues, probably forming ring-like structures, line part of the extracellular mouth and/or the pore wall.

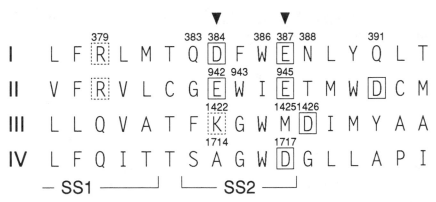

FIGURE 3. Regions encompassing the SS2 segment of the four repeats of rat sodium channel II. The relevant amino acid sequences with one-letter code are aligned,[11] and the positions of the SS1 and SS2 segments are indicated.[15] Negatively charged residues are boxed with solid lines and positively charged residues with broken lines. The numbers of amino acid residues subjected to site-specific mutations are given. The positions of the clusters of residues that have been identified as major determinants of toxin sensitivity are indicated by arrowheads. (From Terlau et al.[16] Reprinted with permission.)

Ion Permeation Properties of Mutated Channels

From a physiological point of view, the calcium channel is distinct from the sodium channel in terms of ion selectivity. However, when the primary structure of the calcium channel was determined, it turned out that the sodium channel and the calcium channel are homologous both in amino acid sequence and in proposed transmembrane topology.[17] If the amino acid sequences in the regions encompassing the short segment SS2 of the four repeats of sodium and calcium channels are aligned, the positions corresponding to one of the two amino acid clusters that determine the toxin sensitivity of the sodium channel are occupied by glutamic acid in all four repeats of calcium channels (FIG. 4). At these positions of the sodium channels, repeats I and II have negatively charged residues, but repeat III has a positively charged residue (K1422) and repeat IV has an uncharged residue (A1714). These differences in charge in the regions likely to form part of the channel lining led to the idea that they might underlie the difference in the ion selectivity of sodium and calcium channels. Therefore we examined ion permeation properties of mutated sodium channels in which glutamic acid was substituted for K1422 (K1422E), for A1714 (A1714E), or both (K1422E·A1714E).[18]

While the strict ion selectivity for Na^+ over K^+, Rb^+, and Cs^+ is a defining property of the wild-type sodium channel, the mutant K1422E was highly permeable to K^+, Rb^+, and Cs^+ as well as Na^+, Li^+, and NH_4^+. When the reversal potential was measured precisely using macro-patch technique with defined ion compositions on both sides of the membrane, the calculated permeability ratio of K^+ relative to Na^+ (P_K/P_{Na}; without taking permeation of Ca^{2+} into account) was 0.69 for the mutant K1422E and 0.15 for the mutant A1714E, whereas the P_K/P_{Na} for the wild type was 0.3. These results indicated that these mutant channels lost the high ion selectivity for Na^+ over K^+ of the wild-type channel.

In addition, current records of the mutant channels showed marked outward rectification, which suggested that the external Ca^{2+} might block the mutant channels much more potently than the wild-type channel. Therefore dependence on extracellular Ca^{2+} of the inward current observed in oocytes expressing the mutant K1422E was examined. As the external Ca^{2+} concentration was decreased, the inward current became larger. When Na^+ in external medium was entirely replaced by Ca^{2+} or Ba^{2+}, large inward currents were observed, indicating the mutant K1422E was highly permeable to Ca^{2+} and Ba^{2+}. Thus the inward current of the mutant K1422E showed an anomalous mole-fraction dependence on external Ca^{2+} concentration. This relationship was very similar to that observed for native calcium channel, although the IC_{50} value of K1422E was higher by two orders of magnitude than those of native calcium channels. The inward current through the mutant A1714E was also blocked by external Ca^{2+}. When the mutants K1422E and A1714E were combined, the resulting double mutant K1422E·A1714E showed even higher sensitivity to Ca^{2+} than the mutant with single amino acid substitutions. The inward current through the double mutant was strongly reduced at 500 µM, but increased inward current was observed with increased Ca^{2+} concentrations. From the IC_{50} value, inward current was estimated to be carried mainly by Ca^{2+}, even at 500 µM.

Repeat I

							384							

		R	L	M	T	Q	D	F	W	E	N	L	Y	Q	379-391
	Brain II	R	L	M	T	Q	D	F	W	E	N	L	Y	Q	379-391
Sodium channel	Brain III	R	L	M	T	Q	D	Y	W	E	N	L	Y	Q	378-390
	Heart I	R	L	M	T	Q	D	C	W	E	R	L	Y	Q	368-380
	Skeletal μI	R	L	M	T	Q	D	Y	W	E	N	L	F	Q	395-407
	Electrophorus	R	L	M	L	Q	D	Y	W	E	N	L	Y	Q	356-368
Calcium channel	Brain BI	Q	C	I	T	M	E	G	W	T	D	L	L	Y	313-325
	Cardiac	Q	C	I	T	M	E	G	W	T	D	V	L	Y	388-400
	Skeletal	Q	C	I	T	M	E	G	W	T	D	V	L	Y	287-299

Repeat II

						942								
Sodium channel	Brain II	R	V	L	C	G	E	W	I	E	T	M	W	D	937-949
	Brain III	R	V	L	C	G	E	W	I	E	T	M	W	D	889-901
	Heart I	R	I	L	C	G	E	W	I	E	T	M	W	D	896-908
	Skeletal μI	R	I	L	C	G	E	W	I	E	T	M	W	D	750-762
	Electrophorus	R	A	L	C	G	E	W	I	E	T	M	W	D	744-756
Calcium channel	Brain BI	Q	I	L	T	G	E	D	W	N	E	V	M	Y	663-675
	Cardiac	Q	I	L	T	G	E	D	W	N	S	V	M	Y	731-743
	Skeletal	Q	V	L	T	G	E	D	W	N	S	V	M	Y	609-621

Repeat III

						1422								
Sodium channel	Brain II	Q	V	A	T	F	K	G	W	M	D	I	M	Y	1417-1429
	Brain III	Q	V	A	T	F	K	G	W	M	D	I	M	Y	1363-1375
	Heart I	Q	V	A	T	F	K	G	W	M	D	I	M	Y	1416-1428
	Skeletal μI	Q	V	A	T	F	K	G	W	M	D	I	M	Y	1232-1244
	Electrophorus	Q	V	S	T	F	K	G	W	M	D	I	M	Y	1208-1220
Calcium channel	Brain BI	T	V	S	T	G	E	G	W	P	Q	V	L	K	1464-1476
	Cardiac	T	V	S	T	F	E	G	W	P	E	L	L	Y	1140-1152
	Skeletal	T	V	S	T	F	E	G	W	P	Q	L	L	Y	1009-1021

Repeat IV

						1714								
Sodium channel	Brain II	Q	I	T	T	S	A	G	W	D	G	L	L	A	1709-1721
	Brain III	Q	I	T	T	S	A	G	W	D	G	L	L	A	1655-1667
	Heart I	Q	I	T	T	S	A	G	W	D	G	L	L	S	1708-1720
	Skeletal μI	E	I	T	T	S	A	G	W	D	G	L	L	N	1524-1536
	Electrophorus	E	I	T	T	S	A	G	W	D	G	L	L	L	1500-1512
Calcium channel	Brain BI	R	S	A	T	G	E	A	W	H	N	I	M	L	1760-1772
	Cardiac	R	C	A	T	G	E	A	W	Q	D	I	M	L	1441-1453
	Skeletal	R	C	A	T	G	E	A	W	Q	E	I	L	L	1318-1330

—SS1—┘　└—SS2—┘

These results indicated that the amino acid residues K1422 and A1714 are critical in determining selectivity of Na^+ over K^+ and also in distinguishing the sodium channel from the calcium channel with respect to ion selectivity. They also suggested that these sites of the sodium channel and corresponding sites of the calcium channel form part of the selectivity filter of these channels.

CONCLUDING REMARKS

The analyses of ion permeation properties of site specifically mutated AChR channels and sodium channels allow us to speculate on the overall plan of the open channels, although direct information of the tertiary structure of the ion channel proteins remains to be obtained. At first, from a methodological point of view, the fact that a large number of site-specific mutations have been successful suggests that channel proteins can usually accommodate a few amino acid substitutions. At least, by systematically introducing mutations, we can distinguish designed local effects of mutations from unexpected global and conformational effects.

The systematic mutational work on the AChR channel shows that the critical part of the open channel is relatively short. It means that there must be a really steep voltage drop across the channel constriction. If we assume 10 mV voltage drop over a length of 1 nm, this voltage gradient corresponds to 100 kV/cm, which is 1,000 times more steep than that used in electrophoresis for nucleotide sequencing. This steep drop of membrane potential should generate a strong driving force for permeating ions. The constriction should be short in length also to be effective in transporting ions.[19]

Current analyses of mutated AChR channels and sodium channels show that ion permeation properties, conductance and ion selectivity, are determined by interactions between a limited number of amino acid residues and permeating ions. The finding that these residues are well conserved among channels from various species supports their importance. These residues are often charged residues. Classical biophysics stressed the importance of the pore size and the electrostatic interaction in determining ion selectivity of the ion channels.[1] Our studies have by and large supported those classical views, demonstrating that both the pore size and the fixed charge are major determinants of the rate and selectivity of ion transport. Negatively charged residues are probably critical in attracting permeating cations into the selectivity constriction, while negatively

---

FIGURE 4. Alignment of the amino acid sequences (in one-letter code) in the regions encompassing the SS2 segment of the four repeats of different sodium and calcium channels. The sequences from top to bottom are: rat brain sodium channel II[11]; rat brain sodium channel III[23]; rat heart I sodium channel[24]; rat skeletal muscle μI sodium channel[25]; *Electrophorus electricus* electroplax sodium channel[26]; rabbit brain calcium channel BI[27]; rabbit cardiac dihydropyridine (DHP)-sensitive calcium channel[28]; and rabbit skeletal DHP-sensitive calcium channel.[17] The numbers of the amino acid residues in each sequence are given on the right-hand side. The positions of the SS1 and SS2 segments are indicated. The positions of the clusters of residues that have been identified as major determinants of toxin sensitivity are indicated by arrowheads. (From Heinemann *et al.*[18] Reprinted with permission.)

charged residues at the mouth contribute to accumulation of cations. Our AChR work also demonstrated the importance of hydrophilic environment. Uncharged polar residues in the channel constriction would contribute to dehydration of permeating cations by replacing some of the water molecules of their hydration shell.[1] We speculate that cooperation of the charged and uncharged polar residues at the channel constriction ensure the high cation selectivity without hampering the high rate of ion transport. The possibility that changes in net charge alter ion permeation properties by affecting local structures cannot be neglected, as our AChR work suggested.

Involvement of a positively charged residue (K1442) in Na^+ selectivity was rather unexpected. Since substitution of glutamic acid for this lysine confers calcium-channel properties on the sodium channel, this lysine may be important in preventing too much Ca^{2+} accumulation. The same may be the case for A1714. Furthermore, electrostatic attraction between aspartic acid in repeat I (D384) and K1422 would result in a smaller pore, presumably being suitable for selection of smaller alkali metal ions (Li^+ and Na^+) over larger ions (K^+, Rb^+, and Cs^+). However, recent reports of the primary sequence of other sodium channels indicate that this lysine is not always conserved.[20,21] Functional studies of these new types of sodium channel are awaited.

Our knowledge of the molecular mechanisms of ion permeation is still very limited and fragmented. Experimental results including those discussed here, however, certainly refine consideration of model systems. More systematic and comparative mutational studies will be necessary for understanding how ions permeate the membrane. The determination of the tertiary structure of channel proteins remains essential.

ACKNOWLEDGMENTS

The author wishes to thank the collaborators who contributed to the work described in this article.

REFERENCES

1. HILLE, B. 1992. Ionic Channels of Excitable Membranes. 2nd edit. Sinauer Associates. Sunderland, MA.
2. NUMA, S. 1989. A molecular view of neurotransmitter receptors and ionic channels. Harvey Lect. **83:** 121–165.
3. IMOTO, K., C. METHFESSEL, B. SAKMANN, M. MISHINA, Y. MORI, T. KONNO, K. FUKUDA, M. KURASAKI, H. BUJO, Y. FUJITA & S. NUMA. 1986. Location of a δ-subunit region determining ion transport through the acetylcholine receptor channel. Nature **324:** 670–674.
4. IMOTO, K., C. BUSCH, B. SAKMANN, M. MISHINA, T. KONNO, J. NAKAI, H. BUJO, Y. MORI, K. FUKUDA & S. NUMA. 1988. Rings of negatively charged amino acids determine the acetylcholine receptor channel conductance. Nature **335:** 645–648.
5. IMOTO, K., T. KONNO, J. NAKAI, F. WANG, M. MISHINA & S. NUMA. 1991. A ring of uncharged polar amino acids as a component of channel constriction in the nicotinic acetylcholine receptor. FEBS Lett. **289:**193–200.
6. ADAMS, D. J., T. M. DWYER & B. HILLE. 1980. The permeability of endplate channels to monovalent and divalent metal cations. J. Gen. Physiol. **75:** 493–510.

7. KONNO, T., C. BUSCH, E. VON KITZING, K. IMOTO, F. WANG, J. KANAI, M. MISHINA, S. NUMA & B. SAKMANN. 1991. Rings of anionic amino acids as structural determinants of ion selectivity in the acetylcholine receptor channel. Proc. R. Soc. Lond. B 244: 69–79.

8. WANG, F. & K. IMOTO. 1992. Pore size and negative charge as structural determinants of permeability in the Torpedo nicotinic acetylcholine receptor channel. Proc. R. Soc. Lond. B 250: 11–17.

9. DWYER, T. M., D. J. ADAMS & B. HILLE. 1980. The permeability of the endplate channel to organic cations in frog muscle. J. Gen. Physiol. 75: 469–492.

10. DANI, J. A. 1989. Open channel structure and ion binding sites of the nicotinic acetylcholine receptor channel. J. Neurosci. 9: 884–892.

11. NODA, M., T. IKEDA, T. KAYANO, II. SUZUKI, H. TAKESHIMA, M. KURASAKI, H TAKAHASHI & S. NUMA. 1986. Existence of distinct sodium channel messenger RNAs in rat brain. Nature 320: 188–192.

12. NODA, M., H. SUZUKI, S. NUMA & W. STÜHMER. 1989. A single point mutation confers tetrodotoxin and saxitoxin insensitivity on the sodium channel II. FEBS Lett. 259: 213–216.

13. PUSCH, M., M. NODA, W. STÜHMER, S. NUMA & F. CONTI. 1991. Single point mutations of the sodium channel drastically reduce the pore permeability without preventing its gating. Eur. Biophys. J. 20: 127–133.

14. GUY, H. R. & P. SEETHARAMULU. 1986. Molecular model of the action potential sodium channel. Proc. Natl. Acad. Sci. USA 83: 508–512.

15. GUY, H. R. & F. CONTI. 1990. Pursuing the structure and function of voltage-gated channels. Trends Neurosci. 13: 201–206.

16. TERLAU, H., S. H. HEINEMANN, W. STÜHMER, M. PUSCH, F. CONTI, K. IMOTO & S. NUMA. 1991. Mapping the site of block by tetrodotoxin and saxitoxin of sodium channel II. FEBS Lett. 293: 93–96.

17. TANABE, T., H. TAKESHIMA, A. MIKAMI, V. FLOCKERZI, H. TAKAHASHI, K. KANGAWA, M. KOJIMA, H. MATSUO, T. HIROSE & S. NUMA. 1987. Primary structure of the receptor for calcium channel blockers from skeletal muscle. Nature 328: 313–318.

18. HEINEMANN, S. H., H. TERLAU, W. STÜHMER, K. IMOTO & S. NUMA. 1992. Calcium channel characteristics conferred on the sodium channel by single mutations. Nature 356: 441–443.

19 LATORRE, R. & C. MILLER. 1983. Conduction and selectivity in potassium channels. J. Membr. Biol. 71: 11–30.

20. GEORGE JR., A. L., T. J. KNITTLE & M. M. TAMKUN. 1992. Molecular cloning of an atypical voltage-gated sodium channel expressed in human heart and uterus: Evidence for a distinct gene family. Proc. Natl. Acad. Sci. USA 89: 4893–4897.

21. SATO, C. & G. MATSUMOTO. 1992. Primary structure of squid sodium channel deduced from the complementary DNA sequence. Biochem. Biophys. Res. Commun. 186: 61–68.

22. NODA, M., H. TAKAHASHI, T. TANABE, M. TOYOSATO, S. KIKYOTANI, Y. FURUTANI, T. HIROSE, H. TAKASHIMA, S. INAYAMA, T. MIYATA & S. NUMA. 1983. Structural homology of Torpedo californica acetylcholine receptor subunits. Nature 302: 528–532.

23. KAYANO, T., M. NODA, V. FLOCKERZI, H. TAKAHASHI & S. NUMA. 1988. Primary structure of rat brain sodium channel III deduced from the cDNA sequence. FEBS Lett. 228: 187–194.

24. ROGART, R. B., L. L. CRIBBS, L. K. MUGLIA, D. D. KEPHART & M. W. KAISER. 1989. Molecular cloning of a putative tetrodotoxin-resistant rat heart Na$^+$ channel isoform. Proc. Natl. Acad. Sci. USA 86: 8170–8174.

25. TRIMMER, J. S., S. S. COOPERMAN, S. A. TOMIKO, J. ZHOU, S. M. CREAN, M. B. BOYLE, R. G. KALLEN, Z. SHENG, R. L. BARCHI, F. J. SIGWORTH, R. H. GOODMAN, W. S. AGNEW & G. MANDEL. 1989. Primary structure and functional expression of a mammalian skeletal muscle sodium channel. Neuron 3: 33–49.

26. NODA, M., S. SHIMIZU, T. TANABE, T. TAKAI, T. KAYANO, T. IKEDA, H. TAKAHASHI, H. NAKAY-
    AMA, Y. KANAOKA, N. MINAMINO, K. KANGAWA, H. MATSUO, M. A. RAFTERY, T. HIROSE,
    S. INAYAMA, H. HAYASHIDA, T. MIYATA & S. NUMA. 1984. Primary structure of *Electrophorus
    electricus* sodium channel deduced from cDNA sequence. Nature 312: 121–127.
27. MORI, Y., T. FRIEDRICH, M.-S. KIM, A. MIKAMI, J. NAKAI, P. RUTH, E. BOSSE, F. HOFMANN,
    V. FLOCKERZI, T. FURUICHI, K. MIKOSHIBA, K. IMOTO, T. TANABE & S. NUMA. 1991. Primary
    structure and functional expression from complementary DNA of a brain calcium channel.
    Nature 350: 398–402.
28. MIKAMI, A., K. IMOTO, T. TANABE, T. NIIDOME, Y. MORI, H. TAKESHIMA, S. NARUMIYA &
    S. NUMA. 1989. Primary structure and functional expression of the cardiac dihydropyridine-
    sensitive calcium channel. Nature 340: 230–233.

Assembly of Potassium Channels[a]

MIN LI, EHUD ISACOFF,[b] YUH NUNG JAN,
AND LILY YEH JAN

Department of Physiology and Biochemistry
Howard Hughes Medical Institute
University of California
San Francisco, California 94143

INTRODUCTION

Voltage-gated potassium channels are present in both excitable and non-excitable cells and serve a variety of biological functions.[2-8] Unlike other channels, e.g., sodium channels and acetylcholine receptor (AChR) channels, a combination of low abundance, high heterogeneity, and lack of high affinity ligands makes the purification and biochemical analysis of potassium channels very difficult. The first potassium channel gene to be isolated was cloned from *Drosophila* and based on a mutant phenotype, the loss of an A-type potassium current in both neurons and muscles.[9-12] Subsequently, a large number of the genes and their splice variants that encode related potassium channels have been isolated from a variety of species and tissues. Electrophysiological studies of the channels expressed in *Xenopus* oocytes have shown that the functional form of the potassium channel is a multimeric protein complex, likely a tetramer.[12,13] The cloned channels are classified into four subfamilies: Shaker, Shaw, Shal, and Shab. Only polypeptides in the same subfamily can form heteromultimeric channels when coexpressed in *Xenopus* oocytes.[14] Their distinct but overlapping expression patterns in mammalian brain are compatible with the idea that the characteristics of the excitability of specific neurons derive from the particular subset of potassium channel genes that they express and the types of heteromultimeric and homomultimeric channels that are thus formed in the cell.[15-18]

Based on sequence comparison, all cloned potassium channels share a common design. Each subunit consists of a single polypeptide that can be divided into three domains: a hydrophobic domain with six putative transmembrane segments flanked by two (amino and carboxyl) cytoplasmic hydrophilic domains. The molecular events involved in the formation of functional channels are poorly understood. Here we discuss experiments indicating that the interaction between amino-terminal domains of subunits is critical for channel assembly and determines the compatibility of polypeptides in the formation of heteromultimeric channels.

[a] This work was supported by the Helen Hay Whitney Foundation (M. L.) and Howard Hughes Medical Institute (L. Y. J. and Y. N. J.).
[b] Present address: Department of Molecular and Cell Biology, 229 Stanley Hall, University of California, Berkeley, CA 94270.

51

Specific Association between the Hydrophilic Amino-Terminal Domains

Comparison of amino acid sequence of potassium channels in the Shaker subfamily reveals a high degree of amino acid sequence similarity within the hydrophilic amino-terminal domains.[18,19] This region is not dispensable since the deletion of this region in ShB (a splice variant *Shaker* gene) eliminates the functional expression.[20] To test whether the hydrophilic domain is involved in subunit assembly, we first asked whether this domain associates and forms multimers. The amino-terminal domain was cloned, expressed in bacteria, and purified. It behaved in gel filtration column as two discrete species corresponding to monomers (36 kD) and multimers (140 kD). The size of the multimer is consistent with that of a tetramer.[1] This indicates that there is a specific homophilic interaction involving the amino-terminal domain.

If the association is physiologically relevant, coexpression of the amino-terminal hydrophilic domain with the full length potassium channel polypeptide should result in association between amino-terminal domains that do not form channels with the full length subunits, thus reducing the number of functional channels formed by the full length polypeptides as homomultimers. To test this hypothesis, the cRNAs that encode either amino-terminal domain or the full length ShB polypeptide were coexpressed in *Xenopus* oocytes. These cRNAs have identical 5' and 3' untranslated sequences and the amount of cRNA injected was adjusted to avoid saturation of the translation machinery of the oocyte. The formation of functional channels was tested by two-electrode voltage clamp. The current amplitude decreased by at least a factor of ten when equal mass of cRNA for the amino-terminal domain and cRNA for ShB was coexpressed.[1] The probable explanation is that the amino-terminal domain co-assembles with full length polypeptides and results in nonfunctional channels.

FIGURE 1. Homophilic association of ShB amino-terminal hydrophilic domains (NShB) revealed by binding ^{32}P-labeled NShB fusion protein to immobilized ShB and NShB. (A) Diagrammatic representation of the fusion protein expressed in bacteria. At the amino-terminus of the fusion protein is a short peptide (the "FLAG"), which is recognized by the commercially available antibody to FLAG (Immunex Corporation), and two heart muscle kinase sites. The transcription was driven by T7 polymerase. Single-letter amino acid codes are: A, Ala; R, Arg; N, Asn; D, Asp; C, Cys; Q, Gln; E, Glu; G, Gly; H, His; I, Ile; L, Leu; K, Lys; M, Met; F, Phe; P, Pro; S, Ser; T, Thr; W, Trp; Y, Tyr; V, Val. (B) Binding of NShB to ShB and NShB, but not to the carboxyl-terminal domain of ShB (CShB). Proteins (5 μg/lane) were fractionated by SDS-PAGE (lanes 1 to 12) and either visualized by Coomassie blue staining (lanes 5 and 6) or transferred onto a nitrocellulose filter (lanes 1 to 4, 7 to 12): 0 and 46 represent total protein of SF9 cells at 0 and 46 hours after infection by the 3A1 strain of recombinant baculovirus carrying ShB cDNA. The labels NShB and CShB above the blot on the right indicate lysates of IPTG (isopropyl β-D-thiogalactoside)-induced (lanes 9 and 11) and noninduced (lanes 10 and 12) bacteria that contain the expression vector for IPTG-induced expression of NShB (amino acids 1 to 227) and CShB (amino acids 479 to 656) fusion protein, respectively. Rabbit antisera to NShB (anti-NShB) and CShB (anti-CShB) were obtained by immunizing the rabbits with purified NShB and CShB fusion protein, respectively. These antibodies (dilution: 1/10,000 and 1/500) were used in immunoblots. They specifically recognize the ShB polypeptide (82 kD) expressed in SF9 cells (lanes 1 to 4), as does ^{32}P-labeled NShB (lanes 7 and 8). The NShB but not the CShB fusion protein is recognized by ^{32}P-labeled NShB (lanes 9 to 12) (From Li *et al.*[1] Reprinted with permission.)

To identify the sequences required for the homophilic interaction of the amino-terminal domain, we used a filter binding assay to test the interaction between ^{32}P-labeled polypeptides and immobilized polypeptides. The development of the filter binding assay is based upon the observation that many unfolded proteins can be refolded into their native conformation either in solution or after immobilization on a solid support. The specific interaction of the ^{32}P-labeled amino-terminal domain, but not the carboxyl-terminal domain, with the full length ShB polypeptides was detected[1] (FIG. 1). The immobilized protein preparation was the total cell lysate. Although the ShB polypeptides constituted less than 2% of the total protein loaded, they account for the only detected interaction with ^{32}P-labeled amino-terminal do-

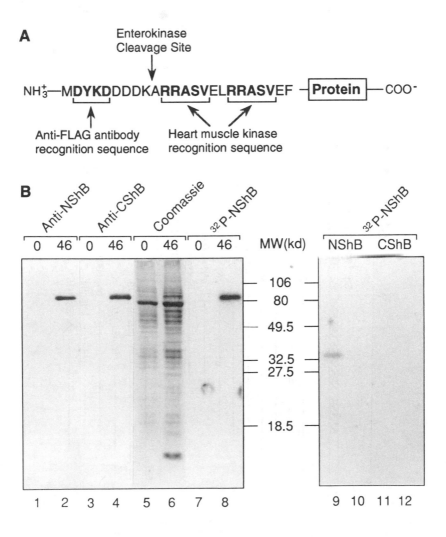

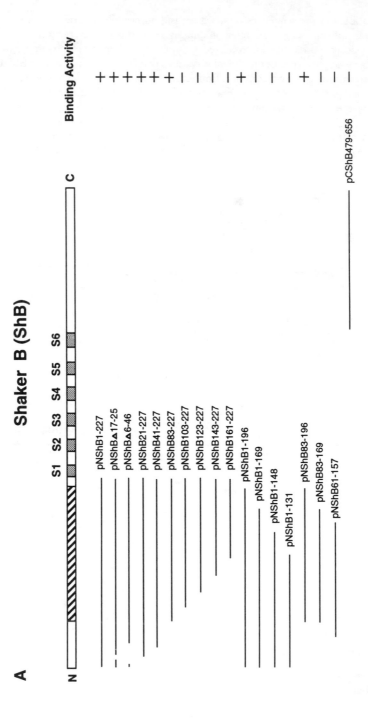

B

	90	100	110	120	130	140	150	160	170	180	190	
ShB	PQHFEPIPHDHDFCERVVINVSGLRFETQLRTLNQFPDTLLGDPARRLRYFDPLRNEYFFDRSRPSFDAILYYCSGGRLLRRPVNPLDVFSEEIKFYELGDQAINKFREDEGF											196
Ak01	NGMGV-GSDYDCS------------------------------N-QK-N--Y--EN-FERY------------											168 (77)
XSha2	--DSYDPEP--EC-----I---------K--S---E------KK-M------------F------------------------------I---------EE-MEI------------											134 (78)
RCK1	SYPRQADHD--EC-----I---------K--A---N------N-KK-M--------------------------------M--------------------EE-ME--Y---------											137 (77)
RCK2	EFQEAEGGGCCSS--L---I--------------SL-------------------N------------------------I-M--R--Q--E-LAA-----C											140 (73)
RCK3	LPPAL-AAGEQ--C--------------K--C--E-------K--M-----------------------------I------I-I------R--Q--EE-ME--Y---											154 (75)
RCK4	GGGGYSSVRYS--C------------MK--A---E-------EK-TQ---------------------------------K---------F-I-T--V--Q--EE-LL--Y---											278 (72)
RCK5	--DTYDPEA--EC-----I---------K--A---E------KK-M----------------------------------K---------------R--EE-MEM----Y											133 (78)
Kv1	EEDQA-QDAGSLHHQ--L--I--------G--A---N------N--M-------------G------------------------S--AD--R--Q--E-MER-----											210 (72)
MBK1	SYPRQADH---EC-----I---------K--A---N------N-KK-M------D----------------------F-------M----------------EE-ME--P---											176 (77)
MK1	SYPRQADHD--EC-----I---------K--A---E------N-KK-M------------------------------------M----------------EE-ME------------											136 (78)
MK2	--DTYDPEA--EC-----I---------K--A---E------KK-M-------------------------------------R--------R--EE-MEM----Y											132 (79)
MK3	-PALPAAGEQDCCG--------------K--C--E-------K------------------------------------I---------I-I------R--Q--EE-ME-----											156 (76)
hPCN1	TVEDQALGTASLHHC--H--I-------G--A---N------N--------------------G-------------S--AD--R--Q--E-MER----											199 (70)
hPCN2	GGGGYSSVRYS--S--------------MK--A---E------EK-TQ------------------------------------F-I-T--V--Q--EE-LL---											275 (74)
hPCN3	-PSLPAAGEQDCCG------------I--I-----I------K--C---E-M---R--M--V------------------I--------I-I------R--Q--EE-ME---											251 (74)
HBK2	DFPEAGGGGCCSS--L--I--------------------S-SL--I--L--R--Q--E-LAA------C											140 (72)

DRK1	RRVRLNVGGLAHEVLWRTLDRLPRTRLGK--DC-TH -S-- QVCDDYSLE- ------HPGA-TS--NF-RT ---HMMEEMCALS--Q-LJYWGIDEIYLESCCQARYH 134 (19)

FIGURE 2. A 114-amino acid fragment within the ShB amino-terminal domain (NShB) is required for the homophilic interaction. This was determined by using ³²P-labeled NShB as a probe to test for binding to various fragments of NShB(B) or by using these fragments as probes to test for their binding to ShB(C). (A) *(Top)*: A diagrammatic representation of the full-length ShB polypeptide. The shaded boxes represent the six putative transmembrane segments. The slashed box located in the hydrophilic amino-terminal domain represents amino acids 83 to 196. *(Bottom)* NShB, CShB, and fragments of NShB are produced by using expression plasmids, which include the coding sequences for the segments marked by the horizontal lines beneath the diagram of ShB. [p, plasmid; N, amino-terminal domain; C, carboxyl-terminal domain; numbers indicate the first and last amino acid residues in the fragment; pNShBΔ17-25 and pNShBΔ6-46 represent the amino-terminal domain (amino acids 1 to 227) with an internal deletion of amino acids 17 to 25 and amino acids 6 to 46, respectively.] (+) indicates that the fragment binds to both ShB and NShB when used either as a probe or as an immobilized substrate. (−) indicates that no binding is detectable in either case. The smallest fragment that shows binding contains amino acids 83 to 196. (B) Sequence conservation of mapped association region among genes in the Shaker subfamily. The sequence required for NShB homophilic association is shown by single letter code. The amino acid sequences of known K⁺ channel genes of the Shaker subfamily from different species (*Aplysia*, *Xenopus*, rat, mouse and human) are aligned with the ShB sequence from *Drosophila*. A rat gene (DRK1, Shab subfamily) is also included at the bottom. (−) indicates that the amino acid at that position is identical to that in the ShB sequence. Spaces in DRK1 sequence indicate gaps introduced. The name of each potassium channel gene is given on the left of the sequence. The position of the last amino acid is identified by the number on the right. The numbers in parentheses indicate the percentage of identity to ShB sequence in the region shown. (From Li *et al.*[1] Reprinted with permission.)

mains. The sensitivity of assay is within 100 picogram range and is partly dependent on immobilization efficiency. The minimum sequence requirement for the interaction was mapped within a fragment of 114 amino acid residues (amino acids 83 to 196 of ShB) (FIG. 2).[1]

Specification of the Formation of Heteromultimeric Channels

Among genes in the Shaker subfamily, there is more than 70% amino acid identity within the 114 amino acid fragment critical for association (FIG. 2, B).[1] Most of the 30% divergence is located on the amino end of the region; it remains to be tested whether most residues between position 83 and 196 are necessary for the homophilic interaction. To test whether the compatibility between the different members of Shaker subfamily in forming heteromultimers is reflected by the ability of their hydrophilic amino-terminal domains to interact, we expressed the amino-terminal domain of a mammalian homolog of ShB, RCK1. Indeed this fusion protein can specifically associate with itself, with the full length ShB polypeptide, and with the ShB amino-terminal domain.[1]

The cloned potassium channel genes have been classified into four subfamilies: Shaker, Shal, Shab, and Shaw. The amino acid identity in the hydrophobic domain is 70% for genes within a given subfamily, while this number drops to about 40% between genes in different subfamilies. Electrophysiological studies have shown that only coexpression of genes from the same subfamily in *Xenopus* oocytes will result in the formation of heteromultimeric channels. To test whether the incompatibility between potassium channels from different subfamilies could be due to incompatible interactions between their hydrophilic amino-terminal domains, we constructed a chimeric cDNA

FIGURE 3. Formation of functional heteromultimeric channels by ShB and a chimera of ShB and DRK1. This chimera, NShBΔ6-46/TmCDRK1, has the hydrophobic core region and the carboxyl-terminal domain of DRK1; the hydrophilic amino-terminal domain of DRK1 (amino acids 1 to 180) is replaced with that of ShB (amino acids 1 to 226), and an internal deletion of amino acids 6 to 46 of ShB is introduced to remove fast inactivation. This is shown schematically in the diagrams in (A through C), ShB: solid line, DRK1: dotted line. The chimera was functionally expressed in *Xenopus* oocytes and produced currents (B) that resembled the DRK1 K$^+$ current (C); in both cases the current activated much more slowly than the ShB K$^+$ current (A). Coexpression of ShB and the chimera gave rise to currents of different waveform (D) from those due to co-expression of ShB and DRK1 (E). (G) The current due to co-expression of ShB and DRK1 (elicited at + 60 mV, solid line) is similar in waveform to that generated by digital addition of ShB and DRK1 currents (dotted lines, with ratios of the two currents indicated on the right of the traces); it matches the simulation of ShB:DRK1 = 1:1.2. (F) The current due to co-expression of ShB and the chimera (at + 60mV, solid line) does not match a simulation at any ratios. The currents were elicited by 85-msec test pulses at + 20 mV, + 40 mV, and + 60 mV from a holding potential of – 100 mV. The top traces in (B) and (C) were generated by 900-msec test pulses at + 60 mV. The interval between test pulses was 3 seconds. The horizontal scale bar is 300 msec for inserted panels in (B) and (C) and 20 msec for all other traces. The vertical scale bar is 0.5 μA for (A), (D), (E); 0.14 μA for (B); 0.24 μA for (C); 1.1 μA for the inserted panel in (B); 1.9 μA for the inserted panel in (C). Each trace shown is representative of records from at least 4 oocytes. (From Li *et al.*[1] Reprinted with permission.)

that carries the ShB (Shaker subfamily) amino-terminal domain and DRK1(Shab subfamily) hydrophobic and carboxyl-terminal domains. This cDNA induced a current similar to that of DRK1. Unlike DRK1, however, the chimeric channel polypeptide was able to associate with ShB polypeptides to form heteromultimers with novel kinetic properties (FIG. 3).

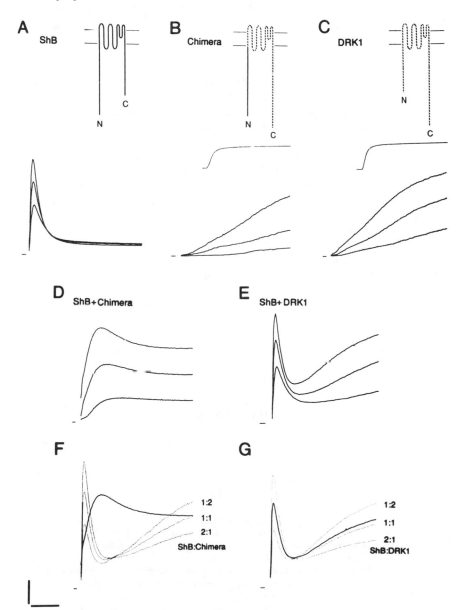

DISCUSSION

Biochemical and physiological studies demonstrate that the highly conserved region in ShB hydrophilic amino-terminal domain is critical for the formation of functional channels. In addition the amino-terminal domain is the determinant for specifying the formation of heteromultimeric channels. Studies from mutagenesis in conjunction with electrophysiological analysis have shown that the hydrophobic core regions of the potassium channel are involved in channel assembly.[20,21] The interaction between subunits in the hydrophobic domain should be critical for preserving the integrity and function of the ion conducting pathway. Such an interaction may also account for the observation that a null mutant of ShB carrying a deletion of the amino-terminal domain suppresses the formation of the functional channels by full length ShB polypeptides. In summary, we propose that there are at least two regions important for subunit interaction and formation of functional potassium channels: (1) a highly conserved hydrophilic sequence located before the first putative transmembrane segment; (2) a less well defined region in the hydrophobic domain. The hydrophobic regions, at least between the Shaker and Shab subfamily genes, are compatible in forming a functional channel, even though their hydrophobic domains show only 40% amino acid identity. The amino-terminal domain of Shaker subfamily genes *per se* can assemble to form multimers and this association may determine the compatibility in the formation of heteromultimeric channels.

The hydrophilic amino-terminal domain of the Shaker potassium channel alone dominantly suppresses the formation of a functional channel by full length polypeptides.[1] Although the detailed mechanisms at the cellular level remain to be studied, this observation itself indicates that cDNA for the hydrophilic association domain is a useful reagent for studying the physiological function of the potassium channels. It should be possible to use this construct to remove the function of endogenous gene product(s). This can be an important complementary approach to the experiments of knockout of potassium channel gene(s) by homologous recombination in embryonic stem (ES) cells.

REFERENCES

1. LI, M., Y. N. JAN & L. Y. JAN. 1992. Specification of subunit assembly by the hydrophilic amino-terminal domain of the Shaker potassium channel. Science 257: 1225–1230.
2. LATORRE, R., R. CORRONADO & C. VERGARA. 1984. Potassium channels gated by voltage and ions. Annu. Rev. Physiol. 46: 485–495.
3. ROGAWSKI, M. A. 1985. The A-current: How ubiquitous a feature of excitable cell is it? Trends Neurosci. 8: 214–219.
4. RUDY, B. 1988. Diversity and ubiquity of potassium channels. Neuroscience 25: 729–749.
5. JAN, L. Y. & Y. N. JAN. 1989. Voltage sensitive ion channels. Cell 56: 13–25.
6. KOLB, H-A. 1990. Potassium channels in excitable and non-excitable cells. Rev. Physiol. Pharmacol. 115: 51–91.
7. HILLE, B. 1991. Ionic Channels of Excitable Membranes. Sinauer Associates. Sunderland, MA.
8. JAN, L. Y. & Y. N. JAN. 1992. Structural elements involved in specific potassium channel functions. Annu. Rev. Physiol. 54: 537–555.
9. PAPAZIAN, D. M., T. L. SCHWARZ, B. L. TEMPEL, Y. N. JAN & L. Y. JAN. 1987. Cloning of

genomic and complementary DNA from *Shaker*, a putative potassium channel gene from *Drosophila*. Science 237: 749–753.

10. KAMB, A., L. E. IVERSON & M. A. TANOUYE. 1987. Molecular characterization of *Shaker*, a *Drosophila* gene that encodes a potassium channel. Cell 50: 405–413.

11. PONGS, O., *et al.* 1987. *Shaker* encodes a family of putative potassium channel proteins in the nervous system of *Drosophila*. EMBO J. 7: 1087–1096.

12. TEMPEL, B. L., D. M. PAPAZIAN, T. L. SCHWARZ, Y. N. JAN & L. Y. JAN. 1987. Sequence of a probable potassium channel component encoded at the *Shaker* locus of *Drosophila*. Science 237: 770–775.

13. MACKINNON, R. 1991. Determination of the subunit stoichiometry of a voltage-activated potassium channel. Nature 350: 232–235.

14. COVARRUBIAS, M., A. WEI & L. SALKOFF. 1991. *Shaker, Shal, Shah,* and *Shaw* express independent potassium current systems. Neuron 7: 763–773.

15. BECKH, S. & O. PONGS. 1990. Members of the RCK potassium channel family are differentially expressed in the rat nervous system. EMBO J. 9: 777–782.

16. TASUR, M-L., M. SHENG, D. H. LOWENSTEIN, Y. N. JAN & L. Y. JAN. 1992. Differential expression of potassium channel mRNAs in the rat brain and down-regulation in the hippocampus following seizures. Neuron 8: 1055–1067.

17. SHENG, M., M-L. TASUR, Y. N. JAN & L. Y. JAN. 1992. Subcellular segregation of two A-type potassium channel proteins in rat central neurons. Neuron 9: 271–284.

18. DREWE, J. A., S. VERMA, G. FRECH & R. H. JOHO. 1992. Distinct spatial and temporal expression patterns of potassium channel mRNAs from different subfamilies. J. Neurosci. 12: 538–548.

19. TEMPEL, B. L., Y. N. JAN & L. Y. JAN. 1988. Cloning of a probable potassium channel gene from mouse brain. Nature 332: 837–839.

20. ISACOFF, E. Y., Y. N. JAN & L. Y. JAN. 1990. Evidence for the formation of heteromultimeric potassium channels in *Xenopus* oocytes. Nature 345: 530–534.

21. VANDONGEN, A. M. J., G. C. FRECH, J. A. DREWE, R. H. JOHO & A. M. BROWN. 1990. Alteration and restoration of potassium channel function by deletions at the N- and C-termini. Neuron. 5: 433–443.

Potassium Channels Cloned from NG108-15 Neuroblastoma-Glioma Hybrid Cells

Functional Expression in *Xenopus* Oocytes and Mammalian Fibroblast Cells[a]

SHIGERU YOKOYAMA, TETSURO KAWAMURA,
YUJI ITO, NAOTO HOSHI, KOH-ICHI ENOMOTO[b]
AND HARUHIRO HIGASHIDA[c]

Department of Biophysics
Neuroinformation Research Institute
Kanazawa University School of Medicine
Kanazawa 920, Ishikawa, Japan
and
[b] Department of Physiology
Shimane Medical University
Izumo 693, Shimane, Japan

INTRODUCTION

Voltage-gated potassium (K^+) channels are membrane proteins responsible for control of cell excitability. A number of different voltage-gated K^+ channels were electrophysiologically identified and were demonstrated to regulate action potential repolarization, modulate firing pattern, and set the resting membrane potential.[1] Recently, the molecular genetic approach has uncovered a diverse group of voltage-gated K^+ channel genes.[2,3] Expression studies of these genes, combined with mutational experiments, have revealed several important structural elements involved in channel activation, inactivation, and ion permeation.[4–6]

In contrast with the extensive research of the structure-function relationship, only a few studies focused on the physiological correlations between the K^+ channels expressed from cloned cDNAs and native channels in a particular cell.[7–10] One major reason is that these cDNAs were isolated from whole brain tissues in most instances.

[a] This work was supported in part by grants from the Japanese Ministry of Education, Science & Culture and by the fund for medical treatment of elderly from Kanazawa University School of Medicine.
[c] Address correspondence Dr. Haruhiro Higashida.

This makes it difficult to compare the currents expressed in heterologous expression systems with those in cells of the native tissue.

To avoid this complexity, we have used a simple mammalian neuronal NG108-15 cell line.[11] This mouse neuroblastoma × rat glioma hybrid cell line expresses well-characterized K+ currents, such as delayed rectifier,[12] slowly inactivating voltage-activated K+ currents,[13] M-current,[12] and Ca^{2+}-activated K+ currents.[12] We have isolated two cDNAs encoding voltage-gated channels from NG108-15 cells: NGK1 and NGK2.[14] NGK1 protein is structurally more closely related to the *Drosophila Shaker* voltage-gated K+ channel gene product, whereas NGK2 protein is more closely related to the *Drosophila Shaw* voltage-gated K+ channel gene product. In the new nomenclature proposed by Chandy et al.,[15] NGK1 and NGK2 are designated as rat Kv1.2 (rKv1.2) and mouse Kv3.1 (mKv3.1), respectively. Subsequently, a second alternatively spliced Kv3.1 transcript encoding different carboxyl terminus has been identified in the rat brain[16] and mouse T lymphocytes.[10] Here we follow the designation by Grissmer et al.[10]: mKv3.1a for NGK2 and mKv3.1b for the second alternatively spliced product.

In the present report, we measured the rKv1.2 and the mKv3.1a currents expressed in two heterologous systems, *Xenopus* oocytes and mammalian fibroblast cells at macroscopic and microscopic levels. Then we compared them with the native voltage-gated K+ channels in NG108-15 cells.

Expression of the mRNAs for the Kv1.2 and Kv3.1 Genes in NG108-15 Cells

RNA preparations from NG108-15 cells were subjected to blot hybridization analysis with probes from rKv1.2 or mKv3.1a cDNAs. The size of the rKv1.2 mRNA in NG108-15 cells was estimated to be ~ 10,000 bases (FIG. 1,A: lane 1), in agreement with the data obtained from the rat brain.[17,18] The size of the mKv3.1a mRNA was estimated to be ~ 7,000 bases.[14] In addition, upon longer exposure to the autoradiogram, poly (A)+ RNA from NG108-15 cells exhibited an additional faint signal corresponding to sizes of ~ 4,000 nucleotides (FIG. 1,A: lane 2). The ~ 4,000-nucleotide RNA species may suggest the presence of the mKv3.1b. The amount of Kv3.1a mRNA appeared to be much larger than that of Kv1.2 mRNA. These data indicate that the mKv3.1a protein may be one of the major components responsible for one or two types of voltage-gated K+ channels seen in NG108-15 cells. We also examined the parental neuroblastoma cell line, N18TG2, and related cells. The ~ 7,000-nucleotide transcript hybridizable with mKv3.1a probe was detected in NL308 mouse neuroblastoma × mouse fibroblast hybrid cells[19] (FIG. 1,B: lane 2), but not in other neuroblastoma or hibrid cells (FIG. 1,B: lanes 1, 3, 5, 6). The ~ 4,000-nucleotide RNA species was not detectable. The level of mKv3.1 mRNA appears to differ in various neuronal cells, although more sensitive experiments should be performed for a detailed analysis.

It has been reported that in the adult rat brain, the ~ 4,500-nucleotide transcript is predominant,[16,23] whereas the ~ 7,000-nucleotide transcript is predominantly expressed in the embryonic and perinatal rat brain.[23] Interestingly, transfection of the human Ha-*ras* oncogene has been shown to induce the predominant expression of the ~ 7,000-nucleotide transcript in the mouse AtT20 cell line, derived from an anterior pituitary tumor.[24] These lines of evidence indicate that the mKv3.1a-

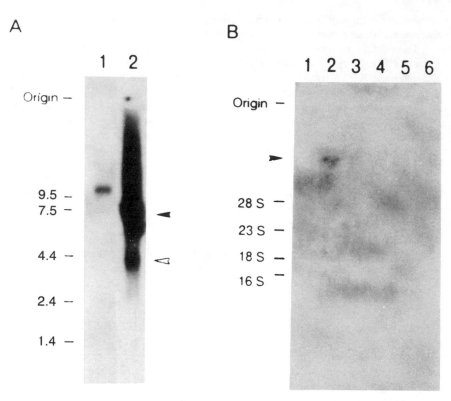

FIGURE 1. Autoradiograms of RNA blot hybridization analysis. The RNA samples used were as follows: (A) 15 μg of Poly (A)⁺RNA from NG108-15 cells. (B) 15 μg of total RNA from (Lane 1) neuroblastoma N18TG2 × B82 fibroblast hybrid NL309[19] (Lane 2) neuroblastoma N18TG2 × B82 fibroblast hybrid NL308,[19] (Lane 3) NL1F,[19] (Lane 4) B82 fibroblast cells,[20] (Lane 5) N4TG neuroblastoma cells,[19] and (Lane 6) N18TG2 neuroblastoma cells.[19] The ~ 7,000-nucleotide hybridizing band is indicated by a closed arrowhead, and the ~ 4,000-nucleotide hybridizing band by an open arrowhead. These samples were electrophoresed on a 1.0% agarose gel containing 2.2 M formamide[21] and transferred[22] to Biodyne nylon membrane (Pall). The hybridization was carried out at 42°C for 24 h in a solution containing 50% formamide, 50 mM sodium phosphate buffer (pH 7.0), 5 × SSC, 250 μg/ml denatured salmon sperm DNA, 0.1% SDS, 0.1% Ficoll, 0.1% polyvinylpyrolidone 0.1% bovine serum albumin and ~ 1.5 × 10⁶ cpm/ml ³²P-labeled probes. The probes were (A, Lane 1) 1.9 kb *Pvu*II/*Eco*RI fragment from λNGK1.23C2,[14] (A, Lane 2) 1.7 kb *Pst*I/*Bam*HI fragment from pSPNGK2[14] and (B) ~ 0.44 kb *Eco*RI fragment from λNGK2.5B1[14] The blots were washed at 60°C in 0.1 × SSC containing 0.1% SDS. The duration of autoradiography at −70°C with an intensifying screen was (A) 5 days and (B) 7 days. The size markers used were (A) RNA ladders from Bethesda Research Laboratories (size in kilobases) and (B) rat and *Escherichia coli* rRNAs.

predominant expression in NG108-15 cell may be closely related to tumor transformation, although functional distinction between Kv3.1a and Kv3.1b is not yet ascertained. We suppose that NG108-15 cells could serve as an excellent system for exploring the character of the Kv3.1a channel in a normal cell environment, in which Kv3.1a and other K⁺ channels are functioning.

Functional Expression from the cDNAs in Xenopus Oocytes

mRNAs specific for the rKv1.2 protein and for the mKv3.1a protein were synthesized by transcription *in vitro* of the cloned cDNAs and were injected into *Xenopus* oocytes. The peak amplitudes of the currents expressed in oocytes injected with rKv1.2 or mKv3.1a mRNAs were proportional to the dose of mRNA (FIG. 2). Larger mRNA doses (12.5 ng/µl) produced larger currents of 3–25 µA. As shown in TABLE 1, the currents elicited by 25 ng/µl mRNA solutions decayed slowly during 1-sec voltage steps to 0 mV, with the decrease in current size being $1.6 \pm 1.6\%$ (mean $\pm$ S.E.M., $n = 5$) of the peak value for the rKv1.2 current and $7.8 \pm 3.6\%$ ($n = 7$) for the mKv3.1a current, respectively. However, when we used 2.5 ng/µl mRNA solutions, the mKv3.1a currents (1–5 µA in amplitude) showed three- to fourfold greater inactivation ($27 \pm 8.8\%$, $n = 11$), whereas little change in inactivation was observed in the rKv1.2 currents ($3.5 \pm 1.6\%$, $n = 6$). At this mRNA concentration, inactivation of the mKv3.1a current varied widely and the maximal inactivation reached 76%; therefore the difference in the size of inactivation between rKv1.2 and mKv3.1 currents became marked. In the subsequent experiments, we used 2.5 ng/µl mRNA solutions, giving 0.10–0.15 ng mRNA per oocyte.

The current families shown in FIGURE 3 are examples obtained from oocytes injected with the rKv1.2- or mKv3.1a-specific mRNA (FIG. 3, A and B). These were evoked by depolarizing test pulses from a holding potential of −60 mV. In oocytes expressing rKv1.2 channels the threshold for current activation was -36 ± 2 mV ($n = 17$). A typical example of current-voltage (I-V) relationships is shown in FIGURE 3(C and E), indicating voltage-activated current. At 0 mV the current amplitude was 1.7 ± 0.3 µA ($n = 13$) with a time to peak of 163 ± 17 msec. Currents in oocytes expressing mKv3.1a channels (FIG. 3, D and F) were activated more positively than those of rKv1.2 (-14 ± 2 mV, $n = 22$), judging from the I-V curves. The mKv3.1a current amplitude at 0 mV was larger (4.9 ± 1.2 µA, $n = 13$) and time to peak was faster (75 ± 9 msec, $n = 19$). Furthermore, the time to peak was voltage dependent, decreasing with increasing depolarization.

The voltage dependence of steady-state inactivation was assessed by inactivating the current with setting for 10 sec to various membrane potentials from −60 mV just before applying a voltage step to 0 mV for 5 sec. FIGURE 4 shows the inactivation curves for rKv1.2 (FIG. 4, A) and mKv3.1a (FIG. 4,B). The half-inactivation potentials for the peak and the residual current at the end of the 5-sec step were similar for both the rKv1.2 current and the mKv3.1a current: rKv1.2 (peak, -29 ± 2 mV; steady state, -27 ± 3 mV, $n = 3$); mKv3.1a (peak, -31 ± 0.6 mV; steady state, -28 ± 4 mV, $n = 4$). The normalized curves are shown for the peak (FIG. 4,C) and steady-state currents (FIG. 4,D), showing complete removal of inactivation between −60 and −90 mV and complete inactivation at 0 mV.

As already demonstrated, there is a difference in the rate (FIG. 2, A and B) and amount (FIG. 2,E and F) of inactivation between the rKv1.2 and mKv3.1a currents. The proportions of the rKv1.2 and mKv3.1a currents that had inactivated at the end of a 5-sec pulse were $46 \pm 6\%$ ($n = 8$) and $84 \pm 3\%$ ($n = 18$), respectively. The inactivation of rKv1.2 currents was slow (FIG. 5,A) and best fit by two exponentials with time constants of 4.4 ± 0.7 sec and 19 ± 2 sec ($n = 4$). mKv3.1a currents inactivated very much faster (FIG. 3, B and FIG. 5, B) (0.37 + 0.04 sec and 1.5 $\pm$

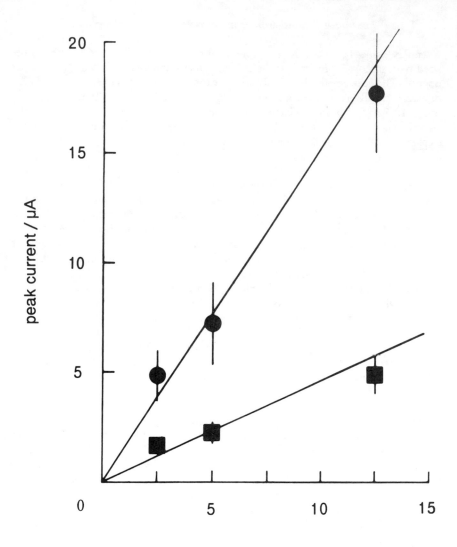

FIGURE 2. Relationships between peak amplitude of outward currents and concentration of mRNA solutions injected into *Xenopus* oocytes. Symbols are mean currents ($\pm$ S.E.M., bars) expressed by rKv1.2-specific mRNA (squares) and mKv3.1a-specific mRNA (circles). mRNAs used were synthesized by transcription *in vitro* of each cDNA as described in Yokoyama *et al.*[14] RNA solution was diluted with distilled water to the indicated concentration from a stock solution (0.25 μg/μl). Outward currents were recorded in response to 500 msec voltage step depolarization to 0 mV from a holding potential of -60 mV in 4–10 oocytes. Though the exact amount mRNA injected into an oocyte is assumed to receive roughly 50 nl, the point at 5 ng/μl RNA represents 0.25 ng mRNA for each oocyte. (Modified from Ito *et al.*[9])

TABLE 1. Effect of mRNA Injection Dose on Inactivation of Outward Current

mRNA (ng/µl)	Inactivation of Outward Current (%)	
	rKv1.2 (*n*)	mKv3.1a (*n*)
25	1.6 ± 1.6 (5)	7.8 ± 3.6 (7)
2.5	3.5 ± 1.6 (6)	27 ± 8.8 (11)

Denuded oocytes isolated from one frog were injected with about 50 nl of 2.5 or 25 ng/µl solution of rKv1.2 or mKv3.1a mRNAs, respectively. All current recordings were made on the fourth day after the injection in the normal frog Ringer solution (in mM): 115 NaCl, 2.5 KCl, 1.8 CaCl$_2$, 10 HEPES, pH adjusted to 7.2 with NaOH. Outward currents were evoked by 5 sec voltage step depolarization to 0 mV from a holding potential of −60 mV. Inactivation was calculated by subtracting currents at 1 sec from the start of depolarization from peak currents. Values are the mean ± S.E.M. in *n* individuals in parentheses.

0.08 sec (*n* = 5)), and with the slowest decay of 19 ± 2 sec (*n* = 5). The same time constants were obtained for rKv1.2 and mKv3.1a currents evoked at different pulse potentials.

TABLE 2 summarizes the characteristics of rKv1.2 and mKv3.1a whole-cell currents recorded from *Xenopus* oocytes together with the transient, slowly inactivating outward whole-cell current reported in NG108-15 cells by Robbins and Sim.[13] The activation and inactivation kinetics of the NG108-15 transient outward current seem to resemble the mKv3.1a current. These results suggest that mKv3.1a channels rather than rKv1.2 channels may function to generate the slowly inactivating current observed in NG108-15 cells. However, there is one difference between the mKv3.1a current and the native current in NG108-15 cells. The tetraethylammonium (TEA) sensitivity of the mKv3.1a current (IC$_{50}$ = 0.034 mM) was lower by around 60-fold in the native current (IC$_{50}$ = 2 mM). This may suggest that the TEA-insensitive rKv1.2 current (IC$_{50}$, > 10 mM) and/or other currents also contribute to the native whole-cell currents.

Functional Expression from cDNAs in Mammalian Cells

To confirm the above expression studies in *Xenopus* oocytes,[9] we have established transformed B82 mouse fibroblast cell lines that stably express the rKv1.2 and mKv3.1a gene products.[25] Each of the rKv1.2 and mKv3.1a cDNAs was ligated to the mammalian expression vector, pKNHneo, containing the simian virus 40 (SV40) early promoters and an aminoglycoside-3'-phosphotransferase gene. The resulting plasmids, pKNGK1 or pKNGK2, were introduced into B82 cells by the cationic liposome-mediated transfection method and stably transformed cell lines were selected by geneticin. Expression of each transcript was identified by RNA blot hybridization analysis (data not shown). In untransformed B82 cells, no hybridization signal was detected either with rKv1.2 probe (data not shown) or with mKv3.1a probe (FIG. 1, B: lane 4).

As observed in mRNA-injected oocytes, B82 fibroblast cells transformed to express rKv1.2[26] or mKv3.1a (FIG. 6, A) exhibited whole-cell outward currents of up to 5–10 nA. Parental B82 cells showed outward currents less than 40 pA.[26] This agrees well with the observation shown in FIGURE 1(B), in that no mRNA was detectable. The

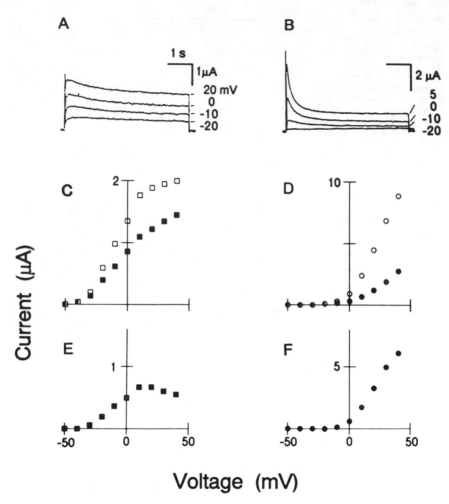

FIGURE 3. Activation of K⁺ currents in *Xenopus* oocytes injected with solutions (2.5 ng/µl) of rKv1.2-specific mRNA (**A, C, E**) and mKv3.1a-specific mRNA (**B, D, F**). (**A** and **B**) Current responses to depolarizing test pulses of 5 sec to indicated potentials from a holding potential of −60 mV in voltage-clamped oocytes with two electrodes. (**C** and **D**) Current-voltage relationship for the outward currents measured at the peak (open symbols) and at the end of the pulse (closed symbols). (**E** and **F**) Plots of currents after subtracting the amplitudes at the end of the pulse from the peak currents, showing the amount of inactivating components. Data of (**A, C, E**) are obtained from the same experiments but (**B, D, F**) are not. (Modified from Ito *et al.*[9])

mKv3.1a current decayed faster and more profoundly than the rKv1.2 current and resembled the transient, slowly inactivating current in NG108-15 cells. The inactivation of the mKv3.1a current was fitted by a single exponential with a time constant of 3.1 sec. The mKv3.1a current activated at potential positive to −20 mV (FIG. 6, A

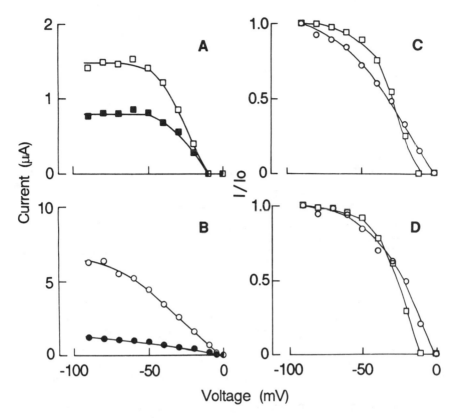

FIGURE 4. Steady-state inactivation of (**A**) rKv1.2 and (**B**) mKv3.1a currents from one oocyte each. Membrane potential was held at -100 mV, prepulsed for 10 sec to the indicated voltages, and stepped to a test pulse of 0 mV (5 sec duration). Peak currents (open symbols) and currents at the end of the pulse (closed symbols) were recorded. The ratio I/I_0 of currents at the peak (**C**) and at the end (**D**) was plotted as a function of the prepulse potentials, where I_0 is the amplitude at -90 mV and I is at test potentials. The voltage of half-maximal inactivation can be obtained from (**C**) and (**D**). Lines were drawn by eye fitting. (Modified from Ito *et al.*[9])

and C), whereas the native current activated at slightly more negative potential than that for the mKv3.1a current (FIG. 6,C).

Furthermore, we measured single-channel currents from the transformed fibroblast cells. FIGURE 7 shows representative single-channel currents flowing through the rKv1.2 and mKv3.1a channels. Although both channels usually activated in response to step depolarization, activation thresholds were different. The rKv1.2 channel was activated at potentials less than -10 mV (FIG. 7, A), whereas the mKv3.1a channel was activated at a more positive potential than -10 mV (FIG. 7, B). Occasionally both channels displayed partial closure states. The open possibilities of the rKv1.2 channel were always higher than those of the mKv3.1a channel. Notably, the mKv3.1a channel exhibited long periods of inactivity at potentials more positive than $+30$ mV. This

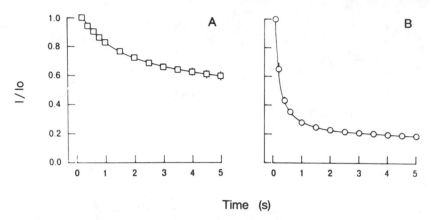

Time (s)

FIGURE 5. Time course of current decay during depolarization for (**A**) rKv1.2 and (**B**) mKv3.1a. Currents evoked by command pulses to 0 mV from −60 mV for 5 sec were normalized to the peak current at the indicated time. Each point represents the mean ± S.E.M. of four rKv1.2 currents and five mKv3.1a currents. (Modified from Ito *et al.*[9])

tendency was not observed in the rKv1.2 channel. The estimated reversal potentials of the rKv1.2 and mKv3.1a channel were −72 mV and −64 mV, respectively.[25] The slope conductances of the single rKv1.2 and mKv3.1a channel were 11 pS and 18 pS, respectively. These values in transformed fibroblasts are smaller than previously reported values measured in the *Xenopus* oocyte expression system: 17.5 ± 0.4 pS (*n* = 5) for rKv1.2 and 26.4 ± 1.7 pS (*n* = 6) for mKv3.1a.[14] The reason is still unknown.

The slowly inactivating outward current was observed in NG108-15 cells, which resembles those observed in mKv3.1a-transformed cells. But the outward currents in NG108-15 cells began to open at slightly more negative potential (Fig. 6, C). To see in detail, we performed single-channel recordings on NG108-15 cells and mKv3.1a-transformed cells. The channels with slope conductance of 21.9 ± 1.7 pS (*n* = 11) were frequently observed from NG108-15 cells (Fig. 8, B), and 20.0 ± 3.3 pS (*n* =

TABLE 2. Electrophysiological and Pharmacological Parameters of K⁺ Currents Expressed in Oocytes with rKv1.2 mRNA and mKv3.1a mRNA, with a Comparison to the Transient Outward Current in NG108-15 Cells

	rKv1.2	mKv3.1a	NG108-15
Peak current at 0 mV (μA)	1.7 ± 0.3 (13)	4.9 ± 1.2 (13)	—
Activation threshold (mV)	−36 ± 2 (17)	−14 ± 2 (22)	−30
Activation time (msec)	163 ± 17 (15)	75 ± 9 (19)	—
Half-inactivation of peak current (mV)	−29 ± 2 (3)	−31 ± 0.6 (4)	−53
Inactivation removed (mV)	−60	−90	−90
Inactivation completed (mV)	−10	0	−10
90% recovery from maximal inhibition (sec)	8.6 ± 1 (3)	30 ± 1 (7)	—
TEA sensitivity (IC₅₀) (mM)	>10	0.034	2

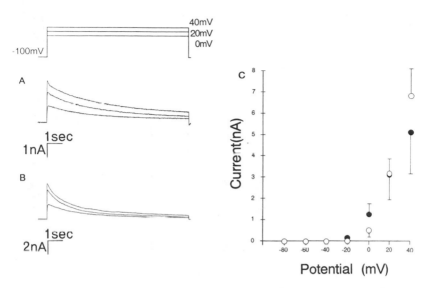

FIGURE 6. Transient, slowly inactivating currents. Records from an NG108-15 cell (**A**) and a mKv3.1a-transformed fibroblast cell (**B**). (**A** and **B**) The membrane potential was held at −100 mV and stepped to 0 mV, 20 mV, and 40 mV for 9.5 sec. Voltage pulse protocols are shown at the top. Records of the potential were low-pass filtered at 300 Hz and sampled at 200 Hz. (**C**) Current-voltage relationship for the outward currents from mKv3.1a-transformed cells (open circles) and NG108-15 cells (closed circles). Currents were measured at the end of the 400 msec step command. Symbols show means of 13 and 16 records from mKv3.1a-transformed cells and NG108-15 cells, respectively. Bars indicate SD. Records of the current were low-pass filtered at 1 kHz and sampled at 5 kHz. Linear leak and capacity transient were subtracted by P/N protocol. Cells were superfused with HEPES-buffered Ringer solution (in mM):134 NaCl, 5.4 KCl, 0.8 MgSO₄, 1.1 NaH₂PO₄, 1.8 CaCl₂, 5 NaH₂CO₃, 5.5 glucose, 10 HEPES, pH adjusted to 7.4 with NaOH. The electrodes contained a solution of the following composition (in mM): 150 KCl, 2 MgCl₂, 0.1 CaCl₂, 1.1 EGTA, 5 HEPES, pH adjusted to 7.2 with KOH. Electrodes had resistances of 3–5 MOhm. Membrane potentials were set at −100 mV using a single-electrode voltage clamp amplifier at switching frequency of 7–10 kHz.

25) from mKv3.1a-transformed cells (Fig. 8, A). This 21 pS channel from NG108-15 cells appears to be identical with that of NG108-15 cells previously described by McGee *et al.*[27] Both the 21 pS channel from NG108-15 cells and the mKv3.1a channel from transformed fibroblasts, exhibited two components of open time at 0 mV with the same order as follows: 0.97 msec and 6.75 msec for the 21 pS channel from NG108-15 cells, 0.69 msec and 4.55 msec for the mKv3.1a channel. These two channels had similar reversal potentials; −61.8 ± 7.4 mV (n = 11) from NG108-15 cells and −64.4 ± 12.5 mV (n = 18) from mKv3.1a-transformed cells. But they had different open probabilities at 0 mV: 0.40 ± 0.14 (n = 11) for the 21 pS channel from NG108-15 cells and 0.15 ± 0.12 (n = 18) for the mKv3.1a channel. Thus, it can be summarized that most of the parameters from these two channels showed similar values except for the voltage dependence of activation. However, we could

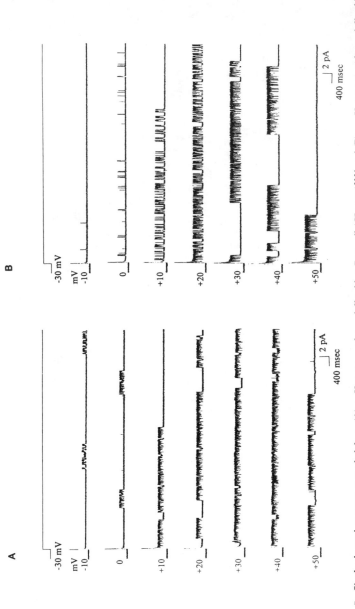

FIGURE 7. Single-channel currents recorded from (A) a rKv1.2-transformed fibroblast B82 cell (CL301)[25] under cell-attached configuration. Voltage pulses were stepped to the voltage indicated at the left of each trace from the holding potential of −30 mV. Outward current upwards. The resting membrane potential was assumed to be −30 mV. The patch pipette contained HEPES-buffered Ringer solution. and (B) a mKv3.1a-transformed fibroblast B82 cell (CL1023)[25,26]

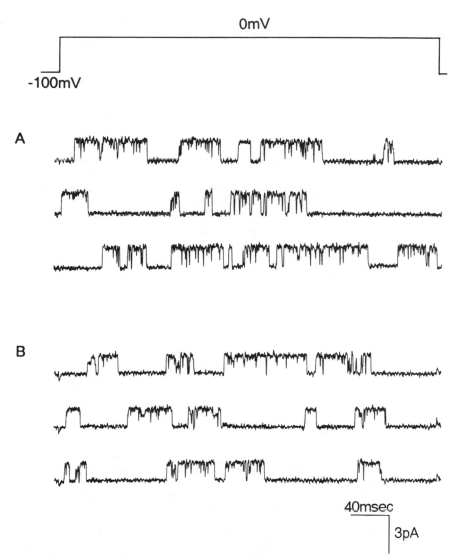

FIGURE 8. Representative single-channel recordings of a mKv3.1a channel from a transformed cell (**A**) and a 21 pS voltage-gated K⁺ channel from an NG108-15 cell (**B**) under cell-attached configuration. The voltage pulse protocol is shown at the top. Records were low-pass filtered at 1 kHz and sampled at 5 kHz. Cells were superfused with a solution of the following composition (in mM): 150 potassium aspartate, 5 MgCl₂, 5 EGTA, 10 glucose, 10 HEPES, pH adjusted to 7.4 with KOH. The patch pipette solution used was the same as in FIG. 7.

not identify rKv1.2-like channels out of 47 successful patches from NG108-15 cells. This reasonably agrees with its low-level mRNA expression.

CONCLUSION

In *Xenopus* oocytes and mammalian fibroblast cells, we expressed the rKv1.2 and mKv3.1a cDNAs and analyzed the channel properties. In both expression systems, the rKv1.2 and mKv3.1a currents exhibited a marked difference. The mKv3.1a current decayed faster and greater than the rKv1.2 current, but much slower than the typical transient, rapidly inactivating A-type current, which occurs within 20–50 msec. The present results show that the mKv3.1a protein is the essential component of the voltage-gated K⁺ channel that generates the transient, slowly inactivating outward current in NG108-15 cells. Recently, many reports of transient, slowly inactivating outward currents have been described in the mammalian brain.[28-30] They display the inactivation time course similar to the mKv3.1a current or the transient outward current in NG108-15 cells. It would be of interest to investigate if the mKv3.1a protein may contribute to such current *in vivo*.

REFERENCES

1. RUDY, B. 1988. Diversity and ubiquity of K channels. Neuroscience **25**: 729–749.
2. JAN, L. Y. & Y. N. JAN. 1990. How might the diversity of potassium channels be generated? Trends Neurosci. **13**: 415–419.
3. PERNEY, T. M. & L. K. KACZMAREK. 1991. The molecular biology of K⁺ channels. Curr. Opinion Cell Biol. **3**: 663–670.
4. MILLER, C. 1991. Annus Mirabilis of potassium channels. Science **252**: 1092–1096.
5. MacKINNON, R. 1991. New insights into the structure and function of potassium channels. Curr. Opinion Neurobiol. **1**: 14–19.
6. JAN, L. Y. & Y. N. JAN. 1992. Structural elements involved in specific K⁺ channel functions. Annu. Rev. Physiol. **54**: 537–555.
7. BAKER, K. & L. SALKOFF. 1990. The Drosophila *Shaker* gene codes for a distinctive K⁺ current in a subset of neurons. Neuron **2**: 129–140.
8. PFAFFINGER, P. J., Y. FURUKAWA, B. ZHAO, D. DUGAN & E. R. KANDEL. 1991. Cloning and expression of an *Aplysia* K⁺ channel and comparison with native *Aplysia* K⁺ currents. J. Neurosci. **11**: 918–927.
9. ITO, Y., S. YOKOYAMA & H. HIGASHIDA. 1992. Potassium channels cloned from neuroblastoma cells display slowly inactivating outward currents in *Xenopus* oocytes. Proc. R. Soc. London Ser. B **248**: 95–101.
10. GRISSMER, S., S. GHANSHANI, B. DETHLEFS, J. D. McPHERSON, J. J. WASMUTH, G. A. GUTMAN, M. D. CAHALAN & K. G. CHANDY. 1992. The *Shaw*-related potassium channel gene, Kv3.1, on human chromosome 11, encodes the type *l* K⁺ channel in T cells. J. Biol. Chem. **267**: 20971–20979.
11. NIRENBERG, M., S. WILSON, H. HIGASHIDA, A. ROTTER, K. KRUEGER, N. BUSIS, R. RAY, J. G. KENIMER & M. ADLER. 1983. Modulation of synapse formation by cyclic adenosine monophosphate. Science **222**: 794–799.
12. BROWN, D. A. & H. HIGASHIDA. 1988. Voltage- and calcium-activated potassium currents in mouse neuroblastoma × rat glioma hybrid cells. J. Physiol. Lond. **397**: 149–165.
13. ROBBINS, J. & J. A. SIM. 1990. A transient outward current in NG108-15 neuroblastoma × glioma hybrid cells. Pflügers Arch. **416**: 130–137.

14. YOKOYAMA S., K. IMOTO, T. KAWAMURA, H. HIGASHIDA, N. IWABE, T. MIYATA & S. NUMA. 1989. Potassium channels from NG108-15 neuroblastoma-glioma hybrid cells: Primary structure and functional expression from cDNAs. FEBS Lett. **259**: 37–42.

15. CHANDY, K. G., J. DOUGLAS, G. A. GUTMAN, L. JAN, R. JOHO, L. KACZMAREK, D. MCKINNON, R. A. NORTH, S. NUMA, L. PHILIPSON, A. B. RIBERA, B. RUDY, L. SALKOFF, R. SWANSON, D. STEINER, M. TANOUYE & B. TEMPEL. 1991. Simplified gene nomenclature. Nature **352**: 26.

16. LUNEAU, C. J., J. B. WILLIAMS, J. MARSHALL, E. S. LEVITAN, C. OLIVA, J. S. SMITH, J. ANTANAVAGE, K. FOLANDER, R. B. STEIN, R. SWANSON, L. K. KACZMAREK & S. A. BUHROW. 1991. Alternative splicing contributes to K⁺ channel diversity in the mammalian central nervous system. Proc. Natl. Acad. Sci. USA **88**: 3932–3936.

17. MCKINNON, D. 1989. Isolation of a cDNA clone coding for a putative second potassium channel indicates the existence of a gene family. J. Biol. Chem. **264**: 8230–8236.

18. BECKH, S. & O. PONGS. 1990. Members of the RCK potassium channel family are differently expressed in the rat nervous system. EMBO J. **9**: 777–782.

19. MINNA, J., D. GLAZER & M. NIRENBERG. 1972. Genetic dissection of neural properties using somatic cell hybrids. Nature New Biol. **235**: 225–231.

20. MINNA, J., P. NELSON, J. PEACOCK, D. GLAZER & G. NIRENBERG. 1971. Genes for neuronal properties expressed in neuroblastoma × L cell hybrids. Proc. Natl. Acad. Sci. USA **68**: 234–239.

21. LEHRACH, H., D. DIAMOND J. M. WOZNEY & H. BOEDTKER. 1977. RNA molecular weight determinations by gel electrophoresis under denaturing conditions, a critical reexamination. Biochemistry **16**: 4743–4751.

22. THOMAS, P. S. 1980. Hybridization of denatured RNA and small DNA fragments transferred to nitrocellulose. Proc. Natl. Acad. Sci. USA **77**: 5201–5205.

23. PERNEY, T. M., J. MARSHALL, K. A. MARTIN, S. HOCKFIELD & L. K. KACZMAREK. 1992. Expression of the mRNAs for the Kv3.1 potassium channel gene in the adult and developing rat brain. J. Neurophysiol. **68**: 756–766.

24. HEMMICK, L. M., T. M. PERNEY, R. E. FLAMM, L. K. KACZMAREK & N. C. BIRNBERG. 1992. Expression of the H-ras oncogene induces potassium conductance and neuron-specific potassium channel mRNAs in the AtT20 cell line. J. Neurosci. **12**: 2007–2014.

25. KAWAMURA, T., S. YOKOYAMA, J. YAMASHITA & H. HIGASHIDA. 1990. Establishment of transformed fibroblast with potassium channel genes, NGK1 and NGK2. Soc. Neurosci. Abstr. **16**: 670

26. WERKMAN, T. R., T. KAWAMURA, S. YOKOYAMA, H. HIGASHIDA & M. A. ROGAWSKI. 1992. Charybdotoxin, dendrotoxin and mast cell degranulating peptide block the voltage-activated K⁺ current of fibroblast cells stably transfected with NGK1 (Kv1.2) K⁺ channel complementary DNA. Neurosci. **50**: 935–946.

27. MCGEE, R., M. S. P. SANSOM & P. N. R. USHERWOOD. 1988. Characterization of a delayed rectifier K⁺ channel in NG108-15 neuroblastoma × glioma cells: gating kinetics and the effects of enrichment of membrane phospholipids with arachidonic acid. J. Membr. Biol. **102**: 21–34.

28. GUSTAFSSON, B., M. GALVAN, P. GRAFE & H. WIGSTRÖM. 1982. A transient outward current in a mammalian central neurone blocked by 4-aminopyridine. Nature **299**: 252–254.

29. GREENE, R. W., H. L. HAAS & P. B. REINER. 1990. Two transient outward currents in histamine neurones of the rat hypothalamus *in vitro*. J. Physiol. Lond. **420**: 149–163.

30. FICKER, E. & U. HEINEMANN. 1992. Slow and fast transient potassium currents in cultured rat hippocampal cells. J. Physiol. Lond. **445**: 431–455.

The Potassium Pore and Its Regulation[a]

A. M. BROWN, J. A. DREWE, H. A. HARTMANN,
M. TAGLIALATELA, M. DE BIASI, K. SOMAN,
AND G. E. KIRSCH[b]

Department of Molecular Physiology and Biophysics
[b]Department of Anesthesiology
Baylor College of Medicine
One Baylor Plaza
Houston, Texas 77030

INTRODUCTION

Voltage-dependent K^+ channels have three principal functional components: the ion conduction pathway or pore, a voltage sensor responsive to changes in membrane potential, and a mechanism coupling the sensor to the pore. The moving parts of the coupling mechanism are thought to be gates. An electromechanical model of voltage-gated K^+ channels depicting these components is shown in FIGURE 1.

This paper deals only with the K^+ pore. A linear sequence has been identified that embodies most of the pore and the experimental evidence supporting this conclusion will be reviewed. Within the pore, residues have been identified that are involved in ionic selectivity and binding. From these results we will present a unique β-barrel model of the pore.

LOCALIZING THE PORE WITHIN THE LINEAR AMINO ACID SEQUENCE OF K^+ CHANNELS

The linker between the fifth and sixth transmembrane segments (S5-S6 loop) is the most conserved region in voltage-dependent K^+ channels.[1] The possibility that a homologous region in Na^+ channels formed the channel pore was raised when a glutamate→arginine mutation in this region greatly reduced blockade by tetrodotoxin (TTX).[2] The first functional evidence in K^+ channels was the observation that open channel block of a Shaker K^+ channel by the peptide toxin, charybdotoxin, was modified by point mutations in the S5-S6 loop.[3] Other point mutations in the linker affected open channel block by the small quaternary ammonium ion tetraethylammonium (TEA).[4] Yellen *et al.*[5] showed that internal TEA blockade could be changed

[a] This work supported in part by National Institutes of Health Grants HL37044, HL39262, HL36930, and NS23877 to A. M. Brown and NS29473 to G. Kirsch.

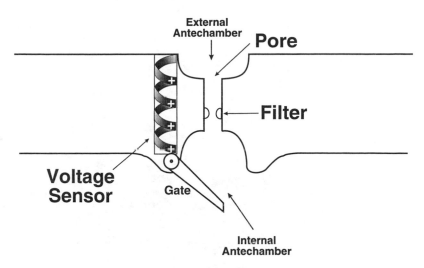

FIGURE 1. Electromechanical model of K⁺ channel.

by a threonine→serine substitution at a position almost midway between mutations producing changes in external TEA blockade. Yool and Schwarz[6] showed that the same mutation involved in blockade by internal TEA produced large increases in the relative permeability of NH_4^+ as did a nearby phenylalanine→serine mutation.

Rather than using point mutations, Hartmann *et al.*[7] used large scale mutagenesis to identify the pore. A DNA sequence thought to embody the pore of two related K⁺ channels with markedly different pore properties was transplanted from one channel to the other. An NGK2-like (Kv3.1)[8] K⁺ channel had a conductance of about 26 pS and was sensitive to external TEA while a delayed rectifier K⁺ channel DRK1 (Kv2.1)[9] had a conductance of about 8 pS and was sensitive to internal TEA. When a stretch of DNA that encoded 21 amino acids was transplanted from NGK2 (Kv3.1) to DRK1 (Kv2.1), the chimeric channel CHM adopted the pore behavior of the parental NGK2-like (Kv3.1) phenotype.

With regards to voltage sensitivity, the behavior of the chimeric channel resembled that of the host phenotype DRK1. The fact that pore properties were mainly exchanged made it likely that the effects were local rather than global and suggests that the voltage sensor, gate, and pore are discrete modular structures.[10]

The transplanted sequence of amino acids included the mutations of Yellen *et al.*[5] and Yool and Schwarz.[6] From considerations of the length of the sequence involved and the positions for blockade by external and internal TEA, the arrangement favored for the pore (P) region was a β-hairpin of 18–20 amino acids (FIG. 2).

FUNCTIONS OF RESIDUES IN THE K⁺ PORE

The transplanted K⁺ pore was studied further by mutational analysis taking advantage of the differences in phenotype and genotype (nine residues) differences between

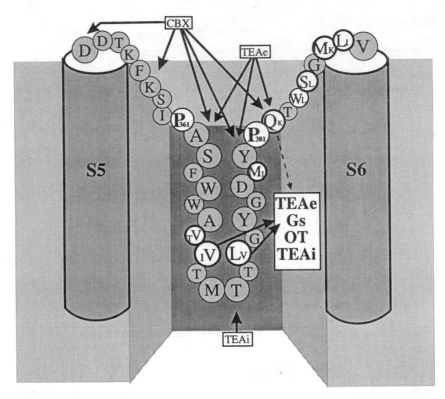

FIGURE 2. Model of a K⁺ pore. The loop between transmembrane segments 5 and 6 of CHM is shown. Nine residues account for the differences between the donor NGK2 and the host DRK1. The arrows at the top and bottom of the figure indicate the residues that align with Shaker residues responsible for blockade by charybdotoxin, external and internal TEA. Also indicated are the residues at positions 369 and 374. The residues determine the differences in pore phenotype between DRK1 and NGK2.

CHM and DRK1. Point reversions in CHM outside the deep pore defined by proline residues at 361 and 381, and downstream or towards the C-terminus from P381 had either no effect in the case of the conservative reversions, or had consequences predicted from a change in side-chain charge in this region affecting the local concentration of K⁺.[11] The substitution Q382K reduced inward and outward current with the reduction in inward current being greater, resulting in a decrease in outward rectification of the single channel current. Similar, but less marked, results were obtained for M387K.

Within the deep pore or tunnel between amino acid residues 361 and 381 there were four differences between CHM and DRK1, three of which were conservative (FIG. 2). The reversions at 368 and 379 had little effect. Surprisingly, the conservative reversions at 369 and 374 introduced novel phenotypes. For both positions, open times were shortened almost tenfold but for V369I the shortening was due to stabilization of an inactivated state that was apparent in the whole cell currents producing P-type inactivation. For L374V, the reduced open time was due to rapid departure from the open state to a closed state and the currents rather than inactivating remained steady during standard test pulses. In L374V, single channel K$^+$ conductance was greatly reduced.

The next surprise was that combined reversions at 369 and 374 restored the host pore phenotype. None of the other five possible double reversions had this effect, suggesting that positions 369 and 374 interacted with each other. The likelihood of interaction among separate subunits was confirmed when it was established that co-injection of V369I and L374V cRNAs produced channels having properties similar to those of the double reversion.[12] This result implies that interaction between positions 369 and 374 occurs within and between subunits.

The substitution L374V in CHM and DRK1 switched the pore from being selective for K$^+$ to being selective for Rb$^+$.[13] To analyze this selectivity filter, extensive substitutions were introduced at position 374.[14] We found that hydrophobic residues favored Rb$^+$ whereas polar residues favored K$^+$. Hydrophobic substitutions enhanced TEA blockade consistent with a hydrophobic component in the interaction between TEA and the pore. Interactions between conducting ions and TEA blockade led to an interpretation in which both the side chains at 374 and the peptide backbone were involved in selectivity and internal TEA blockade.

PORE OR CORE INACTIVATION

Three regions of the amino acid sequence of K$^+$ channels are important for inactivation, the N terminus, the sixth transmembrane segment S6, and the P region. A comparison of the inactivation properties expressed by alternatively spliced Shaker cRNAs showed that differences in the N or C termini were responsible for differences in rates of inactivation.[15]

An N-terminus stretch of 19 residues was shown to be responsible for fast inactivation, and a more distal stretch of residues modified the rate at which inactivation occurred.[16] This gave use to the term N-terminus (N-type) inactivation and the ball-and-chain physical model for the process.

Inactivation was different in alternatively spliced Shaker channels that shared common N termini but differed in C termini (C-type).[15] *Shak* A and B channels with the same N-terminal deletions differed markedly in their inactivation rates.[17]

The S6 residue responsible for the C-type inactivation in Shaker was located near the external mouth of the pore.[17] In lymphocytes, a His at a corresponding position part of the external TEA receptor when protonated produced large changes in inactivation.[18] Thus residues in the hydrophobic core as opposed to the charged, non-hydrophobic N-terminus appeared to contribute to inactivation of the non-"ball" type. Point reversions of a chimeric K$^+$ channel showed that a Val→Ile reversion at a position in the tunnel or deep part of the pore[11] produced a novel phenotype with the unique property of rapid inactivation but unitary conductance was unchanged. Subsequently, a Val→Ser substitution at the same position produced more complete inactivation and marked reduction in conductance, confirming this position to be in the pore.[19]

P-type inactivation differed from C-type inactivation as originally described, because external TEA rather than slowing inactivation, increased channel availability. At higher concentrations external TEA did produce blockade. It was unlikely that the pore mutations introduced a receptor for N-terminus inactivation since internal TEA scaled the current downward but did not slow inactivation.

For inactivation produced by substitutions of pore residues, external K^+ slowed inactivation and increased recovery from inactivation. The results seem to indicate a highly complex situation in which TEA and K^+ receptors near the external mouth of the pore couple to an inactivation process within the pore. There appears to be an external site at which TEA produces blockade, another site at which TEA has an enhancing effect on K^+ currents, possibly the same site as the one responsible for the enhancing effect of K_o, and there may be a site at which increased K_o slows inactivation.

The effects of external TEA can be accounted for by a closed-closed-open-block model including transitions between C_1 and I_4. A fifth state is the blocked state β, produced by higher concentrations of TEA. Thus, we have:

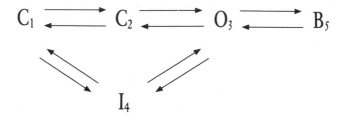

The $I_4 \rightarrow C_1 \rightarrow O_3$ transitions are stabilized by lower concentrations of TEA and enhance currents. The $O_3 \rightarrow B_5$ transition is stabilized by higher concentrations of TEA to produce blockade. The potentiating effects of increased K_o^+ were accounted for similarly while omitting B_5.

P- and N-types of inactivation can be compared by testing the effects of internal TEA injected into oocytes with the method described by Taglialatela et al.[20] For N-type inactivation, internal TEA prolonged the currents whereas for P-type inactivation internal TEA simply reduced the currents.[19]

Both P- and C-types of inactivation involve residues in or near the pores and may share a similar mechanism. They may be distinguished from the N-terminus "ball" type of activation[16] and from deletions of the C-terminus, which in Kv2.1 increased the rate of inactivation.[21] Since the distinction lies between inactivation mechanisms involving the non-conserved cytoplasmic N- and C-termini and inactivation mechanisms involving the conserved hydrophobic core, a more suitable nomenclature would be core and non-core inactivation.

UNIQUE β-BARREL K+ PORE

A β-barrel structure composed of four β-hairpins has been proposed.[5-7,11] Based on these results, two more specific examples of β-barrel K+ pores have been described.[22,23] Neither of these structures predicts the results presented here. Also, they do not incorporate new findings indicating that the residues at 374[14] and 369[19] are at the surface and that position 369 appears to be closer to the external mouth of the pore.[19]

We have developed a unique β-barrel model that deals with the region of the pore where positions 369 and 374 interact. The strands consist of amino acids 364–371 and 372–389, and have a "right-handed" tilt. Intra- and inter strand hydrogen bonding of the backbone as is found in β-sheets are present. Residues 369 and 374 are at the surface and residue 369 is closer to the external mouth than residue 374. The strands are joined to obtain a continuous stretch from 364 to 379 and additional subunits are generated by fourfold rotation around the axis of the β-barrel.

The side chains at 369 and 374 may interact between subunits and within a subunit, residues 369 and 374 are in register for β-sheet hydrogen bonding.

Small energy differences resulting from mutations at 369 and 374 produce changes in TEA affinity and ion conduction and these are explained by changes in van der Waals interactions. Side chains that are too large or have altered flexibility may have unfavorable van der Waals interaction, leading to a less stable barrel, shortened open times, and altered ion conduction. Side chains that are too small produce a pore that is too capacious and Rb+ is preferred over K+.[14] When the side chains are optimal, a unique K+ pore phenotype is expressed.

ACKNOWLEDGMENTS

We thank Judy Breedlove and Marianne Anderson for their secretarial assistance and Derwin Furra for his graphics design.

REFERENCES

1. TEMPEL, B. L., D. M. PAPAZIAN, T. L. SCHWARZ, Y. N. JAN & L. Y. JAN. 1987. Sequence of a probable potassium channel component encoded at the *Shaker* locus of *Drosophila*. Science 237: 770–775.
2. NODA, M., H. SUZUKI, S. NUMA & W. STÜHMER. 1989. A single point mutation confers tetrodotoxin and saxitoxin insensitivity on the sodium channel II. FEBS Lett. 259: 213–216.
3. MACKINNON, R. & C. MILLER. 1989. Mutant potassium channels with altered binding of charybdotoxin, a pore-blocking peptide inhibitor. Science 245: 1382–1385.
4. MACKINNON, R. & G. YELLEN. 1990. Mutations affecting TEA blockade and ion permeation in voltage-activated K+ channels. Science 250: 276–279.
5. YELLEN, G., M. JURMAN, T. ABRAMSON & R. MACKINNON. 1991. Mutations affecting internal TEA blockade identify the probable pore-forming region of a K+ channel. Science 251: 939–942.
6. YOOL, A. J. & T. L. SCHWARZ. 1991. Alteration of ionic selectivity of a K+ channel by mutation of the H5 region. Nature 349: 700–704.
7. HARTMAN, H. A., G. E. KIRSCH, J. A. DREWE, M. TAGLIALATELA, R. H. JOHO & A. M. BROWN.

1991. Exchange of conduction pathways between two related K^+ channels. Science **251**: 942–944.

8. YOKOYAMA, S., K. IMOTO, T. KAWAMURA, H. HIGASHIDA, N. IWABE, T. MIYATA & S. NUMA. 1989. Potassium channels from NG108-15 neuroblastoma-glioma hybrid cells. FEBS Lett. **259**: 37–42.

9. FRECH, G. C., A. M. J. VANDONGEN, G. SCHUSTER, A. M. BROWN & R. H. JOHO. 1989. A novel potassium channel with delayed rectifier properties isolated from rat brain by expression cloning. Nature **340**: 642–645.

10. TAGLIALATELA, M., L. TORO & E. STEFANI. 1992. Novel voltage clamp to record small, fast currents from ion channels expressed in *Xenopus* oocytes. Biophys. J. **61**: 78–82.

11. KIRSCH, G. E., J. A. DREWE, H. A. HARTMANN, M. TAGLIALATELA, M. DE BIASI, A. M. BROWN & R. H. JOHO. 1992. Differences between the deep pores of K^+ channels determined by an interacting pair of non-polar amino acids. Neuron **8**: 499–505.

12. KIRSCH, G. E., J. A. DREWE, M. DEBIASI, H. A. HARTMANN & A. M. BROWN. 1993. Functional interactions between K^+ pore residues located in different subunits. J. Biol. Chem. In press.

13. KIRSCH, G. E., J. A. DREWE, M. TAGLIALATELA, R. H. JOHO, M. DE BIASI, H. A. HARTMANN & A. M. BROWN. 1992. A single non-polar residue in the deep pore of related K^+ channels acts as a $K^+:Rb^+$ conductance switch. Biophys. J. **62**: 136–144.

14. TAGLIALATELA, M., J. A. DREWE, G. E. KIRSCH, M. DEBIASI, H. A. HARTMANN & A. M. BROWN. 1993. Regulation of K^+/Rb^+ selectivity and internal TEA blockade by mutations at a single site in K^+ pores. Pflügers Archiv. **423**: 104–112.

15. TIMPE, L. C., Y. N. JAN & L. Y. JAN. 1988. Four cDNA clones from the *Shaker* locus of Drosophila induce kinetically distinct A-type potassium currents in Xenopus oocytes. Neuron **1**: 659–667.

16. HOSHI, T., W. N. ZAGOTTA & R. W. ALDRICH. 1990. Biophysical and molecular mechanisms of *Shaker* potassium channel inactivation. Science **250**: 533–538.

17. HOSHI, T., W. N. ZAGOTTA & R. W. ALDRICH. 1991. Two types of inactivation in *Shaker* K^+ channels: effects of alterations in the carboxy-terminal region. Neuron **7**: 547–556.

18. BUSCH, A. E., R. S. HURST, R. A. NORTH, J. P. ADELMAN & M. P. KAVANAUGH. 1991. Current inactivation involves a histidine residue in the pore of the rat lymphocyte potassium channel RGK5. Biochem. Biophys. Res. Commun. **179**: 1384–1390.

19. DEBIASI, M., H. A. HARTMANN, J. A. DREWE, M. TAGLIALATELA, A. M. BROWN & G. E. KIRSCH. Inactivation determined by a single site in K^+ pores. Pflügers Archiv. **422**: 354–363.

20. TAGLIALATELA, M., A. M. J. VANDONGEN, J. A. DREWE, R. H. JOHO, A. M. BROWN & G. E. KIRSCH. 1991. Patterns of internal and external tetraethylammonium block in four homologous K^+ channels. Mol. Pharmacol. **40**: 299–307.

21. VANDONGEN, A. M. J., G. C. FRECH, J. A. DREWE, R. H. JOHO & A. M. BROWN. 1990. Alteration and restoration of K^+ channel function by deletions at the N- and C-termini. Neuron **5**: 433–443.

22. BOGUSZ, S. & D. BUSATH. 1992. Is a β-barrel model of the K^+ channel energetically feasible? Biophys. J. **62**: 19–21.

23. DURELL, S. R. & H. R. GUY. 1992. Atomic scale structure and functional models of voltage-gated potassium channels. Biophys. J. Discussions **62**: 243–252.

Structure and Function of Voltage-Dependent Calcium Channels from Muscle[a]

TSUTOMU TANABE,[b] ATSUSHI MIKAMI,
TETSUHIRO NIIDOME, AND SHOSAKU NUMA[c]

bHoward Hughes Medical Institute
Department of Cellular & Molecular Physiology
Boyer Center for Molecular Medicine
Yale University School of Medicine
P.O. Box 9812
New Haven, Connecticut 06536-0812

Departments of Medical Chemistry and Molecular Genetics
Kyoto University Faculty of Medicine
Kyoto 606, Japan

BRETT A. ADAMS AND KURT G. BEAM

Department of Physiology
College of Veterinary Medicine and Biomedical Sciences
Colorado State University
Fort Collins, Colorado 80523

INTRODUCTION

Voltage-dependent calcium channels play important roles in the regulation of a variety of cellular functions, including membrane excitability, muscle contraction, synaptic transmission, and secretion.[1-3] Several types of calcium channels are known to be co-expressed in single cells and different types of cells apparently use these channels for different purposes. We are interested in the structure-function relationships of calcium channels and the molecular basis of specialization of individual types of channels for particular functions. For this purpose we have used skeletal muscle and cardiac dihydropyridine (DHP) receptor cDNAs as important representatives of the voltage-

[a] This work was supported in part by the Ministry of Education, Science and Culture of Japan and by the National Institutes of Health. T. T. is an investigator of the Howard Hughes Medical Institute.
[c] Deceased.

81

dependent calcium channels and have used primary cultured myotubes of skeletal muscle from dysgenic mice as an expression system.

EXPRESSION OF SKELETAL MUSCLE
AND CARDIAC DHP RECEPTOR

Both skeletal muscle and cardiac DHP receptors function as calcium channels and as essential components of excitation-contraction (E-C) coupling but these two kinds of DHP receptors display important functional differences.[4] To understand the molecular basis for these differences, we have injected expression plasmids carrying the entire protein coding sequence of the rabbit skeletal muscle DHP receptor cDNA[5] (pCAC6) or the cardiac DHP receptor cDNA[6] (pCARD1) into nuclei of dysgenic myotubes in primary culture.[4,7]

As a consequence of a mutation of the gene encoding the skeletal muscle DHP receptor,[7,8] depolarization does not cause contraction of dysgenic myotubes.[9-13] However, both pCAC6-injected and pCARD1-injected dysgenic myotubes, examined 2–3 days after injection, displayed spontaneous and electrically evoked contractions. Thus, both skeletal muscle and cardiac DHP receptors restored depolarization-contraction coupling, although the nature of this coupling is different for the two kinds of DHP receptor. Like normal myotubes, dysgenic myotubes injected with pCAC6 underwent electrically evoked contractions in normal rodent Ringer's solution, in Ca^{2+}-free Ringer, and in 0.5 mM Cd^{2+}-containing Ringer. Thus, Ca^{2+} entering across the sarcolemma is not necessary for E-C coupling mediated by the skeletal muscle DHP receptor.[14,15] On the other hand, pCARD1-injected dysgenic myotubes displayed electrically evoked contractions in normal rodent Ringer, but not in Ca^{2+}-free Ringer or in 0.5 mM Cd^{2+}-containing Ringer. Thus, unlike the E-C coupling that is restored in dysgenic myotubes by injection of pCAC6, depolarization-contraction coupling restored in dysgenic myotubes by injection of pCARD1 requires the voltage-dependent entry of Ca^{2+}. Experiments using caffeine-treated myotubes suggested the involvement of Ca^{2+}-induced release of Ca^{2+} from the sarcoplasmic reticulum (SR)[16] in depolarization-contraction coupling in pCARD1 injected myotubes. Thus, it appears that depolarization-contraction coupling in pCARD1-injected dysgenic myotubes mimics coupling in cardiac cells[17-19]: Ca^{2+} entry is required and the Ca^{2+} that enters triggers additional Ca^{2+} release from the SR.

All dysgenic myotubes that had been injected with pCAC6 or pCARD1 and observed to contract expressed L-type calcium current, but the nature of these currents was very different.[1] In pCAC6-injected myotubes, the rate of activation was slow, similar to that observed in normal skeletal muscle.[20,21] In contrast, the L-type calcium current in pCARD1-injected myotubes activated more rapidly, like the L-type current in cardiac muscle.[22,23]

LOCALIZATION OF FUNCTIONAL REGIONS

As described above, injection of an expression plasmid carrying the skeletal muscle DHP receptor cDNA restores both E-C coupling and skeletal L-type calcium current.

The restored coupling resembles normal skeletal muscle E-C coupling, which does not require entry of extracellular Ca^{2+}. By contrast, injection of an expression plasmid carrying the cardiac DHP receptor cDNA into dysgenic myotubes produces cardiac-like L-type current and cardiac-type E-C coupling, which does require entry of extracellular Ca^{2+}. To investigate the molecular basis for these differences in calcium currents and in E-C coupling, various chimeric DHP receptor cDNAs were expressed in dysgenic myotubes.

Regions Critical for Skeletal-type E-C Coupling

In skeletal muscle it has been suggested[24,25] that depolarization causes a molecular rearrangement of a hypothetical structure, the "voltage sensor," and that this rearrangement gates calcium flow across the SR. If the DHP receptor corresponds to this voltage sensor and controls the release of Ca^{2+} from the SR[26] in an electromechanical fashion,[24,25] then discrete regions of the DHP receptor might interact directly (or indirectly through a putative linking protein) with the foot region of the ryanodine receptor[27] to gate the calcium release channel. The main differences in primary structure between the skeletal muscle and cardiac DHP receptors reside in the large, putative cytoplasmic regions, that is, the amino- and carboxy-terminal regions as well as the regions linking repeats I and II and repeats II and III. To determine whether any of these regions are critical for interaction between the DHP receptor and ryanodine receptor, cDNAs encoding chimeric proteins with one or more of the large, putative cytoplasmic regions of the cardiac DHP receptor replaced by corresponding regions of the skeletal muscle DHP receptor were expressed in dysgenic myotubes. Examination of the electrically evoked contraction of these myotubes showed that the putative cytoplasmic region linking repeats II and III of the skeletal muscle DHP receptor is a major determinant site for skeletal-type E-C coupling.[28] Furthermore, analysis of calcium current and intramembrane charge movement indicated that the slow activation of the skeletal muscle L-type channel is not an obligatory consequence of the DHP receptor serving as the voltage sensor for E-C coupling. In addition, participation in direct, skeletal-type E-C coupling does not necessarily cause a DHP receptor to lose its efficiency as a calcium channel.[29]

Regions Critical in Determining Activation Kinetics

For both sodium channels and calcium channels, kinetic analysis of current has suggested that during activation the channel undergoes several distinct conformational changes before reaching the open state.[30-34] Based on the structural characteristics of these channels, it has been postulated that the distinct conformational transitions inferred from kinetic analysis may be equated with conformational changes of the individual structural repeats. As a means of testing this hypothesis, cDNAs encoding chimeric calcium channels in which one or more of the four repeats of the skeletal muscle DHP receptor were replaced by the corresponding repeats derived from the cardiac DHP receptor were expressed in dysgenic myotubes.[35] The current produced by the expression of pSkC15, in which repeat I is skeletal and repeats II–IV are cardiac, activated slowly, resembling the current produced by pCAC6. In contrast, the current produced by pSkC11, in which repeat I is cardiac and repeats II–IV are skeletal, activated rapidly, resembling that for pCARD1. The measured currents could be approximated by a single exponential function, providing a simple measure (τ_{act}) of the activation kinetics for different constructs. Because the rate of activation varies with test potential, τ_{act} was determined in each myotube expressing a cDNA for a test potential at or just above the potential that elicited maximal inward current. For any given plasmid, τ_{act} shows a considerable range of values, but the data clearly fall into two groups with almost no overlap. The only consistent structural feature that distinguishes these two groups of chimeric plasmids is repeat I. For all chimeric plasmids in which repeat I is of skeletal muscle origin, τ_{act} is large, whereas for those in which repeat I is of cardiac origin, τ_{act} is small. Thus a single repeat, the first, governs whether the calcium channel shows slow (skeletal-like) or rapid (cardiac-like) activation. One simple mechanism that can account for the results is that each of the four repeats can independently and reversibly interconvert between resting and activated states, and that the conformational transition of repeat I might be slow. Alternatively, the identity of repeat I might affect the interconversion of other repeats.

Functional Significance of the Carboxyl-terminal Regions of Skeletal Muscle DHP Receptor

Biochemical analysis indicates the presence of two forms of the DHP receptor polypeptide in skeletal muscle: a full-length translation product present as a minor species and a much more abundant form that has a truncated C-terminus.[36-38] This raises the possibility of different roles for these two molecules.[39] To resolve this issue we have constructed a cDNA (pC6Δ1) encoding a protein corresponding to the truncated DHP receptor in skeletal muscle and expressed this cDNA in dysgenic myotubes.[40]

Dysgenic myotubes injected with pC6Δ1 were found to have restored E-C coupling, and the fraction with restored E-C coupling was comparable to that observed for dysgenic myotubes injected with pCAC6. Contractions of pC6Δ1-injected myotubes could be elicited in the absence of external calcium or in the presence of 0.5 mM external cadmium, indicating that the truncated DHP receptor encoded by pC6Δ1 mediates skeletal muscle-type E-C coupling, which does not require the entry of external calcium. Expression of pC6Δ1 restored the intramembrane charge movement that has been implicated as representing voltage-driven conformational changes of the DHP receptor functioning as the voltage sensor for E-C coupling. Thus the truncated DHP receptor is able to function as a voltage sensor for skeletal muscle-type E-C coupling.

Injection of dysgenic myotubes with pC6Δ1 also restored skeletal L-type calcium current with a density, time course of activation, DHP sensitivity, and efficiency of channel function all being very similar to those of L-type current in dysgenic myotubes injected with pCAC6. Thus DHP receptors produced by expression of the full-length or truncated cDNAs seem to differ little in their ability to function as calcium channels. Our results demonstrate that a truncated DHP receptor, which probably corresponds closely in length to the predominant form detected biochemically in skeletal muscle, is able to function not only as a voltage sensor for E-C coupling but also as a calcium channel. Although we cannot conclude that individual DHP receptors function simultaneously as channel and voltage sensor, our results are consistent with the idea that both of these functions are performed in skeletal muscle by a single class of DHP receptors.

CONCLUSIONS

By expressing the cDNAs encoding skeletal muscle, cardiac, chimeric, and truncated DHP receptors in dysgenic myotubes, we have identified several significant regions that determine the specific character of calcium channel molecules. Expression of cDNAs encoding chimeras with one or more of the large, putative cytoplasmic regions of the cardiac DHP receptor replaced by corresponding regions of the skeletal muscle DHP receptor, showed that the putative cytoplasmic region linking repeats II and III is a major determinant of skeletal muscle-type E-C coupling. Expression of cDNAs encoding chimeras in which one or more of the four repeats of the skeletal muscle DHP receptor are replaced by the corresponding repeats derived from the cardiac DHP receptor, showed that the repeat I determines whether the chimeric calcium channel shows slow (skeletal muscle-like) or rapid (cardiac-like) activation. Expression of cDNA encoding a truncated skeletal muscle DHP receptor suggests that a single molecular species of DHP receptors performs dual functions as calcium channel and voltage sensor for E-C coupling.

REFERENCES

1. BEAN, B. P. 1989. Annu. Rev. Physiol. 51: 367–384.
2. TSIEN, R. W. & R. Y. TSIEN. 1990. Annu. Rev. Cell Biol. 6: 715–760.
3. BERTOLINO, M. & R. R. LLINÁS. 1992. Annu. Rev. Pharmacol. Toxicol. 32: 399–421.
4. TANABE, T., A. MIKAMI, S. NUMA & K. G. BEAM. 1990. Nature 344: 451–453.
5. TANABE, T., H. TAKESHIMA, A. MIKAMI, V. FLOCKERZI, H. TAKAHASHI, K. KANGAWA, M. KOJIMA, II. MATSUO, T. HIROSE & S. NUMA. 1987. Nature 328: 313–318.
6. MIKAMI, A., K. IMOTO, T. TANABE, T. NIIDOME, Y. MORI, H. TAKESHIMA, S NARUMIYA & S. NUMA. 1989. Nature 340: 230–233.
7. TANABE, T., K. G. BEAM, J. A. POWELL & S. NUMA. 1988. Nature 336: 134–139.
8. CHAUDHARI, N. 1992. J. Biol. Chem. 267: 25636–25639.
9. POWELL, J. A. & D. M. FAMBROUGH. 1973. J. Cell Physiol. 82: 21–38.
10. KLAUS, M. M., S. P. SCORDILIS, J. M. RAPALUS, R. T. BRIGGS & J. A. POWELL. 1983. Devel. Biol. 99: 152–166.
11. BEAM, K. G., C. M. KNUDSON & J. A. POWELL. 1986. Nature 320: 168–170.
12. BOURNAUD, R. & A. MALLART. 1987. Pflügers Arch. 409: 468–476.
13. RIEGER, F., R. BOURNAUD, T. SHIMAHARA, L. GARCIA, M. PINÇON-RAYMOND, G. ROMEY & M. LAZDUNSKI. 1987. Nature 330: 563–566.
14. ARMSTRONG, C. M., F. M. BEZANILLA & P. HOROWICZ. 1972. Biochim. Biophys. Acta 267: 605–608.
15. CHIARANDINI, D. J., J. A. SÁNCHEZ & E. STEFANI. 1980. J. Physiol. 303: 153–163.
16. ENDO, M. 1977. Physiol. Rev. 57: 71–108.
17. FABIATO, A. 1985. J. Gen. Physiol. 85: 291–320.
18. BEUCKELMANN, D. J. & W. G. WIER. 1988. J. Physiol. 405: 233–255.
19. NÄBAUER, M., G. CALLEWAERT, L. CLEEMANN & M. MORAD. 1989. Science 244: 800–803.
20. SÁNCHEZ, J. A. & E. STEFANI. 1978. J. Physiol. 283: 197–209.

21. DONALDSON, P. L. & K. G. BEAM. 1983. J. Gen. Physiol. 82: 449–468.
22. ISENBERG, G. & U. KLÖCKNER. 1982. Pflügers Arch. 395: 30–41.
23. LEE, K. S. & R. W. TSIEN. 1982. Nature 297: 498–501.
24. SCHNEIDER, M. F. & W. K. CHANDLER. 1973. Nature 242: 244–246.
25. CHANDLER, W. K., R. F. RAKOWSKI & M. F. SCHNEIDER. 1976. J. Physiol. 254: 285–316.
26. RIOS, E. & G. BRUM. 1987. Nature 325: 717–720.
27. FLEISCHER, S. & M. INUI. 1989. Annu. Rev. Biophys. Biophys. Chem. 18: 333–364.
28. TANABE, T., K. G. BEAM, B. A. ADAMS, T. NIIDOME & S. NUMA. 1990. Nature 346: 567–569.
29. ADAMS, B. A., T. TANABE, A. MIKAMI, S. NUMA. & K. G. BEAM. 1990. Nature 346: 569–572.
30. HODGKIN, A. L. & A. F. HUXLEY. 1952. J. Physiol. 117: 500–544.
31. ARMSTRONG, C. M. 1981. Physiol. Rev. 61(3): 644–683.
32. KOSTYUK, P. G., O. A. KRISHTAL & V. I. PIDOPLICHKO. 1981. J. Physiol. 310: 403–421.
33. SÁNCHEZ, J. A. & E. STEFANI. 1983. J. Physiol. 337: 1–17.
34. KEYNES, R. D., N. G. GREEFF & I. C. FORSTER. 1990. Proc. R. Soc. Lond. B 240: 411–423.
35. TANABE, T., B. A. ADAMS, S. NUMA & K. G. BEAM. 1991. Nature 352: 800–803.
36. DE JONGH, K. S., D. K. MERRICK & W. A. CATTERALL. 1989. Proc. Natl. Acad. Sci. USA 86: 8585–8589.
37. LAI, Y., M. J. SEAGAR, M. TAKAHASHI & W. A. CATTERALL. 1990. J. Biol. Chem. 265: 20839–20848.
38. DE JONGH, K. S., C. WARNER, A. A. COLVIN & W. A. CATTERALL. 1991. Proc. Natl. Acad. Sci. USA 88: 10778–10782.
39. CATTERALL, W. A. 1991. Cell 64: 871–874.
40. BEAM, K. G., B. A. ADAMS, T. NIIDOME, S. NUMA & T. TANABE. 1992. Nature 360: 169–171.

Molecular Diversity of Voltage-Dependent Calcium Channel

YASUO MORI, TETSUHIRO NIIDOME,
YOSHIHIKO FUJITA, MICHELLE MYNLIEFF,[a]
ROBERT T. DIRKSEN,[a] KURT G. BEAM,[a]
NAOYUKI IWABE,[b] TAKASHI MIYATA,[b]
DAISUKE FURUTAMA,[c] TEIICHI FURUICHI,[c]
AND KATSUHIKO MIKOSHIBA[c]

Departments of Medical Chemistry and Molecular Genetics
Kyoto University Faculty of Medicine
Kyoto 606-01, Japan

[a]Department of Physiology
College of Veterinary Medicine and Biomedical Sciences
Colorado State University
Fort Collins, Colorado 80523

[b]Department of Biophysics
Kyoto University Faculty of Science
Kyoto 606-01, Japan

[c]Department of Molecular Neurobiology
Institute of Medical Science
University of Tokyo
Tokyo 108, Japan

The influx of calcium through voltage-dependent calcium channels plays a vital role in the regulation of a variety of cellular functions, including membrane excitability, enzyme activity, axonal outgrowth, muscle contraction, and neurotransmitter release. Establishing criteria for distinguishing between the many types of calcium channels is a prerequisite for elucidation of molecular mechanisms that underlie diverse cellular processes.[1]

Biophysical and pharmacological criteria are usually employed as a first step towards identifying the type of channel involved in a particular physiological process. On the basis of electrophysiological and pharmacological properties, at least four types of calcium channel (designated T-, L-, N-, and P-type) have been distinguished (TABLE

1).[1,2] L-type calcium channels are high voltage–activated, sensitive to dihydropyridine (DHP), and show a single-channel conductance of 22–27 pS. They are found in virtually all excitable tissues and in many non-excitable cells. They trigger excitation-contraction coupling in skeletal muscle, heart, and smooth muscle and they control hormone or transmitter release from endocrine cells and some neurons. The N-type channel is a high voltage–activated calcium channel largely restricted to neurons. It differs pharmacologically from L-type in being resistant to DHP and irreversibly blocked by ω-conotoxin (ω-CgTx).[2–5] In addition, ω-CgTx inhibits transmitter release in a variety of mammalian neuronal preparations,[6–9] thus supporting the hypothesis[10] that influx of calcium through N-type calcium channels controls neurotransmitter release. Additionally, N-type calcium channels also appear to play a role in directed migration of immature neurons.[11] P-type calcium channels, first identified in Purkinje cells,[12,13] are found in a variety of neurons.[14] They are high voltage–activated calcium channels selectively blocked by ω-agatoxin-IVA (ω-AgaIVA)[15] or funnel-web spider toxin (FTX),[11] but insensitive to both DHP and ω-CgTx. P-type channels play an essential role in inducing long-term depression. T-type calcium channels are known as low voltage–activated calcium channels and called "fast" because they inactivate rapidly. T-type channels are sensitive to Ni^{2+}, amirolide, and octanol, but they are resistant to DHP and ω-CgTx. They have a single-channel conductance of ~ 8 pS. T-type channels have been implicated in repetitive firing and pacemaker activity in heart and neurons.

However, recent results indicate that on the basis of these criteria it may not be possible to assign unequivocally a measured calcium current to the currently defined T, N, L, or P categories, and that currents of a particular pharmacological specificity may vary considerably in their biophysical properties.[15–21] Thus, it is critical to establish the electrophysiological and pharmacological properties for calcium channels of known molecular identity. Over the last five years, there have been rapid advances in molecular biological approaches to calcium channels. The approaches have been proven to be a powerful way to aid the understanding of the diversity of voltage-dependent calcium channels.

TABLE 1. Functional Classification of Voltage-dependent Calcium Channels

Type	Properties	Function
T type	Low voltage–activated, blocked by low Ni^{2+} and octanol, inactivates rapidly	Pacemaker activity and repetitive firing in heart and neurons
L type	High voltage–activated, blocked by dihydropyridine antagonists and low Cd^{2+}	Excitation-contraction coupling, excitation-secretion coupling in endocrine cells and some neurons
N type	High voltage–activated, blocked by ω-conotoxin and low Cd^{2+}	Trigger neurotransmitter release
P type	High-voltage activated, blocked by ω-AgaIVA, insensitive to dihydropyridine and ω-conotoxin	Trigger neurotransmitter release, calcium spike in some neurons, induces long-term depression

TABLE 2. Molecular Classification of Voltage-dependent Calcium Channels

Numa Class	Snutch Class	Perez-Reyes Class	Primary Tissue Location	Functional Class
Sk	–	1	skeletal muscle	L type
C	C	2	heart, smooth muscle	L type
BIV	D	3	brain, pancreas	L type
BI	A	4	brain	P type
BII	E	–	brain	?
BIII	B	5	brain	N type

STRUCTURAL DIVERSITY AND DIFFERENTIAL EXPRESSION OF CALCIUM CHANNELS

Biochemical and molecular biological approaches began with studies of the L-type calcium channel (the DHP receptor) complex in skeletal muscle.[22] The DHP receptor has been purified from skeletal muscle by several laboratories.[23-28] The receptor is composed of the α_1 subunit (170 kD polypeptide), the α_2 subunit (140 kD polypeptide), the β subunit (55 kD polypeptide), and the γ subunit (33 kD polypeptide). A stoichiometric ratio of 1:1:1:1 was obtained for the α_1, α_2, β, and γ subunits, respectively, which suggests that all four subunits are integral components of the DHP receptor.[29] The δ subunit has been observed as small proteins of 24–33 kD disulfide-linked to α_2 subunits to form α_2/δ.[25] The subunit compositions of calcium channel complexes from other tissues are not exactly the same as those from skeletal muscle.[30-32] One form of L-type calcium channels in brain contains α_1-like subunit in association with α_2/δ-like and β-like subunits. The brain ω-CgTx-sensitive N-type calcium channel contains components homologous to the α_1 subunit,[33-36] the α_2 subunit,[34,36] and the β subunit[36-38] of the DHP receptor. An additional polypeptide of 110 kD[36] or 100 kD[31] is present in N-type or L-type calcium channels, respectively, from brain. Other polypeptides of 36 kD, 28 kD,[39] and 58 kD[40] associated with ω-CgTx receptors have also been reported. Using FTX to form an affinity gel, a protein with an apparent molecular mass of 90–100 kD was isolated from brain.[41]

Molecular biological studies have revealed an even greater diversity than functional studies among calcium channels, arising from multiple genes and alternative splicing of the composing subunits. Recent evidence suggests that α_1 subunits are encoded by a gene family comprising at least six distinct genes (TABLE 2). Amino acid sequences

FIGURE 1. (*Following four pages.*) Alignment of amino acid sequence of the different calcium channels. The six sequences compared (from top to bottom): BIII; BI[62]; BII[64]; rabbit cardiac muscle DHP-sensitive calcium channel (C)[51]; rabbit skeletal muscle DHP-sensitive calcium channel (Sk)[24]; human brain DHP-sensitive calcium channel α_{1D} (D).[60] The sequences for BI and BII are those of the BI-2 and BII-2 isoforms, respectively. Sets of six identical residues at one position are enclosed with solid lines, and sets of six identical or conservative residues[90] with broken lines. The conserved sequences (aligned positions 106–827 and 1333–2131) adopted to infer the phylogenetic tree (FIG. 3) are indicated by arrows. The numbers of the amino acid residues at the right end of the individual lines are given. Segments S1-S6 in each repeat are shown.

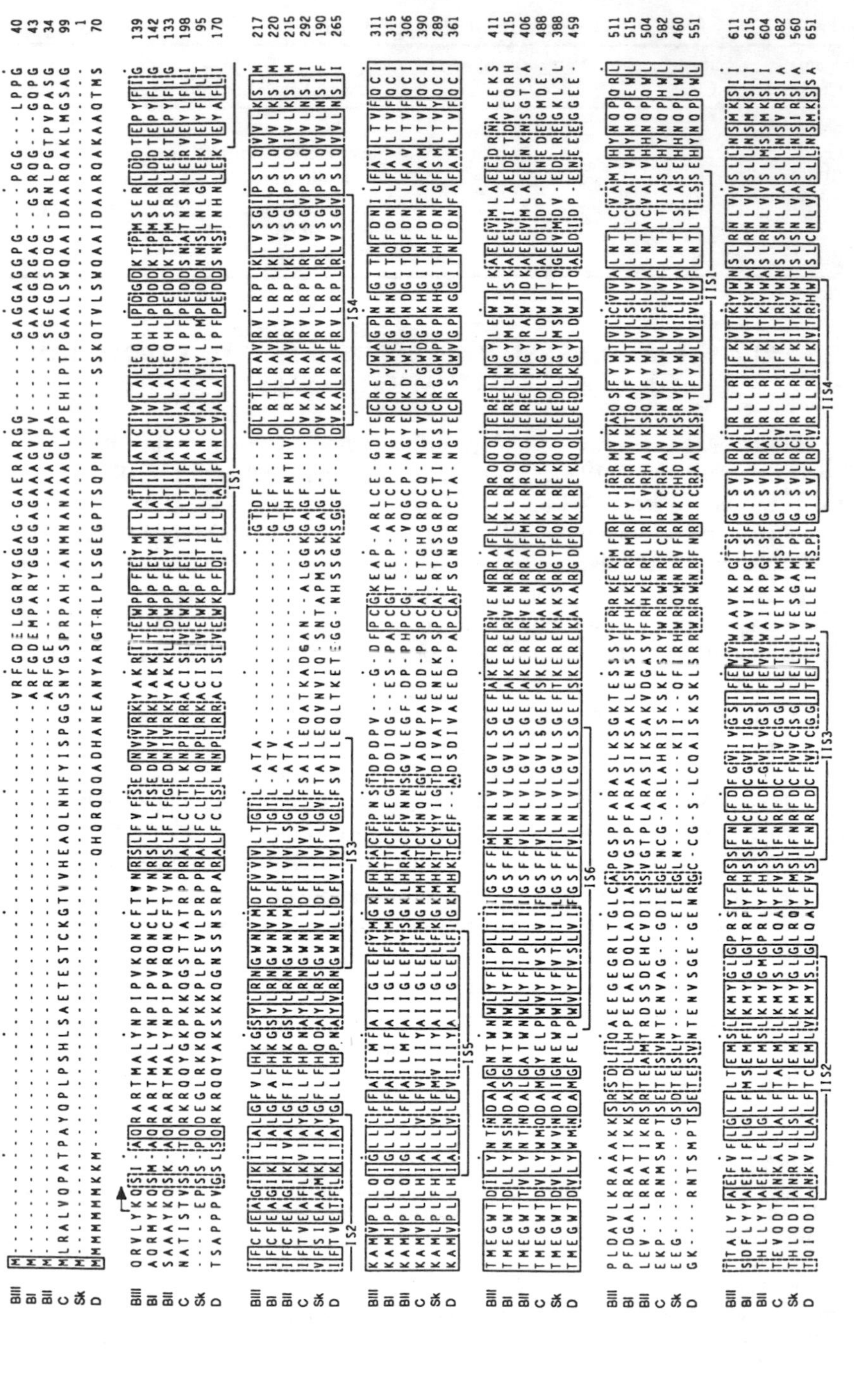

FIG. 1 continued.

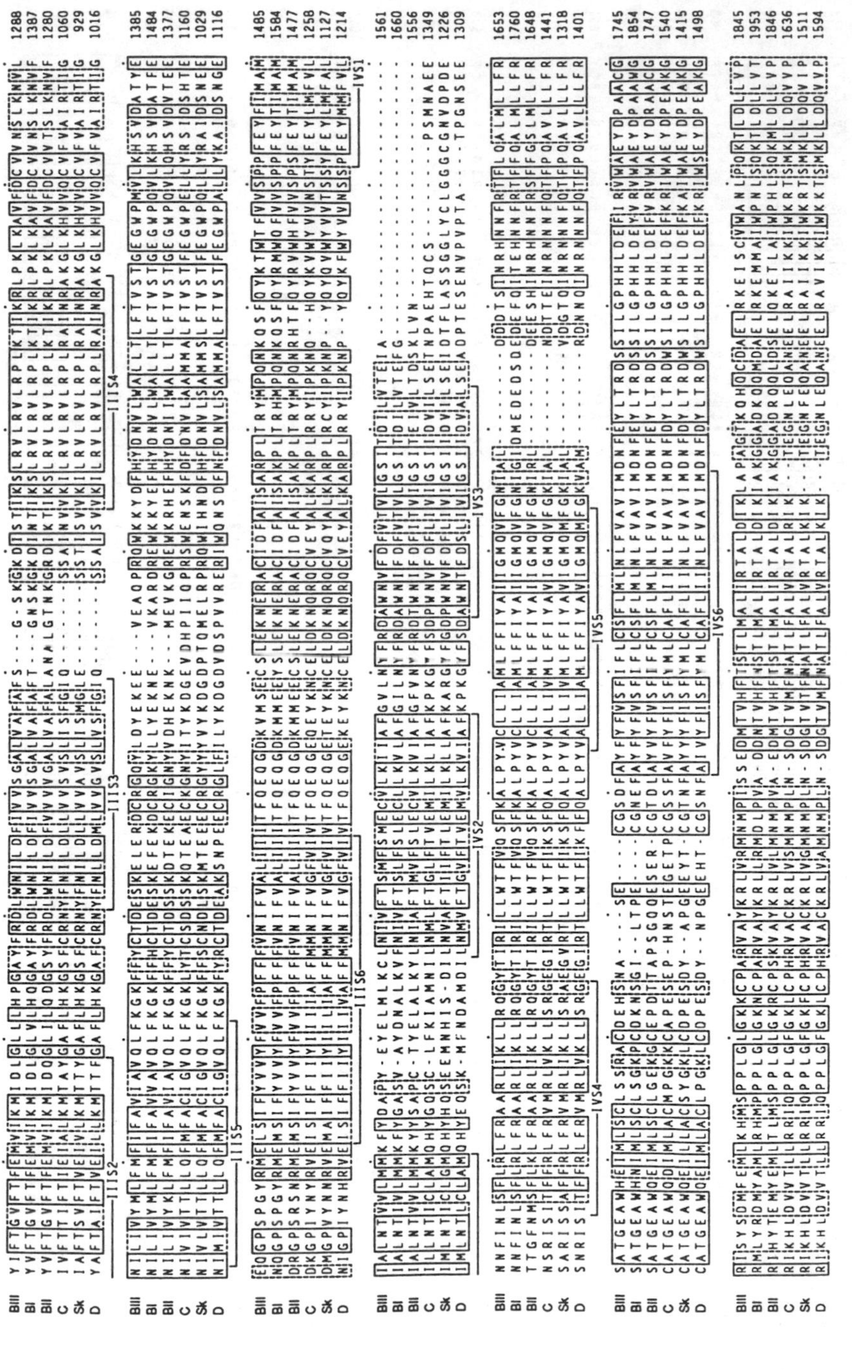

FIG. 1 continued.

of the six classes of calcium channels have been deduced by cloning the sequencing of the cDNAs (FIG. 1). All the α_1 subunits share general structural features with voltage-dependent sodium channels, thus apparently having the same transmembrane topology proposed for sodium channels.[42] They contain four repeating homologous units and each repeat has one positively charged segment (S4), which probably represents a voltage-sensing region,[43] and five hydrophobic segments (S1, S2, S3, S5, and S6). The conserved charged residues in segments S2 and S3 are also retained.[42] The glutamic acid residues in the SS1–SS2 region, which may be critical for ion selectivity of calcium channels,[44] are conserved among the α_1 subunits. Previous cDNA expression studies suggest that the function of the sodium channel can be manifested by the large subunit.[45] Thus the striking structural similarity found between the α_1 subunit and the sodium channel may imply that the α_1 subunit itself is a voltage-dependent channel protein. cDNA expression studies of calcium channels support this notion and are discussed later in this paper.

The skeletal muscle DHP receptor (class Sk calcium channel) cDNA was cloned first of all the α_1 subunits.[24] Using immunochemical techniques, this channel was detected only in skeletal muscle.[46] A prominent transcript of the class Sk channel was detected only in RNA preparations from skeletal muscle by Northern blot analysis.[47] However, successful isolation of class Sk channel cDNAs from brain, pancreatic β-cell–derived HIT cells, and ovary implies that this class cannot be considered as exclusively a skeletal muscle gene.[48] The class Sk channel is expressed in a developmentally regulated fashion, being induced upon myogenic differentiation.[46,49] Splice variants of this class, which differ in the region between S3 and S4 of repeat IV,[48] and, interestingly, cDNAs encoding a two-motif isoform of the class Sk channel have been isolated.[50]

The class C calcium channel is a ubiquitous calcium channel. cDNAs encoding the class C channel were isolated from heart,[51] lung,[52] aorta,[53] brain,[54] skin fibroblast cell line,[55] ovarian cells, and hamster pancreatic β-cell–derived HIT cells.[48] Expression of this class in intestine, stomach, spinal cord, pituitary, adrenal gland, liver, kidney, testes, and spleen was shown by Northern blot analyses and analyses using polymerase chain reaction (PCR).[48,53,54] The class C channel mRNA was also found in early skeletal muscle myotubes.[55] Numerous isoforms of the class C channel are generated by alternative splicing at different sites.[48,52–54,56,57] The sites of structural variation due to alternative splicing are located in the N-terminal portion, the region between segment S5 and S6 of repeat I (IS5 and IS6, respectively), IS6, two regions in the vicinity of IIS6, the region between repeats I and II (I-II loop), the II-III loop, IIIS2, IVS3 and its vicinity, and the carboxy-terminal region. The variability within IVS3 is generated by a developmentally regulated, mutually exclusive splicing mechanism.[56] Differential expression of these variants was shown by PCR analysis of RNAs from different tissues.[54]

Class BIV (or class D) calcium channel cDNAs were isolated from brain, human neuroblastoma IMR32 cell line, pancreatic islet cells, hamster pancreatic β-cell–derived HIT cells, and ovarian cells.[48,58–60] Therefore the BIV channel is called the neuroendocrine type.[48] Expression of this class in pancreatic β cells[59] and in cells from the central nervous system[61] was shown by in situ hybridization. Multiple types of the class BIV channel isoform are generated by alternative splicing.[48,60] Variable regions of this channel are located in IS6, the I-II loop, IVS3 and its vicinity, and the carboxy-terminal region.

Until recently, molecular identification of calcium channel has been limited to L-type calcium channels. The class BI (or class A) calcium channel cDNA is the first cloned non–L-type calcium channel cDNA.[62,63] The presence of multiple isoforms of this class, generated by alternative RNA splicing, has been reported.[62] The sites of variation are located in the II-III loop, the vicinity of IVS6, and the carboxy terminal region. The class BI channel is distributed widely in the brain, being abundant in the cerebellum.[62] Blotting analysis of RNA from the cerebellums of mutant mice with different types of cerebellar degeneration suggested that this channel is expressed in Purkinje cells and granule cells. Expression of this class in the heart[62] and pituitary[63] was also demonstrated.

The class BII (or class E) calcium channel cDNA was also isolated from the brain.[64] Two isoforms, possibly generated by alternative splicing, differ from each other in the carboxy-terminal sequence. The spatial distribution of the class BII channel in the brain is different from that of the class BI channel. This class is abundant in cerebral cortex, hippocampus, and corpus striatum.

Class BIII (or class B) calcium channel cDNAs were cloned from brain[65] and human neuroblastoma IMR32 cell line.[66] Two isoforms, that differ at the carboxy-terminal region, are generated by alternative RNA splicing.[66] Polyclonal antiserum generated against a peptide from the class BIII sequence selectively immunoprecipitates high-affinity ω-CgTx binding sites from forebrain membranes.[65] RNA preparations from different rabbit tissues and from different regions of rabbit brain were subjected to Northern blot analysis with a BIII cDNA probe (data not shown). Of all the tissues studied, only the brain expressed a major hybridizable RNA species (~ 9,500 nucleo-tides in length). A BIII mRNA species of ~ 9,500 nucleotides was present in the cerebral cortex, hippocampus, corpus striatum, midbrain, and cerebellum. An additional mRNA species of ~ 9,300 nucleotides was observed in the striatum and midbrain. This class is expressed also in calcitonin-secreting C-cells and undifferentiated and differentiated rat pheochromocytoma PC-12 cells.[65] Notably, the distribution of BIII mRNA within rabbit brain is rather different from that of rbB-1 mRNA within rat brain.[65] In rabbit, BIII mRNA appears most abundantly expressed in the striatum and midbrain, while in rat, rbB-1 is relatively more abundant in the cerebellum, hippocampus, and thalamus-hypothalamus than in other brain regions. This suggests the possibilities that similar neuronal functions in the two species are carried out by different types of calcium channels or that there are differences in synaptic architecture. Alternatively, differences in developmental age may account for the distinct patterns of tissue distribution.

In situ hybridization histochemistry of the BIII channel revealed widespread but uneven signals throughout rabbit brain (FIG. 2). The BIII mRNA was detectable in the olfactory bulb (the external and internal plexiform layers and the mitral and granular cell layers), anterior olfactory cortex, olfactory tubercle, caudate-putamen, primary olfactory cortex, neocortex, entorhinal cortex, hippocampal formation, amygdaloid nucleus, thalamus-hypothalamus, colliculus, cerebellar cortex, and medulla-pons (the motor trigeminal nucleus, facial nucleus, lateral reticular nucleus, and inferior olivary nucleus). The BI mRNA was found in similar brain regions as the BIII mRNA, but at different expression levels (autoradiography carried out for 18 days and 5 days for BIII and BI probes, respectively, in FIG. 2(A); 38 days and 12 days in FIG. 2(B and C)). Dark-field observation of emulsion-dipped hippocampal sections (FIG. 2, B) showed that BI mRNA expression was prominent in CA3 hippocampal pyramidal cells and

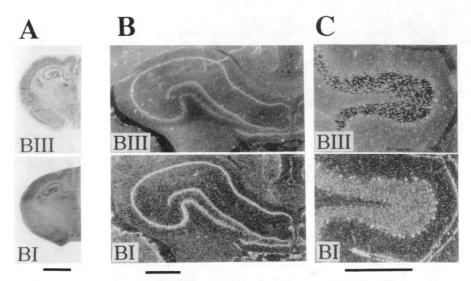

FIGURE 2. Spatial distribution of BIII and BI calcium channel mRNAs determined with *in situ* hybridization histochemistry. Cryosection preparation (12 μm in thickness) and *in situ* hybridization were performed essentially as described previously.[91] (**A**) Film autoradiogram of coronal sections hybridized with [35S]BIII and BI cRNA probes. Bar = 5 mm. The film was Hyper β-max (Amersham, Buckinghamshire, UK). (**B** and **C**) Dark-field observation of emulsion-dipped coronal sections of the hippocampal formation (**B**) and cerebellar lobes (**C**), hybridized with [35S]BIII and BI cRNA probes. (**B**) Bar = 100 μm. (**C**) Bar = 50 μm. The emulsion was NB2 (Kodak, Rochester, NY).

hilar cells, whereas the expression level of BIII mRNA was relatively lower in these neurons than that in CA1-2 pyramidal cells or dentate granular cells. In the cerebellum (Fig. 2, C), BI mRNA was predominantly expressed in the Purkinje cell layer and, at a slightly lower level, in the granular cell layer. This distribution agrees well with the distribution inferred from Northern blot analysis of RNA from cerebellums of mutant mice having different types of cerebellar degeneration.[62] The BIII mRNA was also present in Purkinje cells and in granular cells at a low density. The BI mRNA encodes a calcium channel insensitive to both ω-CgTx and DHP,[62] whereas the BIII mRNA encodes a ω-CgTx-sensitive calcium channel (see below). Thus, differential expression of the BI and BIII channels may underlie the electrophysiological and pharmacological heterogeneity of calcium currents that has been described for neurons of different types.[14]

A phylogenetic tree indicates that the α_1 subunits of voltage-dependent calcium channels can be classified into two subfamilies (Fig. 3). One is the DHP-sensitive L-type calcium channel subfamily consisting of the cardiac muscle, skeletal muscle, and neuroendocrine BIV (α_{1D}) channels. The second is the channel subfamily consisting of the BI, BII, and BIII calcium channels, which are predominantly or exclusively expressed in neuronal tissues. These two subfamilies diverged from a common ancestor, although it remains unclear when the divergence occurred. It is evident that each of

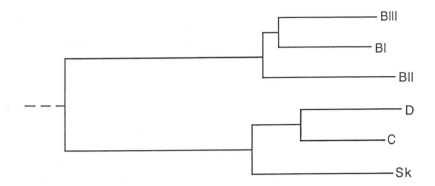

FIGURE 3. Identity (similarity) matrix (**A**) and phylogenetic tree (**B**) for members of the calcium channel family. (**A**) The calculations are based on the alignment in FIGURE 2. Gaps are counted as one substitution regardless of their length. The percentages of overall amino acid identity (or similarity calculated for identical plus conservative residues[90]) of the calcium channel pairs are shown on the upper right side of the diagonal, and the percentages of identity (or similarity) within the conserved sequences (aligned positions 106–827 and 1333–2131 in Fig. 2) on the lower left side. BI and BII represent BI-2 and BII-2 sequences, respectively. (**B**) The neighbor-joining method[92] was used; the conserved sequences were adopted for the calculation. Lengths of horizontal lines are proportional to estimated numbers of amino acid substitutions.

the subfamilies further diverged at about the same time, creating the six currently known members of the calcium channel family. The phylogenetic tree also indicates that the relationship between BIII and BI is closer than those between BIII and BII and between BI and BII. The regions corresponding to the four internal repeats and the short segment between repeats III and IV (III-IV loop) are relatively well conserved. However, the remaining regions, all of which are assigned to the cytoplasmic side of the membrane, are less well conserved. The amino acid sequences of the putative cytoplasmic region between repeats II and III (II-III loop) and the C-terminal region are highly diverged among calcium channels.

HETEROLOGOUS EXPRESSION OF CALCIUM CHANNELS

To relate the molecular classification of calcium channels to their functional classification, expression of mRNAs or cDNAs in foreign cells has proven to be essential. This form of functional reconstitution is termed heterologous expression, and several expression systems have been developed for calcium channels. The most favored expression system for calcium channels is based on the injection of mRNAs into *Xenopus* oocytes, which serves both as decisive proof of successful cloning and as a useful tool for studying structure-function relationships of various types of ion channels. To reconstitute a particular calcium channel subtype, it is desirable to inject the subtype-specific mRNA in oocytes. This has become possible by *in vitro* transcription of cDNAs using the bacteriophage SP6 promoter.[67] The technique has provided an effective approach to the study of what roles the individual subunits, domains, or amino acids play in the operation of calcium channels. Techniques of expression in mammalian cells also offer a useful way to study functional properties of calcium channels.

In *Xenopus* oocytes, mRNA derived from the class C calcium channel cDNA isolated from heart directed the formation of a functional DHP-sensitive calcium channel current.[51] The Ba^{2+} current carried by the channel was high voltage–activated, increased several times in amplitude by BAY K 8644, and virtually abolished by nifedipine and Cd^{2+}. Properties of a splice variant of the class C channel from lung were similar to those of the class C variant from heart in oocytes.[52] These results indicate that the α_1 subunit alone is sufficient to exhibit L-type calcium channel activity. Co-injection of the skeletal muscle α_2 subunit enhanced the peak inward current about three times in amplitude. The increasing effect on the activity of class C channel was observed also for coexpression of the class CaB1 β subunit from skeletal muscle,[68–70] the class CaB2 β subunit from brain[71] and heart,[72] the class CaB3 β subunit from heart,[72] and the γ subunit from skeletal muscle.[68,69] In Chinese hamster ovary (CHO) cells, the class C channel cDNA from smooth muscle alone was sufficient for stable expression of functional calcium channels.[73] The single-channel conductance was 26 pS with 80 mM Ba^{2+} as the charge carrier, corresponding with 24.6 pS in vascular smooth muscle.

The class Sk channel-specific mRNA has failed to direct L-type calcium channel activity in oocytes. Skeletal muscle myotubes from mice with muscular dysgenesis (mdg) has proved to be a useful expression system for various classes of calcium channels. mdg is a fatal autosomal recessive mutation[74] that is expressed in skeletal muscle as a failure of excitation-contraction (E-C) coupling[75] and absence of the slow DHP-sensitive L-type calcium current.[76] Blot hybridization analysis of genomic DNA and skeletal muscle RNA suggested that the mdg mutation alters the structural gene for skeletal muscle DHP receptor.[77] Both functional defects of dysgenic myotubes were restored by microinjection of the class Sk channel (the skeletal muscle DHP receptor) cDNA. Cardiac L-type calcium channel, which activates more rapidly than the skeletal muscle calcium channel, was also expressed in myotubes from mdg mice by microinjection of the class C channel cDNA from heart.[78] E-C coupling restored by injection of the class C cDNA didn't require entry of extracellular Ca^{2+}, in contrast to that restored by injection of the class Sk channel. Thus the behaviors of the expressed calcium channels mirror the physiological situation in skeletal and cardiac muscle.

Skeletal muscle L-type channel was stably expressed in murine L-cells.[79] The class CaB1 β subunit cDNA from skeletal muscle was transfected to examine its functional roles. Coexpression of the β subunit accelerated activation kinetics and increased drug binding sites, but did not increase the currents.[80,81]

The two class BI (class A) channel isoforms (BI-1 and BI-2) were expressed in oocytes.[62] Both isoforms were high voltage–activated; insensitive to Ni^{2+}, nifedipine, and ω-CgTx; and inhibited moderately by Bay K 8644 and strongly by funnel web spider venom and low concentrations of Cd^{2+} with the half-blocking concentrations of 0.5 μM. The single-channel slope conductance of the BI channel was 16 pS with 110 mM Ba^{2+} as the charge carrier. Thus the properties of the BI channel as well as its tissue distribution suggest that the BI channel represents the P-type channel. Oocytes injected with the BI-specific mRNA alone showed only small inward Ba^{2+} current in 40 mM external Ba^{2+}. But when the α_2 subunit and the class CaB1 β subunit from skeletal muscle were coexpressed, the Ba^{2+} current increased by two orders of magnitude. Coexpression of the α_2 or the β subunit alone increased the current to a much smaller extent, the latter being more effective than the former, whereas coexpression of the skeletal muscle γ subunit exerted no significant effect.

The class BIV (class D) channel expressed in oocytes was the DHP-sensitive L-type channel.[60] Coexpression of the neuronal-type splice variant of the class CaB1 β subunit was necessary for the functional expression of the BIV channel, whereas the neuronal-type splice variant of the α_2 subunit played an accessory role that potentiates calcium channel activity. The BIV channel was reversibly blocked by ω-CgTx. This observation, together with the distribution of both the class C and class BIV channels in neurons, may imply that the class BIV represents the neuronal L-type reversibly inhibited by ω-CgTx[17] and the class C represents the neuronal L-type insensitive to ω-CgTx.[16]

To address functional properties of the BIII calcium channel, an expression plasmid carrying the full-length BIII cDNA (pKCRB3) was injected into cultured skeletal muscle myotubes from mice homozygous for the muscular dysgenesis mutation. Dysgenic myotubes endogenously express only a low voltage–activated (T-type) calcium current,[76] and also a very low level of I_{dys}, a high voltage–activated (HVA) calcium current that is blocked by dihydropyridines. Previous studies have shown that dysgenic myotubes injected with cDNAs encoding cardiac and skeletal muscle DHP receptors express high levels of DHP-sensitive calcium current. In these studies, the injected myotubes that were expressing cDNAs encoding DHP receptors could be identified on the basis of contraction in response to electrical stimulation.[77,78] However, in the case of dysgenic myotubes injected with pKCRB3, contractions in response to electrical stimulation were not observed. Thus, 3–4 days after injection of pKCRB3, we randomly assayed the injected myotubes using the whole-cell patch clamp technique to determine whether they were expressing HVA calcium current. Of 43 injected myotubes examined, 14 expressed an appreciable density of HVA calcium current.

An example of currents recorded from a BIII-expressing cell are illustrated in FIGURE 4(A). In this cell, test depolarizations to potentials $>$ – 10 mV elicited a partially inactivating current that became maximal for a test pulse to + 30 mV (FIG. 4, B). In other cells, the maximal current occurred at potentials ranging from + 10 to + 30 mV, with + 20 mV being the most common. This voltage dependence is similar to that previously reported for N-type current.[3,82] The time course of the expressed current, which was similar with either Ca^{2+} or Ba^{2+} as the charge carrier (data not shown), resembles that of both the ensemble average determined for a single N-type channel in a differentiated PC12 cell[16] and the ω-CgTx-sensitive currents produced in HEK293 cells by transfection with cDNA isolated from human neuroblastoma cells.[66] At a test potential for + 20 mV, the magnitude of the peak current averaged 6.89 ± 0.65 pA/pF ($n = 14$). The expressed current was insensitive to dihydropyridine antagonists. In 7 cells, the maximal current after addition of 1 mM (+) PN 200-110 was nearly identical ($110 \pm 4\%$) to the current recorded prior to the addition (run-up of the expressed current, which occurred even in the absence of any solution change, probably accounts for the small increase). It has been shown that depolarizing prepulses, which are too weak to activate N-type current (as defined by sensitivity to ω-CgTx), substantially inactivate the transient component of the current activated by a subsequent, stronger depolarization.[83] FIGURE 4(C) shows that this is also the case for the current present in myotubes expressing pKCRB3 since the transient component of the current elicited by depolarization from -80 to $+20$ mV is substantially reduced when the depolarization to $+20$ mV is preceded by a 1 sec prepulse to -20 mV. On average, 1 sec prepulses to -20 mV and -30 mV reduced the transient component at $+20$ mV by $75.2 \pm 3.1\%$ ($n = 3$) and $62.6 \pm 2.3\%$ ($n = 6$), respectively.

An irreversible block by ω-CgTx is increasingly accepted as a defining attribute of N-type calcium channels.[14,16] FIGURE 4(D) illustrates the effect of 10 mM ω-CgTx on the current in a myotube expressing the BIII calcium channel. This block required several minutes to reach a steady-state and could not be reversed by prolonged washing. For a 300 msec test pulse to $+20$ mV, 10 mM ω-CgTx reduced the peak current, and the current sustained to the end of the test pulse, respectively, by 85.6 ± 4.1

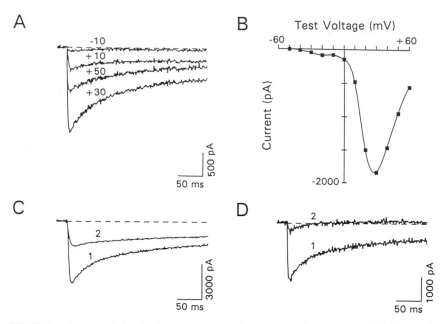

FIGURE 4. N-type whole-cell calcium currents in dysgenic myotubes expressing BIII cDNA. (A) Family of calcium currents elicited by test depolarizations (V_{TEST}) to the indicated potentials (mV) that were applied from a holding potential (HP) of -80 mV. (B) Peak current-voltage relationship for the cell illustrated in (A). The small amount of current at negative test potentials represents the endogenous T-type current (I_{fast}) of dysgenic myotubes.[76] (C) A depolarizing prepulse preferentially inactivates the transient component of expressed N-type current. The test depolarization ($+20$ mV) was applied either directly from the HP of -80 mV (1) or after a 1 sec prepulse to -20 mV (2). (D) Expressed N-type current recorded before (1) and after (2) exposure to ω-CgTx. Between the recordings of traces 1 and 2, the cell was exposed for 5 min to 10 μM ω-CgTx GVIA and then extensively washed with toxin-free solution for another 5 min. The residual current after exposure to ω-CgTx likely represents unblocked T-type current. $V_{TEST} = +40$ mV, HP $= -80$ mV. In both (C) and (D) 1 μM ($+$) PN 200-110 was present throughout the experiment to block any endogenous I_{dys}.[85] All currents were measured with an "external solution" containing (all concentrations are given in mM) 10 $CaCl_2$, 145 Tetraethylammonium-Cl, 0.003 tetrodotoxin, and 10 HEPES (pH 7.4 with CsOH) and an "internal solution" containing 140 Cs-aspartate, 10 Cs_2-EGTA, 5 $MgCl_2$, and 10 HEPES (pH 7.4 with CsOH).

and 87.5 ± 5.8% (n = 5 cells, including 4 that were exposed continuously to 1 mM (+) PN 200-110 to block any endogenous I_{dys}).

In addition to examining whole-cell currents in pKCRB3-injected myotubes, we also examined currents in cell-attached patches. FIGURE 5(A–C) illustrates unitary activity of BIII calcium channel in response to test depolarizations to − 10, + 10, and + 30 mV. Channel openings were relatively rare at a test potential of − 10 mV, became increasingly frequent at + 10 mV, and still more frequent at + 30 mV. Occasionally, as many as three channels were observed to be open simultaneously in this patch (e.g., beginning of second trace in FIG. 5, C). A plot of unitary current amplitude versus membrane potential (FIG. 5, D) yielded a slope conductance of 15 pS for the channels in this patch. Based on a total of five experiments, the average slope conductance was 14.3 ± 2.2 pS, a value similar to previously published values for N-type channels.[3,16,17,84]

As noted above, dysgenic myotubes endogenously express both T-type calcium current and a low density of the DHP-sensitive calcium current I_{dys}. However, it is very unlikely that the calcium channels that produce these endogenous currents are contributing to the unitary activity like that illustrated in FIGURE 5. First, for all the cell-attached recordings, the pipette solution contained 1 mM (+) PN 200-110, a concentration sufficient to produce nearly complete block of I_{dys}.[85] Second, for T-type channels, the unitary slope conductance is considerably smaller (only ~ 10 pS), and activation occurs over a range of potentials >30 mV hyperpolarized, compared to the expressed BIII channels.[4,72,86]

The ensemble averages of unitary BIII currents (FIG. 5) have a time course qualitatively similar to that of whole-cell BIII currents (FIG. 4, A). Each displays both transient and sustained phase. Thus, these data provide strong support for the notion that a single molecular species can produce an N-type calcium current exhibiting both inactivating and non-inactivating components[16] and for identifying the BIII calcium channel as that molecular species.

In initial descriptions, the inactivating and non-inactivating components of HVA calcium current were assigned to N-type and DHP-sensitive L-type channels, respectively.[3,82] However, more recent work has demonstrated that in many cells ω-CgTx blocks both inactivating HVA calcium current and a part of the non-inactivating HVA calcium current that is insensitive to DHP.[16,17,19,21,87] This sensitivity to ω-CgTx suggested the existence of a non-inactivating component of N current, but did not establish whether the non-inactivating and inactivating components arose from the identical channel type. A strong argument that both components do arise from a single type of channel is the demonstration that both components are reconstructed in ensemble averages of individual HVA calcium channels in PC12[16] and rat sympathetic ganglion cells.[87] By a process of exclusion, the recorded individual channels analyzed in both cell types were very likely of the ω-CgTx–sensitive type, but the molecular identity of individual channels is difficult to establish in native neuronal tissues. Thus, our results—that inactivating and non-inactivating components are present in both the unitary and macroscopic currents that are produced by expression of the BIII channel in dysgenic myotubes—is important evidence in support of the conclusion that single N-type channels produce both components of current. Additionally, brief depolarizing prepulses do not provide a useful method of distinguishing N-type channels from other types since a 1 sec prepulse to − 20 mV had a comparatively small effect on

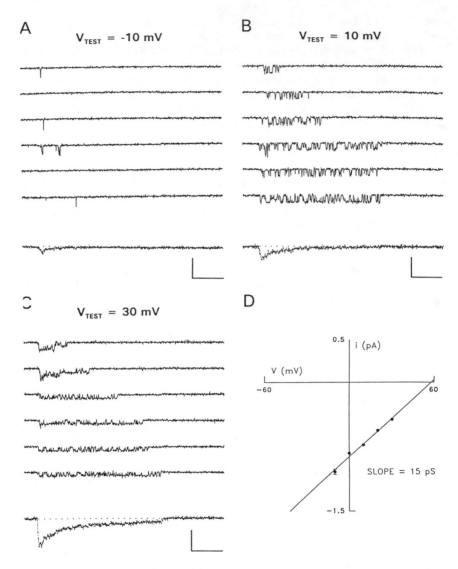

FIGURE 5. Cell-attached patch recordings of N-type calcium channels in a dysgenic myotube expressing BIII cDNA. (A–C) Six representative (non-consecutive) records of unitary activity (above) and the ensemble average (below) obtained from a three-channel patch for each of the indicated test depolarizations, which were 200 msec in duration and applied from HP = − 90 mV. The pipette solution contained (in mM): 110 BaCl₂, 10 HEPES, 0.003 tetrodotoxin, and 0.001 (+) PN 200-110 (pH = 7.4 with tetraethylammonium-OH). The bath solution contained: 145 NaCl, 5 KCl, 2 CaCl₂, 1 MgCl₂ and 10 HEPES (pH = 7.4 with NaOH). Linear components of capacitative and leak current were removed by null subtraction. Ensemble currents are averages of 40 (A), 120 (B), or 100 (C) individual sweeps. Horizontal scale bars: 50 msec. Vertical scale bars: 2 pA (unitary records) or 0.5 pA (ensemble averages). (D) Unitary current-voltage relationship for the same experiment as in (A–C). Each point represents the mean ± s.e.m. of at least 65 individual openings (except at − 10 mV where only 12 openings were observed in 40 sweeps). Data were fit by a linear regression, yielding a slope of 15 pS (R = 0.994).

the maintained component of current through BIII channels while almost completely inactivating the transient component (FIG. 4, C).

The effect of the coexpression of the auxiliary subunits on the macroscopic current properties and pharmacological sensitivities of the class C and class BI channel had not been reported. However, recent works have shown that the auxiliary subunits have pronounced effects on macroscopic characteristics such as drug sensitivity, kinetics, and voltage dependence of activation and inactivation of the class C calcium channels.[68,69,72,88] The presence of the β subunit was required for cAMP-mediated increase of the class C calcium channel activity.[89] It is conceivable that different combinations between multiple α_1 subunits and β subunits comprise heterologous calcium channel complexes and contribute to the functional heterogeneity of calcium channels.

CONCLUSIONS

So far, molecular biological studies of calcium channels have justified the former existing functional classification. Two subfamilies of α_1 subunit, defined from the molecular point of view, correspond to two classes, L-types and non–L-types, which can be distinguished by DHP sensitivity (TABLE 2). The L-type subfamily consists of the class Sk, C, and BIV channels. The expressed class Sk and C channels mirror the physiological situation in skeletal and cardiac muscle, respectively. The class BIV channel is the neuroendocrine calcium channel reversibly blocked by ω-CgTx. On the other hand, the non–L-type subfamily consists of the BI, BII, and BIII channel. The presence of the BI channel in cerebellar Purkinje cells and in granule cells as well as the pharmacological and electrophysiological properties suggest that the BI channel represents the P-type calcium channel in neurons. The BIII calcium channel is the N type calcium channel in neurons. Because the voltage-sensing region S4,[43] the pore-forming region between S5 and S6,[44] and the region between repeats III and IV, involved in inactivation process,[43] are highly conserved among the class BI, BII, and BIII channels, the class BII channel is supposed be a high voltage–activated non–L-type channel. Molecular biological studies also aid to settle ambiguity in functional categorization, i.e., inactivation kinetics of N-type channels and ω-CgTx sensitivities of the N-type and L-type channels in neurons.

In the future, in which direction should the study of voltage-dependent calcium channels head? One possible direction is to disintegrate a calcium channel molecule into structural elements responsible for specific functions, such as Ca^{2+} selectivity, voltage-sensing, and gating. Another direction is to find other proteins associated with calcium channels and to integrate them into subcellular structures, i. e., dendritic spines and presynaptic active zones in neurons, or triads in myotubes. This will be a direct approach to elucidate molecular mechanisms of memory in brain, neurotransmitter release, and excitation-contraction coupling.

REFERENCES

1. TSIEN, R. W. & R. Y. TSIEN. 1990. Calcium channels, stores, and oscillations. Annu. Rev. Cell Biol. 6: 715–760.

2. Tsien, R. W., P. T. Ellinor & W. A. Horne. 1991. Molecular diversity of voltage-dependent Ca²⁺ channels. Trends Pharmacol. Sci. 12: 349–354.

3. Fox, A. P., M. C. Nowycky & R. W. Tsien. 1987. Kinetic and pharmacological properties distinguishing three types of calcium currents in chick sensory neurones. J. Physiol. 394: 149–172.

4. Fox, A. P., M. C. Nowycky & R. W. Tsien. 1987. Single-channel recordings of three types of calcium channels in chick sensory neurones. J. Physiol. 394: 173–200.

5. Sher, E. & F. Clementi. 1991. ω-conotoxin-sensitive voltage-operated calcium channels in vertebrate cells. Neuroscience 42: 301–307.

6. Hirning, L. D., A. P. Fox, E. W. McCleskey, B. M. Olivera, S. A. Thayer, R. J. Miller & R. W. Tsien. 1988. Dominant role of N-type Ca²⁺ channels in evoked release of norepinephrine from sympathetic neurons. Science 239: 57–61.

7. Hofmann, F. & E. Habermann. 1990. Role of ω-conotoxin sensitive calcium channels in inositolphosphate production and noradrenaline release due to potassium depolarization or stimulation with carbachol. Naunyn-Schmiedeberg's Arch. Pharmacol. 341: 200–205.

8. Horne, A. L. & J. A. Kemp. 1991. The effect of ω-conotoxin GVIA on synaptic transmission within the nucleus accumbens and hippocampus of the rat in vitro. Br. J. Pharmacol. 103: 1733–1739.

9. Hong, S. J., K. Tsuji & C. C. Chang. 1992. Inhibition by neosurugatoxin and ω-conotoxin of acetylcholine release and muscle and neuronal nicotinic receptors in mouse neuromuscular junction. Neurosci. 48: 727–735.

10. Smith, S. J. & G. J. Augustine. 1988. Calcium ions, active zones and synaptic transmitter release. Trends Neurosci. 11: 458–464.

11. Komuro, H. & P. Rakic. 1992. Selective role of N-type calcium channels in neuronal migration. Science 257: 806–809.

12. Llinás, R., M. Sugimori, J.-W. Lin & B. Cherksey. 1989. Blocking and isolation of a calcium channel from neurons in mammals and cephalopods utilizing a toxin fraction (FTX) from funnel-web spider poison. Proc. Natl. Acad. Sci. USA 86: 1689–1693.

13. Hillman, D., S. Chen, T. T. Aung, B. Cherksey, M. Sugimori & R. Llinás. 1991. Localization of P-type calcium channels in the central nervous system. Proc. Natl. Acad. Sci. USA 88: 7076–7080.

14. Regan, L. J., D. W. Y. Sah & B. P. Bean. 1991. Ca²⁺ channels in rat central and peripheral neurons: High-threshold current resistant to dihydropyridine blockers and ω-conotoxin. Neuron 6: 269–280.

15. Mintz, I. M., M. E. Adams & B. P. Bean. 1992. P-type calcium channels in rat central and peripheral neurons. Neuron 9: 85–95.

16. Plummer, M. R., D.E. Logothetis & P. Hess. 1989. Elementary properties and pharmacological sensitivities of calcium channels in mammalian peripheral neurons. Neuron 2: 1453–1463.

17. Aosaki, T. & H. Kasai. 1989. Characterization of two kinds of high-voltage-activated Ca-channel currents in chick sensory neurons. Pflügers Arch. 414: 150–156.

18. Swandulla, D., E. Carbone & H. D. Lux. 1991. Do calcium channel classifications account for neuronal calcium channel diversity? Trends Neurosci. 14: 46–51.

19. Artalejo, C. R., R. L. Perlman & A. P. Fox. 1992. ω-Conotoxin GVIA blocks a Ca²⁺ current in bovine chromaffin cells that is not the "classic" N type. Neuron 8: 85–95.

20. Hillyard, D. R., V. D. Monje, I. M. Mintz, B. P. Bean, L. Nadasdi, J. Ramachandran, G. Miljanich, A. Azimi-Zoonooz, J. M. McIntosh, L. J. Cruz, J. S. Imperial & B. M. Olivera. 1992. A new conus peptide ligand for mammalian presynaptic Ca²⁺ channels. Neuron 9: 69–77.

21. Mynlieff, M. & K. G. Beam. 1992. Characterization of voltage-dependent calcium currents in mouse motoneurons. J. Neurophysiol. 68: 85–92.

22. Campbell, K. P., A. T. Leung & A. H. Sharp. 1988. The biochemistry and molecular biology of the dihydropyridine-sensitive calcium channel. Trends Neurosci. 11: 425–430.

23. Leung, A. T., T. Imagawa & K. P. Campbell. 1987. Structural characterization of the 1,4-

dihydropyridine receptor of the voltage-dependent Ca²⁺ channel from rabbit skeletal muscle. J. Biol. Chem. **262**: 7943–7946.

24. TANABE, T., H. TAKESHIMA, A. MIKAMI, V. FLOCKERZI, H. TAKAHASHI, K. KANGAWA, M. KOJIMA, H. MATSUO, T. HIROSE & S. NUMA. 1987. Primary structure of the receptor for calcium channel blockers from skeletal muscle. Nature **328**: 313–318.

25. TAKAHASHI, M., M. J. SEGAR, J. F. JONES, B. F. X. REBER & W. A. CATTERALL. 1987. Subunit structure of dihydropyridine-sensitive calcium channels from skeletal muscle. Proc. Natl. Acad. Sci. USA **84**: 5478–5482.

26. MORTON, M. E. & S. C. FROEHNER. 1987. Monoclonal antibody identifies a 200-kDa subunit of the dihydropyridine-sensitive calcium channel. J. Biol. Chem. **262**: 11904–11907.

27. VAGHY, P. L., J. STRIESSNIG, K. MIWA, H.-G. KNAUS, K. ITAGAKI, E. McKENNA, H. GLOSSMANN & A. SCHWARTZ. 1987. Identification of a novel 1,4-dihydropyridine- and phenylalkylamine-binding polypeptide in calcium channel preparations. J. Biol. Chem. **262**: 14337–14342.

28. HOSEY, M. M., J. BARHANIN, A. SCHMID, S. VANDAELE, J. PTASIENSKI, C. O'CALLAHAN, C. COOPER & M. LAZDUNSKI. 1987. Photoaffinity labelling and phosphorylation of a 165 kilodalton peptide associated with dihydropyridine and phenylalkylamine-sensitive calcium channels. Biochem. Biophys. Res. Commun. **147**: 1137–1145.

29. LEUNG, A. T., T. IMAGAWA, B. BLOCK, C. FRANZINI-ARMSTRONG & K. P. CAMPBELL. 1988. Biochemical and ultrastructure of the 1,4-dihydropyridine receptor from rabbit skeletal muscle. J. Biol. Chem. **263**: 994–1001.

30. TAKAHASHI, M. & W. A. CATTERALL. 1987. Identification of an α subunit of dihydropyridine-sensitive brain calcium channels. Science **236**: 88–91.

31. AHLIJANIAN, M. K., R.E. WESTENBROEK & W. A. CATTERALL. 1990. Subunit structure and localization of dihydropyridine-sensitive calcium channels in mammalian brain, spinal cord and retina. Neuron **4**: 819–832.

32. YOSHIDA, A., M. TAKAHASHI, Y. FUJIMOTO, H. TAKISAWA & T. NAKAMURA. 1990. Molecular characterization of 1,4-dihydropyridine-sensitive calcium channels of chick heart and skeletal muscle. J. Biochem. **107**: 608–612.

33. CRUZ, L. J., D. S. JOHNSON & B. M. OLIVERA. 1987. Characterization of the ω-conotoxin target. Evidence for tissue-specific heterogeneity in calcium channel types. Biochemistry **26**: 820–824.

34. ABE, T. & H. SAISU. 1987. Identification of the receptor for ω-conotoxin in brain. J. Biol. Chem. **262**: 9877–9882.

35. MARQUEZE, B., N. MARTIN-MOUTOT, C. LEVÊQUE & F. COURAUD. 1988. Characterization of the ω-conotoxin-binding molecule in rat brain synaptosomes and cultured neurons. Mol. Pharmacol. **34**: 87–90.

36. McENERY, M. W., A. M. SNOWMAN, A. H. SHARP, M. E. ADAMS & S. H. SNYDER. 1991. Purified ω-conotoxin GVIA receptor of rat brain resembles a dihydropyridine-sensitive L-type calcium channel. Proc. Natl. Acad. Sci. USA **88**: 11095–11099.

37. GLOSSMANN, H. & J. STRIESSNIG. 1990. Molecular properties of calcium channels. Rev. Physiol. Biochem. Pharmacol. **114**: 1–105.

38. SAKAMOTO, J. & K. P. CAMPBELL. 1991. A monoclonal antibody to the β subunit of the skeletal muscle dihydropyridine receptor immunoprecipitates the brain ω-conotoxin GVIA receptor. J. Biol. Chem. **266**: 18914–18919.

39. SAISU, H., K. IBARAKI, T. YAMAGUCHI, Y. SEKINE & T. ABE. 1991. Monoclonal antibodies immunoprecipitating ω-conotoxin-sensitive calcium channel molecules recognize two novel proteins localized in the nervous system. Biochem. Biophys. Res. Commun. **181**: 59–66.

40. TAKAHASHI, M., Y. ARIMATSU, S. FUJITA, Y. FUJIMOTO, S. KONDO, T. HAMA & E. MIYAMOTO. 1991. Protein kinase C and Ca²⁺/calmodulin-dependent protein kinase II phosphorylate a novel 58-kDa protein in synaptic vesicles. Brain Res. **551**: 279–292.

41. CHERKSEY, B.D., M. SUGIMORI & R. R. LLINÁS. 1991. Properties of calcium channels isolated with spider toxin, FTX. Ann. NY Acad. Sci. **635**: 80–89.

42. NUMA, S. & M. NODA. 1986. Molecular structure of sodium channels. Ann. NY Acad. Sci. **479**: 338–355.

43. Stühmer, W., F. Conti, H. Suzuki, X. Wang, M. Noda, N. Yahagi, H. Kubo & S. Numa. 1989. Structural parts involved in activation and inactivation of the sodium channel. Nature 339: 597–603.

44. Heinemann, S. H., H. Terlau, W. Stühmer, K. Imoto & S. Numa. 1992. Calcium channel characteristics conferred on the sodium channel by single mutations. Nature 356: 441–443.

45. Noda, M., T. Ikeda, H. Suzuki, H. Takashima, T. Takahashi, M. Kuno & S. Numa. 1986. Expression of functional sodium channels from cloned cDNA. Nature 322: 826–828.

46. Morton, M. E. & S. C. Froehner. 1989. The α_1 and α_2 polypeptides of the dihydropyridine-sensitive calcium channel differ in developmental expression and tissue distribution. Neuron 2: 1499–1506.

47. Ellis, S., M. E. Williams, N. R. Ways, R. Brenner, A. H. Sharp, A. T. Leung, K. P. Campbell, E. McKenna, J. K. Koch, A. Hui, A. Schwartz & M. M. Harpold. 1988. Sequence and expression of mRNAs encoding the α_1 and α_2 subunits of a DHP-sensitive calcium channel. Science 241: 1661–1664.

48. Perez-Reyes, E., X. Wei, A. Castellano & L. Birnbaumer. 1990. Molecular diversity of L-type calcium channels. J. Biol. Chem. 265: 20430–20436.

49. Varadi, G., J. Orlowski & A. Schwartz. 1989. Developmental regulation of expression of the α_1 and α_2 subunits mRNAs of the voltage-dependent calcium channel in a differentiating myogenic cell line. FEBS Lett. 2: 515–518.

50. Malouf, N. N., D. K. McMahon, C. N. Hainsworth & B. K. Kay. 1992. A two-motif isoform of the major calcium channel subunit in skeletal muscle. Neuron 8: 899–906.

51. Mikami, A., K. Imoto, T. Tanabe, T. Niidome, Y. Mori, H. Takeshima, S. Narumiya & S. Numa. 1989. Primary structure and functional expression of the cardiac dihydropyridine-sensitive calcium channel. Nature 340: 230–233.

52. Biel, M., P. Ruth, E. Bosse, R. Hullin, W. Stühmer, V. Flockerzi & F. Hofmann. 1990. Primary structure and functional expression of a high voltage activated calcium channel from rabbit lung. FEBS Lett. 269: 409–412.

53. Koch, W. J., P. T. Ellinor & A. Schwartz. 1990. cDNA cloning of a dihydropyridine-sensitive calcium channel from rat aorta. J. Biol. Chem. 265: 17786–17791.

54. Snutch, T. P., W. J. Tomlinson, J. P. Leonard & M. M. Gilbert. 1991. Distinct calcium channels are generated by alternative splicing and are differentially expressed in the mammalian CNS. Neuron 7: 45–57.

55. Chaudhari, N. & K. G. Beam. 1991. mRNA for the cardiac calcium channel is expressed during development of skeletal muscle. Soc. Neurosci. Abstr. 17: 772.

56. Diebold, R. J., W. J. Koch, P. T. Ellinor, J.-J. Wang, M. Muthuchamy, D. F. Wieczorek & A. Schwartz. 1992. Mutually exclusive exon splicing of the cardiac calcium channel a1 subunit gene generates developmentally regulated isoforms in the rat heart. Proc. Natl. Acad. Sci. USA 89: 1497–1501.

57. Soldadov, N. M. 1992. Molecular diversity of L-type calcium channel transcripts in human fibroblasts. Proc. Natl. Acad. Sci. USA 89: 4628–4632.

58. Hui, A., P. T. Ellinor, O. Krizanova, J.-J. Wang, R. J. Diebold & A. Schwartz. 1991. Molecular cloning of multiple subtypes of a novel rat brain isoform of the α_1 subunit of the voltage-dependent calcium channel. Neuron 7: 35–44.

59. Seino, S., L. Chen, M. Seino, O. Blondel, J. Takeda, J. H. Johnson & G. I. Bell. 1992. Cloning of the α_1 subunit of a voltage-dependent calcium channel expressed in pancreatic β cells. Proc. Natl. Acad. Sci. USA 89: 584–588.

60. Williams, M. E., D. H. Feldman, A. F. McCue, R. Brenner, G. Velicelebi, S. B. Ellis & M. M. Harpold. 1992. Structure and functional expression of α_1, α_2, and β subunits of a novel human neuronal calcium channel subtype. Neuron 8: 71–84.

61. Chin, H., M. A. Smith, H.-L. Kim & H. Kim. 1992. Expression of dihydropyridine-sensitive brain calcium channels in the rat central nervous system. FEBS Lett. 299: 69–74.

62. Mori, Y., T. Friedrich, M.-S. Kim, A. Mikami, J. Nakai, P. Ruth, E. Bosse, F. Hofmann, V. Flockerzi, T. Furuichi, K. Mikoshiba, K. Imoto, T. Tanabe & S. Numa. 1991. Primary

structure and functional expression from complementary DNA of a brain calcium channel. Nature **350**: 398–402.

63. STARR, T. V. B., W. PRYSTAY & T. P. SNUTCH. 1991. Primary structure of a calcium channel that is highly expressed in the rat cerebellum. Proc. Natl. Acad. Sci. USA **88**: 5621–5625.

64. NIIDOME, T., M.-S. KIM, T. FRIEDRICH & Y. MORI. 1992. Molecular cloning and characterization of a novel calcium channel from rabbit brain. FEBS Lett **308**: 7–13.

65. DUBEL, S. J., T. V. B. STARR, J. HELL, M. K. AHLIJANIAN, J. J. ENYEART, W. A. CATTERALL & T. P. SNUTCH. 1992. Molecular cloning of the α-1 subunit of an ω-conotoxin-sensitive calcium channel. Proc. Natl. Acad. Sci. USA **89**: 5058–5062.

66. WILLIAMS, M. E., P. F. BRUST, D. H. FELDMAN, S. PATTHI, S. SIMERSON, A. MAROUFI, A. F. MCCUE, G. VELICELEBI, S. B. ELLIS & M. M. HARPOLD. 1992. Structure and functional expression of an ω-conotoxin-sensitive human N-type calcium channel. Science **257**: 389–395.

67. MISHINA, M., T. TOBIMATSU, K. IMOTO, K. TANAKA, Y. FUJITA, K. FUKUDA, M. KURASAKI, Y. MORIMOTO, H. TAKAHASHI, T. HIROSE, S. INAYAMA, T. TAKAHASHI, M. KUNO & S. NUMA. 1985. Location of functional regions of acetylcholine receptor α-subunit by site-directed mutagenesis. Nature **313**: 364–369.

68. SINGER, D., M. BIEL, I. LOTAN, V. FLOCKERZI, F. HOFMANN & N. DASCAL. 1991. The roles of the subunits in the function of the calcium channel. Science **253**: 1553–1557.

69. WEI, X., E. PEREZ-REYES, A. E. LACERDA, G. SCHUSTER, A. M. BROWN & L. BIRNBAUMER. 1991. Heterologous regulation of the cardiac Ca²⁺ channel α₁ subunit by skeletal muscle β and γ subunits. J. Biol. Chem. **266**: 21943–21947.

70. ITAGAKI, K., W. J. KOCH, I. BODI, U. KLÖCKNER, D.F. SLISH & A. SCHWARTZ. 1992. Native-type DHP-sensitive calcium channel currents are produced by cloned rat aortic smooth muscle and cardiac α₁ subunits expressed in Xenopus laevis oocytes and are regulated by α₂- and β-subunits. FEBS Lett. **297**: 221–225.

71. PEREZ-REYES, E., A. CASTELLANO, H. S. KIM, P. BERTRAND, E. BAGGSTROM, A. E. LACERDA, X. WEI & L. BIRNBAUMER. 1992. Cloning and expression of a cardiac/brain β subunit of the L-type calcium channel. J. Biol. Chem. **267**: 1792–1797.

72. HULLIN, R., D. SINGER-LAHAT, M. FREICHEL, M. BIEL, N. DASCAL, F. HOFMANN & V. FLOCKERZI. 1992. Calcium channel β subunit heterogeneity: functional expression of cloned cDNA from heart, aorta and brain. EMBO J. **11**: 885–890.

73. BOSSE, E., R. BOTTLENDER, T. KLEPPISCH, J. HESCHELER, A. WELLING, F. HOFMANN & V. FLOCKERZI. 1992. Stable and functional expression of the calcium channel α₁ subunit from smooth muscle in somatic cell lines. EMBO J. **11**: 2033–2038.

74. GLUECKSOHN-WAELSCH, S. 1963. Lethal genes and analysis of differentiation. In higher organisms lethal genes serve as tools for studies of cell differentiation and cell genetics. Science **142**: 1269–1276.

75. POWELL, J. A. & D. M. FAMBROUGH. 1973. Electrical properties of normal and dysgenic mouse skeletal muscle in culture. J. Cell. Physiol. **82**: 21–38.

76. BEAM, K. G., C. M. KNUDSON & J. A. POWELL. 1987. A lethal mutation in mice eliminates the slow calcium current in skeletal muscle cells. Nature **320**: 168–170.

77. TANABE, T., K. G. BEAM, J. A. POWELL & S. NUMA. 1988. Restoration of excitation-contraction coupling and slow calcium current in dysgenic muscle by dihydropyridine receptor complementary DNA. Nature **336**: 134–139.

78. TANABE, T., A. MIKAMI, S. NUMA & K. G. BEAM. 1990. Cardiac-type excitation-contraction coupling in dysgenic skeletal muscle injected with cardiac dihydropyridine receptor cDNA. Nature **344**: 451–453.

79. PEREZ-REYES, E., H. S. KIM, A. E. LACERDA, W. HORNE, X. WEI, D. RAMPE, K. P. CAMPBELL, A. M. BROWN & L. BIRNBAUMER. 1989. Induction of calcium currents by the expression of the α₁-subunit of the dihydropyridine receptor from skeletal muscle. Nature **340**: 233–236.

80. VARADI, G., P. LORY, D. SCHULTZ, M. VARADI & A. SCHWARTZ. 1991. Acceleration of activation and inactivation by the β subunit of the skeletal muscle calcium channel. Nature **352**: 159–162.

81. LACERDA, A. E., H. S. KIM, P. RUTH, E. PEREZ-REYES, V. FLOCKERZI, F. HOFMANN, L. BIRNBAUMER & A. M. BROWN. 1991. Normalization of current kinetics by interaction between the α_1 and β subunits of the skeletal muscle dihydropyridine-sensitive Ca^{2+} channel. Nature 352: 527–530.

82. NOWYCKY, M. C., A. P. FOX & R. W. TSIEN. 1985. Three types of neuronal calcium channel with different calcium agonist sensitivity. Nature 316: 440–443.

83. SCHROEDER, J. E., P. S. FISCHBACH, M. MAMO & E. W. McCLESKEY. 1990. Two components of high-threshold Ca^{2+} current inactivate by different mechanisms. Neuron 5: 445–452.

84. FISHER, R. E., R. GRAY & D. JOHNSTON. 1990. Properties and distribution of single voltage-gated calcium channels in adult hippocampal neurons. J Neurophysiol. 64: 91–104.

85. ADAMS, B. A. & K. G. BEAM. 1989. A novel calcium current in dysgenic skeletal muscle. J. Gen. Physiol. 94: 429–444.

86. DIRKSEN, R. T., T. TANABE & K. BEAM 1993. Single channel analysis of native and expressed skeletal muscle calcium channels. Biophys. J. (In press.)

87. PLUMMER, M. R. & P. HESS. 1991. Reversible uncoupling of inactivation in N-type calcium channels. Nature 351: 657–659.

88. PEREZ-REYES, E., A. CASTELLANO, H. S. KIM, P. BERTRAND, E. BAGGSTROM, A. E. LACERDA, X. WEI & L. BIRNBAUMER. 1992. Cloning and expression of a cardiac/brain β subunit of the L-type calcium channel. J. Biol. Chem. 267: 1792–1797.

89. KLÖCKNER, U., K. ITAGAKI, I. BODI & A. SCHWARTZ. 1992. β-subunit expression is required for cAMP-dependent increase of cloned cardiac and vascular calcium channel currents. Pflügers Arch. 420: 413–415.

90. NODA, M., S. SHIMIZU, T. TANABE, T. TAKAI, T. KAYANO, T. IKEDA, H. TAKAHASHI, H. NAKA-YAMA, Y. KANAOKA, N. MINAMINO, K. KANGAWA, H. MATSUO, M. A. RAFTERY, T. HIROSE, S. INAYAMA, H. HAYASHIDA, T. MIYATA & S. NUMA. 1984. Primary structure of Electrophorus electricus sodium channel deduced from cDNA sequence. Nature 312: 121–127.

91. FURUICHI, T., D. SIMON-CHAZOTTES, I. FUJINO, N. YAMADA, M. HASEGAWA, A. MIYAWAKI, S. YOSHIKAWA, J-L. GUÉNET & K. MIKOSHIBA. 1993. Widespread expression of inositol 1,4,5-triphosphate receptor type 1 gene (Insp3r1) in the mouse central nervous system. Recept. Channels. (In press.)

92. SAITOU, N. & M. NEI. 1987. The neighbor-joining method: a new method for reconstructing phylogenetic trees. Mol. Biol. Evol. 4: 406–425.

Structure and Biology of Inhibitory Glycine Receptors

H. BETZ, D. LANGOSCH, N. RUNDSTRÖM,
J. BORMANN, A. KURYATOV, J. KUHSE,
V. SCHMIEDEN, B. MATZENBACH,
AND J. KIRSCH

Max-Planck-Institut für Hirnforschung
Abteilung Neurochemie
Deutschordenstrasse 46
D-60529 Frankfurt/M.
Federal Republic of Germany

Neurotransmission at chemical synapses is mediated by receptors that transduce transmitter binding into alterations of membrane potential. Receptors containing integral ion channels mediate rapid (in the range of a millisecond or less) transduction events, whereas receptors activating G protein–coupled channels operate at slower time scales (in the range of a millisecond to second). At resting membrane potential, excitation results from cation influx, but inhibition of neuronal firing is generated by increased chloride permeability.

The nicotinic acetylcholine receptor (nAChR) at the neuromuscular junction initiates muscle contraction. Because of its abundance in fish electric organ, it is the best characterized ion channel protein known.[38] The major inhibitory neurotransmitters at central synapses, glycine and γ-aminobutyric acid (GABA), gate chloride channel-forming receptors of similar conductance properties,[1] but distinct pharmacology.[2] The convulsive alkaloid strychnine antagonizes postsynaptic inhibition by glycine, the predominant inhibitory neurotransmitter in brainstem and spinal cord, whereas benzodiazepines and barbiturates modify inhibitory GABA$_A$ receptor responses in many regions of the central nervous system (CNS). This report summarizes recent data on the structure and biology of inhibitory glycine receptors (GlyRs).

PRIMARY STRUCTURE OF GLYCINE RECEPTOR SUBUNITS

The GlyR was the first receptor protein to be isolated from the mammalian central nervous system.[3-5] Affinity-purified GlyR contains two glycosylated integral membrane proteins of 48 kD (α) and 58 kD (β), which form the chloride channel of the receptor. A co-purifying peripheral membrane protein of 93 kD, gephyrin,[3-6] is associated with the cytoplasmic domains of the GlyR and has been implicated in the synaptic localization of the receptor.[7,8] Consistent with this view, gephyrin binds with high affinity to polymerized tubulin,[9] and thus may anchor the GlyR to subsynaptic cytoskeletal structures.

The α and β subunits of the GlyR share remarkable similarities in primary structure, as deduced by cDNA sequencing.[10,11] The putative signal sequences of both polypeptides are clearly different, but the amino terminal regions of the mature proteins display 50 to 70% sequence identity. In the C-terminal half of the polypeptide, four highly conserved hydrophobic segments (M1 to M4) are thought to form membrane-spanning α-helices. This arrangement resembles that of nAChR and GABA$_A$ receptor proteins, suggesting that all channel-forming receptors are composed of subunits sharing a common transmembrane topology.[2] Moreover, significant amino acid sequence homology exists between the subunits of different ligand-gated ion channels. Thus, these receptors constitute a protein superfamily that evolved by gene duplication from a common ancestor early in phylogeny.

GlyR α SUBUNITS FORM HOMO-OLIGOMERIC CHANNELS

GlyR α subunits contain domains for both ligand binding and ion conduction. Indeed, heterologous expression of rat or human GlyR α subunits in *Xenopus* oocytes[12] or mammalian cells[13] creates glycine-gated chloride channels, which are blocked by nanomolar concentrations of strychnine, as is the GlyR in spinal neurons. Moreover, these channels are gated by the agonists glycine, β-alanine, and taurine and show a biphasic desensitization behavior.[12] Thus, GlyR α subunits assemble efficiently into homo-oligomeric channels, which closely resemble the native GlyR in biophysical and pharmacological properties. This finding suggests that homomeric GlyRs may exist *in vivo*, an interpretation supported by biochemical data obtained with embryonic neurons.[14] Moreover, it has been exploited for the characterization of different GlyR α subunit isoforms in heterologous expression studies (see below).

THE GlyR IS A PENTAMERIC PROTEIN

Analysis of the subunit composition of the purified GlyR by cross-linking and sedimentation techniques indicates a pentameric quaternary structure with three α and two β subunits, respectively.[15] Interestingly, this closely resembles the nAChR, which also contains five membrane-spanning subunits.[38] In view of the above-discussed sequence homology and common transmembrane topology of different channel-forming receptor proteins,[2,16] a quasisymmetrical pentameric complex of transmembrane polypeptides around a central ion pore is thought to represent the common quaternary structure of different members of the ligand-gated ion channel superfamily.[2,15,16]

THE LIGAND-BINDING DOMAIN

The extracellular N-terminal domain of GlyR subunits contains two precisely conserved cysteine residues that also are found in nAChR and GABA$_A$ receptor polypeptides and may form a disulfide bridge essential for receptor tertiary structure.[2] Photoaffinity labeling experiments with [³H]strychnine indicate that the ligand binding site of the GlyR resides on the α subunit.[17,18] Based on theoretical considerations[10] and proteolytic cleavage of [³H]strychnine-labeled GlyR preparations,[18,19] a stretch of charged residues preceding the first transmembrane segment has been postulated to be part of the binding pocket. Interestingly, a corresponding region containing two neighboring cysteine residues is known to be important for acetylcholine binding to the α subunits of the nAChR.[38]

Site-directed mutagenesis of human and rat GlyR α-subunit isoforms, combined with heterologous expression in Xenopus oocytes or mammalian cells, has led to a more detailed picture of the ligand binding region. By comparing the pharmacology of different variants of the neonatal α2 subunit, residue 160 was identified as crucial for high-affinity strychnine binding.[20] Moreover, amino acids 200 and 202 preceding transmembrane segment M1 have been shown to represent important determinants of glycine binding,[21] and exchanges of residues 111 and 212 were found to strongly affect the potency of the glycinergic agonists β-alanine and taurine.[22] Based on these data, a model of the ligand pocket of the GlyR has been proposed where agonist binding involves multiple interactions with at least three different domains in the extracellular region of the α subunit.[22] Interestingly, their positions correspond to regions of nAChR α subunits known to be involved in agonist and antagonist binding. These results suggest that a common folding pattern of the extracellular domain of α subunits generates the ligand binding pocket of all members of this receptor superfamily.

THE CHLORIDE CHANNEL-FORMING SEGMENT M2

The transmembrane segments M1 to M3 are highly conserved between GlyR and GABA$_A$ receptor subunits, pointing to their potential importance in chloride channel function.[2,16,23] Segment M2 contains a high content of uncharged polar amino acid residues and therefore is thought to provide the hydrophilic inner lining of the chloride channel. Here, eight consecutive amino acid residues are identical in most GABA$_A$ receptor and GlyR α subunits. Interestingly, transmembrane segment M2 of nAChR proteins is known to be involved in cation transport and channel blocker binding. Segment M2 thus is considered a common structural determinant of ligand-gated ion channel function.[2]

The M2 segments of anion-selective GlyR and GABA$_A$ receptor subunits terminate with positively charged residues both intra- and extracellularly. Patch-clamp data indicate two sequentially occupied anion binding sites in both GlyR and GABA$_A$ receptor channels.[1] The charged residues at the termini of the M2 segments may be the structural correlate of these sites at the presumptive inner and outer mouths of receptor ion channels, and thus may provide their ion selectivity. Consistent with this proposals, a synthetic peptide corresponding to segment M2 of the GlyR α subunit has been shown to produce randomly gated "channels" on incorporation into planar lipid bilayers.[24] Notably, the ion selectivity of these channels was altered by inverting the terminal charges of the peptide.

ROLE OF THE β SUBUNIT

The importance of segment M2 in ion channel formation is corroborated further by studies of GlyRs generated by coexpression of the α and β subunits.[25] Although the resulting chloride channels show no obvious differences in agonist pharmacology and strychnine inhibition, they differ greatly in sensitivity to the channel blocking agent picrotoxinin. This alkaloid is a potent antagonist of both $GABA_A$ receptors and homomeric GlyRs generated by α subunit expression.[12,13] However, the channels coassembled from GlyR α and β subunits display a 50–100 fold reduced sensitivity to picrotoxinin block.[25] Site-directed mutagenesis identified transmembrane segment M2 of the β polypeptide as the molecular determinant of picrotoxinin resistance and showed that its substitution by the corresponding residues of the α subunit generates picrotoxinin-sensitive heteromeric receptors.[25] Moreover, preliminary patch clamp data show that the single channel conductances of heteromeric GlyRs are drastically reduced as compared to those of homomeric α subunit receptors.[39]

GLYCINE RECEPTOR DIVERSITY

Biochemical and cDNA sequence data have established subtype diversity as a general phenomenon for brain neurotransmitter receptors. For the GlyR, subtype heterogeneity was first noted during spinal cord development.[26] There, a neonatal receptor isoform is prevalent at birth, which differs in strychnine binding affinity, immunological properties, molecular weight (49 kD) of its ligand binding subunit,[26] and mRNA size[27] from the adult receptor protein. This neonatal GlyR isoform is abundantly expressed in primary cultures of embryonic spinal cord, a condition that facilitated its biochemical analysis.[14] Pulse-chase experiments indicate that the neonatal GlyR is a metabolically stable protein (half-life approximately 2 days) that may have a homo-oligomeric structure.[14]

Evidence for GlyR heterogeneity also came from DNA sequencing data. By screening cDNA and genomic libraries under conditions of low stringency, variants of the originally isolated GlyR α subunit (now termed α1) have been isolated. The α2 subunit sequences predicted from human[28] and rat[20,29,30] cDNAs display about 80% amino acid identity to their α1 counterparts and correspond to the ligand binding subunits of the neonatal GlyR. In addition, an α3 sequence has been isolated from rat,[31] and clones encoding exons of a fourth variant, α4, have been identified in mouse genomic libraries.[40] Thus, considerable α subunit diversity exists for the GlyR.[32] So far, no variants of the β subunit have been detected (A. Kuryatov, unpublished data).

In situ hybridization experiments with sequence-specific oligonucleotides have revealed a complex picture of the developmental and regional distribution of the different GlyR subunit mRNAs.[32] In the rat, α1 transcripts are found in spinal cord, brain stem, and the colliculi, whereas α2 mRNA is also seen in several forebrain regions including the hippocampus, cerebral cortex, and thalamus.[33] Low levels of α3 subunit mRNA are detected in the olfactory bulb, the hippocampus, and in particular the cerebellum, whereas β subunit transcripts are abundantly expressed throughout the entire brain and spinal cord, suggesting that further GlyR subtypes may exist. GlyR α2 and β transcript accumulate already at embryonic stages, whereas α1 and α3 mRNAs appear only postnatally.[33]

Alternative splicing in addition contributes to GlyR heterogeneity. For the rat α1 subunit, variant cDNAs have been identified that originate from alternate splice acceptor site selection at an exon encoding the cytoplasmic domain adjacent to transmembrane segment M3.[34] The resulting insertion contains eight additional amino acid residues and creates a novel phosphorylation site. Similarly, the rat α2 polypeptide exists in two versions, which originate from alternative use of exon 3 and differ by only two conservative amino acid substitutions in the extracellular region of the polypeptide.[29] S1 nuclease mapping, polymerase chain reaction, and/or in situ hybridization experiments indicate that these splice variants are expressed at all stages of postnatal development.

CONCLUSIONS AND PERSPECTIVES

The currently available structural and functional data on GlyRs provide a detailed view of these ion channel proteins. Moreover, they underline the importance of glycinergic synapses in the control of neuronal activity. The crucial role of GlyRs in regulating diverse motor and sensory functions is supported further by studies on animal mutants. Spastic mice[5,35,36] and myoclonic Poll Hereford cattle[37] display severe motor deficits resulting from GlyR deficiencies that reduce the life-span of the affected animals. GlyR anomalies also may be implicated in human neurological diseases. Interestingly, the GlyR α2 subunit gene in humans has been localized in close vicinity to the Duchenne-Becker muscular dystrophy locus on the X chromosome. Thus, further unraveling of the molecular biology of inhibitory GlyRs may not only contribute to understanding the function and pharmacology of this neuronal channel protein, but also help to elucidate pathogenic mechanisms in animals and man.

ACKNOWLEDGMENTS

Work in the authors' laboratory was supported by Deutsche Forschungsgemeinschaft (SFB 169 and Leibniz-Program), German-Israeli Foundation, and Fonds der Chemischen Industrie. We thank S. Wartha for help with the preparation of this manuscript.

REFERENCES

1. BORMANN, J., O. P. HAMILL & B. SAKMANN. 1987. Mechanism of anion permeation through channels gated by glycine and γ-aminobutyric acid in mouse cultured spinal neurones. J. Physiol. (Lond) 385: 243–286.
2. BETZ, H. 1990. Ligand-gated ion channels in the brain: the amino acid receptor superfamily. Neuron 5: 383–392.
3. PFEIFFER, F., D. GRAHAM & H. BETZ. 1982. Purification by affinity chromatography of the glycine receptor of rat spinal cord. J. Biol. Chem. 257: 818–823.

4. GRAHAM, D., F. PFEIFFER, R. SIMLER & H. BETZ. 1985. Purification and characterization of the glycine receptor of pig spinal cord. Biochemistry **24**: 990–994.
5. BECKER, C. M., I HERMANS-BORGMEYER, B. SCHMITT & H. BETZ. 1986. The glycine receptor deficiency of the mutant mouse spastic: evidence for normal glycine receptor structure and localization. J. Neurosci **6**: 1358–1364.
6. PRIOR, P., B. SCHMITT, G. GRENNINGLOH, I. PRIBILLA, G. MULTHAUP, K. BEYREUTHER, Y. MAULET, P. WERNER, D. LANGOSCH, J. KIRSCH & H. BETZ. 1992. Primary structure and alternative splice variants of gephyrin, a putative glycine receptor-tubulin linker protein. Neuron **8**: 1161–1170.
7. SCHMITT, B., P. KNAUS, C. M. BECKER & H. BETZ. 1987. The M_r 93,000 polypeptide of the postsynaptic glycine receptor is a peripheral membrane protein. Biochemistry **26**: 806–811.
8. TRILLER, A., F. CLUZEAUD, F. PFEIFFER, H. BETZ & H. KORN. 1985. Distribution of glycine receptors at central synapses: an immunoelectron microscopy study. J. Cell Biol. **101**: 683–688.
9. KIRSCH, J., D. LANGOSCH, P. PRIOR, U. Z. LITTAUER, B. SCHMITT & H. BETZ. 1991. The 93 kDa glycine receptor-associated protein binds to tubulin. J. Biol. Chem. **266**: 22242–22245.
10. GRENNINGLOH, G., A. RIENITZ, B. SCHMITT, C. METHFESSEL, M. ZENSEN, K. BEYREUTHER, E. D. GUNDELFINGER & H. BETZ. 1987. The strychnine-binding subunit of the glycine receptor shows homology with nicotinic acetylcholine receptors. Nature **328**: 215–220.
11. GRENNINGLOH, G., I. PRIBILLA, P. PRIOR, G. MULTHAUP, K. BEYREUTHER. O. TALEB & H. BETZ. 1990. Cloning and expression of the β subunit of the inhibitory glycine receptor. Neuron **4**: 963–970.
12. SCHMIEDEN, V., G. GRENNINGLOH, P. R. SCHOFIELD & H. BETZ. 1987. Functional expression in *Xenopus* oocytes of the strychnine binding 48 kd subunit of the glycine receptor. EMBO J **8**: 695–700.
13. SONTHEIMER, H., C. M. BECKER & D. B. PRITCHETT, P. G. SCHOFIELD, G. GRENNINGLOH, H. KETTENMANN, H. BETZ & P. H. SEEBURG 1989. Functional chloride channels by mammalian cell expression of rat glycine receptor subunit. Neuron **2**: 1491–1497.
14. HOCH, W., H. BETZ & C. M. BECKER. 1989. Primary cultures of mouse spinal cord express the neonatal isoform of the inhibitory glycine receptor. Neuron **3**: 339–348.
15. LANGOSCH, D., L. THOMAS & H. BETZ. 1988. Conserved quaternary structure of ligand-gated ion channels: the postsynaptic glycine receptor is a pentamer. Proc. Natl. Acad. Sci. USA **85**: 7394–7398.
16. BETZ, H. 1990. Homology and analogy in transmembrane channel design: lessons from synaptic membrane proteins. Biochemistry **29**: 3591–3599.
17. GRAHAM, D., F. PFEIFFER & H. BETZ. 1981. UV light-induced cross-linking of strychnine to the glycine receptor of rat spinal cord membranes. Biochem. Biophys. Res. Commun. **102**: 1330–1335.
18. GRAHAM, D., F. PFEIFFER & H. BETZ. 1983. Photoaffinity-labelling of the glycine receptor of rat spinal cord. Eur. J. Biochem. **131**: 519–525.
19. RUIZ-GOMEZ, A., E. MORATO, M. GARCIA-CALVO, F. VALDIVIESO & F. JR. MAYER. 1990. Localization of the strychnine binding site on the 48-kilodalton subunit of the glycine receptor. Biochemistry **29**: 7033–7040.
20. KUHSE, J., V. SCHMIEDEN & H. BETZ. 1990. A single amino acid exchange alters the pharmacology of neonatal rat glycine receptor subunit. Neuron **5**: 867–873.
21. VANDENBERG, R. J., C. R. FRENCH, P. H. BARRY, J. SHINE & P. SCHOFIELD. 1992. Antagonism of ligand-gated ion channel receptors: two domains of the glycine receptor α subunit form the strychnine-binding site. Proc. Natl. Acad. Sci. USA **89**: 1765–1769.
22. SCHMIEDEN, V., J. KUHSE & H. BETZ. 1992. Agonist pharmacology of neonatal and adult glycine receptor α subunits: identification of amino acid residues involved in taurine activation. EMBO J. **11**: 2025–2032.
23. UNWIN, N. 1989. The structure of ion channels in membranes of excitable cells. Neuron **3**: 665–676.
24. LANGOSCH, D., K. HARTUNG, E. GRELL, E. BAMBERG & H. BETZ. 1991. Synthetic transmembrane

segments of the inhibitory glycine receptor form ion-selective channels in lipid bilayers. Biochim. Biophys. Acta 1063: 36–44.

25. PRIBILLA, I., T. TAKAGI, D. LANGOSCH, J. BORMANN & H. BETZ. 1992. The atypical M2 segment of the β subunit confers picrotoxinin resistance to inhibitory glycine receptor channels. EMBO J 11: 4305–4311.

26. BECKER, C. M., W. HOCH & H. BETZ. 1988. Glycine receptor heterogeneity in rat spinal cord during postnatal development. EMBO J 7: 3717–3726.

27. AKAGI, T. & R. MILEDI. 1988. Heterogeneity of glycine receptors and their RNAs in rat brain and spinal cord. Science 242: 270–273.

28. GRENNINGLOH, G., V. SCHMIEDEN, P. R. SCHOFIELD, P. H. SEEBURG, T. SIDDIQUE, T. K. MOHANDAS, C.-M. BECKER & H. BETZ. 1990. Alpha subunit variants of the human glycine receptor: primary structures, functional expression and chromosomal localization of the corresponding genes. EMBO J. 9: 771–776.

29. KUHSE, J., A. KURYATOV, Y. MAULET, M. L. MALOSIO, V. SCHMIEDEN & H. BETZ. 1991. Alternative splicing generates two isoforms of the α2 subunit of the inhibitory glycine receptor. FEBS Lett. 283: 73–77.

30. AKAGI, H., K. HIRAI & F. HISHINUMA. 1991. Cloning of a glycine receptor subtype expressed in rat brain and spinal cord during a specific period of neuronal development. FEBS Lett. 281: 160–166.

31. KUHSE, J., V. SCHMIEDEN & H. BETZ. 1990. A novel ligand binding subunit of the rat glycine receptor. J. Biol. Chem. 265: 22317–22320.

32. BETZ, H. 1991. Glycine receptors: heterogeneous and wide-spread in the mammalian brain. Trends Neurosci. 14: 458–461.

33. MALOSIO, M. L., B. MARQUEZE-POUEY, J. KUHSE & H. BETZ. 1991. Widespread expression of glycine receptor subunit mRNAs in the adult and developing rat brain. EMBO J. 10: 2401–2409.

34. MALOSIO, M. L., G. GRENNINGLOH, J. KUHSE, V. SCHMIEDEN, B. SCHMITT, P. PRIOR & H. BETZ. 1991. Alternative splicing generates two variants of the α1 subunit of the inhibitory glycine receptor. J. Biol Chem. 266: 2048–2053.

35. WHITE, F. & A. H. HELLER. 1982. Glycine receptor alteration in the mutant mouse spastic. Nature 298: 655–657.

36. BECKER, C.-M., V. SCHMIEDEN, P. TARRONI, U. STRASSER & H. BETZ. 1992. Isoform-selective deficit of glycine receptors in the mouse mutant spastic. Neuron 8: 283–289.

37. GUNDLACH, A. L., P. R. DODD, C. S. G. GRABARA, W. E. J. WATSON, G. A. R. JOHNSTON, P. A. W. HAPER, J. A. DENNIS & P. J. HEALY. 1988. Deficit of spinal cord glycine/strychnine receptors in inherited myoclonus of Poll Hereford calves. Science 241: 1807–1810.

38. CHANGEUX, J. P. 1990. The nicotinic acetylcholine receptor. Fidia Res. Found. Neurosci. Award Lect. 4: 21–168.

39. BORMANN, N. RUNDSTRÖM, H. BETZ & D. LANGOSCH. 1993. Residues within transmembrane segment M2 determine chloride conductance of glycine receptor homo- and hetero-oligomers. EMBO J. 12: 3729–3737.

40. MATZENBACH, B., Y. MAULET, L. SEFTON, B. COURTIER, P. AVNER, J.-L. GUÉNET & H. BETZ. 1993. Structural analysis of mouse glycine receptor alpha subunit genes. J. Biol. Chem. In press.

Multiplicity, Structure, and Function in GABA$_A$ Receptors

ERIC A. BARNARD,[a,b] MARGARET SUTHERLAND,[b]
SHAHID ZAMAN,[b] MITSUHIKO MATSUMOTO,[b]
NAUSHABA NAYEEM,[b] TIM GREEN,[c]
MARK G. DARLISON,[b] AND ALAN N. BATESON[b]

[a]Molecular Neurobiology Unit
Division of Basic Medical Sciences
Royal Free Hospital School of Medicine
London NW3 2PF, United Kingdom

[b]MRC Molecular Neurobiology Unit

[c]Centre for Protein Engineering
MRC Centre
Hills Road
Cambridge, CB2 2QH, United Kingdom

INTRODUCTION

The application of molecular biology to the study of the vertebrate brain GABA$_A$ receptor has, in the period of only five years from the initial cloning of four of its cDNAs,[1,2] produced a vast body of totally new information and insights for the understanding of this, the major inhibitory neurotransmitter receptor of the brain. Four different subunit types of this ion-channel receptor (α, β, γ, and δ) have thus been recognized and work from several laboratories (for references, see the review in Burt & Kamatchi[3]) has resulted in the reporting, to date, of the cDNA and deduced amino acid sequences for six α subunits ($\alpha1-\alpha6$), four β subunits ($\beta1-\beta4$), three γ subunits ($\gamma1-\gamma3$), a δ subunit, as well as $\rho1$ in the retina,[4] which may substitute for an α subunit there. Each is encoded by a separate gene.

The co-expression of these subunits, in either *Xenopus* oocytes[1,2] or transfected mammalian cells,[5] has permitted the pharmacologies produced by different subunit combinations to be defined. Thus, the $\gamma2$ subunit (but not the $\gamma1$ subunit[6]) confers on the complex the normal response to benzodiazepine (BZ) positive and inverse agonists.[5] We shall describe here how variation of the α subunit type in $\alpha\beta\gamma$ combinations also affects specific pharmacological properties.

Such co-expression studies led to the now generally accepted view that the GABA$_A$ receptors *in vivo* are comprised of combinations of particular isoforms of these subunits. For the BZ-sensitive types this would require, on present knowledge, one of the $\alpha\beta\gamma$

116

combinations (though the stoichiometry there is undefined). However, since also some BZ-insensitive types have been detected, in studies on native GABA$_A$ receptors *in situ*[7] this would indicate that additional combinations exist where the γ is replaced presumably by δ or an extra α or β.

In all, the various possibilities compatible with the properties seen in co-expressions could allow a theoretical total on the order of 3,000 GABA$_A$ receptor subtypes. It is not believed that all, or more than a fraction, of these potential combinations actually exist *in vivo*. Mapping the co-distributions of the subunits by *in situ* hybridization of their mRNAs does not have the resolution needed to determine the actual number of combinations, although there is clearly a complex set of non matching distributions.[8] This demonstrates that high diversity in the receptor subtypes does indeed exist.

What is the reason for the great multiplicity of combinations of these subunits *in vivo*? It is presumed that each natural combination represents one subtype of the GABA$_A$ receptor, which differs from other subtypes in the detailed configuration of one or more of the functional sites of the receptor. Significant functional differences have been detected when the isoform of α in a given combination is changed, as in the case of the benzodiazepine site ligands.[9] Further analysis of this has recently been made and will be presented in this paper. In addition, the programming of neuronal development to provide the switching-on of two different isoform genes in different pathways, or at different stages, may permit their differential regulation even if the two receptors are very similar. A high multiplicity of the receptors may, for both of these reasons, allow GABA to be employed in constructing a variety of complex neuronal circuits, although it is a ubiquitous, single inhibitory transmitter in the nervous system.

If each permitted combination of isoforms indeed represents a different subtype of GABA$_A$ receptor, the pharmacology of this receptor must be far more complex than any that could be analyzed by conventional methods. The different pharmacologies can only be defined by expressing different mixtures of recombinant isoforms. This should preferably be done in permanent cell lines, for constant reference and general availability. Since a comparable level of stable expression of several DNAs is required together, this operation is less straightforward to perform successfully than for single-subunit receptors. Multiple GABA$_A$-receptor subunits can, however, be so expressed when an inducible promoter system is used,[10] when the exogenous receptors are not present on the cells except when being analyzed, so that an ionic imbalance due to channel openings does not accumulate. Such cells, when induced, are suitable for patch-clamping analysis of channel properties, as well as for pharmacological distinctions made both by ligand binding measurements and by recording of currents. That system, initially used with stable αβ combinations,[10] has recently been extended to a stable αβγ cell line.

CHICKEN GABA$_A$ RECEPTOR SUBUNIT cDNAs

The recombinant GABA receptor subunits obtained in this laboratory and elsewhere have hitherto been derived only from mammals (bovine, human, rat, mouse). We have recently gone on to use mammalian GABA$_A$ receptor subunit cDNAs to isolate the corresponding cDNA sequences from the chicken. We have chosen this animal for further molecular biological analysis of GABA$_A$ receptors (*1*) to look for sequence features conserved through the vertebrates; (*2*) because of its amenability to experimental manipulation in relation to gene expression (e.g., for developmental studies); and (*3*) because the original work on the expression of GABA$_A$ receptors in *Xenopus* oocytes showed that poly(A)$^+$ RNA isolated from chick brain induced a particularly strong and robust response[11,12] indicating this as an excellent source (as it has indeed proved) for the cloning of further subtypes.

Using low-stringency hybridization conditions, a series of α-, β-, and γ-subunit clones were isolated from an embryonic chick whole-brain cDNA library. The sequences of the chicken GABA$_A$ receptor mature polypeptides that we have obtained are highly homologous to their mammalian counterparts. For example, the chicken α1 subunit sequence[13] is 98% identical to the corresponding bovine sequence, the chicken β3-subunit sequence[14] is 92% identical to the bovine β3-subunit sequence, and the chicken γ2-subunit sequence[15] is 96% identical to the corresponding rat sequence. This extremely high degree of homology is surprising, considering the large evolutionary distance between avian and mammalian species. The majority of these differences are found in the region of the polypeptide that is presumed to be intracellular, i.e. between transmembrane domains M3 and M4, or near the amino terminus.

Further, we have isolated a novel chicken β4-subunit cDNA[16] not previously known from the mammal. This exhibits only ~72% identity to the mammalian β1-, β2-, and β3-subunit sequences. In contrast, as noted above, other chicken GABA$_A$ receptor subunits are highly homologous (>92% identical) to their precise mammalian counterparts.

Interestingly, DNA sequencing of the β4 subunit cDNA clones described above revealed that two forms of the β4 subunit exist that differ by the presence or absence of an extra four amino acids (Val-Arg-Glu-Gln) in the presumed intracellular loop region.[16] Sequencing of genomic DNA in the region corresponding to the 12 bp insertion, which is found in one of the three cDNA clones isolated and which encodes the additional four amino acids, reveals an intron in the gene at the position corresponding to that where the two different β4 subunit cDNA sequences differ. Two possible 5′ donor splice sites that are separated by 12 bp are found in this region of the gene; use of the more 5′ of these results in a β4 subunit that lacks the four amino acid insertion, while use of the more 3′ of these yields a subunit (designated β4′) that has the extra four amino acids.[17]

Further, a recombinant γ2 subunit was obtained from the chicken, in two forms that differ by the presence or absence of an extra eight amino acids in the presumed intracellular loop region.[18] This is of interest because the shorter form of γ2 has lost one of the consensus sites for serine phosphorylation by protein kinase C. The same alternative splicing, at an equivalent position, was concurrently found in the bovine γ2 mRNA[19] and likewise in the mouse γ2[20] and the actual phosphorylation of this sequence has been demonstrated.[19]

Hence, in the vertebrates in general the GABA$_A$ receptor is encoded by a set of 15 or more genes, and the subunits produced can be further diversified (at least in some cases) by alternative splicing of the gene transcript. Each isoform (α1, α2, etc.) of a given subunit is maintained throughout vertebrate evolution and has an exceptional degree of conservation of its separate identity.

CLONING OF AN INVERTEBRATE GABA~A~ RECEPTOR SUBUNIT cDNA

It is clearly of interest to extend this comparison across the much greater phylogenetic distances to the invertebrates, where GABA$_A$ type receptors have been identified pharmacologically, for example, in much detail on molluscan neurons. There they show behavior in common with mammalian GABA$_A$ receptors, e.g., in anionic selectivity, blockade by bicuculline and by picrotoxin, although some differences can also be found.[21,22] By low-stringency screening of a *Lymnaean* genomic library with the full-length bovine GABA$_A$ receptor β1-subunit cDNA,[1] we were able to isolate a series of clones. Analysis of these has shown that several types of sequence related to vertebrate GABA$_A$ receptor subunits exist in *Lymnaea*. From such β-subunit–like coding sequences we proceeded to the cloning of a full-length *Lymnaean* cDNA. This encodes a polypeptide[23] with a signal prepeptide sequence followed by a mature 54,000-dalton subunit sequence. The latter is much more like the β subunits than the others of mammalian GABA$_A$ receptors, being ~ 50% identical to any of the vertebrate β subunits, suggesting these diversified later in evolution. In its structure (FIG. 1), the molluscan subunit shows the features common to all of the vertebrate subunits, and in particular the four deduced transmembrane domains M1–M4 and the same relative positions of these in the chain. The conservation of the subunit sequence between the mammals and the mollusc is high in M1–M3, but notably lower in M4, perhaps denoting a less important structural role for M4 in the receptor in general. The loop between M3 and M4, which is presumed to be intracellular in the topology of the GABA$_A$ receptor generally,[1] is poorly conserved (FIG. 1), suggesting that much of the loop is needed for its length rather than its specific structure. This loop does, however, contain two overlapping consensus sequences for phosphorylation by cAMP-dependent protein kinase, a type found there in all vertebrate β subunits, so that we deduce that parts of the loop are maintained for functional intracellular interactions.

Also noteworthy is the high conservation (FIG. 1) of the 15-residue "cys loop" structure in the N-terminal region, a feature common to all vertebrate GABA$_A$ receptor

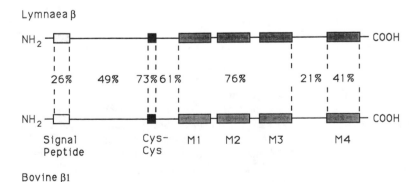

FIGURE 1. The percentages of sequence identity between the *Lymnaea* β subunit and the bovine β1 subunit are shown for the different regions of the polypeptide.

subunits and indeed to the whole of a superfamily of transmitter-gated ion channels to which they belong.[1] All of those residues in this loop, which appear to be important in other GABA receptors,[24] are unchanged in the invertebrate.

There is, in addition, a relatively high degree of conservation of the rest of the large N-terminal (deduced extracellular[1]) domain, which on inspection shows many blocks of complete sequence identity with *all* of the known β subunits.[23] The substitutions that occur outside these blocks represent natural mutations that are compatible with β-subunit function and offer a guide to mutagenesis experiments and their interpretations.

This molluscan subunit is part of a functional GABA$_A$ receptor, since when expressed and assembled in the *Xenopus* oocyte system[25] an anion channel was opened by GABA and muscimol and showed bicuculline blockade.[23] The homo-oligomeric receptor thus evidenced is not thought to be a natural form, since the expression obtainable was low, but a hybrid αβ receptor with high expression could be obtained when the bovine α1 subunit was co-expressed with the molluscan β subunit.[23] This functional association of invertebrate and vertebrate polypeptides, which has not been reported before with neuroreceptors, shows that despite the ~ 50% difference between a mammalian and a molluscan β subunit the invertebrate sequence encodes all of the features necessary for co-operative assembly and function of a β subunit in the mammalian GABA$_A$ receptor.

THE CONSERVATION OF THE GENOMIC STRUCTURE IN GABA RECEPTORS

The constancy of the amino acid sequence for any given subunit isoform (α1 or α2 or etc.) is, as we have seen, exceptionally high in vertebrate phylogeny (>90%), so far as it has been sampled. This supports the concept that each subunit isoform of the GABA$_A$ receptor has a separate and significant functional role. Moreover, construction of the most parsimonious phylogenetic tree for the entire superfamily suggests that the divergence of the GABA$_A$ and acetylcholine receptors arose early in metazoan evolution.[24] It is, therefore, of much interest to see whether the GABA receptor genes show an evolution into subtype-specific exon patterns. Likewise, the proposal that the nicotinic acetylcholine and GABA$_A$ receptors arose in evolution from a common ancestor[1] can be examined further at the level of genomic organization.

Partial structures of some of the genes for GABA receptor isoforms have been determined in our laboratory for representative cases from man and from the chicken, so that the positions of all or some of the introns have been located in each. Thus, the intron/exon pattern was determined fully across the coding sequence for the chicken β4 subunit gene,[17] for most of it in the human α1 and α2 genes and for a few of the boundary positions in the human α5 and γ2 genes.[26] Isolated boundary positions were also found in the course of studies on the chicken γ2 cDNA[18] and the bovine α3 cDNA.[2] Others have reported the boundary positions for the mouse δ subunit gene[26] and recently for a partial human β1 gene structure.[27]

From each of these studies, the same pattern has emerged (FIG. 2). The vertebrate α, β, and γ subunit genes are exceptionally large for receptor genes, ~ 100–200 kb, and their introns are therefore very large. The mouse δ gene,[28] in contrast, is only

13 kb. There are nine exons in all cases (plus a tenth, not always used, found so far only in $\gamma2$, being involved in the alternative splicing to give a variant structure of the intracellular loop, as described above). The positions of the exons are constant or extremely close (in all the cases where they have been determined) to give the fixed pattern shown in FIGURE 2. There the exons (between the arrows) are shown under the corresponding points in the subunit polypeptide. Two exons cover M1–M3, with M4 always separated. The Cys loop in the extracellular region (noted above) is always in a separate small exon. The constancy of the intron positions in vertebrate evolution and in the different subunit types is remarkable.

Further, similar analysis of the gene for the above-described molluscan β subunit[23] shows—as far as has been determined—that the positions of these boundaries are maintained with high precision (FIG. 2). This invariance of the genomic organization in the subunits across the evolutionary distance involved is, again, highly exceptional and reinforces the argument for the very high selection pressure on the subunit structure in the GABA_A receptors.

The genomic organization has also been determined for the nicotinic acetylcholine receptors, for various subunits from muscle receptor from man,[28,29] mouse,[30] and chicken.[31] This can be compared with that for the neuronal nicotinic receptor subunits from chicken[32] and rat.[33] The pattern is constant for all of the neuronal subunits analyzed, whereas the muscle gene pattern varies somewhat with the subunit type (9 exons in α, 12 in the others). The neuronal nicotinic receptors have a simpler genomic organization: the first four exons have the same locations as in the muscle genes, but the fifth exon covers four or seven exons of the latter.

Few features of the pattern are common to GABA_A and nicotinic receptor genes. The boundaries are certainly not in conserved locations between the two series. In contrast, the preservation of the organization of the GABA_A receptor genes through all subunit types, isoforms, and evolution stands out as a feature. Their intronic sequences (where examined) vary, of course, greatly between species, whereas the exons and their boundaries are maintained almost constant for all isoforms. The combination of a high multiplicity of the GABA_A receptor genes and an invariancy of each of them may be required, therefore, as discussed above, for the subunits of such a receptor series employed in the construction of most neuronal pathways.

SUBUNIT FUNCTIONAL DIFFERENCES

The effect of both the γ and the α subunit types on the benzodiazepine pharmacology of the receptor is profound and is discussed by others in this volume. For the interaction with the transmitter, GABA, it appears that each subunit has some form of an agonist site, since homo-oligomeric receptors can form from either α or β or γ subunits expressed singly (albeit very poorly and not with all the native receptor properties) and these respond weakly to GABA to open the channel.[6,34,35] However, in the well-expressed $\alpha\beta\gamma$ combinations the GABA affinity can be much stronger, and there the α subunit isoform present can greatly affect the EC_{50} for GABA.[2,9,36]

In recent studies the modulatory site for neurosteroids has, also, been compared between subunit compositions. Neurosteroids of the pregnanolone series, which potentiate GABA action, show considerable differences in their dose-response curves as

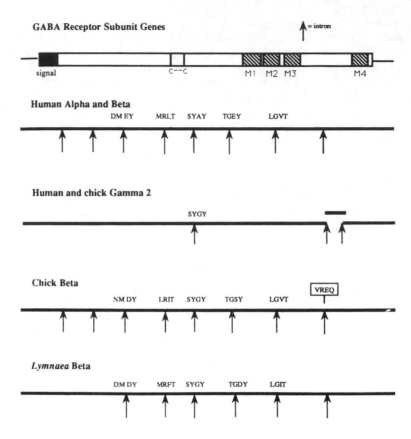

FIGURE 2. Schematic representation of the genomic organization for the subunits of GABA$_A$ receptors. At the top, the encoded polypeptide is shown, with features common to all of the subunits: the positions of the signal peptide, the cysteine-loop region (C−C), and the four membrane-spanning domains (M1 to M4) are shown. For the genes, designated below this, the arrows indicate the positions of the introns relative to the coding regions above. For references, see the text. For the γ2 subunit, only three of the introns have yet been located; two are at equivalent positions to boundaries in the α and β genes, while the extra exon shown (of only 24 nucleotides) has thus far only been detected in the γ2 subunit, given rise to an alternative splicing of the primary gene transcript. In the chicken β4 subunit pattern shown below this, the box shows the additional sequence that can be inserted due to an alternative choice of the splice site. For *Lymnaea*, only six of the introns present have as yet been mapped. Not shown is the mouse δ gene pattern, which is identical to that for the α and β patterns here. Likewise, the two introns located so far in the bovine α3 gene (2) coincide with the fourth and the eighth introns of the human α and β genes. It can be seen that the positions of the introns are identical through all subunits and species examined. The one-letter amino acid code symbols show the codons that are split or (where a space is shown) divided, by the intron. Again, it is clear that even at the level of the insertion in or around a particular codon, there is a remarkable conservation of the pattern through the phyla.

the α subunit is varied.[37] Pregnenolone sulfate, which acts in the inverse manner, inhibiting GABA action, again shows differences in its potency in some cases of $\alpha\beta\gamma$ combinations when the α subunit isoform is varied.[36] Although a neurosteroid is expected to interact to some extent with hydrophobic receptor structures within the membrane, there is nevertheless obviously also some interaction with a site on the α subunit whose sequence varies significantly between the α isoforms.

In summary, functional differences between subunit combinations exist in the agonist site, the benzodiazepine site, and the neurosteroid site. It is proposed that, with sufficiently structure-sensitive pharmacological probing, functional differences will occur between all of the combinations that exist *in vivo*, and this (with presumed natural modulators) will contribute to the need for the great heterogeneity of subtypes of the GABA_A receptor that is now observed.

REFERENCES

1. Schofield, P. R., M. G. Darlison, N. Fujita, et al. 1987. Sequence and functional expression of the GABA-A receptor shows a ligand-gated receptor superfamily. Nature **328:** 221–227.
2. Levitan, E. S., P. R. Schofield, D. R. Burt, et al. 1988. Structural and functional basis for GABA_A receptor heterogeneity. Nature **335:** 76–79.
3. Burt, D. R. & G. L. Kamatchi. 1991. GABA_A receptor subtypes: from pharmacology to molecular biology. FASEB J **5:** 2916–2923.
4. Cutting, G. R., L. L. Bruc, F. O'Hara, et al. 1991. Cloning of the γ-aminobutyric (GABA) ρ_1 cDNA: a GABA receptor subunit highly expressed in the retina. Proc. Natl. Acad. Sci. USA **88:** 2673–2677.
5. Verdoon, T. A., A. Draguhn, S. Ymer, P. H. Seeburg & B. Sakmann. 1990. Functional properties of recombinant rat GABA_A receptors depend upon subunit composition. Neuron **4:** 919–918.
6. Ymer, S., A. Draguhn, W. Wisden, et al. 1990. Structural and functional characterization of the $\gamma 1$ subunit of GABA_A/benzodiazepine receptors. EMBO J. **9:** 261–267.
7. De Blas, A. L., J. Vitorica & P. Friedrich. 1988. Localization of the GABA_A receptor in the rat brain with a monoclonal antibody to the 57,000 M_r peptide of the GABA_A receptor/benzodiazepine receptor/Cl$^-$ channel complex. J. Neurosci. **8:** 602–614.
8. Olsen, R. W. & A. J. Tobin. 1990. Molecular biology of GABA_A receptors. FASEB J. **4:** 1469–1480.
9. Sigal, E. R., Baur, G. Trube, H. Möhler & P. Malherbe. 1990. The effect of subunit composition of rat brain GABA_A receptors on channel function. Neuron **5:** 703–711.
10. Moss, S. J., T. G. Smart, N. M. Porter, et al. 1990. Cloned GABA receptors are maintained in a stable cell line: allosteric and channel properties. Eur. J. Pharm. **189:** 77–88.
11. Smart, T. G., A. Constanti, G. Bilbe, D. A. Brown & E. A. Barnard. 1983. Synthesis of functional chick brain GABA-benzodiazepine-barbiturate/receptor complexes in mRNA-injected *Xenopus* oocytes. Neurosci. Lett. **40:** 55–59.
12. Houamed, K., A. Constanti, T. G. Smart, et al. 1984. Expression of functional GABA, glycine and glutamate receptors in *Xenopus* oocytes injected with rat brain mRNA. Nature **310:** 318–321.
13. Bateson, A. N., R. J. Harvey, W. Wisden, et al. 1991. The chicken GABA_A receptor $\alpha 1$ subunit: cDNA sequence and localization of the corresponding mRNA. Mol. Brain Res. **9:** 333–339.
14. Bateson, A. N., R. J. Harvey, C. C. M. Bloks & M. G. Darlison. 1990. Sequence of the chicken GABA_A receptor $\beta 3$-subunit cDNA. Nucleic Acids Res. **18:** 5557.
15. Glencorse, T. A., A. N. Bateson & M. G. Darlison. 1990. Sequence of the chicken GABA_A receptor $\gamma 2$-subunit cDNA. Nucleic Acids Res. **18:** 7157.

16. BATESON, A. N., A. LASHAM & M. G. DARLISON. 1991. γ-Aminobutyric acid$_A$ receptor heterogeneity is increased by alternative splicing of a novel β-subunit gene transcript. J. Neurochem. 56: 1437–1440.

17. LASHAM, A., E. VREUGDENHIL, A. N. BATESON, E. A. BARNARD & M. G. DARLISON. 1991. Conserved organization of γ-aminobutyric acid$_A$ receptor genes: cloning and analysis of the chicken β4-subunit gene. J. Neurochem. 57: 352–355.

18. GLENCOURSE, T. A., A. N. BATESON & M. G. DARLISON. 1992. Differential localization of two alternatively spliced GABA$_A$ receptor γ2-subunit mRNAs in the chick brain. Eur. J. Neurosci. In press.

19. WHITING, P., R. M. MCKERNAN & L. L. IVERSEN. 1990. Another mechanism for creating diversity in γ-aminobutyrate type A receptors: RNA splicing directs expression of two forms of γ2 subunit, one of which contains a protein kinase C phosphorylation site. Proc. Natl. Acad. Sci. USA 87: 9966–9970.

20. KOFUJI, P., J. B. WANG, S. J. MOSS, R. L. HUGANIR & D. R. BURT. 1991. Generation of two forms of the γ-aminobutyric acid$_A$ receptor γ2-subunit in mice by alternative splicing. J. Neurochem. 56: 713–715.

21. YONGSIRI, A., K. FUNASE, H. TAKEUCHI, K. SHIMAMOTO & Y. OHFUNE. 1988. Classification of GABA receptors in snail neurones. Eur. J. Pharmacol. 155: 239–245.

22. KING, W. M. & D. O. CARPENTER. 1989. Voltage-clamp characterization of Cl conductance gated by GABA and L-glutamate in single neurons of *Aplysia*. J. Neurophysiol. 61: 892–901.

23. HARVEY, R. J., E. VREUGDENHIL, S. H. ZAMAN, N. S. BHANDAL, P. N. R. USHERWOOD, E. A. BARNARD & M. G. DARLISON. 1991. Sequence of a functional invertebrate GABA$_A$ receptor subunit which can form a chimeric receptor with a vertebrate α subunit. EMBO J. 10: 3239–3245.

24. COCKROFT, V. B., D. J. OSGUTHORPE, E. A. BARNARD, A. F. FRIDAY, S. WONNACUTT & G. G. LUNT. 1991. Ligand-gated ion-channels: homology and diversity. Molec. Neurobiol. In Press.

25. BARNARD, E. A., R. MILADI & K. SUMIKAWA. 1982. Translation of exogenous messenger RNA coding for nicotinic acetylcholine receptors produces functional receptors in *Xenopus* oocytes. Proc. R. Soc. Lond. B 215: 241–264.

26. SOMMER, B., A. POUSTKA, N. K. SPURR & P. H. SEEBURG. 1990. The murine GABA$_A$ receptor δ-subunit gene: structure and assignment to human chromosome 1. DNA Cell Biol. 9: 561–568.

27. KIRKNESS, E. F., J. W. KUSIAK, J. T. FLEMING, *et al.* 1991. Isolation, characterization, and localization of human genomic DNA encoding the β1 subunit of the GABA$_A$ receptor. Genomics 10: 985–995.

28. NODA, M., Y. FURUTANI, H. TAKAHASI, *et al.* 1983. Cloning and sequence analysis of calf cDNA and human genomic DNA encoding α-subunit precursor of muscle acetylcholine receptor. Nature 305: 818–823.

29. SHIBAHARA, S., T. KUBO, P. PERSKI, H. TAKAHASHI, M. NODA & S. NUMA. 1985. Cloning and sequence analysis of human genomic DNA encoding γ subunit precursor of muscle acetylcholine receptor. Eur. J. Biochem. 146: 15–22.

30. BUONANNO, A., J. MUDD & J. P. MERLIE. 1989. Isolation and characterization of the β and ε subunit genes of mouse muscle acetylcholine receptor. J. Biol. Chem. 264: 7611–7616.

31. NEF, P., A. MAURON, K. R. STALDER, C. ALLIOD & M. BALLIVET. 1984. Structure, linkage and sequence of the two genes encoding the δ and γ subunits of the nicotinic acetylcholine receptor. Proc. Natl. Acad. Sci. USA 81: 7975–7979.

32. NEF, P., C. ONEYSER, C. ALLIOD, S. COUTURIER & M. BALLIVET. 1988. Genes expressed in the brain define three distinct neuronal nicotinic acetylcholine receptors. EMBO J. 7: 595–601.

33. BOULTER, J., A. O'SHEA-GREENFIELD, R. M. DUVOISIN, *et al.* 1991. α3, α5, and β4: three members of the rat neuronal nicotinic acetylcholine receptor-related gene family form a gene cluster. J. Biol. Chem. 265: 4472–4482.

34. LEVITAN, E. S., L. A. C. BLAIR, V. E. DIONNE & E. BARNARD. 1988. Biophysical and pharmacologi-

cal properties of cloned GABA$_A$ receptor subunits expressed in *Xenopus* oocytes. Neuron **1**: 773–781.

35. BLAIR, L. A. C., E. S. LEVITAN, V. E. DIONNE & E. A. BARNARD. 1988. Single subunits of the GABA$_A$ receptor form ion channels with properties characteristics of the native receptor. Science **242**: 577–579.

36. ZAMAN, S. H., R. SHINGAI, R. J. HARVEY, M. G. DARLISON & E. A. BARNARD. 1992. Effects of subunit types of the recombinant GABA$_A$ receptor on the response to a neurosteroid. Eur. J. Pharmacol. In press.

37. SHINGAI, R., M. L. SUTHERLAND & E. A. BARNARD. 1991. Effects of subunit types of the cloned GABA$_A$ receptor on the response to a neurosteroid. Eur. J. Pharmacol. **206**: 77–80.

Glutamate Receptors in the Central Nervous System

PETER JONAS

Max-Planck-Institut für medizinische Forschung
Abteilung Zellphysiologie
Jahnstrasse 29
W-6900 Heidelberg, Germany

Glutamate is the major excitatory neurotransmitter in the central nervous system (CNS). In general, glutamate directly gates two different types of nonselective cationic channels, designated as AMPA (S-α-amino-3-hydroxy-5-methyl-4-isoxazolepropionic acid)/KA (kainate) and NMDA (N-methyl-D-aspartic acid) glutamate receptors (GluRs). Both types of channels are involved in mediating excitatory synaptic transmission. The present paper reviews the functional properties of native GluR channels characterized using the patch-clamp technique, predominantly applied to cells in brain slices.[1] This method has the advantage that it is possible to record, with high resolution, from visually identified central neurons within an intact synaptic environment. Two different aspects of the functional properties of GluR channels will be addressed here. On the one hand, we[2,3] wanted to mimic synaptic events by fast application of glutamate to outside-out patches excised from neurons in slices. Thereby, it might be possible to find out how GluR channels are gated during a synaptic event and which factors determine time course and amplitude of excitatory postsynaptic currents (EPSCs). On the other hand, we[2,4] wanted to compare the properties of the native GluR channels with the characteristics of cloned and functionally expressed receptors.[5-8] The aim was to propose a likely molecular subunit composition of the native GluR channels in different neuronal (and non-neuronal) cell types. Most of the experiments were done on the cells of the hippocampal trisynaptic circuit: dentate gyrus granule cells, CA3, and CA1 pyramidal cells. The hippocampal formation is of great interest for many neurophysiologists because long-term changes at the level of hippocampal glutamatergic synapses have been demonstrated, which presumably are related to the formation of short-term memory.

It has been reported that in hippocampal as well as other types of neurons, EPSCs consist of an AMPA/KA and a NMDA receptor-mediated component.[9] The two components showed substantial differences in their kinetic properties. The rise time of the AMPA/KA component was very fast, about 500 μsec under the most favorable voltage-clamp conditions obtained so far.[10] The decay time constant was also comparatively fast, 3.2–9.5 msec for dentate gyrus granule cells,[11] 3.0–6.6 msec for mossy fiber synapses on CA3 pyramidal cells,[10] and 4–8 msec for CA1 pyramidal cells.[12]

However, the exact value of the EPSC decay time constant is influenced by the quality of the voltage clamp, and hence the real conductance changes at the postsynaptic membrane are probably faster than the currents recorded with somatic patch pipettes.[10]

In contrast, the kinetics of the NMDA component of the EPSC were more than one order of magnitude slower. The rise time was about 10 msec; the decay was slow, with time constants between 50 and 250 msec.[11,12] Evidence has been obtained that both types of channels may be co-localized at the same postsynaptic density.[13,14] If this is the case, the question arises: How is it possible that the same synaptically released glutamate pulse activates two current components with so strikingly different kinetic properties?

FAST APPLICATION OF GLUTAMATE TO MEMBRANE PATCHES: MIMICKING EPSCs

A common feature of all ligand-gated ion channels is that they show desensitization, i.e. current decrease in the maintained presence of the agonist. To study their kinetic features quantitatively, it is therefore essential to apply the agonist very rapidly. At present, this is only possible with outside-out membrane patches, usually obtained from the soma of different cell types. Fast application techniques originally developed by Franke et al.[15] were modified as described by Jonas and Sakmann[2] and Colquhoun et al.[3] Briefly, a patch pipette with an outside-out membrane patch was positioned near the tip of a double-barrelled application pipette pulled from theta glass tubing. The application pipette was then moved rapidly with a piezo-electric element, and the patch thereby crossed the interface between the solutions flowing out of the two barrels (FIG. 1). Solution changes occurred within 100 µsec on an open patch pipette and within 200 µsec on an intact membrane patch.[3] Pulses of glutamate as brief as 1 msec could be applied reliably, making it possible to mimic synaptic release from a presynaptic terminal. The comparison of currents activated by fast application of glutamate with synaptic currents requires the assumption that postsynaptic GluR channels and channels in somatic membrane patches have identical functional properties.

COACTIVATION OF AMPA/KA AND NMDA RECEPTORS BY FAST APPLICATION OF GLUTAMATE IN THE SAME MEMBRANE PATCH

Outside-out patches isolated from the soma of hippocampal neurons in slices contained a high density of AMPA/KA as well as NMDA receptor channels. Appropriate experimental conditions can be chosen to activate either one of the two components separately or both simultaneously.

The AMPA/KA GluR channels can be investigated in isolation at negative membrane potentials in glycine-free extracellular solutions with 1 mM Mg. Glycine strongly potentiates NMDA responses or is even an essential cofactor[16]; the absence of glycine therefore minimizes NMDA receptor activation. Mg ions are known to block NMDA type GluR channels at negative membrane potentials.

The NMDA receptor channels can be investigated in the presence of glycine (10

FIGURE 1. Fast application of agonists to excised outside-out patches isolated from the soma of hippocampal neurons in brain slices. The two barrels of the application pipette were perfused with physiological extracellular solution and with 10% physiological extracellular solution, respectively. The sharp interface between the two solutions, due to their different refractory indices, is clearly visible. The diameter of the application pipette tip was approximately 200 μm. (From Colquhoun et al.[3] With permission.)

μM) and in Mg-free extracellular solutions. To study these channels in isolation, 10 μM 6-cyano-7-nitroquinoxaline-2,3-dione (CNQX), a selective AMPA/KA type GluR channel antagonist, was added to the external solution.

AMPA/KA and NMDA receptor channels are activated simultaneously in the same membrane patch in glycine-containing, Mg-free extracellular solution (FIG. 2). There are striking differences in kinetics between the AMPA/KA and the NMDA component activated by a brief (10 msec) pulse of a high concentration (1 mM) of glutamate. The AMPA/KA component showed a rapid rise (less than 500 μsec) and also a rapid decay, partly due to desensitization during the pulse, partly due to channel closure after the end of the pulse (see below). The NMDA component, in contrast, rose and decayed much more slowly. The rise time was 10 msec, and the decay time constant was more than 100 msec. FIGURE 2 shows that a brief glutamate pulse can qualitatively mimic the fast AMPA/KA as well as the slow NMDA component of an EPSC, suggesting that different gating properties of the two types of channels are responsible for the different time courses. Variation of the length of the glutamate pulse within the range 1 to 100 msec strongly influenced the decay kinetics of the AMPA/KA component (see below), but did not noticeably change the decay time

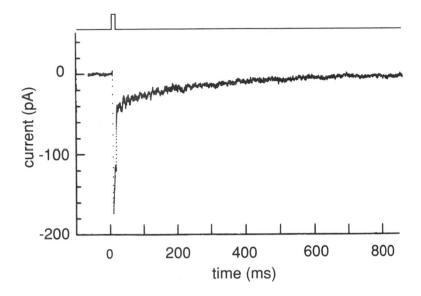

FIGURE 2. AMPA/KA and NMDA components of current activated by glutamate pulses. CA1 pyramidal cell patch, membrane potential was − 50 mV. 10 msec pulse of 1 mM glutamate, five single traces were averaged. Physiological extracellular solution contained 10 μM glycine and was nominally Mg free. Filter frequency 1 kHz. Note that the NMDA component, represented by the slow tail following the termination of the pulse, showed much slower kinetics as compared to the AMPA/KA component.

course of the NMDA component (data not shown). This suggests that the AMPA/KA GluR channel (rather than the NMDA GluR channel) would be suitable to "monitor" the rapid time course of transmitter concentration in the synaptic cleft during an EPSC. Therefore, the dependence of decay kinetics of AMPA/KA receptor-mediated currents on the length and the concentration of the glutamate pulse was studied.

TIME COURSE OF CURRENTS MEDIATED BY AMPA/KA TYPE CHANNELS ACTIVATED BY GLUTAMATE PULSES OF DIFFERENT LENGTHS

The decay time course of the AMPA/KA receptor-mediated component was strongly dependent on the length of the glutamate pulse (FIG. 3). For 1 msec pulses of 1 mM glutamate, the decay time constant (designated as "offset τ") was between 2.3 msec and 3.0 msec for the three principal neurons of the hippocampal circuit.[3] For 100 msec pulses of 1 mM glutamate, the decay time constant (designated as "desensitization τ") was between 9.3 and 11.3 msec. Thus, offset kinetics were by a factor of about four faster than desensitization. For a pulse of intermediate length (10 msec), the current initially decayed slowly as for the long pulse; when the agonist was removed, it became faster, and the decay time constant was then similar to the offset τ (FIG. 3).

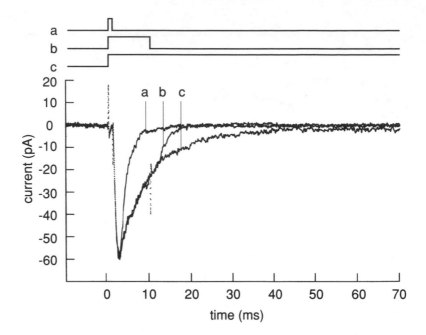

FIGURE 3. Comparison of offset time constant and desensitization time constant of the AMPA/ KA component. CA3 pyramidal cell patch, membrane potential was – 50 mV. Responses to 1 msec, 10 msec, and 100 msec pulses of 1 mM glutamate are superimposed; single traces. Extracellular solution was glycine free and contained 1 mM Mg. Filter frequency 3 kHz. Note that the decay at the end of the 10 msec pulse is faster than the initial decay. (From Colquhoun *et al.*[3] With permission.)

The offset τ was not measurably dependent on the concentration of glutamate during the pulse.[3] The desensitization τ decreased with increasing glutamate concentration in the range from 100 μM to 3 mM,[2] although the concentration dependence was rather weak. Therefore, offset kinetics were faster than desensitization over a wide range of glutamate concentrations. Comparison of offset and desensitization τ with the decay of synaptic currents in hippocampal neurons shows that the fastest measured EPSC decay time constants are very close to the values of the offset τ. Even the slowest EPSCs decay more rapidly than the desensitization τ. This implies that the glutamate pulse in the synaptic cleft during excitatory synaptic transmission must be relatively brief, about 1 msec or only slightly longer than this. Such a result would be expected if there is free diffusion of transmitter in the synaptic cleft.[17]

DESENSITIZATION OF AMPA/KA TYPE GʟᴜR CHANNELS: BRIEF-PULSE DESENSITIZATION AND EQUILIBRIUM DESENSITIZATION

The results described in the previous section suggested that offset kinetics rather than desensitization may determine the decay time course of an EPSC. However, this does not mean that desensitization of GluR channels does not play a role in excitatory synaptic transmission, because a fraction of channels could enter the desensitized state(s) during or after a brief pulse of glutamate applied to the receptors. To quantify the amount of this "brief-pulse desensitization," double-pulse experiments were performed, one 1-msec glutamate pulse being followed by a second one after variable intervals. The second pulse thus tested the amount of desensitization caused by the first one. For intervals longer than 20 msec, the response to the second pulse was smaller as compared to the first. Back-extrapolation (towards brief recovery intervals) revealed that about 50% of GluR channels of the AMPA/KA type were desensitized after a 1 msec pulse of 1 mM glutamate. The recovery from brief-pulse desensitization occurred with a time constant of about 50 msec in CA3 and CA1 pyramidal cell patches and with two time constants of 33 and 450 msec in dentate gyrus granule cell patches.[3] Slow recovery from desensitization might thus play a role in synaptic transmission when the presynaptic neurons fire at high frequencies. Under these conditions, desensitization of postsynaptic GluRs may contribute to synaptic depression.[18]

It has been reported that the concentration of ambient glutamate in the cerebrospinal fluid is between 1 and 3 μM.[19,20] Moreover, thermodynamic considerations suggest that the equilibrium concentration of extracellular glutamate generated by glutamate uptake mechanisms is in the same range.[21] Since desensitization occurs in the maintained presence of glutamate, it is possible that *in vivo* a significant fraction of GluR channels might be desensitized by ambient glutamate. To address this question experimentally, prepulse experiments were performed. One msec pulses of 1 mM glutamate were preceded by 30 sec prepulses of micromolar concentrations of glutamate, and it was tested if the response with the prepulse was reduced as compared to the response without prepulse. Surprisingly low concentrations of glutamate caused desensitization of AMPA/KA type GluR channels at equilibrium; the half-maximal inhibitory concentrations were between 2.4 and 9.6 μM for the three principal types of neurons of the hippocampal circuit.[3] Therefore, under physiological conditions, a considerable fraction of AMPA/KA type GluR channels may be desensitized. Moreover, during ischemia the ambient glutamate concentration in the CNS rises to values as high as 10 μM,[19] and a major fraction of AMPA/KA type GluR channels would become desensitized. This suggests a (probably protective) role of desensitization under pathophysiological conditions.

MOLECULAR IDENTIFICATION OF NATIVE GluR CHANNELS

The functional properties of native GluR channels described above show many similarities to recombinant receptors of the AMPA/KA family (assembled from GluR-1 to -4 or, equivalently, GluR-A to -D subunits) and the NMDA family (assembled from NR1 + NR2A to NR2D subunits), respectively.[22–24] To obtain further insight into the molecular composition of native receptors, it was necessary to correlate electrophysiological properties of native channels, functional properties of recombinant channels, and subunit expression patterns. This task was, however, complicated by the fact that many subunits are coexpressed in different cell types and that they may form mosaics of heteromeric channels. Moreover, subunits of one family often cannot be distinguished electrophysiologically because they have similar functional properties. It is therefore only possible to identify dominant subunits.

Among all recombinant GluR channels characterized so far, the most striking subunit-specific differences have been found for the AMPA/KA-GluR-B subunit as compared to other subunits of the AMPA/KA family. One parameter of channels assembled in expression systems that has been studied extensively was the shape of the macroscopic steady-state current-voltage relationship (I-V). Homomeric channels assembled from AMPA/KA-GluR-B channels are characterized by outward rectification, whereas channels assembled from AMPA/KA-GluR-A, -C, or -D subunits have doubly-rectifying I-V relations.[7] Another well-characterized parameter of the ionic channels formed by these subunits is the Ca permeability. Homomeric channels assembled from AMPA/KA-GluR-B subunits have low Ca permeability ($P_{Ca}/P_{Cs} < 0.1$), whereas channels assembled from AMPA/KA-GluR-A, -C, or -D subunits show high Ca permeability ($P_{Ca}/P_{Cs} > 1$).[8] In heteromeric combinations, the GluR-B subunit dominates these functional properties.[8,22,23]

HIPPOCAMPAL NEURONS: "AMPA/KA-GluR-B CELL TYPES"

In situ hybridization demonstrated that hippocampal neurons strongly express GluR subunits of the AMPA/KA family,[6] and that all of them express the GluR-B subunit. We therefore determined rectification properties and Ca permeability of AMPA/KA GluR channels in the three principal hippocampal neurons of rat brain slices. FIGURE 4 shows, as an example, results from CA3 pyramidal cell patches. With physiological extracellular solution on the outer side of the membrane, the steady-state current activated by KA showed outward rectification and reversed at potentials close to 0 mV, as expected for a nonselective cationic conductance. With a high-Ca solution on the outer side of the membrane, the reversal potential shifted to negative values, indicating that the Ca permeability is small ($P_{Ca}/P_{Cs} < 0.1$).[2,3] Similar results were obtained for dentate gyrus granule cell and CA1 pyramidal cell patches. This suggests that the GluR-B subunit—as predicted from the properties of recombinant channels and the expression pattern—dominates the functional properties of the native channels in these cells. It also seems likely that there is only limited Ca influx through postsynaptic AMPA/KA receptors during excitatory synaptic transmission if somatic and postsynaptic receptors have the same subunit composition. In contrast, doubly-rectifying I-V relations and high Ca permeability have been found in a subset of cultured hippocampal neurons,[25,26] but the origin and the expression pattern of these cells is not known yet.

BERGMANN GLIAL CELLS: "AMPA/KA-NON-GluR-B CELL TYPE"

Uniquely in the whole CNS, Bergmann glial cells in the cerebellum lack the expression of the AMPA/KA-GluR-B subunit. The first evidence that Bergmann glial cells do not express the GluR-B subunit, but only the GluR-A and GluR-D subunit, came from *in situ* hybridization studies.[4,6] Recently, studies with subunit-specific antibodies confirmed these findings.[27] *In situ* hybridization and antibody staining made clear predictions for the electrophysiological properties. This allowed us to test the

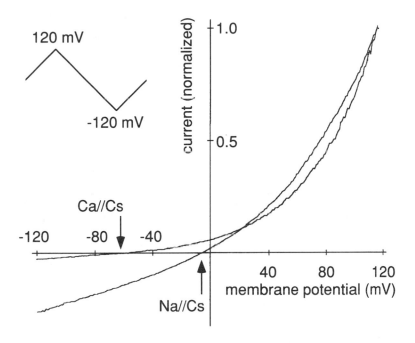

FIGURE 4. Outward rectification and low Ca permeability of AMPA/KA type GluR channels in hippocampal neurons. Ramp (2 sec) from -120 to $+120$ mV. Physiological extracellular solution and 100 mM $CaCl_2$, respectively. Data from two different CA3 cell patches; current was normalized to the value at $+120$ mV. The reversal potential in physiological extracellular solution was close to 0 mV, the reversal potential in 100 mM $CaCl_2$ was -62 mV. Filter frequency 1.5 kHz. (Modified from Jonas & Sakmann.[2])

hypothesis that the native channels are assembled from subunits of the AMPA/KA family in a more rigorous way than previously possible. We therefore studied GluR channels in cultured fusiform cerebellar glial cells presumably derived from Bergmann glia.[4,28] It was demonstrated that these cultured cells lacked the expression of the AMPA/KA-GluR-B subunit, as did the native cells.

The electrophysiological properties of the GluR channels in these cerebellar glial cells were characterized by a doubly-rectifying I-V relation and a high Ca permeability $(P_{Ca}/P_{Cs} = 1.44)$.[4] The properties of the KA-activated current were almost indistinguishable from that of a GluR-A/D heteromeric recombinant channel. Influx of Ca ions into Bergmann glial cells after application of KA has also been demonstrated using imaging techniques.[29] The functional role of these Ca-permeable AMPA/KA receptors is, however, still unclear. It is conceivable that glutamate released from excitatory synapses made by the parallel-fiber nerve endings on Purkinje cells could spill over from the synaptic cleft and activate GluR channels on Bergmann glial cells, which sheathe these synapses. Thereby, second messenger cascades in glial cells could be triggered.

The results presented in this paper strongly suggested that native AMPA/KA type

GluR channels were assembled from subunits of the GluR-A to -D family and that the presence or absence of the GluR-B subunit determined their likely most important functional properties: rectification and Ca permeability.

CONCLUSIONS

Fast application techniques allow one to mimic release of glutamate from presynaptic terminals. Brief pulses of a high concentration of glutamate applied to excised membrane patches can simulate fast AMPA/KA as well as slow NMDA receptor-mediated components of EPSCs. It is likely that closure of AMPA/KA type channels after removal of glutamate rather than desensitization determines the decay time course of an EPSC. Brief-pulse desensitization and equilibrium desensitization may, however, play a role in setting the amplitude of synaptic currents. The presence of the AMPA/KA-GluR-B subunit seems to be a critical factor for the functional properties of native AMPA/KA type GluR channels. Hippocampal neurons express the GluR-B subunit; AMPA/KA type GluRs in these cells show outwardly rectifying steady-state I-V relations and low Ca permeability. Bergmann glial cells do not express the GluR-B subunit; AMPA/KA type GluRs in these cells show doubly-rectifying I-V relations and high Ca permeability. The differential expression of the AMPA/KA-GluR-B subunit may thus determine the Ca permeability of AMPA/KA receptors in various types of cells in the CNS.

ACKNOWLEDGMENTS

I thank Prof. B. Sakmann for generous support and Drs. M. Häusser and A. Villarroel for critically reading the manuscript.

REFERENCES

1. EDWARDS, F. A., A. KONNERTH, B. SAKMANN & T. TAKAHASHI. 1989. A thin slice preparation for patch clamp recordings from neurones of the mammalian central nervous system. Pflügers Arch. 414: 600–612.
2. JONAS, P. & B. SAKMANN. 1992. Glutamate receptor channels in isolated patches from CA1 and CA3 pyramidal cells of rat hippocampal slices. J. Physiol. 455: 143–171.
3. COLQUHOUN, D., P. JONAS & B. SAKMANN. 1992. Action of brief pulses of glutamate on AMPA/kainate receptors in patches from different neurones of rat hippocampal slices. J. Physiol. 458: 261–287.
4. BURNASHEV, N., A. KHODOROVA, P. JONAS, P. J. HELM, W. WISDEN, H. MONYER, P. H. SEEBURG & B. SAKMANN. 1992. Calcium-permeable AMPA-kainate receptors in fusiform cerebellar glial cells. Science 256: 1566–1570.
5. HOLLMANN, M., A. O'SHEA-GREENFIELD, S. W. ROGERS & S. HEINEMANN. 1989. Cloning by functional expression of a member of the glutamate receptor family. Nature 342: 643–648.
6. KEINÄNEN, K., W. WISDEN, B. SOMMER, P. WERNER, A. HERB, T. A. VERDOORN, B. SAKMANN & P. H. SEEBURG. 1990. A family of AMPA-selective glutamate receptors. Science 249: 556–560.
7. VERDOORN, T. A., N. BURNASHEV, H. MONYER, P. H. SEEBURG & B. SAKMANN. 1991. Structural determinants of ion flow through recombinant glutamate receptor channels. Science 252: 1715–1718.

8. BURNASHEV, N., H. MONYER, P. H. SEEBURG & B. SAKMANN. 1992. Divalent ion permeability of AMPA receptor channels is dominated by the edited form of a single subunit. Neuron 8: 189–198.

9. DINGLEDINE, R., L. M. BOLAND, N. L. CHAMBERLIN, K. KAWASAKI, N. W. KLECKNER, S. F. TRAYNELIS & T. A. VERDOORN. 1988. Amino acid receptors and uptake systems in the mammalian central nervous system. Crit. Rev. Neurobiol. 4: 1–96.

10. JONAS, P., G. MAJOR & B. SAKMANN. 1993. Quantal analysis of unitary EPSCs at the mossy fiber synapse on CA3 pyramidal cells of rat hippocampus. J. Physiol. In press.

11. KELLER, B. U., A. KONNERTH & Y. YAARI. 1991. Patch clamp analysis of excitatory synaptic currents in granule cells of rat hippocampus. J. Physiol. 435: 275–293.

12. HESTRIN, S., R. A. NICOLL, D. J. PERKEL & P. SAH. 1990. Analysis of excitatory synaptic action in pyramidal cells using whole-cell recording from rat hippocampal slices. J. Physiol. 422: 203–225.

13. BEKKERS, J. M. & C. F. STEVENS. 1989. NMDA and nonNMDA receptors are co-localized at individual excitatory synapses in cultured rat hippocampus. Nature 341: 230–233.

14. STERN, P., F. A. EDWARDS & B. SAKMANN. 1992. Fast and slow components of unitary EPSCs on stellate cells elicited by focal stimulation in slices of rat visual cortex. J. Physiol. 449: 247–278.

15. FRANKE, C., H. HATT & J. DUDEL. 1987. Liquid filament switch for ultra-fast exchanges of solutions at excised patches of synaptic membrane of crayfish muscle. Neurosci. Lett. 77: 199–204.

16. JOHNSON, J. W. & P. ASCHER. 1987. Glycine potentiates the NMDA response in cultured mouse brain neurons. Nature 325: 529–531.

17. ECCLES, J. C. & J. C. JAEGER. 1958. The relationship between the mode of operation and the dimensions of the junctional regions at synapses and motor end-organs. Proc. R. Soc. Lond. B 148: 38–56.

18. LARKMAN, A., K. STRATFORD & J. JACK. 1991. Quantal analysis of excitatory synaptic action and depression in hippocampal slices. Nature 350: 344–347.

19. BENVENISTE, H., J. DREJER, A. SCHOUSBOE & N. H. DIEMER. 1984. Elevation of the extracellular concentrations of glutamate and aspartate in rat hippocampus during transient cerebral ischemia monitored by intracerebral microdialysis. J. Neurochemistry 43: 1369–1374.

20. LERMA, J., A. S. HERRANZ, O. HERRERAS, V. ABRAIRA & R. MARTIN DEL RIO. 1986. In vivo determination of extracellular concentration of amino acids in the rat hippocampus. A method based on brain dialysis and computerized analysis. Brain Res. 384: 145–155.

21. NICHOLLS, D. & D. ATTWELL. 1990. The release and uptake of excitatory amino acids. Trends Pharmacol. Sci. 11: 462–468.

22. NAKANISHI, N., N. A. SHNEIDER & R. AXEL. 1990. A family of glutamate receptor genes: evidence for the formation of heteromultimeric receptors with distinct channel properties. Neuron 5: 569–581.

23. SOMMER, B. & P. H. SEEBURG. 1992. Glutamate receptor channels: novel properties and new clones. Trends Pharmacol. Sci. 13: 291–296.

24. MONYER, H., R. SPRENGEL, R. SCHOEPFER, A. HERB, M. HIGUCHI, H. LOMELI, N. BURNASHEV, B. SAKMANN & P. H. SEEBURG. 1992. Heteromeric NMDA receptors: Molecular and functional distinction of subtypes. Science 256: 1217–1221.

25. IINO, M., S. OZAWA & K. TSUZUKI. 1990. Permeation of calcium through excitatory amino acid receptor channels in cultured rat hippocampal neurones. J. Physiol. 424: 151–165.

26. OZAWA, S., M. IINO & K. TSUZUKI. 1991. Two types of kainate response in cultured rat hippocampal neurons. J. Neurophysiol. 66: 2–11.

27. PETRALIA, R. S. & R. J. WENTHOLD. 1992. Light and electron immunocytochemical localization of AMPA-selective glutamate receptors in the rat brain. J. Comp. Neurol. 318: 329–354.

28. ORTEGA, A., N. ESHHAR & V. I. TEICHBERG. 1991. Properties of kainate receptor channels on cultured Bergmann glia. Neuroscience 41: 335–349.

29. MÜLLER, T., T. MÖLLER, T. BERGER, J. SCHNITZER & H. KETTENMANN. 1992. Calcium entry through kainate receptors and resulting potassium-channel blockade in Bergmann glial cells. Science 256: 1563–1566.

Molecular and Functional Diversity of the NMDA Receptor Channel[a]

MASAYOSHI MISHINA, HISASHI MORI,
KAZUAKI ARAKI, ETSUKO KUSHIYA,
HIROYUKI MEGURO, TATSUYA KUTSUWADA,
NOBUKO KASHIWABUCHI, KAZUTAKA IKEDA,
MICHIAKI NAGASAWA, MAKOTO YAMAZAKI,
HISASHI MASAKI, TOMOHIRO YAMAKURA,
TAKAO MORITA, AND KENJI SAKIMURA

Department of Neuropharmacology
Brain Research Institute
Niigata University
Asahimachi 1
Niigata 951, Japan

INTRODUCTION

The glutamate receptor (GluR) channel plays a key role in brain function by mediating most of the fast excitatory synaptic transmission in the central nervous system.[1] The heterogeneity of the GluR channel was suggested by electrophysiological and pharmacological studies, which led to the classification of the GluR channel into the α-amino-3-hydroxy-5-methyl-4-isoxazole propionic acid (AMPA), kainate, and N-methyl-D-aspartate (NMDA) receptors.[1,2] The NMDA receptor channel is highly permeable to Ca^{2+} but is blocked by Mg^{2+} in a voltage-dependent manner. These characteristics are essential for the NMDA receptor channel to mediate the induction of long-term potentiation of synaptic efficacy, which is thought to underlie memory acquisition and learning.[3] Furthermore, cumulative evidence suggests the involvement of the NMDA receptor channel in experience-dependent synaptic plasticity in the developing brain.[4] In addition, abnormal activation of the NMDA receptor channel may trigger neuronal cell death observed in various acute and chronic brain disorders.[5,6]

Recently, the molecular entity of the NMDA receptor channel has been revealed by cloning and expression of the subunit cDNAs. The NMDA receptor channel

[a] This work was supported in part by research grants from the Ministry of Education, Science and Culture of Japan, the Institute of Physical and Chemical Research, the Ministry of Health and Welfare of Japan, and the Mochida Memorial Foundation.

subunits show significant amino acid sequence homology with the subunits of the AMPA- or kainate-selective GluR channel and thus are the members of the GluR channel subunit family. Clear differences in pharmacological and electrophysiological properties and regulation have been found among diverse NMDA receptor channel subunits. Furthermore, distinct distributions of the NMDA receptor channel subunit mRNAs in the brain have been visualized by in situ hybridization analyses using cloned cDNAs as probes. These studies have clearly shown the molecular and functional diversity of the NMDA receptor channel and provide solid bases for future investigation of higher brain functions and neurological disorders.

MOLECULAR DIVERSITY

In an attempt to understand the molecular basis of higher brain functions, we have identified numbers of mouse GluR channel subunits by successive screening of mouse brain cDNA libraries under low stringency conditions with α1 and α2 subunit cDNAs[7] and newly identified subunit cDNAs as probes.[8-14] The entire primary structures of these subunits were deduced by nucleotide sequence analysis of the cloned cDNAs (FIG. 1). We have classified these subunits into six subfamilies according to the amino acid sequence identity (TABLE 1); the rat counterparts of the α, β, γ, ε, and ζ subunits are differently named by the groups of Heinemann,[15-19] Seeburg,[20-23] and Nakanishi.[24] Amino acid sequence identity is as high as ~ 40% to ~ 70% within a subfamily, but is as low as ~ 10% to ~ 40% between subfamilies. The presence of variants produced most likely by alternative splicing is noted for some subunits.

GluR channel subunits possess four hydrophobic putative transmembrane segments (M1-M4) and a putative signal peptide at the amino terminus. These structural characteristics are common for neurotransmitter-gated ion channels, the nicotinic acetylcholine receptor channel, the GABA receptor channel, and the glycine receptor channel. Segment M2 is assumed to constitute the inner wall of the ion channel, analogous to the acetylcholine receptor channel. Highly conserved regions among GluR channel subunits are putative transmembrane segments, the region preceding segment M1 and the regions between segments M3 and M4. The region preceding segment M1 contains potential N-glycosylation sites and putative agonist binding sites.[7,25] Point mutations introduced into these regions have been found to affect dose-response relationships for agonists of the AMPA-selective GluR channel.[26,27] The putative cytoplasmic domain between segments M3 and M4 contains potential phosphorylation sites for Ca^{2+}/calmodulin-dependent protein kinase and protein kinase C, which have been suggested to mediate the induction and maintainance of long-term potentiation.[28]

Functional properties of the respective subunits were examined after expression in Xenopus oocytes by injection of the subunit-specific mRNAs synthesized in vitro using cloned cDNAs as templates. The α1 subunit alone forms functional GluR channels, but the current responses are much larger when the α1 and α2 subunits are expressed together.[7,15,16,20] The increases in channel activity and Hill coefficient values for the dose-response curves for agonists indicate a positive cooperative interaction between the α1 and α2 subunits. The apparent affinities for agonists of the α1/α2 heteromeric channel as well as the α1 homomeric channel are quisqualate > AMPA > L-glutamate > kainate. Thus, the α subfamily represents the AMPA-selective GluR channel. The

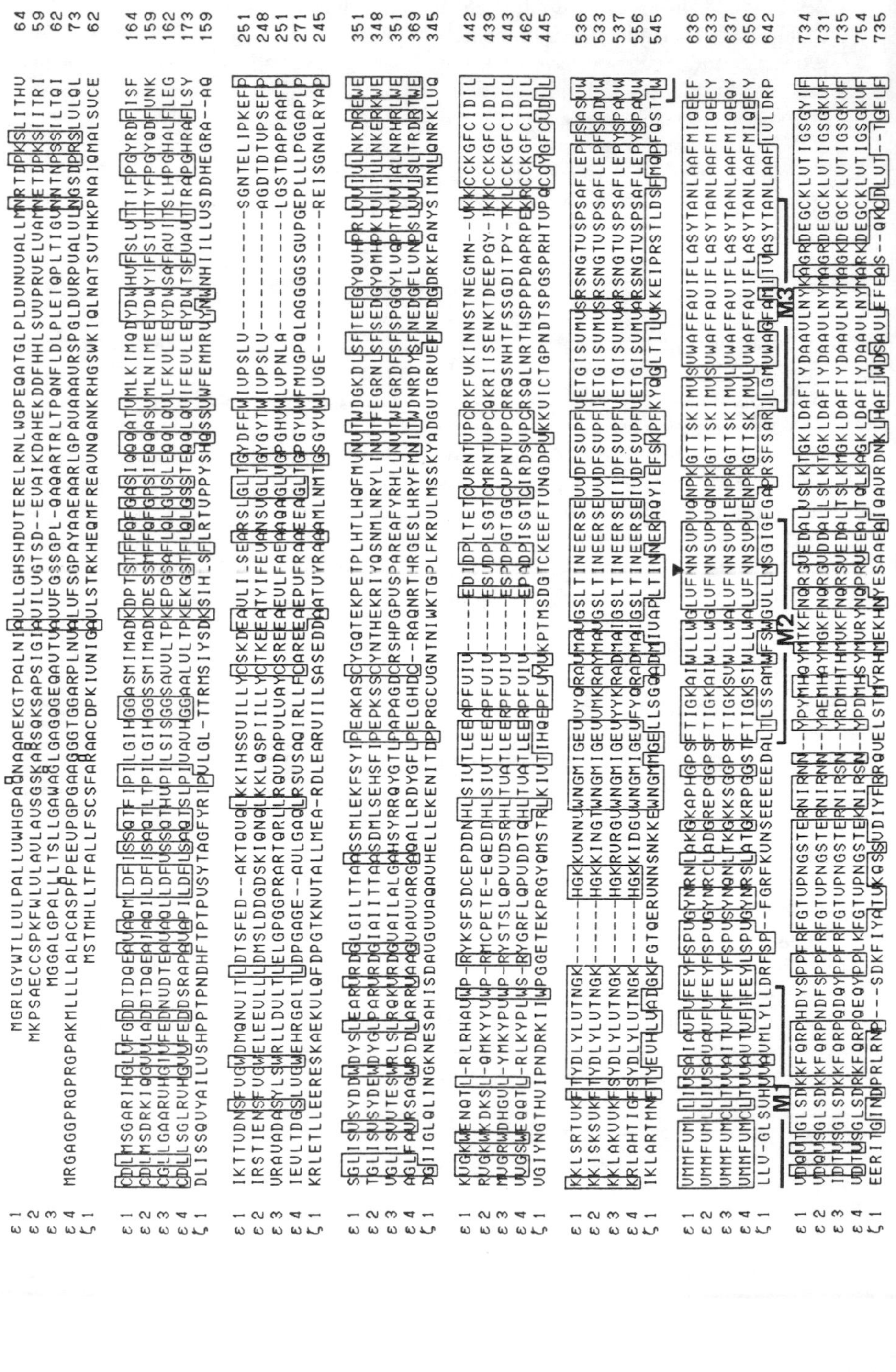
M1
M2
M3

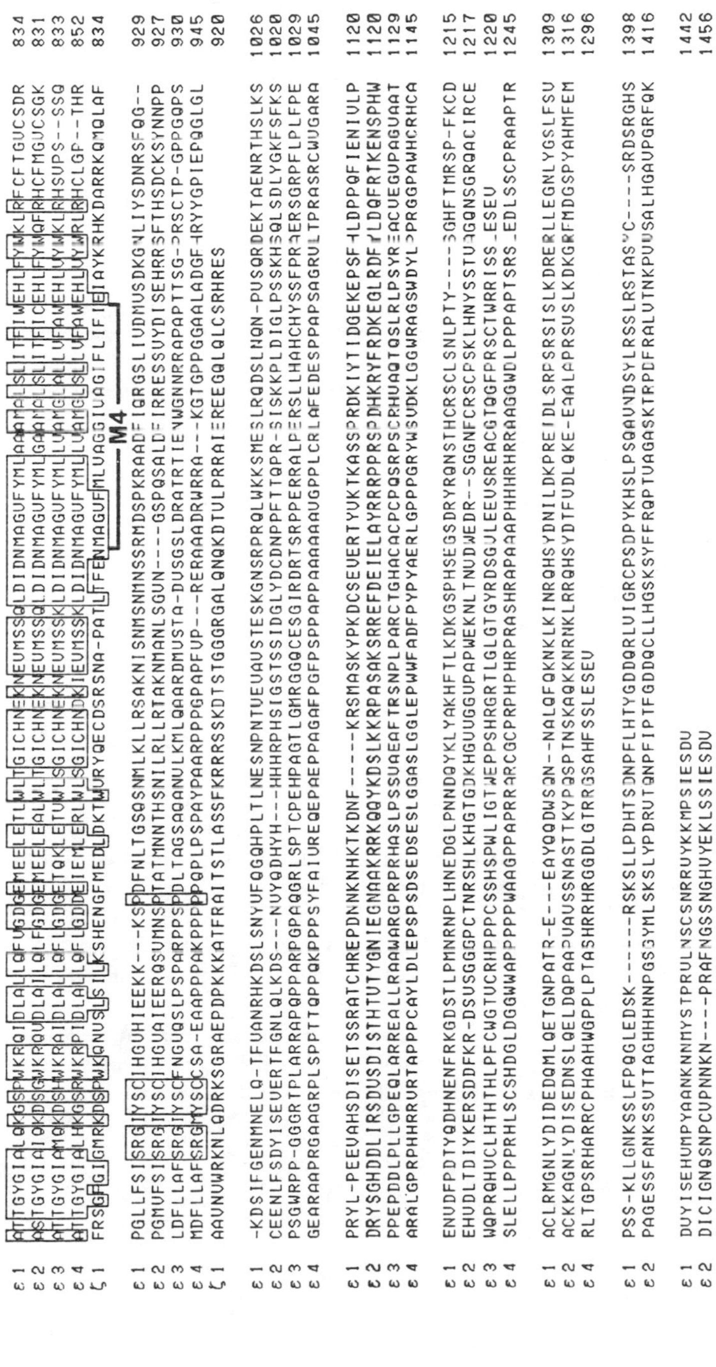

FIGURE 1. Alignment of the deduced amino acid sequences of the ε1, ε2, ε3, ε4, and ζ1 subunits of the mouse NMDA receptor channel. Amino acid residues are numbered beginning with the amino terminal residue of the proposed mature subunit (*open squares*). Numbers of the amino acid residues at the end of the individual lines are given. Sets of identical amino acid residues among five subunits or four ε subunits are enclosed. The putative transmembrane segments (M1-M4) are indicated. Asparagine residues involved in Mg^{2+} block are marked with an arrowhead. (From references 11-14.)

TABLE 1. Subunit Families of the Rodent GluR Channels

| Subfamily | Subunit | | Function |
	Mouse	Rat	
α	α1, α2, α3, α4	GluR 1, 2, 3, 4 (GluR A, B, C, D)	AMPA-selective
β	β1, β2, β3	GluR 5, 6, 7	Kainate-selective
γ	γ1, γ2	KA-1, KA-2	Kainate-selective
δ	δ1	–	Unknown
ε	ε1, ε2, ε3, ε4	NR2A, 2B, 2C	NMDA
ζ	ζ1	NMDAR1	NMDA

β2 subunit forms homomeric channels responsive to L-glutamate, kainate, and quisqualate.[8,9,18] The γ2 subunit, when expressed together with the β2 subunit, forms GluR channels selective for kainate.[9,22] The order of potency is kainate > L-glutamate ≈ quisqualate. Co-expression of the γ2 subunit increases channel activity several fold and the apparent affinity for L-glutamate and quisqualate, compared with the β2 subunit expressed alone. Therefore, the members of the β and γ subfamilies constitute the kainate-selective GluR channel.

The ζ1 subunit forms homomeric channels responsive to L-glutamate, L-aspartate, NMDA, and quisqualate in the presence of glycine.[10,24] However, highly active NMDA receptor channels are observed only when the ζ1 subunit is expressed together with one of four ε subunits, suggesting the heteromeric nature of the NMDA receptor channel (FIG. 2).[12-14,23] The responses to 10 μM L-glutamate plus 10 μM glycine of the ε1/ζ1, ε2/ζ1, ε3/ζ1, and ε4/ζ1 heteromeric channels were suppressed by 100 μM D-2-amino-5-phosphonovalerate (APV), a specific competitive antagonist of the NMDA receptor, and 30 μM 7-chlorokynurenate (7CK), reported to be a competitive antagonist for the glycine modulatory site of the NMDA receptor (FIG. 3). The responses of the ε1/ζ1, ε2/ζ1, ε3/ζ1, and ε4/ζ1 channels were inhibited also by non-competitive antagonists of the NMDA receptor channel, such as 100 μM Mg^{2+}, 100 μM Zn^{2+}, and 1 μM (+)-MK-801. The effects of the channel blockers were apparently weaker for the ε3/ζ1 and ε4/ζ1 channels than for the ε1/ζ1 and ε2/ζ1 channels. The ε1/ζ1, ε2/ζ1, ε3/ζ1, and ε4/ζ1 channels exhibited clear inward currents in Na^+- and K^+-free Ringer's solution containing 20 mM Ca^{2+}, whereas a marginal outward current was observed in control Na^+- and K^+-free Ringer's solution, indicating that the heteromeric channels are permeable to Ca^{2+}. The finding of the presence of four ε subunits indicates the diversity of the NMDA receptor channel at the molecular level.

FUNCTIONAL DIVERSITY

The question then arises whether the molecular diversity of the NMDA receptor channel results in the functional diversity. Pharmacological properties of the ε1/ζ1, ε2/ζ1, ε3/ζ1, and ε4/ζ1 channels were examined by analyzing the dose-response

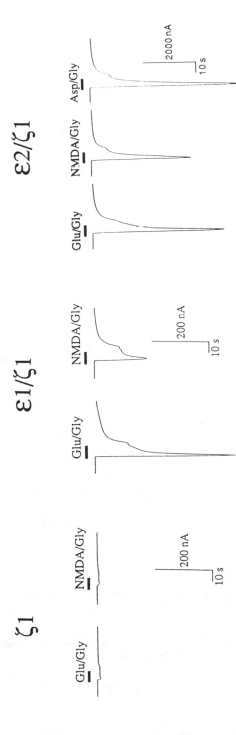

FIGURE 2. Current responses of the ζ1, ε1/ζ1, and ε2/ζ1 channels to 10 μM L-glutamate plus 10 μM glycine and 100 μM NMDA plus 10 μM glycine. The homomeric or heteromeric NMDA receptor channels were expressed in *Xenopus* oocytes by injecting the subunit-specific mRNAs synthesized *in vitro* from cloned cDNAs. Current responses were measured in normal frog Ringer's solution at −70 mV membrane potential. Inward current is downward. (Modified from references 12 and 13.)

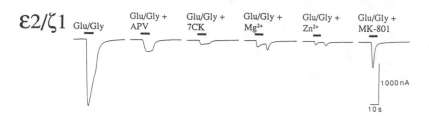

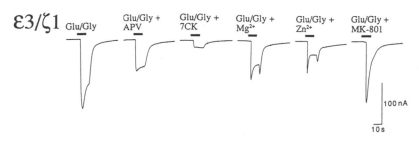

FIGURE 3. Effects of competitive and non-competitive antagonists on the current responses of the ε2/ζ1 and ε3/ζ1 channels to 10 μM L-glutamate plus 10 μM glycine. (Modified from reference 13.)

relationships for L-glutamate and glycine in Ba^{2+}-Ringer's solution to minimize the effect of secondarily activated Ca^{2+}-dependent Cl^- currents (FIG. 4). The EC_{50} values for L-glutamate were 1.7 μM, 0.8 μM, 0.7 μM, and 0.4 μM for the ε1/ζ1, ε2/ζ1, ε3/ζ1, and ε4/ζ1 channels, respectively, whereas those for glycine were 2.1 μM, 0.3 μM, 0.2 μM, and 0.09 μM, respectively. Furthermore, the sensitivities of the ε1/ζ1, ε2/ζ1, ε3/ζ1, and ε4/ζ1 channels to competitive antagonists were compared at the agonist concentrations 10 times as high as the EC_{50} values (FIG. 5). The sensitivity to APV is ε1/ζ1 > ε2/ζ1 > ε3/ζ1 > ε4/ζ1 channels, whereas the sensitivity to 7CK is ε3/ζ1 > ε2/ζ1 > ε1/ζ1 ≈ ε4/ζ1 channels. These results suggest that pharmacologically distinct NMDA receptor channels can be formed depending on the constituting ε subunit. In accord with these findings, the regional variation in ligand-binding properties of the NMDA receptor has been noted in the brain.[29]

Mg^{2+} at the concentrations of 0.1 mM and 1 mM exerted an inhibitory effect on the ε/ζ heteromeric NMDA receptor channels in a voltage-dependent manner (FIG. 6). A clear difference, however, was found among the heteromeric channels. The ε3/ζ1 channel was rather resistant to Mg^{2+} blocking, being active even at −70 mV and −100 mV membrane potentials in the presence of 1 mM Mg^{2+}, whereas both the ε1/ζ1 and ε2/ζ1 channels were strongly suppressed under these conditions.

DISTINCT DISTRIBUTIONS OF THE SUBUNIT mRNAs IN THE BRAIN

Distributions of the subunit mRNAs in the adult mouse brain were examined by *in situ* hybridization analyses.[12,13,30] The ε1 subunit mRNA is distributed widely in

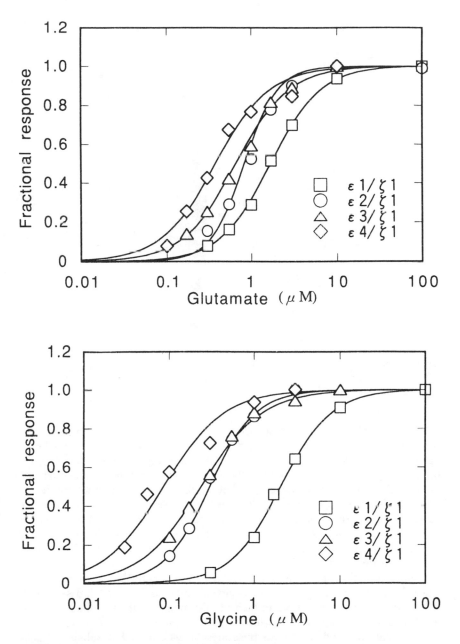

FIGURE 4. Dose-response curves for L-glutamate in the presence of 10 μM glycine (top) and for glycine in the presence of 10 μM L-glutamate (bottom) of the ε1/ζ1, ε2/ζ1, ε3/ζ1, and ε4/ζ1 channels. (From references 13 and 14.)

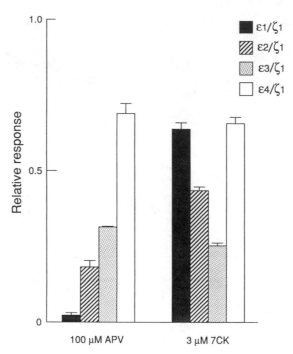

FIGURE 5. Effects of APV and 7CK on the responses to L-glutamate plus glycine of the $\varepsilon1/\zeta1$, $\varepsilon2/\zeta1$, $\varepsilon3/\zeta1$, and $\varepsilon4/\zeta1$ channels. The agonist concentrations were ~ tenfold of the EC_{50} values of the respective heteromeric channels. (From references 13 and 14.)

the brain, but the level of expression is higher in the cerebral cortex and the hippocampal formation. In contrast, the $\varepsilon2$ subunit mRNA is expressed selectively in the forebrain. High levels of expression are observed in the cerebral cortex, the hippocampal formation, the septum, the caudate-putamen, the olfactory bulb, and the thalamus. The $\varepsilon3$ subunit mRNA is found predominantly in the cerebellum. Strong expression is observed in the granule cell layer of the cerebellum, while weak expression is detected in the olfactory bulb and the thalamus. Low levels of the $\varepsilon4$ subunit mRNA are found in the midbrain, the thalamus, and the olfactory bulb. In contrast to the characteristic distributions of the four ε subunit mRNAs, the $\zeta1$ subunit mRNA distribution is ubiquitous in the brain.

Furthermore, the expressions of the respective ε subunit mRNAs are differentially regulated during development.[30] In the embryonic brain, only the $\varepsilon2$ and $\varepsilon4$ subunit mRNAs are expressed. In contrast to the wide distribution of the $\varepsilon2$ subunit mRNA, the $\varepsilon4$ subunit mRNA is found exclusively in the diencephalon and the brainstem. During the first two weeks after birth, the expression patterns of the ε subunit mRNAs change drastically. The $\varepsilon1$ subunit mRNA appears in the entire brain and the $\varepsilon3$ subunit mRNA in the cerebellum. In contrast, the expression of the $\varepsilon2$ subunit mRNA becomes restricted in the forebrain and that of the $\varepsilon4$ subunit mRNA is almost diminished. The $\zeta1$ subunit mRNA is ubiquitously expressed in the brain throughout

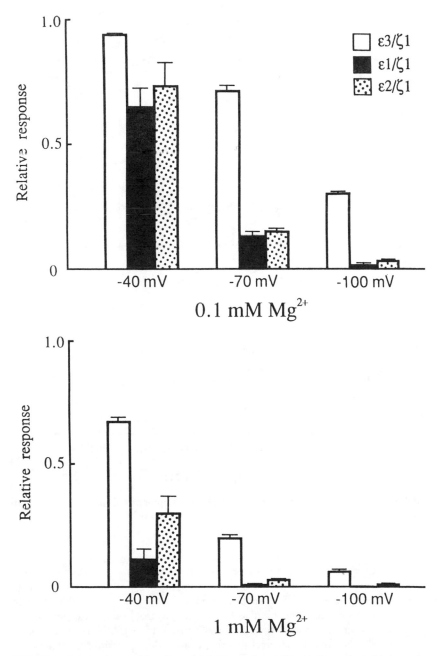

FIGURE 6. Effects of 0.1 mM and 1 mM Mg^{2+} on the $\varepsilon 1/\zeta 1$, $\varepsilon 2/\zeta 1$, and $\varepsilon 3/\zeta 1$ channels at various membrane potentials. (From reference 13.)

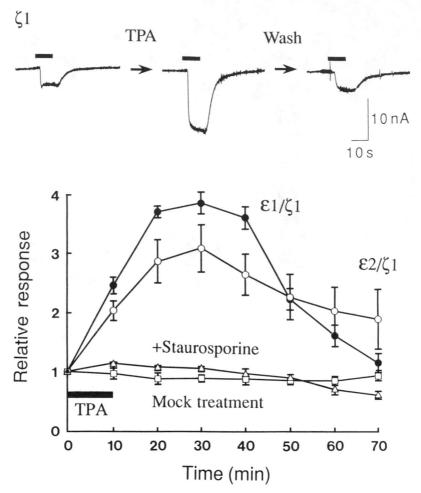

FIGURE 7. Effect of TPA treatment on the $\zeta 1$ homomeric (top) and $\varepsilon 1/\zeta 1$ and $\varepsilon 2/\zeta 1$ heteromeric NMDA receptor channels (bottom). (From references 11 and 13.)

developmental stages. These results suggest that the subunit composition of the NMDA receptor channel alters drastically during development, which would result in changes of the functional properties of the NMDA receptor channel. In accord with this notion, it has been reported that the pharmacological and electrophysiological properties of the NMDA receptor channel, such as the sensitivity to Mg^{2+} block, the apparent affinity for glycine, and the channel kinetics, change during brain development.[31-33] A precedent for such developmental regulation is the changes of the functional properties the muscle acetylcholine receptor channel based on the switch of the subunit composition during neuromuscular junction formation.[34]

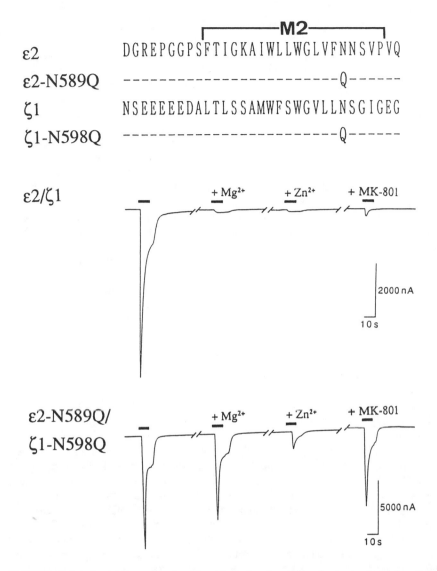

FIGURE 8. Substitution mutations introduced into the ε2 and ζ1 subunits (top) and the effects of the mutations on the sensitivities to non-competitive antagonists (bottom). Amino acid sequences of the mutagenized region encompassing putative transmembrane segment M2 are shown. Bars represent residues identical to those above in the aligned position. The heteromeric ε2/ζ1 and ε2-N589Q/ζ1-N598Q channels were expressed in *Xenopus* oocytes and current responses to 10 μM L-glutamate plus 10 μM glycine were measured in normal frog Ringer's solution at − 70 mV membrane potential. Effects of non-competitive antagonists were examined in Ringer's solution containing 1 mM MgCl₂, 100 μM ZnCl₂, or 1 μM (+)-MK-801. The sensitivity to MK-801 was evaluated after two successive applications. (Modified from reference 36.)

REGULATION OF THE CHANNEL ACTIVITY

The $\zeta 1$ homomeric NMDA receptor channel expressed in *Xenopus* oocytes is transiently potentiated by treatment with 1 μM 12-*O*-tetradecanoyl phorbol 13-acetate (TPA) (FIG. 7). This enhancing effect cannot be nonspecific because the TPA treatment exerts no appreciable effect on the current response of the AMPA-selective $\alpha 1/\alpha 2$ GluR channel. Furthermore, the effect of TPA is abolished by the simultaneous treatment with 5 μM staurosporine, a protein kinase inhibitor.

The effects on heteromeric channels were examined by measuring responses to 10 μM L-glutamate plus 10 μM glycine in Ba^{2+}-Ringer's solution. Treatment of oocytes with 1 μM TPA for 10 min potentiated the responses of the $\epsilon 2/\zeta 1$ and $\epsilon 1/\zeta 1$ heteromeric channels (FIG. 7), but not the response of the $\epsilon 3/\zeta 1$ channel. No significant enhancement of the channel activity was observed by mock treatment or TPA treatment in the presence of 5 μM staurosporine. Failure of potentiation in the presence of staurosporine suggests the involvement of protein kinases in the TPA effect. These observations raise an intriguing possibility that NMDA receptor channels can be positively modulated by protein kinases. The level of NMDA receptor channel activity may hence be regulated by various stimuli leading to activation of protein kinases, for example through G protein-coupled receptors. Such modulation may play a role in determining the threshold of the induction of long-term potentiation, since Ca^{2+} entry through the NMDA receptor channel triggers the persistent change in the efficacy of synaptic transmission.[1,3]

IDENTIFICATION OF THE Mg^{2+} BLOCK SITE

The AMPA-selective $\alpha 1/\alpha 2$ GluR channel is essentially impermeable to Ca^{2+}, like most of native AMPA receptor channels, whereas the $\alpha 1$ homomeric channel is permeable to Ca^{2+}.[26,35] Thus, the $\alpha 2$ subunit regulates the ion selectivity of the AMPA-selective GluR channel. Inspection of the amino acid sequence of the $\alpha 2$ subunit reveals the unique presence of arginine 586 in putative channel-forming segment M2. To examine whether this positively charged residue plays a role in selective cation permeation through the channel, we introduced a point mutation into the $\alpha 2$ subunit to substitute glutamate for arginine (mutation $\alpha 2$-R586Q).[26] The heteromeric channel composed of the wild-type $\alpha 1$ subunit and the mutant $\alpha 2$-R586Q subunit is highly permeable to Ca^{2+}. These results show that arginine 586 in segment M2 of the $\alpha 2$ subunit is a critical determinant of the Ca^{2+} permeability of the AMPA-selective GluR channel.

Voltage-dependent Mg^{2+} block is the key to the depolarization-dependent activation of the NMDA receptor channel, which is the basis of an activity-dependent change of synaptic efficacy, thought to underlie memory, learning, and development. All subunits of the NMDA receptor channel possess asparagine in putative transmembrane segment M2 at the position corresponding to glutamine/arginine of the α subunits that determine the Ca^{2+} permeability of the AMPA-selective GluR channel (FIG. 1). To test the idea that this asparagine residue causes the NMDA receptor channel to be sensitive to voltage-dependent Mg^{2+} block, we introduced substitution mutations into the $\epsilon 2$ and $\zeta 1$ subunits of the NMDA receptor channel (FIG. 8).

The wild-type $\varepsilon 2/\zeta 1$ channel is almost completely suppressed by the physiological concentration of Mg^{2+} in a voltage-dependent manner. Replacement by glutamine of the asparagine in either or both of the subunits strongly reduces the sensitivity to Mg^{2+} block of the $\varepsilon 2/\zeta 1$ heteromeric NMDA receptor channel (FIG. 8). Repetitive application of 1 μM MK-801 suppresses the wild-type $\varepsilon 2/\zeta 1$ channel. However, the heteromeric channel with the mutation on both subunits becomes resistant to MK-801. The sensitivity to Zn^{2+} is partly reduced by the point mutations.[36] These findings suggest that the conserved asparagine in segment M2 constitutes a Mg^{2+} block site and that the MK-801 site overlaps the Mg^{2+} site. There is strong evidence that Mg^{2+} produces a voltage dependent block of the channel by binding a site deep within the ionophore,[37] thus our results are consistent with the view that segment M2 constitutes the transmembrane ion channel of the NMDA receptor channel. Because the $\varepsilon 3/\zeta 1$ and $\varepsilon 4/\zeta 1$ channels are less sensitive to Mg^{2+} block than the $\varepsilon 1/\zeta 1$ and $\varepsilon 2/\zeta 1$ channels and the asparagine is conserved among these subunits, there are probably additional residues involved in interaction with Mg^{2+}. The observation that the sensitivity to Zn^{2+} is only slightly affected by the substitutions at the Mg^{2+} site is consistent with the proposal that Mg^{2+} and Zn^{2+} act on different sites of the NMDA receptor channel.[38] Although the main effect of Zn^{2+} on the NMDA receptor channel is voltage-insensitive, a small portion of the Zn^{2+} block is voltage dependent.[39,40] The slight decrease in the sensitivity to Zn^{2+} of the mutant channels can be explained if the voltage-dependent block by Zn^{2+} is caused by binding to the Mg^{2+} site.

CONCLUSION

We have identified more than a dozen subunits of the mouse GluR channel by molecular cloning. These subunits can be classified into six subfamilies according to the amino acid sequence homology. Functional analysis of GluR channel subunits expressed from the cloned cDNAs shows that this classification purely based on the structural similarity corresponds very well to the pharmacological properties (TABLE 1). When the amino acid sequences of the GluR channel subunits are compared, the asparagine constituting a Mg^{2+} block site of the NMDA receptor channel corresponds in position to the glutamine and arginine that determine the Ca^{2+} permeability of the AMPA-selective GluR channel.[26,41-43] This suggests that the channel structure motif is similar between the NMDA and non-NMDA receptor channels.

We have demonstrated that NMDA receptor channels with high activity are formed by co-expression of the distantly related ε and ζ subunits. The four ε subunits are homologous in primary structure, but are clearly distinct in distribution, functional properties, and regulation. In contrast to the wide distribution of the $\varepsilon 1$ and $\zeta 1$ subunit mRNAs in the brain, the $\varepsilon 2$ subunit mRNA is expressed selectively in the forebrain. The $\varepsilon 3$ subunit mRNA is found predominantly in the cerebellum, whereas the $\varepsilon 4$ subunit mRNA is weakly expressed in the brainstem and the diencephalon. It is remarkable that in the forebrain the distribution patterns of the $\varepsilon 1$ and $\varepsilon 2$ subunit mRNAs coincide well with those of the antagonist-preferring and agonist-preferring forms of the NMDA receptor channel,[34] respectively. Of the ε/ζ heteromeric NMDA receptor channels, the $\varepsilon 4/\zeta 1$ channel exhibits the highest apparent affinities for

L-glutamate and glycine, and the $\varepsilon1/\zeta1$ channel the lowest affinities. Furthermore, the $\varepsilon1/\zeta1$ channel is most sensitive to APV, whereas the $\varepsilon3/\zeta1$ channel is most sensitive to 7CK. The $\varepsilon4/\zeta1$ channel is least sensitive to both antagonists. Remarkably, the $\varepsilon3/\zeta1$ and $\varepsilon4/\zeta1$ channels are less sensitive to Mg^{2+} block than the $\varepsilon1/\zeta1$ and $\varepsilon2/\zeta1$ channels. Furthermore, we have found that the $\varepsilon1/\zeta1$ and $\varepsilon2/\zeta1$ channels, but not the $\varepsilon3/\zeta1$ and $\varepsilon4/\zeta1$ channels, can be positively modulated by TPA treatment. These findings suggest that the functional properties of the NMDA receptor channel are critically determined by the constituting ε subunit, and thus the molecular diversity of the ε subunit family underlies the functional heterogeneity of the NMDA receptor channel.

Because the activity-dependent change in synaptic efficacy mediated by the NMDA receptor channel is found in forebrain regions such as the hippocampus and the cerebral cortex,[1,3,4] it is likely that the $\varepsilon1$ and $\varepsilon2$ subunits expressed in the forebrain play a role in the synaptic plasticity. In accord with this assignment, the $\varepsilon1/\zeta1$ and $\varepsilon2/\zeta1$ channels are highly sensitive to voltage-dependent Mg^{2+} block, which is essential for the NMDA receptor channel to mediate the induction of the activity-dependent long-term potentiation of synaptic efficacy. It is of interest that the $\varepsilon1/\zeta1$ and $\varepsilon2/\zeta1$ channels, but not the $\varepsilon3/\zeta1$ and $\varepsilon4/\zeta1$ channels, are positively regulated by TPA treatment. Thus, modulation of the $\varepsilon1/\zeta1$ and $\varepsilon2/\zeta1$ channels by protein kinases may represent an important regulatory mechanism of the threshold of the induction of long-term potentiation, because Ca^{2+} entry through the NMDA receptor channel triggers the persistent change in the efficacy of synaptic transmission. Furthermore, the $\varepsilon2/\zeta1$ channel expressed from early stages of development would play an important role in experience-dependent synaptic plasticity in the developing brain. The predominant distribution of the $\varepsilon3$ subunit mRNA in the cerebellar granule cell layer and the weak sensitivity of the $\varepsilon3/\zeta1$ channel to Mg^{2+} block suggest that the $\varepsilon3$ subunit is a key component of cerebellar NMDA receptor channels mediating mossy fiber-granule cell synaptic transmission. Similarly, the $\varepsilon4$ subunit may mediate synaptic transmission at early stages of development. The proposed physiological roles of the respective ε subunits of the NMDA receptor channel could be tested *in vivo* through transgenic mouse techniques including gene targeting. Our cDNA clones of the mouse NMDA receptor channel subunits will provide valuable tools for future investigation of higher brain functions and brain disorders at the molecular level.

ACKNOWLEDGMENTS

We thank Misses Yuko Shibuya and Rie Natsume for help in the preparation of the manuscript.

REFERENCES

1. MAYER, M. L. & G. L. WESTBROOK. 1987. Prog. Neurobiol. **28:** 197–276.
2. MONAGHAN, D. T., R. J. BRIDGES & C. W. COTMAN. 1989. Annu. Rev. Pharmacol. Toxicol. **29:** 365–402.
3. COLLINGRIDGE, G. L. & T. V. P. BLISS. 1987. Trends Neurosci. **10:** 288–293.

4. McDonald, J. W. & M. V. Johnston. 1990. Brain Res. Rev. 15: 41–70.
5. Choi, D. W. 1988. Neuron 1: 623–634.
6. Olney, J. W. 1990. Annu. Rev. Pharmacol. Toxicol. 30: 47–71.
7. Sakimura, K., H. Bujo, E. Kushiya, K. Araki, M. Yamazaki, M. Yamazaki, H. Meguro, A. Warashina, S. Numa & M. Mishina. 1990. FEBS Lett. 272: 73–80.
8. Morita, T., K. Sakimura, E. Kushiya, M. Yamazaki, H. Meguro, K. Araki, T. Abe, J. K. Mori & M. Mishina. 1992. Mol. Brain Res. 14: 143–146.
9. Sakimura, K., T. Morita, E. Kushiya & M. Mishiina. 1992. Neuron 8: 267–274.
10. Yamazaki, M., K. Araki, A. Shibata & M. Mishina. 1992. Biochem. Biophys. Res. Commun. 183: 886–892.
11. Yamazaki, M., H. Mori, K. Araki, K. J Mori & M. Mishina. 1992. FEBS Lett. 300: 39–45.
12. Meguro, H., H. Mori, K. Araki, E. Kushiya, T. Kustuwada, M. Yamazaki, T. Kumanishi, M. Arakawa, K. Sakimura & M. Mishina. 1992. Nature 357:70–74.
13. Kustuwada, T., N. Kashiwabuchi, H. Mori, K. Sakimura, E. Kushiya, K. Araki, H. Meguro, H. Masaki, T. Kumanishi, M. Arakawa & M. Mishina. 1992. Nature 358: 36–41.
14. Ikeda, K., M. Nagasawa, H. Mori, K. Araki, K. Sakimura, M. Watanabe, Y. Inoue & M. Mishina. 1992. FEBS Lett. 313: 34–38.
15. Hollmann, M., A. O'Shea-Greenfield, S. W. Rogers & S. Heinemann. 1989. Nature 342: 643–648.
16. Boulter, J., M. Hollmann, A. O'Shea-Greenfield, M. Hartley, E. S. Deneris, C. Maro & S. Heinemann. 1990. Science 249: 1033–1037.
17. Bettler, B., J. Boulter, I. Hermans-Borgmeyer, A. O'Shea-Greenfield, E. S. Deneris, C. Moll, U. Borgmeyer, M. Hollmann & S. Heinemann. 1990. Neuron 5: 583–595.
18. Egebjerg, J., B. Bettler, I. Hermans-Borgmeyer & S. Heinemann. 1991. Nature 351: 745–748.
19. Bettler, B., J. Egebjerg, G. Sharma, G. Pecht, I. Hermans-Borgmeyer, C. Molli, C. F. Stevens, & S. Heinemann. 1992. Neuron 8: 257–265.
20. Keinänen, K., W. Wisden, B. Sommer, P. Werner, A. Herb, T. A. Verdoorn, B. Sakmann & P. H. Seeburg. 1990. Science 249: 556–560.
21. Werner, P., M. Voigt, K. Keinänen, W. Wisden & P. H. Seeburg. 1991. Nature 351: 742–744.
22. Herb, A., N. Burnashev, P. Werner, B. Sakmann, W. Wisden & P. H. Seeburg. 1992. Neuron 8: 775–785.
23. Monyer, H., R. Sprengel, R. Schoepfer, A. Herb, M. Higuchi, H. Lomeli, N. Burnashev, B. Sakmann & P. H. Seeburg. 1992. Science 256: 1217–1221.
24. Moriyoshi, K., M. Masu, T. Ishii, R. Shigemoto, N. Mizuno & S. Nakanishi. 1991. Nature 354: 31–37.
25. Nakanishi, N., N. A. Schneider & R. Axel. 1990. Neuron 5: 569–581.
26. Mishina, M., K. Sakimura, H. Mori, E. Kushiya, M. Harabayashi, S. Uchino & K. Nagahari. 1991. Biochem. Biophys. Res. Commun. 180: 813–821.
27. Uchino, S., K. Sakimura, K. Nagahari & M. Mishina. 1992. FEBS Lett. 308: 253–257.
28. Hu, G.-Y., Ø. Hvalby, S. I. Walaas, K. A. Albert, P. Skjeflo, P. Anderson & P. Greenyard. 1987. Nature 328: 426–429.
29. Monaghan, D. T. & K. J. Anderson. 1991. In Excitatory Amino Acids and Synaptic Transmission. H. Wheal & A. Thomson, Eds.: 33–45. Academic Press. San Diego.
30. Watanabe, M., Y. Inoue, K. Sakimura & M. Mishina. 1992. NeuroReport 3: 1138–1140.
31. Ben-Ari, Y., E. Cherubini & K. Krnjevic. 1988. Neurosci. Lett. 94: 88–92.
32. Kleckner, N. W. & R. Dingledine. 1991. Mol. Brain Res. 11: 151–159.
33. Hestrin, S. 1992. Nature 357: 686–689.
34. Mishina, M., T. Kakai, K. Imoto, M. Noda, T. Takahashi, S. Numa, C. Methfessel & B. Sakmann. 1986. Nature 321: 406–411.
35. Hollmenn, M., M. Hartley & S. Heinemann. 1991. Science 252: 851–853.
36. Mori, H., H. Masaki, T. Yamakura & M. Mishina. 1992. Nature 358: 673–675.

37. Ascher, P. & L. Nowak. 1988. J. Physiol. **399**: 247–266.
38. Westbrook, G. L. & M. L. Mayer. 1987. Nature **328**: 640–643.
39. Christine, C. W. & D. W. Chai. 1990. J. Neurosci. **10**: 108–116.
40. Legendre, P. & G. L. Westbrook. 1990. J. Physiol. **429**: 429–449.
41. Hume, R. I., R. Dingledine & S. Heinemann. 1991. Science **253**: 1028–1031.
42. Burnashev, N., H. Monyer, P. H. Seeburg & B. Sakmann. 1992. Neuron **8**: 189–198.
43. Burnashev, N., R. Schoepher, H. Monyer, J. P. Ruppersberg, W. Gunther, P. H. Seeburg & B. Sakmann. 1992. Science **257**: 1415–1419.

Molecular Characterization of NMDA and Metabotropic Glutamate Receptors

MASAYUKI MASU, YOSHIAKI NAKAJIMA,
KOKI MORIYOSHI, TAKAHIRO ISHII,
CHIHIRO AKAZAWA,
AND SHIGETADA NAKANASHI

Institute for Immunology
Kyoto University Faculty of Medicine
Kyoto 606, Japan

INTRODUCTION

Glutamate is a major excitatory neurotransmitter and plays an important role in many neuronal functions in the central nervous system (CNS).[1,2] The glutamate functions include the ordinary synaptic transmission, the modification of synaptic connection during development, and the modulation of transmission efficacy during plastic changes in the adult brain.[1,2] Glutamate is thus thought to be involved in many higher brain functions such as learning and memory acquisition. In contrast, overactivation of glutamate receptors under pathological conditions causes neurodegeneration and neuronal cell death and may also lead to some slowly progressive neurodegenerative disorders such as Huntington's disease and Alzheimer's disease.[2,3]

The diverse functions of glutamate neurotransmission are mediated by a variety of glutamate receptors that can be categorized into two distinct groups termed ionotropic and metabotropic receptors (mGluRs).[1,2] The ionotropic receptors can be subdivided into N-methyl-D-aspartate (NMDA) receptors and α-amino-3-hydroxy-5-methyl-4-isoxazolepropionate (AMPA)/kainate receptors. They contain glutamate-gated, integral cation channels in their molecules. The metabotropic receptors are coupled to the modulation of intracellular second messengers through GTP-binding proteins.[1,2,4] We have been working on molecular characterization of the NMDA receptors and mGluRs and have revealed the existence of many subtypes or subunits of these receptors in the CNS. This article deals with our recent studies concerning the molecular diversity of the glutamate receptors and discusses some new aspects in this field of research.

PHYSIOLOGICAL IMPLICATIONS OF THE NMDA RECEPTOR

The NMDA receptor plays a key role in many neuronal functions in the CNS.[2] This receptor is essential for inducing long-term potentiation (LTP), a long-lasting and activity-dependent enhancement of transmission efficacy thought to underlie fundamental processes for learning and memory.[5,6] This receptor also plays a crucial role in inducing neurodegeneration and neuronal cell death under pathological conditions such as cerebral ischemia and epilepsy.[3]

The NMDA receptor has several characteristic properties that clearly distinguish it from other types of glutamate receptors.[2] The integral channel of the NMDA receptor is highly permeable to Ca^{2+}, and the increased intracellular Ca^{2+} in neuronal cells is thought to be responsible for evoking glutamate-mediated neuronal plasticity and neurotoxicity. This receptor also shows several unique features, including voltage-dependent Mg^{2+} blockade, Zn^{2+} inhibition, the modulation by glycine, and channel blockade by selective channel blockers such as MK-801. The voltage-dependent Mg^{2+} blockade of the NMDA receptor channel is postulated to be a key switch in controlling the induction of LTP during tetanic stimuli.[5,6]

MOLECULAR CLONING OF NMDA RECEPTORS

Using a *Xenopus* oocyte expression system combined with electrophysiology, we first isolated and characterized a key subunit of the rat NMDA receptor (NMDAR1).[7] The cloned receptor expressed in *Xenopus* oocytes faithfully reproduces the characteristic properties of the NMDA receptor and shows a high Ca^{2+} permeability and voltage-dependent Mg^{2+} blockade. The pharmacological profile is also in good agreement with that reported for the NMDA receptor in the brain. Therefore the single protein encoded by NMDAR1 forms a functional receptor-channel complex possessing all pharmacological and electrophysiological properties characteristic of the NMDA receptor in neuronal cells.[7] The major isoform of NMDAR1 consists of 938 amino acid residues but there are at least seven isoforms of NMDAR1 generated by alternative splicing.[8]

The above characterization of the homomeric NMDAR1 receptor, however, also showed a low channel activity of the cloned receptor and some disparities between the expression sites of NMDAR1 mRNA and the radioligand binding sites of the NMDA receptor in the brain.[7] We extended molecular screening of the NMDA receptors and identified four additional cDNA clones encoding different subunits of the NMDA receptor, termed NMDAR2A–NMDAR2D.[9] The molecular cloning of these subunits has also been reported from the laboratories of Mishina and Seeburg.[10-12]

The NMDAR2A–NMDAR2D subunits share only 15% amino acid sequence homology with NMDAR1 but show 40–50% homology with one another. They consist of about 1,250–1,500 amino acid residues and have a peculiar large carboxyl-terminal extension following the transmembrane segments.[9-12] All NMDAR subunits are thought to comprise four transmembrane segments and belong to the ligand-gated ion channel family.[9-12]

None of the NMDAR2 subunits evoked any appreciable electrophysiological responses after agonist application in homomeric expression in *Xenopus* oocytes. However, the combined expression of the NMDAR2 subunits with the NMDAR1 subunit markedly potentiated the responses to agonists.[9-12] The combination of NMDAR1 with different NMDAR2 subunits also conferred a functional variability in the electrophysiological and pharmacological properties. Depending on the subunit combination, the affinity for agonists and antagonists and the sensitivity to Mg^{2+} blockade are different from one another.[9,10,12] Therefore, NMDAR1 serves as a key subunit necessary for the NMDA receptor activity, while the individual NMDAR2 subunits are involved in potentiating the glutamate response and producing the functional variability by forming the heteromeric configuration of the NMDA receptor-channel complex (TABLE 1).

STRUCTURE-FUNCTION RELATIONSHIP OF NMDA RECEPTOR CHANNEL

All the NMDAR subunits can fit into a transmembrane model of the ligand-gated ion channels, in which the second transmembrane (TM II) segment is thought to be involved in lining a channel pore.[13] To investigate the structural features that control Ca^{2+} permeation and channel blockade of the NMDA receptor, we systematically changed single amino acids in the vicinity of TM II segment of NMDAR1 and examined the channel properties of the resultant mutants in combined expression with the wild-type NMDAR2A in *Xenopus* oocytes.[14] In the mutants examined, the substitution of the asparagine in TM II segment by either glutamine or arginine dramatically reduced Ca^{2+} permeability and the sensitivity to the blockades by Mg^{2+} and MK-801.[14-16] These mutations also reduced the inhibitory effects of Zn^{2+} and an antidepressant desipramine.[14] All NMDA receptor subunits contain asparagine at the corresponding positions of TM II segments.[9] Therefore an asparagine ring could be formed at the central part of the channel pore of the NMDA receptor and may provide a constriction that controls both the Ca^{2+} permeation and the blockades by Mg^{2+} and other cationic channel blockers.

EXPRESSION OF NMDAR SUBUNITS

In situ hybridization analysis indicated that the NMDAR1 mRNA is expressed in almost all the neuronal cells in both central and peripheral nervous systems (FIG. 1).[7,17] The mRNAs for different NMDAR2 subunits show overlapping but different expression patterns in the brain.[9,10,12] The NMDAR2A mRNA is prominently expressed in the cerebral cortex and hippocampus, while the NMDAR2B mRNA is distributed throughout the forebrain region. The NMDAR2C mRNA predominates in the cerebellum and the NMDAR2D mRNA is expressed in the diencephalic/lower brain stem regions (FIG. 1).[9,10,12] It has been reported that the pharmacological and physiological properties of the NMDA receptors vary in different brain regions.[18] Thus, the anatomical and functional differences of the NMDAR2 subunits provide the molecular basis for the heterogeneity of the NMDA receptors.

The timing of the appearance of each NMDAR subunit mRNA during development is also different from subunit to subunit.[19] NMDAR1 appears early in the

TABLE 1. The NMDA Receptor Family

	NMDAR1	NMDAR2
Isoform/subtype	Seven isoforms generated by alternative splicing	Four subtypes encoded by different genes
Amino acid residues	938, 959, 901, 922, 885, 922, and 906 for NMDAR1A to NMDAR1G, respectively	1464, 1482, 1250, and 1323 for NMDAR2A to NMDAR2D, respectively
Structure	Four transmembrane domains Asparagine residue in TMII[a] segment	Four transmembrane domains Asparagine residue in TMII segment Large carboxyl-terminal extension Low sequence homology ($\sim 15\%$) with NMDAR1
Function	Forms homomeric channel Essential for the NMDA receptor activity Possesses all characteristic properties of the NMDA receptor; high Ca^{2+} permeability, voltage-dependent Mg^{2+} block, glycine potentiation, and Zn^{2+} inhibition	Forms heteromeric channel in the coexpression with NMDAR1 Potentiates NMDA receptor activity Confers functional variabilities in the sensitivity to agonists and antagonists, and Mg^{2+} blockade
Expression	Expressed abundantly in almost all neuronal cells Appears early in the embryonic stage	Expressed differently and restrictedly Ontogenetically controlled

[a] TMII: second transmembrane.

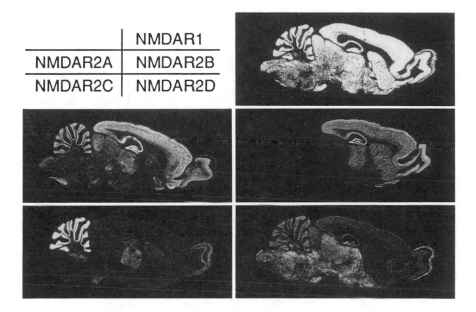

	NMDAR1
NMDAR2A	NMDAR2B
NMDAR2C	NMDAR2D

FIGURE 1. Distribution of mRNAs for the NMDAR1 and NMDAR2 subunits in the adult rat brain. Negative film images of *in situ* hybridization of parasagittal sections are shown. Individual NMDAR subunits show different expression patterns in the brain.

embryonic stage and maintains thereafter a very high level of expression throughout the brain regions. NMDAR2B and NMDAR2D also appear in the embryonic brain and show a broad range of expression during the neonatal stage. Their expressions then gradually decline and converge to more limited portions. NMDAR2A and NMDAR2C, on the other hand, gradually appear after birth and reach a plateau of expression in the adult brain.[19] The different ontogenic change in the expression of the different NMDA receptor subunits may subserve specific functions during neuronal development, but its physiological meaning remains to be elucidated.

MOLECULAR CLONING OF METABOTROPIC GLUTAMATE RECEPTORS

Our molecular cloning studies have now revealed that there are at least six different subtypes of the mGluR family (TABLE 2).[20-24] These six subtypes, termed mGluR1–mGluR6, possess a common structural architecture with a large extracellular amino-terminal domain preceding the seven transmembrane segments and are highly homologous in both their transmembrane segments and extracellular domains.[21,22,24] The high conservation of cysteine residues in the extracellular domains is also striking and is thought to be crucial for the structural formation of the mGluRs.[21,22,24] The mGluRs do not show any sequence homology with other members of G protein-coupled receptors and represent a novel family of G protein-coupled receptors.[20,25] The six

TABLE 2. The Metabotropic Glutamate Receptor Family

	Subgroup 1		Subgroup 2		Subgroup 3	
Subtype	mGluR1	mGluR5	mGluR2	mGluR3	mGluR4	mGluR6
Amino acid residues	1199 (mGluR1α) 906 (mGluR1β)	1171	872	879	912	871
Effector	IP_3/Ca^{2+}↑	IP_3/Ca^{2+}↑	cAMP↓	cAMP↓	cAMP↓	cAMP↓
Agonist selectivity	QA>Glu≥Ibo >tACPD	QA>Glu≥Ibo >tACPD	Glu≥tACPD >Ibo>QA	Glu≥tACPD >Ibo>QA	AP4>Glu≥SOP	AP4>SOP>Glu
Main expression sites	Cerebellum (Purkinje cell) DG CA2-3 OB (Mitral and tufted cells)	Cerebral cortex Striatum DG CA1-3 OB (Internal granular layer)	Cerebellum (Golgi cell) Cerebral cortex DG AOB	Cerebral cortex Striatum DG Thalamus (Reticular nucleus) Glial cells	Cerebellum (Granule cell) Thalamus OB (Internal granular layer)	Retina (Inner nuclear layer)

IP_3, inositol trisphosphate; cAMP, adenosine 3′,5′-monophosphate; QA, quisqualate; Glu, L-glutamate; Ibo, ibotenate; tACPD, trans-1-aminocyclopentane-1,3-dicarboxylate; AP4, L-2-amino-4-phosphonobutyrate; SOP, L-serine-O-phosphate; DG, granule cells in hippocampal dentate gyrus; CA, pyramidal cells in hippocampal CA subregions; OB, olfactory bulb; AOB, accessory olfactory bulb.

receptor subtypes can be divided into three subgroups according to their sequence similarities; mGluR1/mGluR5, mGluR2/mGluR3, and mGluR4/mGluR6. They show about 60–70% homology within the same subgroups and about 40% homology between different subgroups (TABLE 2).

FUNCTIONAL ANALYSES OF THE CLONED mGluRs

To investigate signal transduction mechanisms and pharmacological properties of individual mGluR subtypes, we established cell lines permanently expressing each mGluR subtype by DNA transfection and then examined the modulation of the second messengers after agonist stimulation.

mGluR1 and mGluR5 stimulate inositol trisphosphate (IP3) formation and lead to the intracellular Ca^{2+} mobilization.[22,26] The agonist selectivity is virtually the same between these two receptor subtypes and the rank order of agonist potencies is quisqualate > glutamate ≥ ibotenate > trans-1-aminocyclopentane-1,3-dicarboxylate (tACPD).[22,26]

All other subtypes, on the other hand, mediate the inhibition of the forskolin-stimulated accumulation of intracellular adenosine 3',5'-monophosphate (cAMP) in agonist-dependent manner.[21,23,24] mGluR2 and mGluR3 resemble each other in their agonist selectivity and the rank order of agonist potencies is glutamate ≥ tACPD > ibotenate > quisqualate.[21,23] mGluR4 and mGluR6 are totally different from mGluR2 and mGluR3 in their pharmacological properties. These subtypes potently react with L-2-amino-4-phosphonobutyrate (L-AP4) in a stereoselective manner and also respond to L-serine-O-phosphate (L-SOP).[23,24] Therefore the mGluR subtypes can be classified into three subgroups according to not only their sequence similarities but also their signal transduction pathways and agonist selectivities (TABLE 2).

L-AP4 has been shown to suppress synaptic transmission by inhibiting glutamate release at presynaptic sites in many brain regions.[27] The agonist selectivity of mGluR4 corresponds well to that of the putative AP4 receptor.[23] Furthermore, tACPD, which is a potent agonist for mGluR2/mGluR3, has also been reported to act suppressively at the presynaptic site.[28] Therefore, mGluR2, mGluR3, and mGluR4 appear to represent presynaptic autoreceptors and may play an important role in controlling glutamate transmission.

EXPRESSION OF METABOTROPIC GLUTAMATE RECEPTORS

The mRNAs for different mGluR subtypes show overlapping but distinct expression patterns in the rat brain (FIG. 2).[20–24,29,30] The mGluR1 mRNA is highly expressed in the cerebellum and the hippocampal dentate gyrus and pyramidal cells of CA2-3 regions.[20,29] The mGluR5 mRNA is prominently expressed in the cerebral cortex, striatum, olfactory bulb, and pyramidal cells throughout the hippocampal CA1-3 regions. The expression of this mRNA is very low in the cerebellum.[22] The mGluR2 mRNA is highly expressed in cerebellar Golgi cells, the accessory olfactory bulb, and granule cells of the hippocampal dentate gyrus.[21,30] The mGluR3 mRNA is prominently expressed in the cerebral cortex, striatum, hippocampal dentate gyrus, and thalamus.

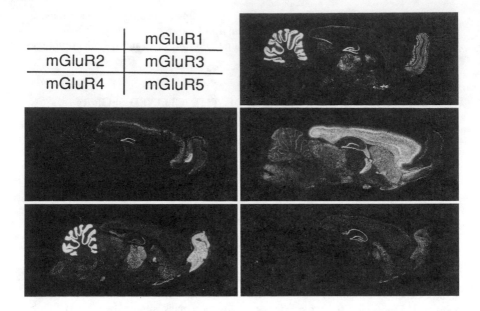

	mGluR1
mGluR2	mGluR3
mGluR4	mGluR5

FIGURE 2. Distribution of mRNAs for the mGluR1–mGluR5 subunits in the adult brain. Negative film images of *in situ* hybridization of parasagittal sections are shown. Each mGluR subunit shows a characteristic expression pattern in the brain.

The characteristic feature of this mRNA is its expression in both glial cells and neuronal cells.[23] The mGluR4 mRNA is more restrictedly expressed and is confined to granule cells of the cerebellum, the thalamus, and olfactory bulb.[23]

RETINAL L-AP4-SENSITIVE mGluR

Several lines of evidence suggested some regulatory functions of metabotropic glutamate receptors in the CNS.[31-33] However, many of the physiological roles of mGluRs still remain to be elucidated. Recent electrophysiological studies indicated that the mGluR that is sensitive to L-AP4 plays an important role in the synaptic transmission between photoreceptor cells and ON-bipolar cells in the visual system through the coupling to the guanosine 3′,5′-monophosphate (cGMP) cascade.[34-36] We thus focused on the putative L-AP4–sensitive receptor in the retina.

By molecular screening of a rat retinal cDNA library, we isolated a novel subtype of the metabotropic glutamate receptor, termed mGluR6.[24] When mGluR6 is transfected and expressed in Chinese hamster ovary (CHO) cells, this receptor inhibits the forskolin-stimulated cAMP formation with an agonist selectivity to L-AP4 and L-serine-O-phosphate.[24] The rank order of agonist potency is thus in good agreement with that reported for the mGluR in retinal ON-bipolar cells. RNA blot hybridization analysis indicated that mGluR6 mRNA is exclusively expressed in the retina.[24] Furthermore, *in situ* hybridization analysis revealed the distribution of mGluR6 mRNA confined to the outer zone of the inner nuclear layer of the retina, where ON-bipolar cells are localized.[24] In contrast to our observation that mGluR6 is coupled to the inhibitory cAMP cascade, the L-AP4–sensitive mGluR in ON-bipolar cells is linked to the cGMP cascade. However, it is also known that under a heterologous expression system where preferred G proteins or subsequent effectors are absent, a transfected receptor can be coupled to a different signal transduction pathway. Therefore, the difference in the signal transduction of the bipolar cell receptor and mGluR6 may be due to the lack of appropriate cGMP signal transduction machinery in CHO cells.

On the basis of our results and the previous reports from other laboratories, the following model can be proposed as a mechanism mediated by the mGluR6 function in the visual transduction system (FIG. 3). When light reaches eyes, intracellular concentration of cGMP in photoreceptor cells decreases as a result of the stimulation of cGMP phosphodiesterase through the activation of transducin. The decrease in cGMP concentrations, in turn, leads to the closure of the cGMP-gated cation channels and hyperpolarizes photoreceptor cells. This hyperpolarization then results in the reduction of glutamate release. Under this situation the mGluR6–G protein–phosphodiesterase system in ON-bipolar cells becomes inactive and increases intracellular concentration of cGMP. The cGMP-gated cation channel is then opened and depolarizes the ON-bipolar cells. This depolarization releases glutamate from ON-bipolar cells and excites ganglion cells. It is thus conceivable that different subtypes of mGluRs are linked to distinct intracellular effectors and may play differential roles in glutamate transmission.

SUMMARY

Our molecular studies have revealed the existence of a large number of different subunits or subtypes for the NMDA and metabotropic glutamate receptors. The individual receptors show functional variabilities and distinct expression patterns in the CNS. The NMDA receptors belong to the ligand-gated ion channel family and consist of a key subunit NMDAR1 and four accessory subunits NMDAR2A–NMDAR2D. The combination of NMDAR1 and NMDAR2 in heteromeric configurations potentiates glutamate response and produces a functional variability. All the NMDAR subunits have an asparagine residue at the corresponding position of the second transmembrane segments, and these residues are thought to be responsible for controlling Ca^{2+} permeation and the channel blockade by Mg^{2+} and cationic channel blockers. Individual NMDAR subunit mRNAs are different in their expression patterns during development and in the adult brain. The mGluR family consists of at least six different subtypes. These subtypes are divided into three subgroups according to their sequence similarities, signal transduction mechanisms, and pharmacological properties. Although their physiological roles largely remain to be elucidated, the retinal L-AP4–sensitive mGluR may have a specific function that mediates excitatory neurotransmission in the visual system. It is thus undoubtedly important to investigate specific functions of different combinations of the NMDA receptor subunits and different subtypes of mGluRs and to explore the molecular mechanisms of glutamate receptor–mediated neuronal plasticity and neurotoxicity.

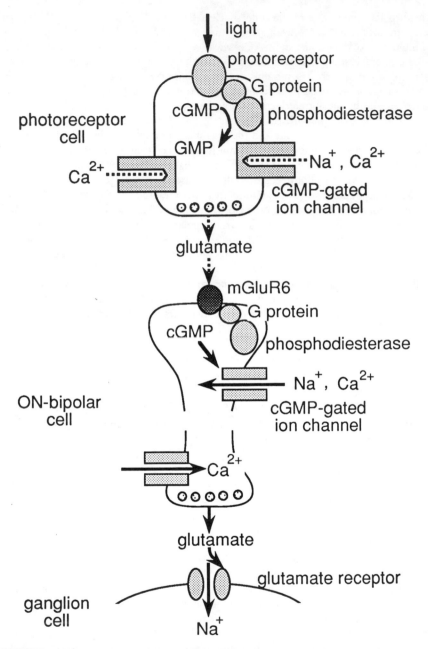

FIGURE 3. Schematic illustration of the hypothetical signal transduction in the visual system. One subtype of metabotropic glutamate receptors, mGluR6, is thought to be localized in ON-bipolar cells and involved in the synaptic transmission between photoreceptor cells and ON-bipolar cells through the coupling to the cyclic GMP cascade.

ACKNOWLEDGMENT

We are grateful to Drs. Shigemoto and Mizuno, Department of Morphological Brain Science, Kyoto University Faculty of Medicine, for collaboration in morphological studies.

REFERENCES

1. NAKANISHI, S. 1992. Molecular diversity of glutamate receptors and implications for brain function. Science 258: 597–603.
2. MONAGHAN, D. T., R. J. BRIDGES & C. W. COTMAN. 1989. The excitatory amino acid receptors: Their classes, pharmacology, and distinct properties in the function of the central nervous system. Annu. Rev. Pharmacol. Toxicol. 29: 365–402.
3. MELDRUM, B. & J. GARTHWAITE. 1990. Excitatory amino acid neurotoxicity and neurodegenerative disease. Trends Pharmacol. Sci. 11: 379–387.
4. SCHOEPP, D., J. BOCKAERT & F. SLADECZEK. 1990. Pharmacological and functional characteristics of metabotropic excitatory amino acid receptors. Trends Pharmacol. Sci. 11: 508–515.
5. COLLINGRIDGE, G. L., & W. SINGER. 1990. Excitatory amino acid receptors and synaptic plasticity. Trends Pharmacol. Sci. 11: 290–296.
6. MADISON, D. V., R. C. MALENKA & R. A. NICOLL. 1991. Mechanisms underlying long-term potentiation of synaptic transmission. Annu. Rev. Neurosci. 14: 379–397.
7. MORIYOSHI, K., M. MASU, T. ISHII, R. SHIGEMOTO, N. MIZUNO & S. NAKANISHI. 1991. Molecular cloning and characterization of the rat NMDA receptor. Nature 354: 31–37.
8. SUGIHARA, H., K. MORIYOSHI, T. ISHII, M. MASU & S. NAKANISHI. 1992. Structures and properties of seven isoforms of the NMDA receptor generated by alternative splicing. Biochem. Biophys. Res. Commun. 185: 826–832.
9. ISHII, T., K. MORIYOSHI, H. SUGIHARA, K. SAKURADA, H. KADOTANI, M. YOKOI, C. AKAZAWA, R. SHIGEMOTO, N. MIZUNO, M. MASU & S. NAKANISHI. 1993. Molecular characterization of the family of N-methyl-D-aspartate receptor subunits. J. Biol. Chem. 268: 2836–2843.
10. KUTSUWADA, T., N. KASHIWABUCHI, H. MORI, K. SAKIMURA, E. KUSHIYA, K. ARAKI, H. MEGURO, H. MASAKI, T. KUMANISHI, M. ARAKAWA & M. MISHINA. 1992. Molecular diversity of the NMDA receptor channel. Nature 358: 36–41.
11. MEGURO, H., H. MORI, K. ARAKI, E. KUSHIYA, T. KUTSUWADA, M. YAMAZAKI, T. KUMANISHI, M. ARAKAWA, K. SAKIMURA & M. MISHINA. 1992. Functional characterization of a heteromeric NMDA receptor channel expressed from cloned cDNAs. Nature 357: 70–74.
12. MONYER, H., R. SPRENGEL, R. SCHOEPFER, A. HERB, M. HIGUCHI, H. LOMELI, N. BURNASHEV, B. SAKMANN & P. H. SEEBURG. 1992. Heteromeric NMDA receptors: Molecular and functional distinction of subtypes. Science 256: 1217–1221.
13. BETZ, H. 1990. Ligand-gated ion channels in the brain: The amino acid receptor superfamily. Neuron 5: 383–392.
14. SAKURADA K., M. MASU & S. NAKANISHI. 1993. Alteration of Ca^{2+} permeability and sensitivity to Mg^{2+} and channel blockers by a single amino acid substitution in the N-methyl-D-aspartate receptor. J. Biol. Chem. 268: 410–415.
15. BURNASHEV, N., R. SCHOEPFER, H. MONYER, J. P. RUPPERSBERG, W. GÜNTHER, P. H. SEEBURG & B. SAKMANN. 1992. Control by asparagine residues of calcium permeability and magnesium blockade in the NMDA receptor. Science 257: 1415–1419.
16. MORI, H., H. MASAKI, T. YAMAKURA & M. MISHINA. 1992. Identification by mutagenesis of a Mg^{2+}-block site of the NMDA receptor channel. Nature 358: 673–675.
17. SHIGEMOTO, R., H. OHISHI, S. NAKANISHI & N. MIZUNO. 1992. Expression of the mRNA for the rat NMDA receptor (NMDAR1) in the sensory and autonomic ganglion neurons. Neurosci. Lett. 144: 229–232.
18. MONAGHAN, D. T. & K. J. ANDERSON 1991. Heterogeneity and organization of excitatory

amino acid receptors and transporters. *In* Excitatory Amino Acids and Synaptic Transmission. H. W. Wheal & A. M. Thomson, Eds.: 33–54. Academic Press. San Diego.

19. Akazawa, C., R. Shigemoto, S. Nakanishi & N. Mizuno. Manuscript in preparation.

20. Masu, M., Y. Tanabe, K. Tsuchida, R. Shigemoto & S. Nakanishi. 1991. Sequence and expression of a metabotropic glutamate receptor. Nature **349**: 760–765.

21. Tanabe, Y., M. Masu, T. Ishii, R. Shigemoto & S. Nakanishi. 1992. A family of metabotropic glutamate receptors. Neuron **8**: 169–179.

22. Abe, T., H. Sugihara, H. Nawa, R. Shigemoto, N. Mizuno & S. Nakanishi. 1992. Molecular characterization of a novel metabotropic glutamate receptor mGluR5 coupled to inositol phosphate/Ca^{2+} signal transduction. J. Biol. Chem. **267**: 13361–13368.

23. Tanabe, Y., A. Nomura, M. Masu, R. Shigemoto, N. Mizuno & S. Nakanishi. 1993. Signal transduction, pharmacological properties, and expression patterns of two rat metabotropic glutamate receptors, mGluR3 and mGluR4. J. Neurosci. **13**: 1372–1378.

24. Nakajima, Y., H. Iwakabe, C. Akazawa, H. Nawa, R. Shigemoto, N. Mizuno & S. Nakanishi. Molecular characterization of a novel retinal metabotropic glutamate receptor mGluR6 with a high agonist selectivity for L-2-amino-4-phosphonobutyrate. J. Biol. Chem. **268**: 11868–11873.

25. Houamed, K. M., J. L. Kuijper, T. L. Gilbert, B. A. Haldeman, P. J. O'Hara, E. R. Mulvihill, W. Almers & F. S. Hagen. 1991. Cloning, expression, and gene structure of a G protein-coupled glutamate receptor from rat brain. Science **252**: 1318–1321.

26. Aramori, I. & S. Nakanishi. 1992. Signal transduction and pharmacological characteristics of a metabotropic glutamate receptor, mGluR1, in transfected CHO cells. Neuron **8**: 757–765.

27. Forsythe, I. D. & J. D. Clements. 1990. Presynaptic glutamate receptors depress excitatory monosynaptic transmission between mouse hippocampal neurones. J. Physiol. (London) **429**: 1–16.

28. Baskys, A. & R. C. Malenka. 1991. Agonists at metabotropic glutamate receptors presynaptically inhibit EPSCs in neonatal rat hippocampus. J. Physiol. (London) **444**: 687–701.

29. Shigemoto, R., S. Nakanishi & N. Mizuno. 1992. Distribution of the mRNA for a metabotropic glutamate receptor (mGluR1) in the central nervous system: An in situ hybridization study in adult and developing rat. J. Comp. Neurol. **322**: 121–135.

30. Ohishi, H., R. Shigemoto, S. Nakanishi & N. Mizuno. 1993. Distribution of the messenger RNA for a metabotropic glutamate receptor, mGluR2, in the central nervous system of the rat. Neuroscience **53**: 1009–1018.

31. Anwyl, R. 1991. The role of the metabotropic receptor in synaptic plasticity. Trends Pharmacol. Sci. **12**: 324–326.

32. Baskys, A. 1992. Metabotropic receptors and 'slow' excitatory actions of glutamate agonists in the hippocampus. Trends Neurosci. **15**: 92–96.

33. Miller, R. J. 1991. Metabotropic excitatory amino acid receptors reveal their true colors. Trends Pharmacol. Sci. **12**: 365–367.

34. Shiller, P. H. 1992. The ON and OFF channels of the visual system. Trends Neurosci. **15**: 86–92.

35. Nawy, S. & C. E. Jahr. 1990. Suppression by glutamate of cGMP-activated conductance in retinal bipolar cells. Nature **346**: 269–271.

36. Shields, R. A. & G. Falk. 1990. Glutamate receptors of rod bipolar cells are linked to a cyclic GMP cascade via a G-protein. Proc. R. Soc. Lond. Ser. B **242**: 91–94.

Primary Structure and Expression from cDNAs of the Ryanodine Receptor[a]

HIROSHI TAKESHIMA

International Institute for Advanced Studies
Shimadzu N-80
1 Nishinokyo-Kuwabara-cho, Nakagyo-ku
Kyoto 604, Japan

INTRODUCTION

The characterization of the ryanodine receptor/calcium release channel has progressed physiologically and biochemically in recent years.[1,2] Skeletal muscle sarcoplasmic reticulum (SR) provides the richest source of the ryanodine receptor. The purified ryanodine receptor with a monomeric relative molecular mass (M_r) of 400–450 kD is thought to form a homotetrameric complex and is morphologically identified with the "foot" structure, which spans the gap between the SR and transverse tubule membranes. When reconstituted into a planar lipid bilayer the receptor has been shown to function as a calcium-release channel. The channel properties and subcellular distribution of the ryanodine receptor suggest its involvement in the SR calcium release that occurs during excitation-contraction (E-C) coupling in skeletal muscle.

Recently, ryanodine receptors were also purified from heart and brain and their functional properties were investigated.[2–4] These results, together with evidence from physiological experiments, suggest that ryanodine receptors take part in calcium signaling in muscle cells and neurons. This paper reviews our experiments on three types of ryanodine receptor using recombinant DNAs.

SKELETAL MUSCLE RYANODINE RECEPTOR

After purification of the ryanodine receptor from rabbit skeletal muscle SR, the cDNA encoding this protein was cloned on the basis of the partial amino acid sequences.[5] Sequence analysis of the cDNA revealed that the rabbit skeletal muscle ryanodine receptor is composed of 5,037 amino acid residues and the calculated M_r is 565 kD (FIG. 1). The deduced amino acid sequence of the ryanodine receptor was analyzed for local hydropathicity and predicted secondary structure (FIG. 2). The

[a]This investigation was supported in part by research grants from Ministry of Education, Science and Culture of Japan and the Institute of Physical and Chemical Research.

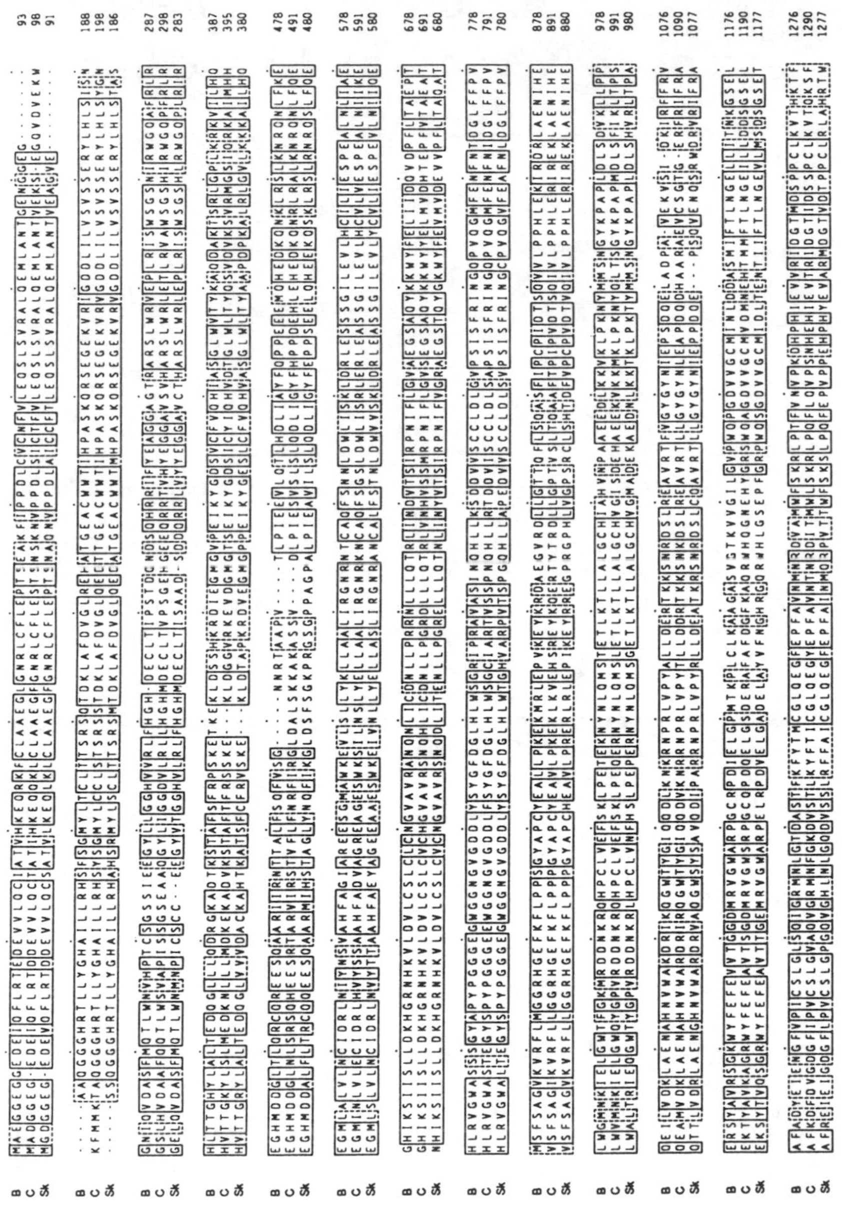

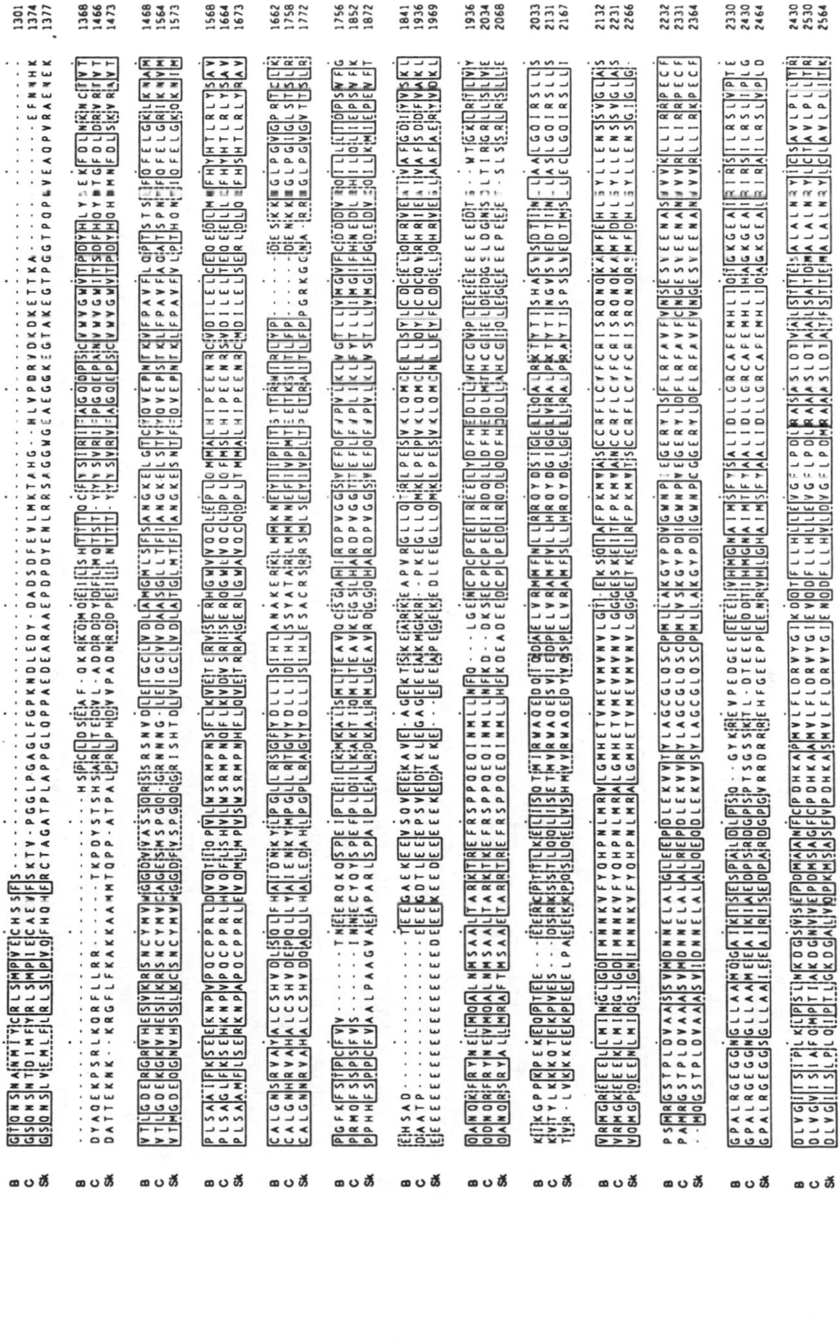

FIGURE 1. (A) Legend on page 170.

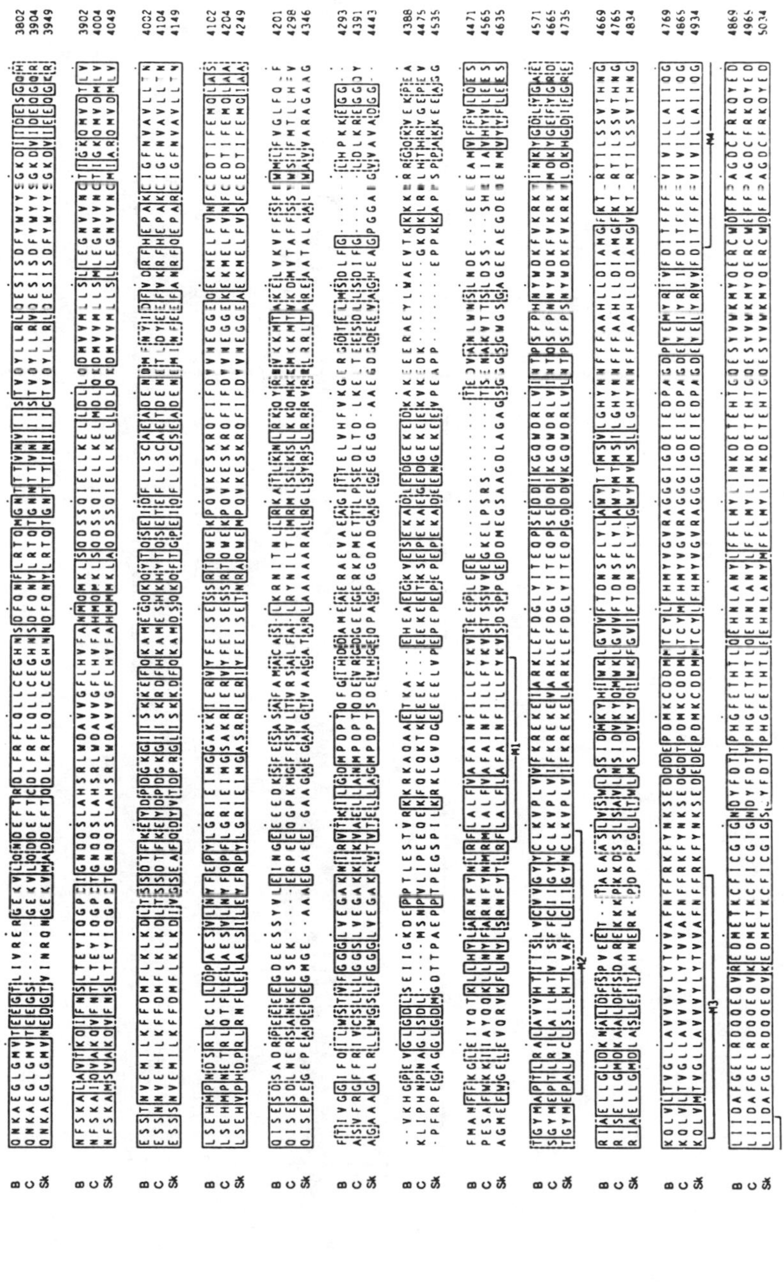

FIGURE 1. (B) Legend on page 170.

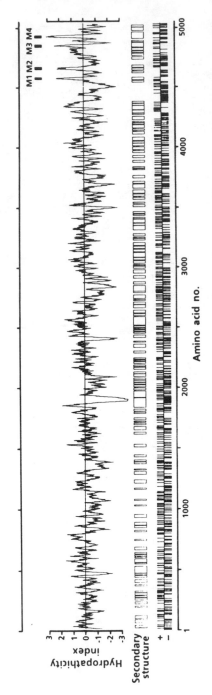

FIGURE 1. (See pages 166–169.) Alignment of the amino acid sequences of the rabbit brain (B), the rabbit cardiac (C), and the rabbit skeletal muscle (Sk) ryanodine receptor. The sequence data have been taken from Takeshima et al.,[5] Nakai et al.,[8] and Hakamata et al.[10] The sets of three identical residues at one position are enclosed with solid lines, and sets of three identical or conservative residues with broken lines. Gaps (-) have been inserted to achieve maximum homology. The putative transmembrane segments M1–M4 are indicated. (From Hakamata et al.[10] Reprinted with permission.)

FIGURE 2. Hydropathicity profile and predicted secondary structures of the skeletal muscle ryanodine receptor. The positions of the putative transmembrane segments M1–M4 are indicated by filled boxes. (From Takeshima et al.[5] Reprinted with permission.)

receptor molecule has four highly hydrophobic segments with predicted secondary structure in its C-terminal tenth. These segments (M1–M4), each comprising about 20 amino acid residues, presumably represent transmembrane α-helices. The ryanodine receptor does not possess a hydrophobic amino-terminal sequence indicative of the signal sequence. This may indicate that the portion preceding segment M1, which constitutes nine-tenths of the receptor molecule, is located on the cytoplasmic side of the SR membrane. The C-terminal region of the ryanodine receptor, including segments M3 and M4, shows remarkable amino acid sequence similarity to the corresponding region of the inositol-1,4,5-triphosphate receptor, which functions as a calcium channel of endoplasmic reticulum.[6] Thus the C-terminal region of both the receptors may be important in forming intracellular membrane channels. These observations suggest that the ryanodine receptor molecule consists of two main parts: the C-terminal channel region and the large cytoplasmic region that apparently corresponds to the "foot" structure (FIG. 3).

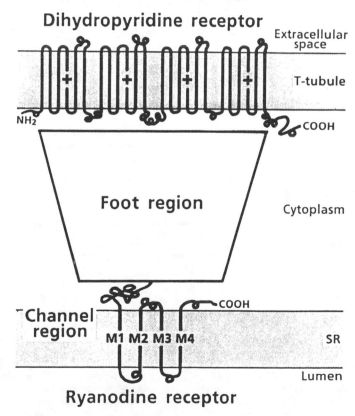

FIGURE 3. Proposed transmembrane topology and molecular architecture of the ryanodine receptor, together with the DHP receptor in the triad junction of skeletal muscle. (From Takeshima *et al.*[5] Reprinted with permission.)

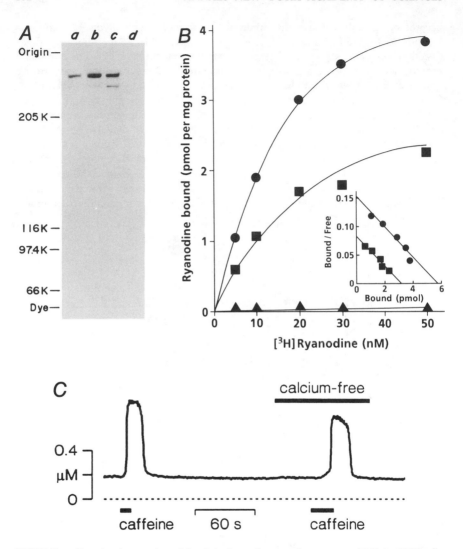

FIGURE 4. Functional expression of the skeletal muscle ryanodine receptor cDNA in CHO cells. (A) Immunoblotting analysis of the heavy SR fraction from rabbit skeletal muscle (lane *a*) and membrane preparations from C7311 cells (lane *b*), C798 cells (lane *c*), and non-transfected CHO cells (lane *d*), using monoclonal antibody against the rabbit skeletal muscle ryanodine receptor. (B) Binding of [³H]ryanodine to membrane preparations from C7311 cells (●), C798 cells (■), and non-transfected CHO cells (▲); the inset shows Scatchard plots of the data. (C) Caffeine-induced calcium release in fura-2–AM-loaded C7311 cell. Application of caffeine (10 mM) was indicated by the bar and the second application was performed after replacing the normal saline by the nominally calcium-free saline. (A and B: From Takeshima *et al.*[5] C: From Penner *et al.*[7] Reprinted with permission.)

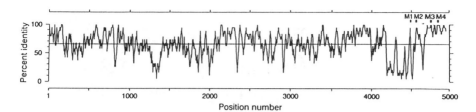

FIGURE 5. Profile of local amino acid sequence identity between the rabbit cardiac and skeletal muscle ryanodine receptors. The horizontal line indicates the overall percent sequence identity between the two ryanodine receptors. The positions of the putative transmembrane segments M1–M4 are shown by filled boxes. (From Nakai et al.[8] Reprinted with permission.)

To examine whether the functional calcium-release channel is formed by the ryanodine receptor cDNA, expression plasmids were constructed that carry the entire protein-coding sequence of the ryanodine receptor cDNA, linked to the SV40 early promoter, and the neomycin-resistance marker gene.[5,7] Chinese hamster ovary (CHO) cells were transfected with the expression plasmids and G418-resistant clones that were shown by RNA blot hybridization analysis to promote ryanodine receptor mRNA were selected. Immunoblotting analysis of membrane preparations using antibody against the ryanodine receptor revealed that the transformed clones produced a protein indistinguishable in M_r from the ryanodine receptor in skeletal muscle SR (FIG. 4, A). FIGURE 4 (B) shows that membrane preparations from the transformed clones were capable of binding ryanodine and Scatchard analysis yields an apparent dissociation constant (K_d) of 19 nM, which is in the same range as the K_d measured with skeletal muscle SR membrane. Fura-2 measurements were performed to study the calcium-release properties of the transformed clones (FIG. 4, C). We used caffeine, a well-known activator of the Ca^{2+}-induced Ca^{2+} release in skeletal muscle. Calcium responses to caffeine were observed in the transformed cells but not in non-transfected cells. The responses were clearly due to release of calcium from intracellular stores, since similar responses were observed after removing calcium from bath solution. Ryanodine also induced calcium release in the transformed clones.[7] These results indicate that functional calcium release channels are formed by expression of the ryanodine receptor cDNA.

CARDIAC RYANODINE RECEPTOR

A rabbit cardiac cDNA library was screened by cross-hybridization with a skeletal muscle ryanodine receptor cDNA probe and the cardiac ryanodine receptor cDNA was isolated. The rabbit cardiac ryanodine receptor, deduced from the cDNA sequence,[8] is homologous in amino acid sequence (67% identity) and shares characteristic structural features with the skeletal muscle counterpart (FIG. 1). FIGURE 5 shows the profile of local amino acid sequence identity between the cardiac and skeletal muscle ryanodine receptors. The C-terminal region that encompasses segments M3 and M4 is highly

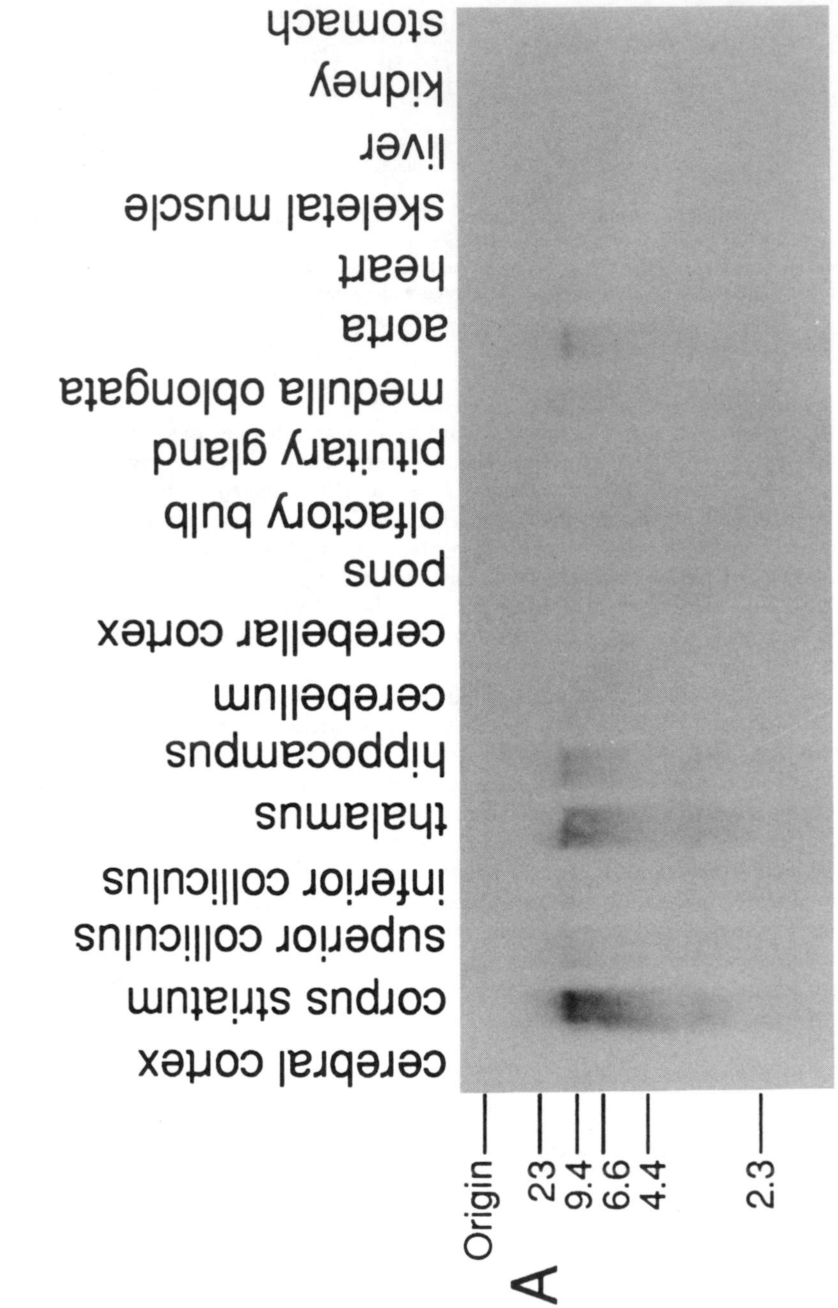
A
Origin
23
9.4
6.6
4.4
2.3
cerebral cortex
corpus striatum
superior colliculus
inferior colliculus
thalamus
hippocampus
cerebellum
cerebellar cortex
pons
olfactory bulb
pituitary gland
medulla oblongata
aorta
heart
skeletal muscle
liver
kidney
stomach

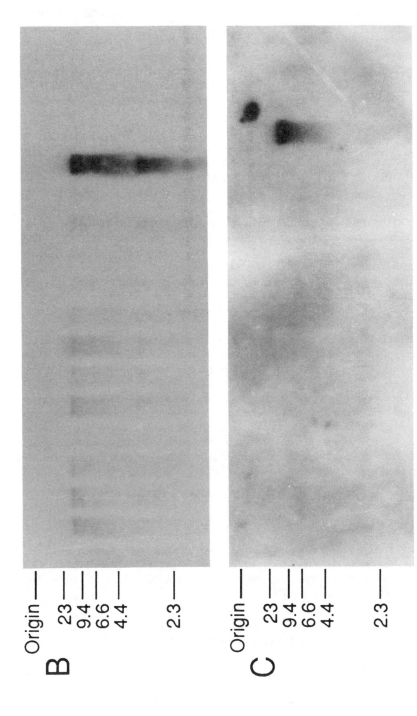

FIGURE 6. Autoradiograms of blot hybridization analysis of RNA from different regions of brain and different tissues of rabbit with cDNA probes for the brain (A), cardiac (B), and skeletal muscle (C) ryanodine receptor mRNAs. (From Hakamata et al.[10] Reprinted with permission.)

conserved, whereas other regions, for example, the region immediately preceding segment M1 and the region around position 1350 are rather divergent. The SR calcium release channel is activated or inhibited by various modulators, including calcium ion, adenine nucleotides, caffeine, calmodulin, and ryanodine. Some candidates for modulator binding sites were discussed in comparison between the amino acid sequence of the skeletal muscle receptor and that of the cardiac receptor elsewhere.[8] And functional calcium release channels were expressed from the cardiac ryanodine receptor cDNA in *Xenopus* oocytes and CHO cells.[8,9]

BRAIN COUNTERPART OF THE RYANODINE RECEPTOR AND DISTRIBUTION OF THE RYANODINE RECEPTORS

A brain cDNA library was screened for ryanodine receptor related protein with probes from the skeletal muscle and cardiac ryanodine receptor cDNAs and the cDNA of a novel brain counterpart of the ryanodine receptor was isolated recently.[10] The cDNA carries an open reading frame that encodes a sequence of 4,872 amino acids and the deduced protein is homologous in amino acid sequence and shares characteristic structural features with the skeletal muscle and cardiac ryanodine receptors (FIG. 1). These results indicate that this protein functions as a calcium release channel and involves in calcium signaling in neurons. However physiological and pharmacological properties of this protein remain to be investigated.

The distribution of the three mRNA types (designated as skeletal muscle, cardiac, and brain types) was examined using RNA blot hybridization analysis (FIG. 6).[8,10] The skeletal muscle ryanodine receptor mRNA is detected only in skeletal muscle. The cardiac ryanodine receptor mRNA is distributed throughout the brain and smooth muscles. On the other hand, the brain ryanodine receptor mRNA is found abundantly in restricted areas of the brain (corpus striatum, thalamus, and hippocampus) and also in aorta. Abundant experimental data have suggested that the major species of the ryanodine receptor in brain is the product of the cardiac ryanodine receptor gene. However different distribution of the ryanodine receptor types may suggest that calcium signaling is divergent in regions of the brain.

CONCLUSIONS

The primary structures of the ryanodine receptors from skeletal muscle and heart have been deduced by cloning and sequencing the cDNAs. Expression of the cDNAs yielded functional calcium release channels. A third ryanodine receptor species has been identified by cloning the cDNA from brain.

ACKNOWLEDGMENTS

I wish to thank the many collaborators who have contributed to the work described here.

REFERENCES

1. ENDO, M. 1977. Calcium release from the sarcoplasmic reticulum. Physiol. Rev. 57: 71–108.
2. FLEISCHER, S. & M. INUI. 1989. Biochemistry and biophysics of excitation-contraction coupling. Annu. Rev. Biophys. Biophys. Chem. 18: 333–364.
3. ELLISMAN, M. H., T, J. DEERINCK, Y. OUYANG, C. F. BECK, S. J. TANKSLEY, P. D. WALTON, J. A. AIREY, & J. L. SUTKO. 1990. Identification and localization of ryanodine binding proteins in the avian central nervous system. Neuron 5: 135–146.
4. McPHERSON, P. S., Y-K. KIM, H. VALDIVIA, C. M. KNUDSON, H. TAKEKURA, C. FRANZINI-ARMSTRONG, R. CORONADO & K. P. CAMPBELL. 1991. The brain ryanodine receptor: A caffeine-sensitive calcium release channel. Neuron 7: 17–25.
5. TAKESHIMA, H., S. NISHIMURA, T. MATSUMOTO, H. ISHIDA, K. KANGAWA, N. MINAMINO, H. MATSUO, M. UEDA, M. HANAOKA, T. HIROSE & S. NUMA. 1989. Primary structure and expression from complementary DNA of skeletal muscle ryanodine receptor. Nature 339: 439–445.
6. FURUICHI, T., S. YOSHIKAWA, A. MIYAWAKI, K. WADA, N. MAEDA & K. MIKOSHIBA. 1989. Primary structure and functional expression of the inositol 1,4,5-triphosphate-binding protein P_{400}. Nature 342: 32–38.
7. PENNER, R., E. NEHER, H. TAKESHIMA, S. NISHIMURA & S. NUMA. 1989. Functional expression of the calcium release channel from skeletal muscle ryanodine receptor cDNA. FEBS Lett. 259: 217–221.
8. NAKAI, J., T. IMAGAWA, Y. HAKAMATA, M. SHIGEKAWA, H. TAKESHIMA & S. NUMA. 1990. Primary structure and functional expression from cDNA of the cardiac ryanodine receptor/calcium release channel. FEBS Lett. 271: 167–177.
9. IMAGAWA, T., J. NAKAI, H. TAKESHIMA, Y. NAKASAKI & M. SIGEKAWA. 1992. Expression of Ca^{2+}-induced Ca^{2+} release channel activity from cardiac ryanodine receptor cDNA in Chinese hamster ovary cells. J. Biochem. (Tokyo) 112: 508–513.
10. HAKAMATA, Y., J. NAKAI, H. TAKESHIMA & K. IMOTO. 1992. Primary structure and distribution of a novel ryanodine receptor/calcium release channel from rabbit brain. FEBS Lett. 312: 229–235.

Structure and Function of Inositol 1,4,5-Trisphosphate Receptor

K. MIKOSHIBA,[a,b] T. FURUICHI,[a] A. MIYAWAKI,[a]
S. YOSHIKAWA,[a,c] S. NAKADE,[c] T. MICHIKAWA,[a]
T. NAKAGAWA,[a] H. OKANO,[a] S. KUME,[a,c]
A. MUTO,[a,c] J. ARUGA,[b] N. YAMADA,[a,c]
Y. HAMANAKA,[a,c] I. FUJINO,[a,c] M. KOBAYASHI[a]

[a]Department of Molecular Neurobiology
The Institute of Medical Science
University of Tokyo
Tokyo 108, Japan

[b]Department of Molecular Neurobiology Laboratory
Tsukuba Life Science Center
The Institute of Physical and Chemical Research (RIKEN)
Ibaragi 305, Japan

[c]Division of Regulation of Macromolecular Function
Institute for Protein Research
Osaka University
Osaka 565, Japan

INTRODUCTION

Many external signals utilize the phosphoinositide signaling pathway to regulate cellular function. The phosphoinositide (PI) turnover system acts as a signal-transducing cascade in the central nervous system.[1,2] Many of the receptors coupled with G protein activate phospholipase C (PLC), which hydrolyzes phosphatidylinositol 4,5-bisphosphate generating inositol 1,4,5-trisphosphate (InsP$_3$) and 1,2-diacylglycerol (DAG). DAG is known to activate protein kinase C.[3] InsP$_3$ is now generally accepted to work as an intracellular second messenger.[2] InsP$_3$ binds to the specific InsP$_3$ receptor (InsP$_3$-R) that releases Ca^{2+} from intracellular storage sites.[4] The released Ca^{2+} modulates various Ca^{2+}-associated proteins, such as calmodulin and Ca^{2+}/calmodulin-dependent protein kinase II (CaM kinase II).

InsP$_3$-R was originally characterized in 1979 as a protein P$_{400}$ present in the cerebellum in normal mice but not in cerebellar Purkinje-cell–deficient mutant mice[5] and also as PCPP-260[6] and GP-A[7] long before the importance of InsP$_3$ was recognized as a second messenger to release Ca^{2+}. InsP$_3$-binding protein and P$_{400}$ protein were shown to be identical immunologically by using specific monoclonal antibodies.[8] Recently, the primary structure of InsP$_3$-R has been determined through cDNA cloning from rodents.[9-11] Here, we describe the molecular properties of InsP$_3$-R and the role of InsP$_3$-R in Ca^{2+} signaling.

P$_{400}$ PROTEIN AND PURKINJE-CELL–DEFICIENT MUTANT MOUSE

The mouse cerebellar cortex contains five types of neurons. One type, the Purkinje cell, plays an important role in information processing, since the sole output from the cerebellar cortex is the axons of Purkinje cells.[12] SDS-polyacrylamide gel electrophoresis has revealed that a membrane glycoprotein P$_{400}$ (M_r 250 kD) capable of binding to concanavalin A (ConA) was found to be greatly decreased in Purkinje-cell–deficient mutant mice such as pcd, nervous (FIG. 1), and Lurcher.[5,13] P$_{400}$ was also decreased in the staggerer mutant, which has poor dendritic arborization lacking tertiary branched spines that serve as the sites of synapse of the Purkinje cells. The weaver mutant, which lacks only granule cells in the cerebellum, contains a slight increase in the protein content (FIG. 1).[14] This increased content is presumably due to a relative increase in the proportion of Purkinje cells in the cerebellum caused by the death of granule cells. P$_{400}$ was labeled by [^{14}C]leucine *in vivo* and was one of the major proteins phosphorylated. These results suggested that P$_{400}$ has an active function in the Purkinje cells.[5,14]

Purification and Characterization of P$_{400}$

P$_{400}$ is localized in the submicrosomal fraction (P31 fraction) of the mouse cerebellum, solubilized by a solution containing 4% Zwittergent 3-14 and 4 M guanidinium chloride, and purified by Sepharose CL-4B and ConA column into homogeneity. Endo-β-N-acetylglucosaminidase F digestion of P$_{400}$ revealed that P$_{400}$ has asparagine-linked oligosaccharide chains.[15]

The purified InsP$_3$-R was phosphorylated *in vitro* by the catalytic subunit of cAMP-dependent protein kinase (PKA) and by the Ca^{2+}/calmodulin-dependent protein kinase II (CaM kinase II).[16] InsP$_3$-R was phosphorylated by incubating the cultured cerebellar cells with [^{32}P]orthophosphate. Three monoclonal antibodies, 4C11, 10A6, and 18A10, were obtained by Western blot screening. P$_{400}$ concentration increased as the Purkinje cells developed, correlating with the dendritic arborization of the Purkinje cells.[15,17]

Identification of P$_{400}$ as an InsP$_3$ Receptor

InsP$_3$-R was isolated in 1988 from the rat cerebellum and showed the following properties: ConA binding, heparin binding, phosphorylation by PKA, M_r 260 kD, and localization in the cerebellum.[18] These properties were consistent with those of P$_{400}$ (or PCPP-260 or GP-A), which is abundant in the cerebellum. In order to confirm the identity of P$_{400}$ as InsP$_3$-R, we purified the InsP$_3$-R from the mouse cerebellum

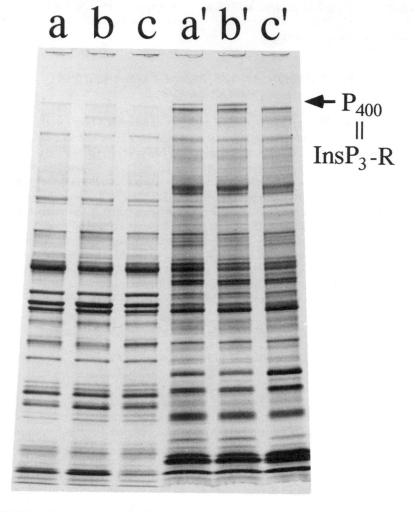

FIGURE 1. SDS-polyacrylamide gel electrophoretic analysis of the protein profile of the cerebellum (a, a') from wild type control, (b, b') from weaver, and (c, c') from nervous mutant mice.[5] Water-soluble fractions are a, b, c, and membrane fractions are a', b', c'. High molecular weight protein, P_{400}, is almost missing in the nervous mutant mouse.

with $InsP_3$ binding as a marker: solubilizing it with Triton X-100 followed by sequential column chromatography by DE52, heparin-agarose, lentil lectin-Sepharose, and hydroxylapatite. The immunoblot analysis revealed that P_{400} and $InsP_3$-R co-migrate at each purification step.[8] All three monoclonal antibodies (mAbs) against P_{400} reacted with the purified mouse $InsP_3$-R. Monoclonal antibody 18A10 immunoprecipitated the $InsP_3$ binding activity. These results demonstrated that P_{400} is identical to the $InsP_3$-R.[8]

cDNA Cloning of the InsP₃ Receptor

We isolated cDNA of InsP₃-R from mouse cerebellar cDNA libraries constructed in phage lambda gt11 expression vectors, using the three mAbs. On the basis of the cDNA sequence, the InsP₃-R is predicted to comprise 2,749 amino acids (M_r 313 kD). From hydropathy profiles, the receptor is predicted to have membrane-spanning domains that cluster around the C-terminus.[9] We have recently found that our previous model was incorrect with regard to the transmembrane topology in which the receptor traverses the membrane an odd number of times. According to the immunoelectron microscopic observations using immunogold (FIG. 2), we revised our model so that the receptor traverses the membrane an even number of times — six times (M1–M6).[19] Thus, a large hydrophilic N-terminal region (2,275 amino acids, 83%) and a short hydrophilic C-terminal region (160 amino acids, 5.8%) are localized on the cytoplasmic side of the membrane. InsP₃-R showed a fragmentary sequence homology with the ryanodine receptor. The transmembrane region and successive C-terminal region of the InsP₃-R showed a striking homology with the ryanodine receptor.[9,10,20]

Recently, novel InsP₃-R subtypes different from the originally cloned receptor were reported.[21-23] We, therefore, call the originally cloned receptor Type I InsP₃-R.

Biochemical Properties of the InsP₃ Receptor

InsP₃ binding activity in the cerebellum is 100–300 times greater than that observed in peripheral tissue.[24] The binding affinities of the purified receptors from cerebellum and aorta smooth muscle are different (K_d 83–100 nM for the cerebellum and 2.4 nM for the smooth muscle), which suggests that there are distinct types of receptors.[25] Here, we focus on the Type I InsP₃-R.

The subunit structure of the InsP₃-R was examined by crosslinking experiments. Agarose-PAGE analysis after crosslinkage of the purified receptor from the cerebellum revealed four distinct bands of M_r 320, 650, 1,000, and 1,250. The same pattern of crosslinking was found with microsome-bound receptor. To investigate whether the native receptor exists in a covalently or non-covalently coupled state, we denatured the purified InsP₃-R using SDS under reducing conditions: the receptor showed an M_r of 320 K in agarose-PAGE. The tetrameric structure of InsP₃-R was slowly destroyed by SDS under non-reducing condition. It is therefore proposed that the native cerebellar InsP₃-R exists as a non-covalently coupled homotetramer of M_r 320,000 subunits.[26]

Deletion analysis suggested that formation of the tetrameric InsP₃-R complex involves the transmembrane domains and/or successive C-termini and that InsP₃ binding is independent of the intermolecular conformation.[27,28]

InsP₃ RECEPTOR AS A Ca²⁺ CHANNEL

InsP₃ Receptor in Reconstituted Lipid Layers

Purified rat InsP₃-R reconstituted into lipid vesicles mediates ⁴⁵Ca²⁺ flux.[4] Purified mouse InsP₃-R in planar lipid bilayers showed InsP₃-induced cation selective channel

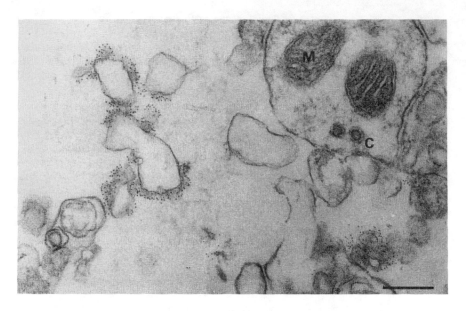

FIGURE 2. Immunogold staining of the InsP$_3$-R in the fuzzy structures associated with the cytoplasmic surface of the smooth vesicotubular structures.[19] These are presumably derived from the dendritic spine apparatus. M: mitochondrion, C: coated vesicule. Scale bar: 0.2 μm.

activity (Ca^{2+} conductance, 26 pS in 54 mM Ca^{2+}; Na$^+$ conductance, 21 pS in 100–500 mM asymmetric Na$^+$ solutions) showing several subconductance states (FIG. 3). Addition of ATP in the presence of InsP$_3$ generated large conductance currents, probably resulting from a change in the full open conductance level or a shift to a greater conductance state. The same high conductance level was occasionally observed even in the absence of ATP. It is likely that ATP modifies the channel to allow it to reach a state of greater conductance. Scatchard analysis of [α-^{32}P]ATP binding to the InsP$_3$-R purified from mouse cerebellum indicated that there is a single ATP-binding site with a K_d value of 17 μM and a B_{max} of 2.3 pmol/μg of protein.[26] Scatchard analysis of binding of tritiated InsP$_3$ to the purified receptor gave a B_{max} value of 2.1 pmol/μg of protein.[8] Three nucleotide-binding consensus sequences (Gly-X-Gly-X-X-Gly) are found in the N-terminal cytoplasmic domain of the mouse InsP$_3$-R but actual site has not been determined.[9] The binding was selective for adenine nucleotides and the affinity is in the order ATP > ADP >> AMP.[26] 100 μM AMP-PCP, a nonhydrolyzable analogue of ATP, increased the open probability of the InsP$_3$-gated channels of the aortic sarcoplasmic reticulum twofold.[29] The receptors from cerebellum and smooth muscle are regulated similarly by adenine nucleotides. In the liposome system, ATP stimulated Ca^{2+} flux from lipid vesicles containing cerebellar InsP$_3$-R in a concentration-dependent manner. ATP increased InsP$_3$ induced Ca^{2+} flux at 1–10 μM, but the effect diminished between 0.1 and 1.0 mM.[30] However, in the reconstituted InsP$_3$-R into planar lipid bilayers the channel opening was most effectively stimulated at 0.6 mM ATP,[26] increased at 1–10 μM ATP, but decreased at 0.1–1 mM

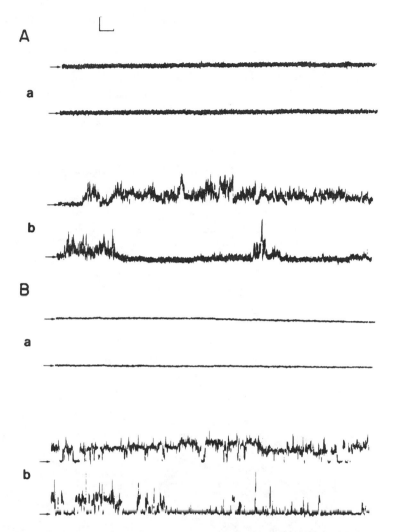

FIGURE 3. Recordings of single-channel currents mediated by the purified InsP₃-R incorporated into planar lipid bilayers.[26] (A) Ca²⁺ currents were recorded after the addition of 2 μg of the InsP₃-R to one chamber termed cis, containing 125 mM Tris, 250 mM HEPES (pH 7.4) and 0.1 μM free Ca²⁺. The other chamber, designated trans, contained 53 mM Ca(OH)₂ and 250 mM HEPES (pH 7.4). (a) No fluctuation in current was observed before the addition of InsP₃. (b) Channel opening was observed after addition of 4.8 μM InsP₃ to the cis chamber. (B) Na⁺ currents mediated by the InsP₃-R were recorded by asymmetric NaCl solutions. The cis chamber contained 0.1 M NaCl, 0.1 μM free Ca²⁺, and 5 mM Tris-HEPES (pH 7.4). Recordings are (a) before and (b) after addition of 4.8 μM InsP₃ to the cis chamber. Vertical calibration: 0.5 pA, horizontal calibration: 5 sec. Arrows to the left indicate the closed state.

ATP.[30] The discrepancy may be due to the difference in the composition of lipids or buffer. Ca^{2+} inhibits InsP₃ binding to microsomal fractions of the cerebellum, but sensitivity to Ca^{2+} is lost in the purified receptor, suggesting some regulatory molecules present in microsomal fractions.[18] Calmodulin antagonists W7, W13, and CGS-934313 inhibited InsP₃-induced Ca^{2+} mobilization in rat liver epithelial cells.[31] It was found that InsP₃-R binds to calmodulin in the presence of Ca^{2+}, but the addition of calmodulin did not affect InsP₃ binding to the cerebellar InsP₃-R.[8] Various regulatory elements of the InsP₃-R are schematically summarized in FIGURE 4.

Expressed InsP₃ Receptor in Fibroblast Cell Has Ca^{2+} Releasing Activity

We transfected neuroblastoma/glioma hybrid cell line NG108-15 and fibroblast L cell with InsP₃-R cDNA with a β-actin promoter at the 5' end to examine whether the clone encodes an InsP₃-binding sequence and a Ca^{2+} release channel.[9,32] The cells contain an endogenous protein immunoreactive against the three mAbs. The M_r of the endogenous InsP₃-R was smaller than that of the receptor from mouse cerebellum (type I InsP₃-R). The protein expression of the type I InsP₃-R derived from the cDNA is apparently coupled with the elevation of InsP₃-binding activity. The expressed protein

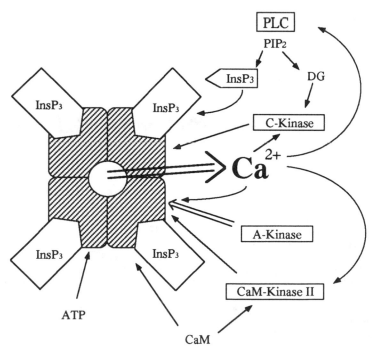

FIGURE 4. Interactions of the InsP₃-R (shaded) with enzymes of other cell-signaling systems. IP₃: inositol 1,4,5-trisphosphate.

displayed high affinity and specificity for $InsP_3$ and a high binding capacity as does the $InsP_3$-R in cerebellar microsomes.[32]

The stable transformants that express the type I $InsP_3$-R have been established in an L-fibroblast cell line (FIG. 5). Direct evidence for Ca^{2+} releasing activity of the receptor was demonstrated by type I receptor expressed in natural membranes of non-neuronal cells. $InsP_3$ releases only a fraction of the Ca^{2+} within cells, suggesting that only some of the Ca^{2+} pools are sensitive to $InsP_3$. It is believed that $InsP_3$-sensitive pools possess $InsP_3$-R. Ca^{2+} ion release experiments revealed that the $InsP_3$-sensitive Ca^{2+} pools in L-fibroblast transformant are larger in size than those in non-transfected cells (FIG. 6). The expression of type I receptor in L-fibroblast may cause some $InsP_3$-insensitive Ca^{2+} pools in L-fibroblasts to be converted into $InsP_3$-sensitive Ca^{2+} pool in the transfected cells. The increase in maximal Ca^{2+} release suggests that the cDNA-derived $InsP_3$-R are distributed not only into $InsP_3$-sensitive pools, but also into $InsP_3$-insensitive pools in transfected cells. The EC_{50} value for $InsP_3$-induced Ca^{2+} release in the transfected cells is about 10-fold lower than that in the non-transfected cells. The greater sensitivity of the transformant may be a result of several factors. First, the measured EC_{50} value is not necessarily a reflection of the actual K_m value for $InsP_3$-induced Ca^{2+} release. It probably results from the greatly increased number of $InsP_3$-R. Second, cDNA-derived $InsP_3$-R may differ from endogenous L-fibroblast receptors in the relative effectiveness of coupling between receptor occupancy and channel opening. Third, the increased sensitivity may be due to a considerably decreased relative effectiveness of proteins that normally control the activity of the receptor, such as protein kinase A.[32]

LOCALIZATION OF THE InsP₃ RECEPTOR IN THE CENTRAL NERVOUS SYSTEM AND PERIPHERAL TISSUES

Both immunohistochemistry (FIG. 7) and *in situ* hybridization showed that the $InsP_3$-R is mostly localized in cerebellar cortex in which Purkinje cells are predominant

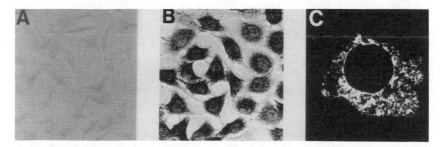

FIGURE 5. $InsP_3$-R expressed after transfection of $InsP_3$-R cDNA into L-fibroblasts.[32] The figure shows immunohistochemical staining of the receptor on L-fibroblasts (using anti-P_{400} monoclonal antibody 4C11, visualized with fluorescein isothiocyanate-labeled goat anti-rabbit IgG) (A) before and (B) after transfection. (C) A confocal section of a single transfected cell strongly fluorescently labeled. Original magnifications A: ×210, B: ×250, C: ×100.

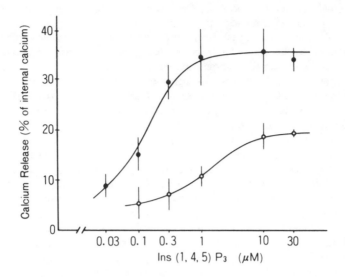

FIGURE 6. Dose-response curves for InsP₃-R-induced Ca²⁺ release from membrane fractions of InsP₃-R L-fibroblast cells expressing the InsP₃-R (closed circles) and from fibroblasts in which the expression vector alone had been transfected (open circles).[32] Results are shown as % of the Ca²⁺ content before addition of InsP₃ and are mean values ± SD from six assays. The membrane fractions were incubated in medium containing ⁴⁵Ca²⁺ and EGTA was added to buffer free Ca²⁺ to 200–400 nM. Ca²⁺ uptake was initiated by addition of ATP (5 mM) and an ATP-regenerating system (creatine phosphate and creatine kinase). After 17 min incubation at 30°C, InsP₃ was added. Samples were filtrated at 16 min and 17 min 40 sec, and the amount of the ⁴⁵Ca²⁺ bound to filters was assayed by liquid scintillation counting. A slight difference in Ca²⁺-pumping activity was detected between the two cell types. In each experiment, the level of ⁴⁵Ca²⁺ accumulated after 17 min in the membrane fractions from cells expressing the receptor was about 85% of that in the fractions from cells containing only the expression vector.

sites.[15,17,33,34] InsP₃-R was widely localized throughout brain—cerebral cortex, nucleus accumbens septi, anterior olfactory nucleus, caudate-putamen, cerebellar nuclei, and in relatively lower concentrations in the amygdaloid cortex, prepiriform cortex, dentate gyrus, olfactory tubercle, precommissural hippocampus, hypothalamus, substantia nigra and pons.[34] These localizations agree well with the sites for [³H]InsP₃ binding.[35] InsP₃-R mRNA and proteins were also located in peripheral tissue such as thymus, heart, lung, liver, spleen, kidney, uterus, oviduct, and testis.[15,33] In lung sections, bronchioles and arteries were labeled. These results may relate to recent finding that endothelin released from endothelial cells and also from renal and tracheal epithelial cells acts as a potent vasoconstrictor and bronchoconstrictor by stimulating the PI turnover system.[36] Strong signals were observed in the smooth muscle cell of oviduct and uterus: the tunica muscularis of the oviduct, and myometrium of the uterus.[33] InsP₃-induced Ca²⁺ release has been shown to play an important role in muscle contraction.[37,38] Our finding that the InsP₃ receptors are plentiful in smooth muscle may

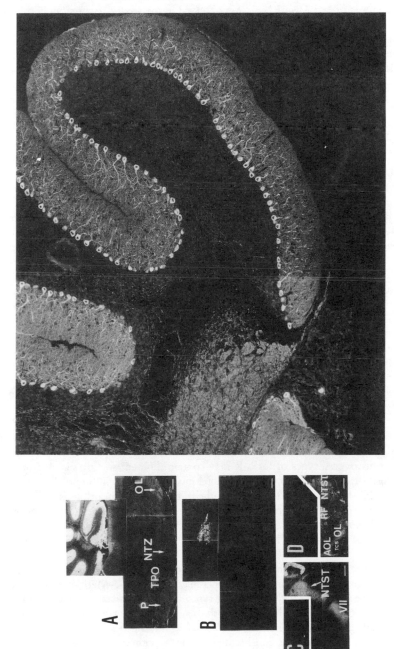

FIGURE 7. Immunohistochemical staining of the Purkinje cells by the antibody against type I InsP₃-R.³⁴ Purkinje cells are strongly stained.

suggest a function of Ca^{2+} suppliers for the activation of smooth muscle contraction. Large amounts of InsP$_3$-R mRNA were observed in the secondary oocytes within the Graafian follicles.[33]

Subcellular localization of the InsP$_3$-R in mouse cerebellar Purkinje cells was studied by immunogold technique with the mAbs.[19] The InsP$_3$-R was localized on the smooth endoplasmic reticulum (ER) (especially on the stacks of flattened smooth ER, subsurface cisterns, and spine apparatus, and less so on the rough ER and the outer nuclear membrane). Receptors could not be localized on the plasmalemma, synaptic densities, mitochondria, or Golgi apparatus. However, recently InsP$_3$-R like immunoreactivity was found on the surface membrane.[39]

LOCALIZATION OF InsP$_3$ RECEPTOR–LIKE PROTEIN IN PLASMALEMMAL CAVEOLAE

Immunolocalization study has been successful in the case of the Purkinje cell where P_{400} is concentrated in an extraordinary degree.[15,17] Judging from the intercellular differences concerning the physiological role of Ca^{2+}, it is plausible that cells in other tissues, especially non-excitable cells, may show a totally different distribution of InsP$_3$-Rs. We found that the immunocytochemical labeling with a mAb raised to P_{400} (mAb4C11) was localized in the plasma membrane of the endothelium, smooth muscle cell, and keratinocyte.[39] Surface biotinylation experiments showed that the antibody recognizes the plasmalemmal protein of 240 kD. Another monoclonal anti-P_{400} antibody (mAb18A10) did not show positive labeling in the cytochemical experiment, but in the surface biotinylation experiments it also bound to the 240 kD protein on the cell surface. The third anti-P_{400} antibody (mAb 10A6) did not label the plasma membrane nor recognize the cell surface protein in immunochemistry.[39] Because the all three mAbs recognize the type I InsP$_3$-R, the plasmalemmal 240 kD protein is possibly a product of a distinct gene from the type I InsP$_3$-R, but the 240 kD protein is likely to show structural homology to InsP$_3$-R. The presence of the type I InsP$_3$-R in the plasma membrane has been thought unlikely since proteins localized in both ER and the plasmalemma have not been known. However, type II or III InsP$_3$-R or other yet discovered types of InsP$_3$-R may contain different sorting signals from the type I InsP$_3$-R and be targeted to the plasma membrane. Our immunolabeling study with mAb4C11 also showed that the endothelial ER and the smooth muscle sarcoplasmic reticulum (SR) are different in reactivity: the former was not labeled positively while the latter was decorated by the antibody.[39] The result indicates that InsP$_3$-R responsible for the intracellular discharge of Ca^{2+} may be diversified in various cell types.

It is interesting to note that the 240 kD protein was localized exclusively in the caveola and not distributed in the entire plasma membrane. 5-Methyltetrahydrofolic acid (the folate receptor),[40] caveolin (a v-src tyrosine kinase substrate),[41] and the 240-kD protein are reported to be localized to the caveola in normal cells. Although the function of the 240 kD protein is not known at present, the localization of an InsP$_3$-R–like protein in the caveola is consistent with the hypothesis that the plasmalemmal differentiation is involved in regulation of cytosolic Ca^{2+} concentration. As indicated from morphological studies, the caveola is closely associated with ER in the endothelium and SR in the smooth muscle, and thus may be related to the Ca^{2+} storage function of the latter organelles. It is essential to elucidate the function of the 240 kD protein.[39]

HETEROGENEITY OF InsP₃ RECEPTOR

Heterogeneity due to alternative splicing was found in the rat,[27,42] human,[42] and mouse[43] InsP₃-R. One alternative splicing in the InsP₃ binding site is a 45-nucleotide sequence coding for 15 amino acids (named SI) (FIG. 8).[43] In the mouse brain, a relative quantity of an mRNA containing the SI is high in the cerebral cortex (88%) and hippocampus (69%), whereas an mRNA lacking the SI is dominant in the cerebellum (85%) and spinal cord (75%). The InsP₃-R containing the SI domain (InsP₃-RSI) shows a peak at P12 and the InsP₃R lacking the SI domain (InsP₃-RSI⁻) develops later than InsP₃-RSI but gradually increases to adult stage. The change of splicing pattern in the InsP₃ binding region may be involved in the development and neuronal function of the cerebellar Purkinje cells, probably by changing the binding affinity of the receptor for InsP₃. In peripheral tissues, the ratio of InsP₃-RSI and InsP₃-RSI⁻ forms differs from tissue to tissue.[43]

Another alternatively splicing segment (SII) is a 120 nucleotide sequence (40 amino acids) between two PKA phosphorylation sites (FIG. 8).[42,43] Within the SII segment, there exist three more subsegments, A, B, and C coding for 23, 1, and 16 amino acids, respectively. We detected the following four splicing variants, i.e., SII (A + B + C, 40 amino acids), SIIB⁻ (A + C, 39 amino acids), SIIBC⁻ (A, 23 amino acids), and SIIABC⁻ (complete deletion of A, B, and C, 40 amino acid deletion). In the mouse

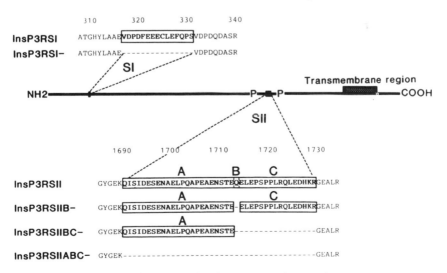

FIGURE 8. Heterogeneity of InsP₃-R produced from RNA splicing of the SI and SII regions.[43] SI is localized near the InsP₃-binding site. SII is located at the center of two phosphorylation sites, indicated by P.

central nervous system, the SIIB⁻ subtype is predominant (50–54%) and the SIIABC⁻ is present at lower levels. The SIIABC⁻ is a predominant splicing subtype in spinal cord (54%). In the peripheral tissues, we detected only SIIABC⁻ subtype mRNA. Thus, the SII, SIIB⁻, and SIIBC⁻ subtypes are brain specific.[43]

As mentioned above, the SII segment is located between the two PKA phosphorylation sites. Because the PKA phosphorylation regulates the Ca^{2+} releasing activity, the splicing at the SII segment will play an important role for the regulation of the function of InsP₃-R.[42] A schematic model of the InsP₃-R is shown in FIGURE 9.

InsP₃ RECEPTOR IN *DROSOPHILA MELANOGASTER*

In insects, InsP₃ is thought to act as a second messenger in the sensory signal transduction of vision and olfaction. It is well established that cGMP and cGMP-dependent phosphodiesterase function in vertebrate visual transduction.[44] The transduction reaction cascade in the fly retina is thought to consist of the following successive events: excitation of the photoreceptor rhodopsin by light stimuli, activation of a G protein, and subsequent activation of phospholipase C, resulting in the production of InsP₃.[45] InsP₃-induced Ca^{2+} release appears to lead to depolarization of the photoreceptor cells.[46] In vertebrate olfactory transduction, cAMP is a potential intracellular messenger that opens the cAMP-gated ion channel involved in depolarization of olfactory receptor cells.[47,48] Adenylate cyclase concentrations in antennae from the cockroach were very low. The cockroach pheromone, periplanone B, stimulated PLC, resulting in accumulation of InsP₃ in the antennae.[49,50] It is thus likely that InsP₃ is also involved in olfactory transduction in insects. We isolated a InsP₃ cDNA clone from

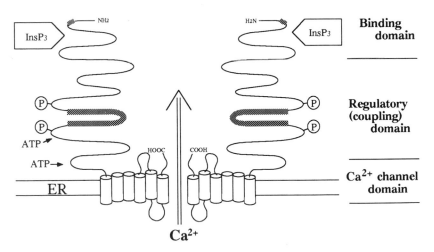

FIGURE 9. Schematic model of InsP₃-R showing transmembrane topology. One half of the homotetrameric structure is depicted. RNA splicing sites are shaped. Both N- and C-terminals are on the cytoplasmic side. Number of transmembrane sites is six. P: phosphorylation sites, ATP: ATP-binding sites.

Drosophila melanogaster.[51] The polypeptide encoded by the cDNA was functionally expressed and showed characteristic InsP$_3$-binding activity. It was expressed throughout development but predominantly in the adult. Localization of the InsP$_3$-R mRNA in adult tissues suggests strong expression in the retina and antennae, indicating the involvement of the InsP$_3$-R in visual and olfactory transduction. In addition, the InsP$_3$-R mRNA is abundant in the legs and thorax, which are primarily muscular system sites. Such localization is consistent with the quantitatively predominant sites for [^{3}H]InsP$_3$ binding in *Drosophila* and the fleshfly (*Boettcherisca peregrina*).[51]

ROLE OF InsP$_3$ RECEPTOR ON Ca^{2+} SIGNALING AND EGG ACTIVATION

Contribution of InsP$_3$ Receptor to the Mechanism of Ca^{2+} Wave and Ca^{2+} Oscillation

The concentration of cytoplasmic free Ca^{2+} increases in various stimulated cells in a wave (Ca^{2+} wave) and in periodic transients (Ca^{2+} oscillations).[52] These phenomena are explained by InsP$_3$-induced Ca^{2+} release (IICR) and Ca^{2+}-induced Ca^{2+} release (CICR) from separate intracellular stores.[53] However, detailed and decisive evidence is lacking. A dramatic, transient increase in the intracellular Ca^{2+} concentration occurs at fertilization in the eggs as a Ca^{2+} wave. Ca^{2+} transient is due to release of intracellular Ca^{2+} and is required for exocytosis of cortical granules to prevent polyspermy and for cell cycle progression.[54,55] Fertilized hamster eggs exhibit repetitive Ca^{2+} transients.[56] IICR is suggested to occur in fertilized eggs of the sea urchin, frog, and hamster and CICR has been detected in sea urchin eggs but it has been difficult to obtain direct evidence for operation of IICR or CICR after eliminating IICR in functioning cells under physiological conditions since specific blocker were not available. IICR and CICR are mediated by the InsP$_3$-R and ryanodine receptor, respectively.[2] Among the three monoclonal antibodies we raised, 18A10 mAb, which recognizes an epitope close to the proposed Ca^{2+} channel region in the COOH terminus of the receptor protein, inhibits IICR in mouse cerebellar microsomes. The mAbs 4C11 and 10A6, which recognize the NH$_2$ terminal and middle regions, respectively, block neither InsP$_3$ binding nor Ca^{2+} release in microsomes.[57] We tested whether 18A10 could block Ca^{2+} release induced by injection of InsP$_3$ or by sperm in hamster eggs. 18A10 antibody completely blocked sperm-induced Ca^{2+} waves and Ca^{2+} oscillations (FIG. 10).[58] The results indicate that Ca^{2+} release in fertilized hamster eggs is mediated solely by the InsP$_3$-R. Together with the evidence that the injection of 18A10 mAb in eggs also suppressed both Ca^{2+}-induced Ca^{2+} wave and Ca^{2+} oscillation, it is indicated that Ca^{2+} release in fertilized hamster eggs is mediated not by CICR but solely by the InsP$_3$-R and Ca^{2+}-sensitized IICR.

InsP$_3$ Receptor and Egg Activation

There is evidence that activation of *Xenopus* eggs is mediated by Ca^{2+} release through activation of a putative receptor of InsP$_3$ on the ER,[59-64] however little is known

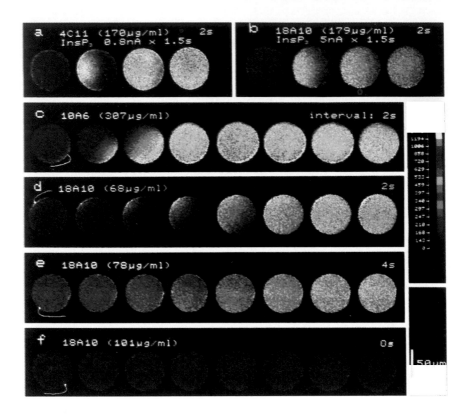

FIGURE 10. Pseudo-colored Ca^{2+} images from the rising phase of Ca^{2+} transients.[58] (a and b) IICR in the presence of 4C11 or 18A10 with injection pulse of $InsP_3$ indicated. (c to f). The first rise in $[Ca^{2+}]$ in fertilized eggs treated with MAb. Sperm were drawn on the computer display. Each image is a combination of F_{340}/F_{360} images accumulated during 0.5 sec intervals every 2 sec in (a to d), 4 sec in (e), and 8 sec in (f).

about the role of $InsP_3$-R in the egg activation. We cloned the $InsP_3$-R expressed in *Xenopus* oocytes and eggs.[65] Primary structure analysis indicated that the cloned cDNA encodes an $InsP_3$-dependent Ca^{2+} channel. Injection of a sequence-specific antisense oligonucleotide of the $InsP_3$-R blocked $InsP_3$-responsive egg activation (cortical contraction), as well as expression of the $InsP_3$-R (Fig. 11).[65] Immunocytochemical staining with an antibody against the $InsP_3$-R fusion protein revealed polarized distribution of the receptor in the cytoplasm of the animal hemisphere in a well-organized ER-like structure and intensive localization in the perinuclear region of stage VI immature oocytes. Dramatic redistribution of the $InsP_3$-R took place during meiotic maturation with relevance to the reorganizations of organelles. $InsP_3$-R was densely localized in the cytoplasm of the animal hemisphere and cortical region of both hemispheres in ovulated unfertilized eggs. After fertilization, $InsP_3$-R changed its distribution drastically in the cortical region. These results imply the predominant role of the $InsP_3$-R in both the formation and propagation of Ca^{2+} waves and activation of egg at fertilization.[65]

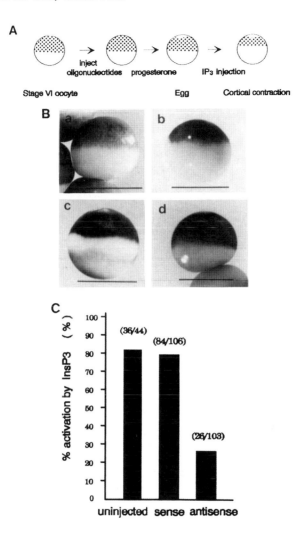

FIGURE 11. Effects of antisense oligonucleotide injection upon InsP₃-responsive egg activation.[65] **(A)** Scheme to show InsP₃ injection results in cortical contraction. **(B)** (a) Before InsP₃ injection. (b) Cortical contraction occurs after InsP₃ injection. (c) InsP₃ induced cortical contraction is inhibited in the presence of antisense nucleotides, while it occurs in the presence of sense nucleotides. **(C)** Antisense blocks InsP₃-responsive egg activation (cortical contraction). Antisense oligonucleotide complementary to the 30 nucleotide sequences of the 5′ flanking and translation start site (141–170, 5′-AACTAGACATCTTGTCTGACATTGCTGCAG-3′) or the corresponding sense oligonucleotide (5′-CTGCAGCAATGTCAGACAAGATGTCTAGTT-3′) was microinjected into fully grown stage V1 oocytes. Injected oocytes meiotically matured with 5 μg/ml progesterone were assayed for InsP₃-responsive cortical contraction. The percentage of eggs undergoing cortical contraction upon InsP₃ injection is shown. Control eggs that received a microinjection of buffer 0.1 mM HEPES (pH 7.8), 10 μM EGTA showed 4.7% (2 of 43) activation.

REFERENCES

1. BERRIDGE, M. J. & R. F. IRVINE. 1989. Inositol phosphates and cell signalling. Nature 341: 197–205.
2. BERRIDGE, M. J. 1993. Inositol trisphosphate and calcium signalling. Nature 361: 315–325.
3. NISHIZUKA, Y. 1988. The molecular heterogeneity of protein kinase C and its implications for cellular regulation. Nature 334: 661–665.
4. FERRIS, C. D., R. L. HUGANIR, S. SUPATTAPONE & S. H. SNYDER. 1989. Purified inositol 1,4,5-trisphosphate receptor mediates calcium flux in reconstituted lipid vesicles. Nature 342: 87–89.
5. MIKOSHIBA, K., M. HUCHET & J. P. CHANGEUX. 1979. Biochemical and immunological studies on the P_{400} protein, a protein characteristic of the Purkinje cell from mouse and rat cerebellum. Dev. Neurosci. 2: 254–275.
6. WALAAS, S. I., A. C. NAIRN & P. GREENGARD. 1986. PCPP-260, a Purkinje cell-specific cyclic AMP-regulated membrane phosphoprotein of Mr 260,000. J. Neurosci. 6: 954–961.
7. GROSWALD, D. E. & P. T. KELLY. 1984. Evidence that a cerebellum-enriched, synaptic junction glycoprotein is related to fodrin and resists extraction with triton in a calcium-dependent manner. J. Neurochem. 42: 534–546.
8. MAEDA, N., M. NIINOBE & K. MIKOSHIBA. 1990. A cerebellar Purkinje cell marker P_{400} protein is an inositol 1,4,5-trisphosphate (InsP₃) receptor protein. Purification and characterization of InsP₃ receptor complex. EMBO J. 9: 61–67.
9. FURUICHI, T., S. YOSHIKAWA, A. MIYAWAKI, K. WADA, N. MAEDA & K. MIKOSHIBA. 1989. Primary structure and functional expression of the inositol 1,4,5-trisphosphate-binding protein P_{400}. Nature 342: 32–38.
10. MIGNERY, G. A., T. S. SUDHOF, K. TAKEI & P. DE CAMILLI. 1989. Putative receptor for inositol 1,4,5-trisphosphate similar to ryanodine receptor. Nature 342: 192–195.
11. MIGNERY, G. A., C. L. NEWTON, B. T. ARCHER III & T. C. SUDHOF. 1990. Structure and expression of the rat inositol 1,4,5-trisphosphate receptor. J. Biol. Chem. 265: 12679–12685.
12. ITO, M. 1984. The Cerebellum and Neural Control. Raven Press. New York.
13. MALLET, J., M. HUCHET, R. POUGEOIS & J. P. CHANGEUX. 1976. Anatomical, physiological and biochemical studies on the cerebellum from mutant mice. III. Protein differences associated with the weaver, staggerer and nervous mutations. Brain Res. 103: 291–312.
14. MIKOSHIBA, K., H. OKANO & Y. TSUKADA. 1985. P_{400} protein characteristic to Purkinje cells and related proteins in cerebella from neuropathological mutant mice: autoradiographic study by ¹⁴C-leucine and phosphorylation. Dev. Neurosci. 7: 179–187.
15. MAEDA, N., M. NIINOBE, K. NAKAHIRA & K. MIKOSHIBA. 1988. Purification and characterization of P_{400} protein, a glycoprotein characteristic of Purkinje cell, from mouse cerebellum. J. Neurochem. 51: 1724–1730.
16. YAMAMOTO, H., N. MAEDA, M. NIINOBE, E. MIYAMOTO & K. MIKOSHIBA. 1989. Phosphorylation of P_{400} protein by cyclic AMP-dependent protein kinase and Ca^{2+}/calmodulin-dependent protein kinase II. J. Neurochem. 53: 917–923.
17. MAEDA, N., M. NIINOBE, Y. INOUE & K. MIKOSHIBA. 1989. Developmental expression and intracellular location of P_{400} protein characteristic of Purkinje cells in the mouse cerebellum. Dev. Biol. 133: 67–76.
18. SUPATTAPONE, S., P. F. WORLEY, J. M. BARABAN & S. H. SNYDER. 1988. Solubilization, purification, and characterization of an inositol trisphosphate receptor. J. Biol. Chem. 263: 1530–1534.
19. OTSU, H., A. YAMAMOTO, N. MAEDA, K. MIKOSHIBA & Y. TASHIRO. 1990. Immunogold localization of inositol 1,4,5-trisphosphate (InsP₃) receptor in mouse cerebellar Purkinje cells using three monoclonal antibodies. Cell Struct. Funct. 15: 163–173.
20. TAKESHIMA, H., S. NISHIMURA, T. MATSUMOTO, H. ISHIDA, K. KANGAWA, N. MINAMINO, H. MATSUO, M. UEDA, M. HANAOKA, T. HIROSE & S. NUMA. 1989. Primary structure and expression from complementary DNA of skeletal muscle ryanodine receptor. Nature 339: 439–445.

21. SUDHOF, T. C., C. L. NEWTON, B. T. ARCHER III, Y. A. USHKARYOV & G. A. MIGNERY. 1991. Structure of a novel InsP$_3$ receptor. EMBO J. **10:** 3199–3206.
22. ROSS, C. A., S. K. DANOFF, M. J. SCHELL, S. H. SNYDER & A. ULLRICH. 1992. Three additional inositol 1,4,5-trisphosphate receptors: molecular cloning and differential localization in brain and peripheral tissues. Proc. Natl. Acad. Sci. USA **89:** 4265–4269.
23. BLONDEL, O., J. TAKEDA, H. JANSSEN, S. SEINO & G. I. BELL. 1993. Sequence and functional characterization of a third inositol trisphosphate receptor subtype, IP3R-3, expressed in pancreatic islets, kidney, gastrointestinal tract, and other tissues. J. Biol. Chem. **268:** 11356–11363.
24. WORLEY, P. F., J. M. BARABAN, J. S. COLVIN & S. H. SNYDER. 1987. Inositol trisphosphate receptor localization in brain: variable stoichiometry with protein kinase C. Nature **325:** 159–161.
25. CHADWICK, C. C., A. SAITO & S. FLEISCHER. 1990. Isolation and characterization of the inositol trisphosphate receptor from smooth muscle. Proc. Natl. Acad. Sci. USA **87:** 2132–2136.
26. MAEDA, N., T. KAWASAKI, S. NAKADE, N. YOKOTA, T. TAGUCHI, M. KASAI & K. MIKOSHIBA. 1991. Structural and functional characterization of inositol 1,4,5-trisphosphate receptor channel from mouse cerebellum. J. Biol. Chem. **266:** 1109–1116.
27. MIGNERY, G. A. & T. C. SUDHOF. 1990. The ligand binding site and transduction mechanism in the inositol 1,4,5-triphosphate receptor. EMBO J. **9:** 3893–3898.
28. MIYAWAKI, A., T. FURUICHI, Y. RYOU, S. YOSHIKAWA, T. NAKAGAWA, T. SAITOH & K. MIKOSHIBA. 1991. Structure-function relationships of the mouse inositol 1,4,5-trisphosphate receptor. Proc. Natl. Acad. Sci. USA **88:** 4911–4915.
29. EHRLICH, B. E. & J. WATRAS. 1988. Inositol 1,4,5-trisphosphate activates a channel for smooth muscle sarcoplasmic reticulum. Nature **336:** 583–586.
30. FERRIS, C. D., R. L. HUGANIR & S. H. SNYDER. 1990. Calcium flux mediated by purified inositol 1,4,5-trisphosphate receptor in reconstituted lipid vesicles is allosterically regulated by adenine nucleotides. Proc. Natl. Acad. Sci. USA **87:** 2147–2151.
31. HILL, T. D., R. CAMPOS-GONZALEZ, H. KINDMARK & A. L. BOYNTON. 1988. Inhibition of inositol trisphosphate-stimulated calcium mobilization by calmodulin antagonists in rat liver epithelial cells. J. Biol. Chem. **263:** 16479–16484.
32. MIYAWAKI, A., T. FURUICHI, N. MAEDA & K. MIKOSHIBA. 1990. Expressed cerebellar-type inositol 1,4,5-trisphosphate receptor, P$_{400}$, has calcium release activity in a fibroblast L cell line. Neuron **5:** 11–18.
33. FURUICHI, T., C. SHIOTA & K. MIKOSHIBA. 1990. Distribution of inositol 1,4,5-trisphosphate receptor mRNA in mouse tissues. FEBS Lett. **267:** 85–88.
34. NAKANISHI, S., N. MAEDA & K. MIKOSHIBA. 1991. Immunohistochemical localization of an inositol 1,4,5-trisphosphate receptor, P$_{400}$, in neural tissue: studies in developing and adult mouse brain. J. Neurosci. **11:** 2075–2086.
35. WORLEY, P. F., J. M. BARABAN & S. H. SNYDER. 1989. Inositol 1,4,5-trisphosphate receptor binding: autoradiographic localization in rat brain. J. Neurosci. **9:** 339–346.
36. HIGHSMITH, R. F., K. BLACKBURN & D. J. SCHMIDT. 1992. Endothelin and calcium dynamics in vascular smooth muscle. Annu. Rev. Physiol. **54:** 257–277.
37. SUEMATSU, E., M. HIRATA, T. HASHIMOTO & H. KURIYAMA. 1984. Inositol 1,4,5-trisphosphate releases Ca^{2+} from intracellular store sites in skinned single cells of porcine coronary artery. Biochem. Biophys. Res. Commun. **120:** 481–485.
38. WALKER, J. W., A. V. SOMLYO, Y. E. GOLDMAN, A. P. SOMLYO & D. R. TRENTHAM. 1987. Kinetics of smooth and skeletal muscle activation by laser pulse photolysis of caged inositol 1,4,5-trisphosphate. Nature **327:** 249–252.
39. FUJIMOTO, T., S. NAKADE, A. MIYAWAKI, K. MIKOSHIBA & K. OGAWA. 1992. Localization of inositol 1,4,5-trisphosphate receptor-like protein in plasmalemmal caveolae. J. Cell Biol. **119:** 1507–1513.
40. ROTHBERG, K. G., Y. S. YING, J. F. KOLHOUSE, B. A. KAMEN & R. G. ANDERSON. 1990. The glycophospholipid-linked folate receptor internalizes folate without entering the clathrin-coated pit endocytic pathway. J. Cell Biol. **110:** 637–649.

41. ROTHBERG, K. G., J. E. HEUSER, W. C. DONZELL, Y. S. YING, J. R. GLENNEY & R. G. ANDERSON. 1992. Caveolin, a protein component of caveolae membrane coats. Cell **68**: 673–682.
42. DANOFF, S. K., C. D. FERRIS, C. DONATH, G. A. FISCHER, S. MUNEMITSU, A. ULLRICH, S. H. SNYDER & C. A. ROSS. 1991. Inositol 1,4,5-trisphosphate receptors: distinct neuronal and nonneuronal forms derived by alternative splicing differ in phosphorylation. Proc. Natl. Acad. Sci. USA **88**: 2951–2955.
43. NAKAGAWA, T., N. OKANO, T. FURUICHI, J. ARUGA & K. MIKOSHIBA. 1991. The subtypes of the mouse inositol 1,4,5-trisphosphate receptor are expressed in a tissue-specific and developmentally specific manner. Proc. Natl. Acad. Sci. USA **88**: 6244–6248.
44. STRYER, L. 1986. Cyclic GMP cascade of vision. Annu. Rev. Neurosci. **9**: 87–119.
45. DEVARY, O., O. HEICHAL, A. BLUMENFELD, D. CASSEL, E. SUSS, S. BARASH, C. T. RUBINSTEIN, B. MINKE & Z. SELINGER. 1987. Coupling of photoexcited rhodopsin to inositol phospholipid hydrolysis in fly photoreceptors. Proc. Natl. Acad. Sci. USA **84**: 6939–6943.
46. MONTELL, C. 1989. Molecular genetics of Drosophila vision. Bioessays **11**: 43–48.
47. PACE, U., E. HANSKI, Y. SALOMON & D. LANCET. 1985. Odorant-sensitive adenylate cyclase may mediate olfactory reception. Nature **316**: 255–258.
48. SKLAR, P. B., R. R. ANHOLT & S. H. SNYDER. 1986. The odorant-sensitive adenylate cyclase of olfactory receptor cells. Differential stimulation by distinct classes of odorants. J. Biol. Chem. **261**: 15538–15543.
49. BOEKHOFF, I., J. STROTMANN, K. RAMING, E. TAREILUS & H. BREER. 1990. Odorant-sensitive phospholipase C in insect antennae. Cell Signal **2**: 49–56.
50. BREER, H., I. BOEKHOFF & E. TAREILUS. 1990. Rapid kinetics of second messenger formation in olfactory transduction. Nature **345**: 65–68.
51. YOSHIKAWA, S., T. TANIMURA, A. MIYAWAKI, M. NAKAMURA, M. YUZAKI, T. FURUICHI & K. MIKOSHIBA. 1992. Molecular cloning and characterization of the inositol 1,4,5-trisphosphate receptor in Drosophila melanogaster. J. Biol. Chem. **267**: 16613–16619.
52. BERRIDGE, M. J. & A. GALIONE. 1988. Cytosolic calcium oscillators. FASEB J. **2**: 3074–3082.
53. BERRIDGE, M. J. 1990. Calcium oscillations. J. Biol. Chem. **265**: 9583–9586.
54. WINKLER, M. M., R. A. STEINHARDT, J. L. GRAINGER & L. MINNING. 1980. Dual ionic controls for the activation of protein synthesis at fertilization. Nature **287**: 558–560.
55. JAFFE, L. F. 1983. Sources of calcium in egg activation: a review and hypothesis. Dev. Biol. **99**: 265–276.
56. MIYAZAKI, S., N. HASHIMOTO, Y. YOSHIMOTO, T. KISHIMOTO, Y. IGUSA & Y. HIRAMOTO. 1986. Temporal and spatial dynamics of the periodic increase in intracellular free calcium at fertilization of golden hamster eggs. Dev. Biol. **118**: 259–267.
57. NAKADE, S., N. MAEDA & K. MIKOSHIBA. 1991. Involvement of the C-terminus of the inositol 1,4,5-trisphosphate receptor in Ca^{2+} release analysed using region-specific monoclonal antibodies. Biochem. J. **27**: 125–131.
58. MIYAZAKI, S., M. YUZAKI, K. NAKADA, H. SHIRAKAWA, S. NAKANISHI, S. NAKADE & K. MIKOSHIBA. 1992. Block of Ca^{2+} wave and Ca^{2+} oscillation by antibody to the inositol 1,4,5-trisphosphate receptor in fertilized hamster eggs. Science **257**: 251–255.
59. BUSA, W. B., J. E. FERGUSON, S. K. JOSEPH, J. R. WILLIAMSON & R. NUCCITELLI. 1985. Activation of frog (Xenopus laevis) eggs by inositol trisphosphate. I. Characterization of Ca^{2+} release from intracellular stores. J. Cell Biol. **101**: 677–682.
60. BUSA, W. B. & R. NUCCITELLI. 1985. An elevated free cytosolic Ca^{2+} wave follows fertilization in eggs of the frog, Xenopus laevis. J. Cell Biol. **100**: 1325–1329.
61. KLINE, D., L. SIMONCINI, G. MANDEL, R. A. MAUE, R. T. KADO & L. A. JAFFE. 1988. Fertilization events induced by neurotransmitters after injection of mRNA in Xenopus eggs. Science **241**: 464–467.
62. KLINE, D. 1988. Calcium-dependent events at fertilization of the frog egg: injection of a calcium buffer blocks ion channel opening, exocytosis, and formation of pronuclei. Dev. Biol. **126**: 346–361.
63. HAN, J. K. & R. NUCCITELLI. 1990. Inositol 1,4,5-trisphosphate-induced calcium release in the organelle layers of the stratified, intact egg of Xenopus laevis. J. Cell Biol. **110**: 1103–1110.

64. DeLisle, S. & M. J. Welsh. 1992. Inositol trisphosphate is required for the propagation of calcium waves in Xenopus oocytes. J. Biol. Chem. 267: 7963–7969.
65. Kume, S., A. Muto, J. Aruga, T. Nakagawa, T. Michikawa, T. Furuichi, S. Nakade, H. Okano & K. Mikoshiba. 1993. The Xenopus IP3 receptor: structure, function, and localization in oocytes and eggs. Cell 73: 555–570.

Ion Channels and Calcium Signaling in Mast Cells[a]

MARKUS HOTH, CRISTINA FASOLATO, AND REINHOLD PENNER

Department of Membrane Biophysics
Max-Planck-Institute for Biophysical Chemistry
Am Fassberg
D-3400 Göttingen, Germany

INTRODUCTION

In response to an external stimulus mast cell granules fuse with the plasma membrane and release substances like histamine, serotonin, and heparin, which can lead to different hypersensitivity reactions such as asthma and allergies.[1] In basophils and mast cells, secretion can be induced by oligomerization of specific IgE receptor by the corresponding antigen. Other physiological stimuli include substance P and tachykinins; mast cells are often clustered around neuropeptide-secreting nerve endings. Secretion can also be induced by non-physiological agents such as compound 48/80 or mastoparan, which are thought to bypass the receptor level and directly activate membrane G proteins.[2] The degranulation mediated by IgE has been reported not to require opening of ion channels,[3] but nevertheless, there is evidence that antigenic stimuli require extracellular calcium[1,4] and ion channels may be involved in stimulus-secretion coupling in mast cells.[5-8]

Calcium plays an important role in the stimulus-secretion coupling of neurons, exocrine and endocrine cells, and also in cells of the immune system. Changes in the intracellular calcium concentration, $[Ca^{2+}]_i$, in response to receptor stimulation usually show a biphasic behavior: an initial Ca^{2+} spike followed by a sustained plateau phase. The former is mainly caused by release of Ca^{2+} from internal stores in response to inositol 1,4,5-trisphosphate production ($InsP_3$), the latter is mainly due to Ca^{2+} entry across the plasma membrane. In mast cells, $[Ca^{2+}]_i$ is believed to have at least a modulatory effect on the signal transduction cascade that leads to secretion. It was shown that the transient increase in $[Ca^{2+}]_i$ is neither sufficient[6,9] nor necessary[6,10,11] to trigger exocytosis. However, a sustained increase in the basal calcium concentration enhances the rate of secretion when combined with an additional stimulus.[12-15] Recent studies

[a] This work supported by Deutsche Forschungsgemeinschaft, Sonderforschungsbereich 236, Hermann- und Lilly-Schilling-Stiftung (R.P.), and the European Molecular Biology Organization (EMBO) (C.F.).
Address correspondence to: Dr. Reinhold Penner, Department of Membrane Biophysics, Max-Planck-Institute for Biophysical Chemistry, Am Fassberg, D-3400 Göttingen, Germany.

have revealed the mechanisms underlying calcium influx in mast cells. Similar mechanisms designed to maintain elevated plateaus of $[Ca^{2+}]_i$ might also be expressed in other cell types.

In mast cells and many other non-excitable cells, there appear to exist two main pathways for Ca^{2+} influx. One mechanism, known as "capacitative" Ca^{2+} entry,[16] is linked to the filling state of Ca^{2+} stores and, upon depletion of cellular Ca^{2+} pools, results in activation of a highly Ca^{2+}-selective current (I_{CRAC} = Calcium Release-Activated Calcium current).[17,18] Another mechanism is provided by nonspecific cation channels, which may be classified as receptor- or second messenger-activated channels.[19] Both Ca^{2+} influx mechanisms appear to be voltage-independent and would provide larger Ca^{2+} entry at hyperpolarizing membrane potentials. Hyperpolarization of the cells could be mediated by potassium or chloride channels, both of which are also found in mast cells.[5,20-22] Furthermore, these channels might also be involved in the degranulation process in a more direct way, as indicated by biochemical studies.[7,8]

CALCIUM RELEASE-ACTIVATED CALCIUM CURRENT

In rat peritoneal mast cells, depletion of internal Ca^{2+} stores activates a Ca^{2+} current.[17,18] This calcium release-activated calcium current (I_{CRAC}) can be activated by at least three different experimental procedures that all result in store depletion ($InsP_3$, ionomycin, Ca^{2+} chelators). Since these procedures share no apparent common mechanism, except for depleting Ca^{2+} stores, it seems unlikely that they involve a direct gating of the current by inositol phosphates or the released Ca^{2+}. This contrasts with previous findings in Jurkat cells[23] and recent work in lobster olfactory receptor neurons[24] where $InsP_3$ appears to activate Ca^{2+} permeable channels directly.

I_{CRAC} is a very selective Ca^{2+} influx pathway with a permeability ratio $P_{Ca^{2+}}/P_{M+}$ ($P_{calcium}/P_{monovalents}$) similar to that of voltage-operated Ca^{2+} channels.[18] Since I_{CRAC} is activated by depletion of Ca^{2+} stores and since it is highly selective for Ca^{2+} over monovalents, it is very likely that I_{CRAC} is the long-sought after Ca^{2+} current that is responsible for "capacitative" Ca^{2+} entry, which had been postulated to exist in a variety of nonexcitable cells. In fact, I_{CRAC} appears to be widely distributed, being found in almost all non-excitable cells that we have tested so far (FIG. 1), including rat peritoneal mast cells, RBL-2H3 (rat basophilic leukemia cells, a mucosal mast cell line), hepatocytes, dissociated thyrocytes, Swiss 3T3 fibroblasts, and HL-60 cells (a human leukemia cell line). Recent evidence suggests that a current with almost identical properties as I_{CRAC} is also present in MDCK cells (an epithelial cell line from kidney),[25] in Jurkat cells[26] (A. Zweifach and R. S. Lewis, personal communication), and in A431 cells (A. Lückhoff and D. Clapham, personal communication). Interestingly, the latter two happen to be cells in which a direct gating of cation channels by $InsP_3$ has also been reported. FIGURE 1 depicts I_{CRAC} in different cell types at 0 mV (left panel) and over the whole voltage range (right panel). In these examples, store depletion was achieved by internal perfusion of $InsP_3$.

In non-excitable cells, Ca^{2+} influx following store depletion is often studied using Mn^{2+} as a Ca^{2+} tracer, taking advantage of its ability to permeate through Ca^{2+} channels and to quench Fura-2 fluorescence.[27,28] We have recently found that I_{CRAC} conducts a small but measurable Mn^{2+} current.[29] In the presence of intracellular BAPTA [1,2-bis(2-aminophenoxy)ethane-N,N,N',N'-tetraacetic acid], a Mn^{2+} current

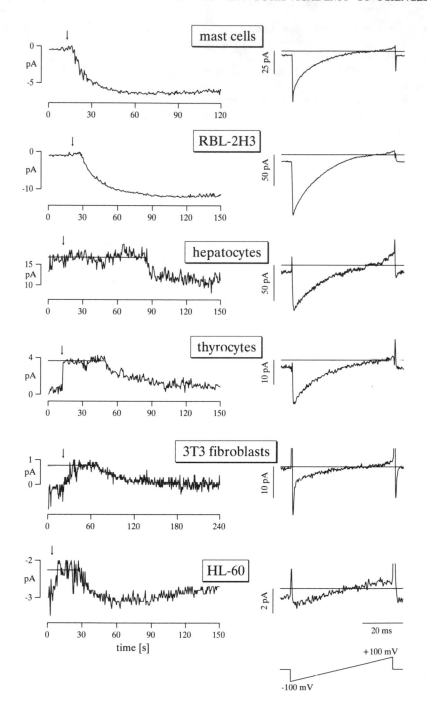

through I_{CRAC} was recorded in isotonic $MnCl_2$ (100 mM). Its amplitude is 10% of that measured in a solution containing 10 mM Ca^{2+}. However, there was no significant quench of Fura-2 fluorescence due to the presence of intracellular BAPTA. A strong quench of Fura-2 fluorescence could be measured after store depletion when omitting intracellular BAPTA, so that all the incoming Mn^{2+} is captured by the fluorescent dye. These findings further strengthen the hypothesis that I_{CRAC} is a ubiquitous mechanism whereby cells accomplish Ca^{2+} influx to refill depleted calcium stores. In mast cells, I_{CRAC} is responsible for the largest part of the sustained Ca^{2+} plateau following receptor stimulation.[30]

NON-SELECTIVE CATION CHANNELS

In response to receptor stimulation (e.g., with compound 48/80) at least two different types of non-selective cation channels can be activated in rat peritoneal mast cells. The first one is a cation channel of 50 pS unitary conductance that is responsible for small whole cell currents (5–50 pA at a holding potential of –40 mV). Channel activation is likely mediated by a G protein as GTPγS mimicks and GDPβS inhibits receptor-mediated activation of these channels. Channel activity is subject to negative feedback inhibition through protein kinase C and high $[Ca^{2+}]_i$.[31,32] Although activation of 50 pS channels is often associated with Ca^{2+} mobilization from intracellular stores, the 50 pS channel is not directly activated by either Ca^{2+} or $InsP_3$.

FIGURE 2 depicts membrane currents measured in the whole cell configuration of the patch-clamp technique and changes in $[Ca^{2+}]_i$ measured with the fluorescent dye Fura-2 during application of 48/80. These experiments were carried out in the presence of intracellularly applied heparin to prevent $InsP_3$-mediated Ca^{2+} influx through I_{CRAC}. The figure demonstrates the relationship between the size of the 50 pS currents and the resulting changes in $[Ca^{2+}]_i$ at different external calcium concentrations for individual cells and the mean relationship. Although the change in $[Ca^{2+}]_i$ elicited by activation

---

FIGURE 1. Activation of I_{CRAC} in different non-excitable cells. The left panel depicts the temporal pattern of activation of an inward current recorded at 0 mV holding potential during perfusion with the standard pipette solution supplemented with $InsP_3$ (Amersham, 10 μM) and the Ca^{2+} chelator EGTA (10 mM). Establishment of the whole cell mode of the patch clamp technique is indicated by the arrow. Immediately after breaking the patch, voltage ramps from –100 mV to +100 mV (duration 50 msec) were applied. The right panel shows these voltage ramps after activation of the inward current corrected by voltage ramps before activation of the inward current. For details of the pulse protocol see Hoth and Penner.[17] **Methods and solutions.** For details see von zur Mühlet *et al.*[11] and Hoth and Penner.[17] Patch-clamp experiments were done in the tight-seal whole-cell configuration[48] at 23–27°C in standard Ringer's solution containing (in mM): NaCl 140, KCl 2.8, $CaCl_2$ 10, $MgCl_2$ 2, glucose 11, HEPES-NaOH 10, pH 7.2. Sylgard-coated patch pipettes had resistances between 2–5 MΩ after filling with standard internal solution which contained (in mM): K-glutamate 145, NaCl 8, $MgCl_2$ 1, Mg-ATP 0.5, HEPES-KOH 10, pH 7.2. Fura-2 pentapotassium salt (Molecular Probes) was regularly added to the internal solution (100 μM). Extracellular solution changes (in case of the application of compound 48/80) were made by local application from a wide-tipped micropipette. The $[Ca^{2+}]_i$ was monitored (using the fluorescent dye Fura-2) with a photo-multiplier-based system.[49]

of 50 pS channels increased with increasing external calcium, the amount of current declined with increasing calcium (FIG. 2, C). This behavior was also observed in a single cell when alternately perfused with Ca^{2+}-free and Ca^{2+}-containing medium. This effect of decreasing current amplitude with increased $[Ca^{2+}]_o$ arises from a decrease in the overall channel open probability accompanied by a minor reduction in the slope conductance.[30] A further attenuation of cation currents is due to the inhibitory effect of elevated $[Ca^{2+}]_i$.[32] Thus, cation channels are tightly regulated by Ca^{2+} ions, exerting efficient negative feedback control on Ca^{2+} influx through 50 pS cation channels.

In about 50% of the cells prolonged applications (tens of seconds) of compound 48/80 evoke a very large inward current, even at resting or buffered $[Ca^{2+}]_i$, probably through non-selective cation channels.[29] The current activates abruptly, in bursts, and only rarely returns to the prestimulus level. The nature and physiological role of this current is still unknown; a cation current this large is expected to cause cell death. Indeed a large conductance, characterized by weak cation selectivity, has been observed in the presence of extracellular ATP (ATP^{4-}). Low concentrations of ATP stimulate Ca^{2+} entry and exocytosis while high doses induce large pores responsible for cell lysis

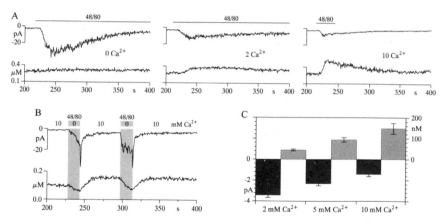

FIGURE 2. Relation among current through 50 pS cation channels and $[Ca^{2+}]_i$. **(A)** Examples of membrane current (upper traces) activated by application of compound 48/80 (5 μg/ml), and the simultaneously determined $[Ca^{2+}]_i$ (lower traces) from three different rat peritoneal mast cells at the indicated concentrations of external calcium. The duration of secretagogue application is indicated by the line above the traces. Heparin (low molecular weight, Sigma, 500 μg/ml) was present in the pipette solution in all three cells to suppress I_{CRAC} (heparin prevents Ca^{2+} release from internal stores through $InsP_3$ which is transiently produced after stimulation with compound 48/80). The holding potential in each case was −40 mV. The inward current elicited by compound 48/80 is due to activation of non-specific cation channels (50 pS channels). **(B)** Examples of membrane current (upper trace) and $[Ca^{2+}]_i$ for a rat peritoneal mast cell bathed in 10 mM external Ca^{2+}. Compound 48/80 was applied in a 0-Ca^{2+} Ringer's solution at the time indicated by the shaded regions. The pipette solution contained 500 μg/ml heparin, and the holding potential was −40 mV. **(C)** Summary of the average increase in membrane current (black bars) and $[Ca^{2+}]_i$ (shaded bars) elicited by compound 48/80 at 2, 5, and 10 mM external Ca^{2+} concentration. The vertical lines indicate ±SEM (n = 16–29). (A: From Fasolato et al.[30] Reprinted with permission.)

and death.[33] Lindau and Fernandez[22] have described another nonselective cation current in rat peritoneal mast cells with a conductance of about 30 pS which, in contrast to the 50 pS channel, the large cation current, and the ATP-gated channel, is activated by an increase in $[Ca^{2+}]_i$. Janiszewski et al. reported that substance P activates large transient currents in RBL-2H3 cells,[34] reaching hundreds of pA, with some cation selectivity. The response was strictly dependent on extracellular Ca^{2+} and could be mimicked by Ca^{2+} ionophores. It is possible that this may reflect positive feedback of Ca^{2+} influx through one influx pathway, which then maintains the activation of cation channels by $[Ca^{2+}]_i$. Evidence for more than one cation channel has also been found in rat peritoneal mast cells by Kuno and Kimura using noise analysis.[35]

Theoretically, all non-selective cation channels could play a role in Ca^{2+} influx provided the current amplitudes are large enough. We have determined the Ca^{2+} selectivity of the 50 pS channels in peritoneal mast cells and found a permeability ratio of P_{Ca2+}/P_{M+} of around 2 in physiological solutions with 2 mM Ca^{2+} (taking activity coefficients into account). This means that 3–4% of the current through 50 pS channels is carried by Ca^{2+} ions. Since inward currents following 48/80 stimulation average less than 4 pA at -40 mV (FIG. 2,C), 50 pS cation channels can at best account for one third of the Ca^{2+} influx necessary to sustain the Ca^{2+} plateaus typically observed in rat peritoneal mast cells.[30]

During the last year, evidence has accumulated that inositol 1,3,4,5-tetrakisphosphate (InsP$_4$) may be involved in Ca^{2+} influx.[36,37] Although InsP$_4$ is unable to activate Ca^{2+} influx by itself, it may enhance Ca^{2+} entry in conjunction with other factors. Thus, in endothelial cells, InsP$_4$ increases the open probability of Ca^{2+}-activated cation channels,[38] whereas in Xenopus oocytes[39] and lacrimal gland cells[40] the additional presence of InsP$_3$ is required to produce the synergistic enhancement of Ca^{2+} influx by InsP$_4$. In a variety of other cell types no actions of InsP$_4$ could be detected. In rat peritoneal mast cells and RBL-2H3 cells, we found neither an effect of InsP$_4$ on I_{CRAC} nor interference with non-selective cation channels.[17,31]

OTHER CHANNELS

In addition to the Ca^{2+}-permeable channels described in the previous two sections, chloride (Cl$^-$) channels and voltage-activated potassium (K$^+$) channels are found in mast cells.[3,20–22] These may be activated and modulated following receptor stimulation or secondary to changes in membrane potential. One of the major functions of the conductances may rest on their ability to set the membrane potential to hyperpolarized levels to support Ca^{2+} influx through voltage-independent cation and calcium currents.

In rat peritoneal mast cells, externally applied secretagogues activate a slowly developing Cl$^-$ current.[21,31] This delayed outward-rectifying current can also be activated by internally applied adenosine-3',5'-cyclic monophosphate (cAMP) or guanosine 5'-O-3-thiotriphosphate (GTPγS) as well as elevated $[Ca^{2+}]_i$. The effect of Ca^{2+} is slow and incomplete however, suggesting that the current is not due to Ca^{2+}-activated Cl$^-$ channels, such as those observed in lacrimal gland cells or Xenopus oocytes. Moreover, with elevated cAMP levels, current activation also occurs in the presence of 2 mM EGTA. The current is reduced by the chloride-channel blocker 4,4'-diisothiocyano-2,2'-stilbenedisulfonate (DIDS). The single-channel conductance of this chloride channel was estimated to a lower limit of 1–2 pS using noise analysis.[21] Since activation of chloride currents would hyperpolarize the membrane potential, it could serve to provide a larger driving force for Ca^{2+} entry via the I_{CRAC} mechanism or through non-selective cation channels. Chloride channels could therefore play a supporting role in the degranulation of mast cells.

Chloride channels can also be found in rat basophilic leukemia (RBL) cells where they are activated by cross-linking of IgE receptors and have a slope conductance of 32 pS.[7] The open-probability increases with depolarizing potentials and the channels are blocked by the Cl^- channel blocker 5-nitro-2-(3-phenylpropylamino) benzoic acid (NPPB) and by the antiallergic drug cromolyn, both in the µM range. NPPB not only inhibits the Cl^- channel but also the serotonin release of these cells with almost the same dose-response relationship. Whether this effect is due to a reduction in driving force for Ca^{2+} influx resulting from the block of Cl^- channels or to some other process controlled by Cl^- channels remains to be determined. It is clear however that depolarizing RBL cells results in impairment of mediator release.[4]

Two main types of K^+ channels have so far been identified in RBL-2H3 cells.[20,22,41] The major resting conductance of this cell line is an inwardly rectifying K^+ channel. This channel seems to be responsible for setting the membrane potential of these cells to negative values (ranging between −50 and −90 mV), which provides a large driving force for Ca^{2+} influx. In physiological K^+ concentrations, the single channel conductance is around 2–3 pS.[22] An increase in $[Ca^{2+}]_i$ leads to a decrease in the open probability of this inward rectifier without affecting the single channel conductance.[41] McCloskey and Cahalan[20] found that this K^+ channel is controlled by a pertussis-sensitive G-protein. Lewis et al.[42] showed that injection of messenger RNA derived from RBL-2H3 cells into Xenopus oocytes resulted in the expression of an inwardly rectifying potassium channel, but cloning of inwardly rectifying potassium channels has so far succeeded only in plant cells.[43] RBL cells also possess another type of K^+ channel (outward rectifier type), which is modulated by non-hydrolyzable GTP analogs.[20] Different GTP-binding proteins seem to regulate the two different potassium channel types such that during activation of these G-proteins, the inward rectifier closes and the outward rectifier opens.

In rat peritoneal mast cells K^+ channel activity was reported by Matthews et al.[21] Less than 5% of the cell preparations showed large conductance, Ca^{2+} and voltage-dependent channels. The reversal potential of the current was more negative than that of the delayed Cl^- current and was affected by changes in external K^+. In a few rare cases, one may detect outward currents in mast cells with the kinetic behavior of an outward rectifying slow-inactivating K^+ channel (Fig. 3). This current may not be seen when clamping the cell to the usual holding potential of 0 mV since it is inactivated at this potential. Changing the holding potential to negative values (−70 mV) lets the channels recover from inactivation and therefore they can be activated by depolarizing voltage pulses. Since this current is observed so rarely, it has not been characterized in great detail. Pharmacological evidence also supports the existence of K^+ channels in the membrane of rat peritoneal mast cells, since K^+ channel blockers induce histamine release.[8]

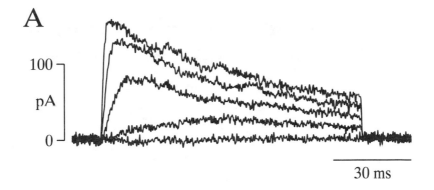

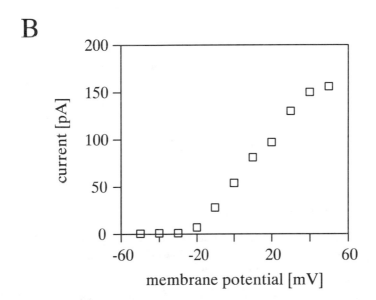

FIGURE 3. Outward currents in rat peritoneal mast cells. (**A**) Outward currents in response to depolarizing voltage pulses (ranging from − 50 to + 50 mV, 20 mM increment) from a holding potential of − 70 mV. (**B**) Current-voltage relationship plotting the peak-current derived from the experiment shown in **A** as a function of the membrane potential.

ION CHANNELS AND MAST CELL ACTIVATION

A simplified overview of the signal transduction pathways, ionic conductances, and second messengers involved in mast cell activation is shown in FIGURE 4. An agonist, such as substance P or compound 48/80, may activate more than one pathway, leading to phospholipase C (PLC) activation, Ca^{2+} release, and Ca^{2+} influx both through the selective pathway (I_{CRAC}) and the non-selective one (50 pS channels). Increased

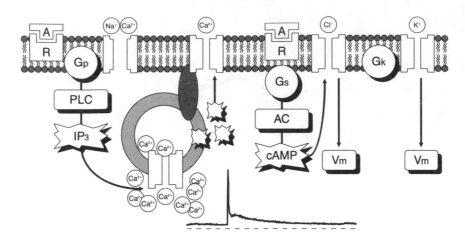

FIGURE 4. Scheme of the different types of channels found in mast cells. From left to right: Non-selective cation channels activated following agonist (A) -receptor (R) binding via a G-protein (Gp). At the same time, InsP$_3$ is produced through activation of phospholipase C (PLC). InsP$_3$ depletes internal Ca^{2+} stores, which leads to activation of I$_{CRAC}$. The signal transduction mechanisms that lead to activation of I$_{CRAC}$ after store depletion are presently unknown. Cl$^-$ channels can be activated by cAMP, which is increased following stimulation of a G-protein (Gs) that activates adenylate cyclase (AC). Voltage-activated K$^+$ channels may be regulated by G-proteins (Gk). Both Cl$^-$ and K$^+$ channels can be responsible for negative membrane potentials (Vm). At the bottom of the figure a typical [Ca^{2+}]$_i$ trace in response to a stimulus like compound 48/80 is shown. The fast Ca^{2+} transient is followed by a sustained plateau phase due to Ca^{2+} influx across the plasma membrane.

levels of [Ca^{2+}]$_i$ and diacylglycerol support the secretory response through the protein kinase C pathway. The same agonist can also activate the delayed Cl$^-$ conductance which, in non voltage-clamped cells, leads to membrane hyperpolarization with further increases of calcium and cation fluxes. It should be noted however that cAMP levels remain unchanged or decrease during 48/80 stimulation but increase with antigen as a stimulus. Since high levels of cAMP also reduce secretion by an unknown mechanism when applied through the patch pipette,[10] the size and the timing of activation of these transduction pathways may explain the ability of different agonists to induce or to suppress the secretory response.[44]

A further level of complexity arises when one considers that the granule content, once secreted, may further activate mast cells. The best known example is displayed by ATP. Mast cells secrete ATP together with histamine[45] and both ATP receptors, coupled to phospholipase C, and ATP-gated channels have been described in mast cells.[33] However, the role of ATP on secretion is still unclear since cellular responses to extracellular ATP, either released by immunocompetent cells or by nerve terminals, can range from [Ca^{2+}]$_i$ rises and membrane depolarization to cell death.[46]

Even within the same type of non-excitable cells, striking differences can be found in two tissue variants: the peritoneal mast cells of the connective tissue type and the basophilic mast cells of the mucosal type. Peritoneal mast cells, under resting conditions, have a very small whole-cell conductance with a resting membrane potential around

0 mV since their dominant conductance, the delayed Cl^- current, is silent. By contrast, the high negative resting potential of the RBL cells is governed by the inwardly rectifying K^+ current.[22] Both cell types are also endowed with the same highly selective Ca^{2+} current activated by store depletion and with a still undefined number of less selective cation pathways. These different channel equipments are likely involved in defining the secretory properties of the two cell types. While in peritoneal mast cells secretion occurs in a few seconds with dramatic morphological changes,[12] a slow release, lasting 20–30 minutes characterizes the secretory process in RBL-2H3 cells.[47]

ACKNOWLEDGMENTS

We would like to thank our colleagues who generously provided us with some of their unpublished results.

REFERENCES

1. METZGER, H., G. ALCARAZ, R. HOHMAN, J.-P. KINET, V. PRIBLUDA & R. QUARTO. 1986. The receptor with high affinity for immunoglobulin E. Ann. Rev. Immunol. **4**: 419–470.
2. MOUSLI, M., C. BRONNER, Y. LANDRY, J. BOCKAERT & B. ROUOT. 1990. Direct activation of GTP-binding regulatory proteins (G-proteins) by substance P and compound 48/80. FEBS Lett. **259**: 260–262.
3. LINDAU, M. & J. M. FERNANDEZ. 1986. IgE-mediated degranulation of mast cells does not require openings of ion channels. Nature **319**: 150–153.
4. MOHR, F. C. & C. FEWTRELL. 1987. The relative contributions of extracellular and intracellular calcium to secretion from tumor mast cells. J. Biol. Chem. **262**: 10638–10643.
5. JANISZEWSKI, J., J. D. HUIZINGA & M. G. BLENNERHASSETT. 1992. Mast cell ionic channels: significance for stimulus-secretion coupling. Can. J. Physiol. Pharmacol. **70**: 1–7.
6. NEHER, E. 1991. Ion influx as a transduction signal in mast cells. Int Arch. Allergy Appl. Immunol. **94**: 47–50.
7. ROMANIN, C., M. REINSPRECHT, I. PECHT & H. SCHINDLER. 1991. Immunological activated chloride channels involved in degranulation of rat mucosal mast cells. EMBO J. **10**: 3603–3608.
8. ELENO, N., L. BOTANA & J. ESPINOSA. 1990. K-channel blocking drugs induce histamine release and ^{45}Ca uptake in isolated mast cells. Int. Arch. Allergy Appl. Immunol. **92**: 162–167.
9. NEHER, E. & W. ALMERS. 1986. Fast calcium transients in rat peritoneal mast cells are not sufficient to trigger exocytosis. EMBO J. **5**: 51–53.
10. PENNER, R. 1988. Multiple signaling pathways control stimulus-secretion coupling in rat peritoneal mast cells. Proc. Natl. Acad. Sci. USA **85**: 9856–9860.
11. VON ZUR MÜHLEN, F., F. ECKSTEIN & R. PENNER. 1991. Guanosine 5'-[β-thio]triphosphate selectively activates calcium signaling in mast cells. Proc. Natl. Acad. Sci. USA **88**: 926–930.
12. PENNER, R. & E. NEHER. 1989. Stimulus-secretion coupling in mast cells. *In* Secretion and Its Control. C. M. Armstrong & G. S. Oxford, Eds.: 296–310. Academic Press. New York.
13. HIDE, M. & M. A. BEAVEN. 1991. Calcium influx in a rat mast cell (RBL-2H3) line. J. Biol. Chem. **266**: 15221–15229.
14. BERTELSEN, H. & T. JOHANSEN. 1991. Dual effect of magnesium on compound 48/80-induced histamine secretion from rat peritoneal mast cells. Pharmacol. Toxicol. **69**: 28–33.
15. NEHER, E. 1988. The influence of intracellular calcium concentration on degranulation of dialysed mast cells from rat peritoneum. J. Physiol. Lond. **395**: 193–214.
16. PUTNEY, J. W. 1990. Capacitative calcium entry revisited. Cell Calcium. **11**: 611–624.

17. Hoth, M. & R. Penner. 1992. Depletion of intracellular calcium stores activates a calcium current in mast cells. Nature 355: 353–356.
18. Hoth, M. & R. Penner. 1993. Calcium release-activated calcium current (I_{CRAC}) in rat mast cells. J. Physiol. Lond. 465: 359–386.
19. Tsien, R. W. & R. Y. Tsien. 1990. Calcium channels, stores, and oscillations. Annu. Rev. Cell Biol. 6: 715–760.
20. McCloskey, M. A. & M. D. Cahalan. G protein control of potassium channel activity in a mast cell line. J. Gen. Physiol. 95: 205–227.
21. Matthews, G., E. Neher & R. Penner. 1989. Chloride conductance activated by external agonists and internal messengers in rat peritoneal mast cells. J. Physiol. Lond. 418: 131–144.
22. Lindau, M. & J. M. Fernandez. 1986. A patch clamp study of histamine-secreting cells. J. Gen. Physiol. 88: 349–368.
23. Kuno, M. & P. Gardner. 1987. Ion channels activated by inositol 1,4,5-trisphosphate in plasma membrane of human T-lymphocytes. Nature 326: 301–304.
24. Fadool, D. A. & B. W. Ache. 1992. Plasma membrane inositol 1,4,5-trisphosphate-activated channels mediate signal transduction in lobster olfactory receptor neurones. Neuron 9: 907–918.
25. Dietl, P. 1993. Depletion of intracellular calcium stores activates a lanthanum inhibitable calcium current in an epithelial cell line (MDCK). Biophys. J. Abstr. 64: A77.
26. McDonald, T. V., B. A. Premack & P. Gardner. 1993. Flash photolysis of caged inositol 1,4,5-trisphosphate activates plasma membrane calcium current in human T cells. J. Biol. Chem. In press.
27. Hallam, T. J. & T. J. Rink. 1985. Agonists stimulate divalent cation channels in the plasma membrane of platelets. FEBS Lett. 186: 175–179.
28. Meldolesi, J., E. Clementi, C. Fasolato, D. Zacchetti & T. Pozzan. 1991. Ca^{2+} influx following receptor activation. Trends Pharmacol. Sci. 12: 289–292.
29. Fasolato, C., M. Hoth & R. Penner. 1993. Multiple mechanisms of manganese-induced quenching of Fura-2 fluorescence in rat mast cells. Pflügers Arch. 286: 3889–3896.
30. Fasolato, C., M. Hoth, G. Matthews & R. Penner. 1993. Ca^{2+} and Mn^{2+} influx through receptor-mediated activation of non-specific cation channels in mast cells. Proc. Natl. Acad. Sci. USA. 90: 3068–3072.
31. Penner, R., G. Matthews & E. Neher. 1988. Regulation of calcium influx by second messengers in rat mast cells. Nature 334: 499–504.
32. Matthews, G., E. Neher & R. Penner. 1989. Second messenger-activated calcium influx of rat peritoneal mast cells. J. Physiol. Lond. 418: 105–130.
33. Tatham, P. E. R. & M. Lindau. 1990. ATP-induced pore formation in the plasma membrane of rat peritoneal mast cells. J. Gen. Physiol. 95: 459–467.
34. Janiszewski, J., J. Bienenstock & M. G. Blennerhassett. 1992. Substance P induces whole cell current transients in RBL-2H3 cells. Am. J. Physiol. 263: C736–C742.
35. Kuno, M. & M. Kimura. 1992. Noise of secretagogue-induced inward currents dependent on extracellular calcium in rat mast cells. J. Membr. Biol. 128: 53–61.
36. Petersen, O. H. 1989. Does inositol tetrakisphosphate play a role in the receptor-mediated control of calcium mobilization? Cell Calcium 10: 375–383.
37. Irvine, R. F. 1992. Inositol phosphates and Ca^{2+} entry: toward a proliferation or simplification? FASEB J. 6: 3085–3091.
38. Lückhoff, A. & D. Clapham. 1992. Inositol 1,3,4,5-tetrakisphosphate activates an endothelial Ca^{2+}-permeable channel. Nature 355: 356–358.
39. DeLisle, S., D. Pittet, B. V. L. Potter, P. D. Lew & M. J. Welsh. 1992. InsP₃ and Ins(1,3,4,5)P₄ act in synergy to stimulate influx of extracellular Ca^{2+} in Xenopus oocytes. Am. J. Physiol. 262: C1456–C1463.
40. Smith, P. M. 1992. Ins(1,3,4,5)P₄ promotes sustained activation of the Ca^{2+}-dependent Cl^- current in isolated mouse lacrimal cells. Biochem. J. 283: 27–30.

41. MUKAI, M., I. KYOGOKU & M. KUNO. 1992. Calcium-dependent inactivation of inwardly rectifying K⁺ channel in a tumor mast cell line. Am. J. Physiol. **262**: C84–C90.

42. LEWIS, D. L., S. R. IKEDA, D. ARYEE & R. H. JOHO. 1991. Expression of an inwardly rectifying K⁺ channel from rat basophilic leukemia cell mRNA in Xenopus oocytes. FEBS Lett. **290**: 17–21.

43. SCHACHTMAN, D. P., J. I. SCHRÖDER, W. J. LUCAS, J. A. ANDERSON & R. F. GABER. 1992. Expression of an inward-rectifying potassium channel by the Arabidopsis KAT1 cDNA. Science **258**: 1654–1658.

44. ALI, H., R. J. R. CUNHA-MELO, W. F. SAUL & M. A. BEAVEN. 1990. Activation of phospholipase C via adenosine receptors provides synergistic signals for secretion in antigen-stimulated RBL-2H3 cells. J Biol. Chem. **265**: 745–753.

45. OSIPCHUK, Y. & M. CAHALAN. 1992. Cell-to-cell spread of calcium signals mediated by ATP receptors in mast cells. Nature **359**: 241–244.

46. ZANOVELLO, P., V. BRONTE, A. ROSATO, P. PIZZO & F. DI VIRGILIO. 1990. Responses of mouse lymphocytes to extracellular ATP. J. Immunol. **145**: 1545–1550.

47. MCGIVNEY, A., Y. MORITA, F. T. CREWS, F. HIRATA, J. AXELROD & R. P. SIRIGANIAN. 1981. Phospholipase activation in the IgE-mediated and Ca²⁺ ionophore A23187-induced release of histamine from rat basophilic leukemia cells. Arch. Biochem. Biophys. **212**: 572–580.

48. HAMILL, O. P., A. MARTY, E. NEHER, B. SAKMANN & F. J. SIGWORTH. 1981. Improved patch-clamp techniques for high-resolution current recording from cells and cell-free membrane patches. Pflügers Arch **391**: 85–100.

49. NEHER, E. 1989. Combined fura-2 and patch clamp measurements in rat peritoneal mast cells. In Neuromuscular Junction. L. C. Sellin, R. Libelius & S. Thesleff, Eds.: 65–76. Elsevier. Amsterdam.

Molecular Basis of the Muscarinic Acetylcholine Receptor[a]

TAI KUBO[b]

International Institute for Advanced Studies
Kyoto 604, Japan

INTRODUCTION

Muscarinic acetylcholine receptors (mAChRs) mediate a variety of the actions of acetylcholine in the central and peripheral nervous systems. The mechanisms of the actions on the cellular level include inhibition of adenylate cyclase, breakdown of phosphoinositides, and modulation of potassium channels, through G proteins.[1] Accumulating evidence has shown heterogeneity among mAChRs on the basis of different binding properties for agonists and antagonists in various tissues. This article deals with the structure of the mAChR, which was revealed by cloning and sequencing of the receptors, and attempts to elucidate the molecular basis of the functional heterogeneity of the mAChR using DNA expression systems.

PRIMARY STRUCTURE OF mAChRs

The mAChR was purified to homogeneity from a membrane fraction of porcine cerebrum using ligand affinity chromatography.[2] The purified protein was digested with trypsin and the resulting peptides were fractionated by reversed-phase HPLC. Five peptides (I–V) were isolated and analyzed for amino acid sequence with the use of gas-phase sequencer (peptide I; ELAALQGSETPGK, peptide II; MPMVDPEAQA-PAK, peptide III; TFSLVK, peptide IV; EPVANQEPVSPXLVQG, peptide V; DDEITQDENTVXXSL.). Two clones (pmACR84 and pmACR60) from a porcine cerebrum cDNA library hybridizable to the peptide I and peptide II probes, were analyzed further for nucleotide sequence.[3] The analysis revealed that the clones contain an open reading frame of 460 amino acid residues including the peptide I–III sequences. Thus the amino acid sequence of porcine cerebrum mAChR was deduced from the cDNA sequences and was designated as m1.[3]

The peptides IV and V, however, were not included in the m1 sequence. This

---

[a] This work supported in part by research grants from the Ministry of Education, Science and Culture of Japan, the Institute of Physical and Chemical Research, and the Japanese Foundation of Metabolism and Diseases.

[b] Correspondence address to: Tai Kubo, Ph.D., International Institute for Advanced Studies, Shimadz N-80, 1 Nishinokyo-Kuwabara-cho, Nakagyo-ku, Kyoto 604, Japan.

finding raised the question of whether these tryptic peptides are derived from a different mAChR protein. An oligodeoxyribonucleotide corresponding to the carboxy-terminal sequence QDENT of peptide V was extended by reverse transcriptase, using porcine cerebral poly(A)$^+$RNA as template.[4] After the single-stranded cDNA was converted to double-stranded followed by cloning into plasmid DNA, the primer-extended cDNA library was screened by a synthetic probe corresponding to the amino-terminal sequence DDEIT of the peptide V. One of the positive clones (pmACR423) was found to encode the amino acid sequences for both peptides IV and V in the same reading frame. Because RNA blot hybridization analysis indicated that porcine atrium contained a relatively large amount of RNA hybridizable with a cDNA probe derived from pmACR423, isolation of cDNA clones that cover the rest of the protein-coding region was completed by screening porcine atrium cDNA libraries.[4] The mAChR encoded by these clones is named m2. Cloning of porcine m2 was also reported by Peralta *et al.*[5] The primary structures of three additional mAChR species, designated as m3, m4, and m5, have subsequently been deduced from the nucleotide sequences of the cloned cDNAs or genomic DNAs.[6-12] Rat m2 cDNA,[13] mouse m1 gene,[14] chicken m2 gene,[15] and *Drosophila* mAChR cDNA and/or gene[16,17] have also been cloned.

In FIGURE 1 porcine m1, m2, m3, and rat m4 sequences are aligned. The hydropathy profiles of the mAChRs are similar and suggest the presence of seven transmembrane segments (I–VII). Primary structures of the mAChRs are also homologous to one another, except that the amino terminal region and the region between segments V and VI (i3) are divergent; some sequence similarity between the subtypes in the i3 region will be discussed later. There are a couple of potential N-glycosylation sites in the amino terminus and several potential sites of phosphorylation by a cAMP-dependent protein kinase in the segment i3 and/or the carboxy terminus. Charged amino acid residues present in these segments are two aspartic acid residues (position 114 and 148 in FIG. 1) that are conserved in all the mAChRs. Involvement of the latter aspartic acid residue in ligand binding to the mAChRs has been suggested,[18] as in the case of the β-adrenergic receptor.[19] These structural features are generally shared by rhodopsin and the β-adrenergic receptor. This suggests that the mAChRs have a transmembrane topology similar to that of bacteriorhodopsin.[3]

PHARMACOLOGICAL CHARACTERIZATION OF MOLECULARLY DEFINED mAChR SUBTYPES

To examine whether these cloned receptors bind muscarinic receptor ligands and whether they have differences in ligand-binding properties, mRNAs specific for the individual mAChRs were synthesized by transcription *in vitro* of the respective cDNAs or genomic DNAs (containing no introns in the protein-coding region) and were injected into *Xenopus* oocytes to yield functional mAChRs.[3,9,20,21] The ligand-binding properties of the four mAChRs expressed in oocytes were examined using cell extracts (TABLE 1). The apparent dissociation constant (K_d) for (−)-[³H]quinuclidinyl benzilate (QNB), estimated by Scatchard analysis, is similar for all the mAChRs (80–130 pM). The apparent K_d values for selective antagonists were obtained by measuring displacement of (−)-[³H]QNB binding by increasing concentrations of the antagonists. Pirenzepine, selective for the M_1 subtype,[22] shows the highest binding affinity for m1 (K_d −

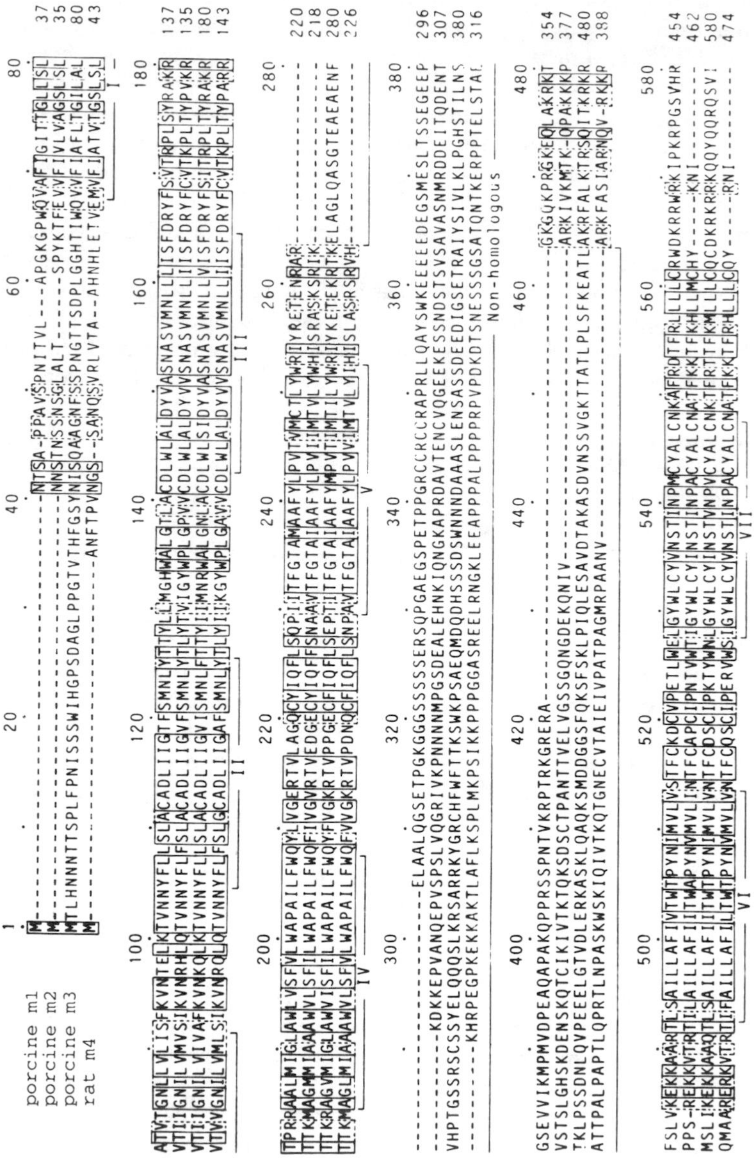

porcine m1
porcine m2
porcine m3
rat m4

TPSRQC----	460
GATR------	466
FHKRVPEQAL	590
GTAR------	478

18 nM). m3 and m4 exhibited intermediate affinities for pirenzepine (K_d = 120–180 nM), and m2 the lowest affinity (K_d = 660 nM). These three classes of affinity for pirenzepine are similar to those reported originally for mAChRs in brain (cerebral cortex and hippocampus), glands, and heart.[22] AF-DX 116, selective for the M_2 cardiac subtype,[23] showed a higher binding affinity for m2 (K_d = 0.73 µM) than for the other mAChR species (K_d = 2.3–3.1 µM). Hexahydrosiladifenidol, selective for the M_2 glandular subtype relative to the M_2 cardiac subtype,[24,25] exhibited the highest binding affinity for m3 (K_d = 4.0–4.4 nM). The affinity for this antagonist decreased in the order of m4 (K_d = 20 nM), m1 (K_d = 51 nM), and m2 (K_d = 280 nM). These results indicate that m1, m2, and m3 correspond most closely to the pharmacologically defined M_1, M_2 cardiac, and M_2 glandular subtypes, respectively. Tissue distribution of the mRNAs encoding these mAChR species also supports this idea. FIGURE 2 shows RNA blot hybridization analysis with specific probes for mAChR species.[26] All four mAChR mRNAs are present in cerebrum, whereas only m2 mRNA is found in heart. Exocrine glands contain both m1 and m3 mRNAs, whereas smooth muscles contain both m2 and m3 mRNAs. Thus the mAChR heterogeneity in tissues with respect to antagonist binding can be accounted for by the presence of molecularly distinct mAChRs or various combinations of them.

CELLULAR RESPONSES MEDIATED BY mAChR SUBTYPES

Various cellular responses, such as the inhibition of adenylyl cyclase activity, the stimulation of phosphoinositide hydrolysis, and the regulation of potassium current in myocardium through G proteins,[27] are known to be induced by mAChR agonists. Coupling of molecularly defined mAChR subtypes with distinct effector systems was investigated using *Xenopus* oocytes and NG108-15 neuroblastoma-glioma hybrid cells[28] as transient and stable expression systems, respectively.[3,20,21,29,30]

Responses in Xenopus Oocytes

FIGURE 3 shows typical ACh responses observed in *Xenopus* oocytes injected with the mRNA specific for m1, m2, m3, or m4. Those are whole-cell currents under

---

FIGURE 1. Alignment of the amino acid sequences of porcine m1, porcine m2, porcine m3, and rat m4. The sequence data have been taken from Kubo *et al.*,[3,4] Akiba *et al.*,[9] and Bonner *et al.*;[6] the amino-terminal sequence of rat m4 has been completed from our unpublished data. Sets of four identical residues at one position are enclosed with solid lines and sets of four identical or conserved residues at one position with broken lines. Conservative amino acid substitutions are defined as pairs of residues belonging to one of the following groups: S, T, P, A, and G; N, D, E, and Q; H, R, and K; M, I, L, and V; F, Y, and W. The nonhomologous sequences that cannot be aligned are also shown. Amino acid residues are numbered beginning with the initiating methionine, and numbers of the residues at the right-hand end of individual lines are given. Positions in the aligned sequences including gaps (−) and in the nonhomologous sequences are numbered beginning with that of the initiating methionine. The putative transmembrane segments I–VII are indicated; the terminals of these segments have been tentatively assigned. The amino acid difference resulting from a nucleotide difference found between the individual clones is Ser at position 336 (in aligned sequences) of m2.

TABLE 1. Apparent Dissociation Constants (K_d) of mAChR Subtypes for Antagonists[a]

Antagonist	K_d (M)				
	m1 (porcine)	m2 (porcine)	m3 (porcine)	m3 (rat)	m4 (rat)
(−)-[³H]QNB	8.4×10^{-11}	1.3×10^{-10}	1.2×10^{-10}	1.0×10^{-10}	9.5×10^{-11}
Pirenzepine	1.8×10^{-8}	6.6×10^{-7}	1.8×10^{-7}	1.3×10^{-7}	1.2×10^{-7}
AF-DX 116	2.5×10^{-6}	7.3×10^{-7}	2.3×10^{-6}	3.1×10^{-6}	2.3×10^{-6}
Hexahydrosiladifenidol	5.1×10^{-8}	2.8×10^{-7}	4.4×10^{-9}	4.0×10^{-9}	2.0×10^{-8}

[a] Data from Akiba et al.[9]

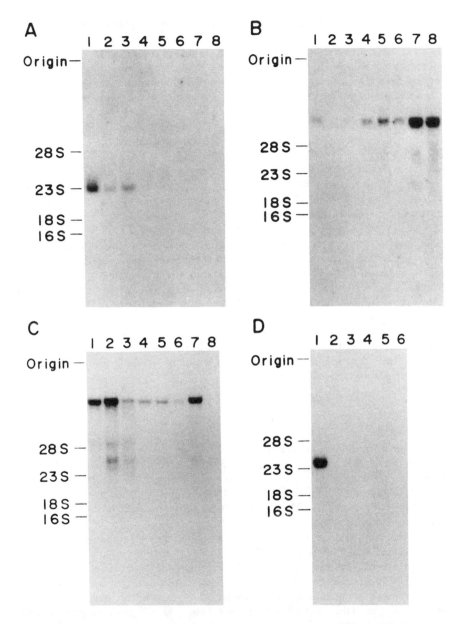

FIGURE 2. Autoradiograms of blot hybridization analysis of poly(A)⁺RNA from porcine (A–C) and rat tissues (D) using probes specific for the mRNAs encoding porcine m1 (A), porcine m2 (B), porcine m3 (C), or rat m4 (D). (A–C) Analysis of poly(A)⁺RNA (15 μg each) from porcine cerebrum (lane 1), lacrimal gland (lane 2), parotid gland (lane 3), small intestine (lane 4), large intestine (lane 5), trachea (lane 6), urinary bladder (lane 7), and atrium (lane 8). The rat RNA species hybridizable with the rat m3-specific probe showed a similar tissue distribution. (D) Analysis of poly(A)⁺RNA (15 μg each) from rat cerebrum (lane 1), submandibular gland (lane 2), small intestine (lane 3), trachea (lane 4), urinary bladder (lane 5), and heart (lane 6). (From Maeda et al.[26] Reprinted with permission.)

voltage clamp at a holding potential of -70 mV. Oocytes implanted with m1 or m3 exhibited an oscillatory inward current (FIG. 3, A and C). In contrast, a typical ACh response induced in oocytes implanted with m2 or m4 comprised an initial smooth inward current followed by an oscillatory component (FIG. 3, B and D). The oscillatory current evoked by activation of m2 or m4 varied in amplitude among the oocytes tested and was undetected in some of them. The latency of the ACh response in m2- or m4-implanted oocytes was shorter than in m1- or m3-implanted oocytes, the former being mostly attributable to the dead-space time in the perfusion system (~ 7 sec). The current response in m1- or m3-implanted oocytes occurred after an additional delay (2–3 sec).

The reversal potential of the oscillatory current induced by activation of each of the four mAChR subtypes, obtained in Ringer's solution, was around -25 mV, which is close to the equilibrium potential of Cl^- in *Xenopus* oocytes.[31] This suggests that the oscillatory current is carried mainly by Cl^-. In contrast, the smooth current mediated by m2 or m4 was reversed in polarity at a potential around 10 mV. This value does not correspond to the equilibrium potential of any single species of ions. Reversal potential measurements in different media performed with EGTA-loaded oocytes suggest that the smooth current evoked by activation of m2 or m4 is carried principally by Na^+ and K^+. Intracellular injection of the Ca^{2+}-chelating agent EGTA almost completely abolished the ACh-activated current in m1- or m3-implanted oocytes, leaving only a small long-lasting inward current. The oscillatory current in m2- or m4-implanted oocytes similarly disappeared after this treatment, whereas the smooth component was virtually unaffected. It is most likely that the activation of the Cl^- current mediated by m1 and m3 results from phosphoinositide hydrolysis leading to Ca^{2+} release from intracellular stores.

Responses in NG108-15 Cells

NG108-15 cells were transfected with cDNA or genomic DNA encoding each of the four mAChR subtypes, using an expression vector carrying the SV40 early gene promoter and the neomycin-resistance marker gene.[29] The expression of each mAChR subtype was confirmed by $(-)$-[^{3}H]QNB binding assay and by blot hybridization analysis of total cellular RNA using mAChR subtype-specific probes. NG108-15 cells endogenously contain the m4 mRNA.[7,29]

The transformed clones were tested for electrophysiological response to ACh under voltage clamp at a holding potential of -30 mV.[29] m1- or m3-transformed cells exhibited a similar response to ACh, which comprised an initial outward current followed by a sustained inward current. The initial outward current was accompanied by an increase in input conductance, as measured by short voltage steps (FIG. 4). The reversal potential of the outward current was -73 to -85 mV, and the current-voltage (*I-V*) relation was apparently linear. The outward current response was reduced or abolished by apamin (0.4 μM) or (+)-tubocurarine (0.2 mM), but was insensitive to tetraethylammonium (1 mM). These results indicate that the initial outward current is principally attributable to activation of a subclass of Ca^{2+}-dependent K^+ currents. Unlike the initial outward current, the secondary inward current induced by ACh was usually accompanied by a decrease in input conductance and by a reduction in

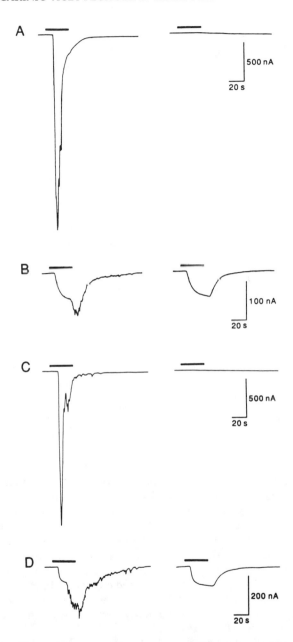

FIGURE 3. ACh-activated currents in *Xenopus* oocytes injected with the mRNA specific for porcine m1 (**A**), porcine m2 (**B**), porcine m3 (**C**), or rat m4 (**D**). Whole-cell currents activated by bath application of 1 µM (**A, B, D**) or 10 nM ACh (**C**) were recorded under voltage clamp at a holding potential of −70 mV before (left) and after (right) intracellular injection of EGTA. Inward current is downward. The duration of ACh application is indicated by bars without taking into account the dead-space time in the perfusion system (∼ 7 sec). The ACh responses in oocytes injected with the rat m3-specific mRNA were similar to the records in (**C**). (**A and B**: From Fukuda *et al.*[20] **C** and **D**: From Bujo *et al.*[21] Reprinted with permission.)

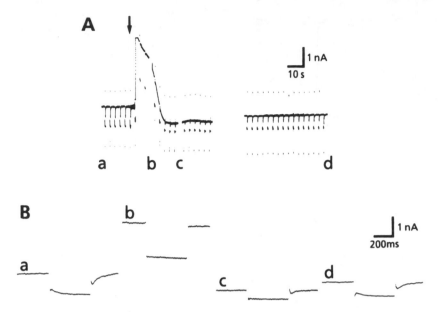

FIGURE 4. ACh response in m1-transformed NG108-15 cells. **(A)** Whole-cell currents activated by applying 3 μl of a 0.1 mM ACh solution at the time indicated by the arrow were recorded under voltage clamp at a holding potential of − 30 mV. Inward current is downward. Repetitive downward deflections are current transients produced by hyperpolarizing steps of 20 mV for 400 msec applied every 3 sec to measure input conductance. An interval of ∼ 5 min separates the trace into two parts, and the trace is interrupted by brief periods (∼ 1 sec) of faster recording. The recorder saturated during the outward current response. **(B)** Expanded records of current transients obtained at the times indicated in A: (a) before applying ACh, (b) during the ACh-induced outward current, (c) during the subsequent inward current, (d) after partial recovery. (From Fukuda *et al.*[29] Reprinted with permission.)

the time-dependent inward current relaxation observed during hyperpolarizing voltage steps (Fig. 4). The steady-state *I-V* curve that was obtained while the sustained inward current was present showed a reduction in the outward rectification in a potential range less negative than approximately − 70 mV (tested up to − 20 mV), where the M current[32,33] is activated. These data indicate that the secondary inward current results primarily from inhibition of the M current, which is known to be present in NG108-15 cells. The initial outward current and the secondary inward current were observed in most of the m1- and m3-transformed cells tested. In contrast, the percentages of responsive cells among the m2- or m4-transformed cells tested, despite their significant amount of (−)-[³H]QNB binding, were not higher than those of nontransfected or vector-transformed control cells.

The effect of carbamylcholine on the formation of [³H]inositol phosphate was examined using [³H]inositol-labeled cells.[29] In m1- or m3-transformed cells, a four- to sevenfold increase in the release of total inositol phosphates occurred in response to carbamylcholine stimulation, as compared with control values obtained without stimulation. But in m2- or m4-transformed cells as well as in nontransfected or vector-transformed cells, no appreciable increase in inositol phosphate release was observed. Studies with the fluorescent indicator dye fra-2 showed that the m1 and m3 expressed in NG108-15 cells, unlike m2 or m4, mediate release of Ca^{2+} from intracellular stores.[30]

Stable expression of the four mAChR subtypes in other mammalian cell lines also shows that m1 and m3 are coupled with stimulation of phosphoinositide hydrolysis,[17,34,35] whereas m2 and m4 are linked mainly with inhibition of adenylate cyclase.[17,34,36] m5 has been shown to be coupled with phosphoinositide hydrolysis efficiently.[10,11] These findings, together with the expression studies using *Xenopus* oocytes suggest that these subtypes are selectively coupled with different effector systems, albeit not exclusively. Of interest in this context is that m3 has a sensitivity to agonist about one order of magnitude higher than that of m1 in mediating phosphoinositide hydrolysis and intracellular Ca^{2+} release in NG108-15 cells as well as the Ca^{2+}-dependent Cl^- current response in *Xenopus* oocytes.[21,30] This is probably attributable, at least partly, to its higher agonist-binding affinity of m3 than that of m1.[21]

LOCATION OF AN mAChR DOMAIN INVOLVED IN SELECTIVE EFFECTOR COUPLING

To localize the region of the mAChR molecules responsible for selective coupling with different effector systems, chimeric mAChR molecules with different combinations of m1 and m2 were expressed in *Xenopus* oocytes from the corresponding cDNA constructs.[37] FIGURE 5(A) schematically shows the structures of different chimeric mAChRs in which corresponding portions of porcine m1 and m2 are replaced with each other. FIGURE 5(B) shows the ACh-induced current responses at − 70 mV membrane potential observed in oocytes implanted with the different chimeric mAChRs and with the parental m1 and m2. Each ACh-induced current was also examined for the effect of EGTA. The ACh-activated inward current mediated by the chimeric mAChR MC2, MC8, or MC10 was oscillatory in nature and was almost completely abolished by EGTA. On the other hand, a typical ACh response mediated by the chimeric mAChR MC4 or MC9 comprised an initial smooth inward current that was virtually unaffected by intracellular injection of EGTA, followed by an oscillatory component that disappeared after this treatment. The average peak inward currents activated by 1 μM ACh were compared taking into account the receptor density, measured by (−)-[³H]N-methylscopolamine (NMS) binding on the cell surface (FIG. 5,C). The normalized current amplitudes for m1, MC2, MC8, and MC10 were about two orders of magnitude larger than that for MC4, MC9, and m2. Thus, MC2, MC8, and MC10, in which the third putative cytoplasmic portion (i3) is derived from m1, mediate an ACh response similar to that mediated by m1.[3,20] In contrast, MC4 and MC9, which share this portion with m2, elicit an ACh response similar to that mediated by m2.[20]

Thus, it is concluded that the region of the mAChR molecules comprising the carboxy-terminal third of the proposed transmembrane segment V and the following putative cytoplasmic portion before the proposed transmembrane segment VI contains a determinant of selective coupling with different effector systems. This is consistent

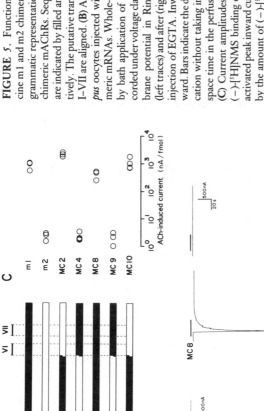

FIGURE 5. Functional properties of porcine m1 and m2 chimeric mAChRs. (A) Diagrammatic representation of the structures of chimeric mAChRs. Sequences of m1 and m2 are indicated by filled and open boxes, respectively. The putative transmembrane segments I–VII are aligned. (B) ACh responses in *Xenopus* oocytes injected with the respective chimeric mRNAs. Whole-cell currents activated by bath application of 1 μM ACh were recorded under voltage clamp at −70 mV membrane potential in Ringer's solution before (left traces) and after (right traces) intracellular injection of EGTA. Inward current is downward. Bars indicate the duration of ACh application without taking into account the dead-space time in the perfusion system (~6 sec). (C) Current amplitudes per unit amount of (−)-[³H]NMS binding on cell surface. ACh-activated peak inward current was normalized by the amount of (−)-[³H]NMS binding per oocyte expressing each chimeric or parental mAChR. (A and B: From Kubo *et al.*[37] Reprinted with permission.)

FIGURE 6. Alignment of amino- and carboxy-terminal sequences of the third cytoplasmic loop of porcine m1, m2, m3, rat m4, and rat m5. The sequence data have been taken from Kubo *et al.*,[3,4] Akiba *et al.*,[9] and Bonner *et al.*[6,10] The carboxy-terminal portion of the putative transmembrane segment V and the amino-terminal portion of the segment VI are indicated. Sets of five identical residues at one position are marked by (#). Residues identical only among m1, m3, and m5 or only between m2 and m4 are marked by (+).

with other studies that used chimeras or mutations, and may be generally valid for G protein-coupled receptors.[38-40] Although the amino acid sequence of this third cytoplasmic loop is divergent among the mAChR subtypes,[3-17] some sequence homology is noted in this portion among m1, m3, and m5 as well as m2 and m4, especially in the both amino- and carboxy-terminus of the cytoplasmic loop (FIG. 6). Recent studies have focused on these regions.[41-44] It has been shown by using chimeras that nine amino acid residues of m3 and no more than 21 amino acid residues of m2 at the amino-terminal portion of the third cytoplasmic loop are critical for selective effector coupling of m3 and m2, respectively.[44]

CONCLUSION

The existence of multiple mAChR subtypes originated from distinct genes has been shown by cloning and sequencing analysis. The antagonist binding properties of the individual subtypes expressed from the cloned DNAs, together with the differential tissue distribution of the mRNAs, indicate that the mAChR heterogeneity in tissues with respect to antagonist binding can be accounted for by the presence of distinct mAChRs in various combinations. Using the DNA expression systems in *Xenopus* oocytes and NG108-15 neuroblastoma-glioma hybrid cells, it has been shown that molecularly defined mAChR subtypes are selectively coupled with different effector system, albeit not exclusively. Functional analysis of chimeric receptors composed of m1 and m2, together with the recent progress by other groups, indicates that the amino-terminal sequences in the third cytoplasmic loop between the putative transmembrane segments V and VI are a possible determinant of selective coupling with different effector systems.

ACKNOWLEDGMENTS

I would like to thank the many collaborators who have contributed to the work described here.

REFERENCES

1. NATHANSON, N. M. 1987. Annu. Rev. Neurosci. 10: 195–236.
2. HAGA, K. & T. HAGA. 1985. J. Biol. Chem. 260: 7927–7935.
3. KUBO, T., K. FUKUDA, A. MIKAMI, A. MAEDA, H. TAKAHASHI, M. MISHINA, T. HAGA, K. HAGA, A. ICHIYAMA, K. KANGAWA, M. KOJIMA, H. MATSUO, T. HIROSE & S. NUMA. 1986. Nature 323: 411–416.
4. KUBO, T., A. MAEDA, K. SUGIMOTO, I. AKIBA, A. MIKAMI, H. TAKAHASHI, T. HAGA, K. HAGA, A. ICHIYAMA, K. KANGAWA, H. MATSUO, T. HIROSE & S. NUMA. 1986. FEBS Lett. 209: 367–372.
5. PERALTA, E. G., J. W. WINSLOW, G. L. PETERSON, D. H. SMITH, A. ASHKENAZI, J. RAMACHANDRAN, M. I. SCHIMERLIK & D. J. CAPON. 1987. Science 236: 600–605.
6. BONNER, T. I., N. J. BUCKLEY, A. C. YOUNG & M. R. BRANN. 1987. Science 237: 527–532, 1556, 1628.

7. PERALTA, E. G., A. ASHKENAZI, J. W. WINSLOW, D. H. SMITH, J. RAMACHANDRAN & D. J. CAPON. 1987. EMBO J. **6**: 3923–3929.
8. BRAUN, T., P. R. SCHOFIELD, B. D. SHIVERS, D. B. PRITCHETT & P. H. SEEBURG. 1987. Biochem. Biophys. Res. Commun. **149**: 125–132.
9. AKIBA, I., T. KUBO, A. MAEDA, H. BUJO, J. NAKAI, M. MISHINA & S. NUMA. 1988. FEBS Lett. **235**: 257–261.
10. BONNER, T. I., A. C. YOUNG, M. R. BRANN & N. J. BUCKLEY. 1988. Neuron **1**: 403–410.
11. LIAO, C-F., A. N. THEMMEN, R. JOHO, C. BARBERIS, M. BIRNBAUMER & L. BIRNBAUMER. 1989. J. Biol. Chem. **264**: 7328–7337.
12. TIETJE, K. M., P. S. GOLDMAN & N. M. NATHANSON. 1990. J. Biol. Chem. **265**: 2828–2834.
13. GOCAYNE, J., D. A. ROBINSON, M. G. FITZGERALD, F-Z. CHUNG, A. R. KERLAVAGE, K-U. LENTES, J. LAI, C-D. WANG, C. M. FRASER & J. C. VENTER. 1987. Proc. Natl. Acad. Sci. USA **84**: 8296–8300.
14. SHAPIRO, R. A., N. M. SCHERER, B. A. HABECKER, E. M. SUBERS & N. M. NATHANSON. 1988. J. Biol. Chem. **263**: 18397–18403.
15. TIETJE, K. M. & N. M. NATHANSON. 1991. J. Biol. Chem. **266**: 17382–17387.
16. ONAI, T., M. G. FITZGERALD, S. ARAKAWA, J. D. GOCAYNE, D. A. URQUHART, L. M. HALL, C. M. FRASER, W. A. McCOMBIE & J. C. VENTER. 1989. FEBS Lett. **255**: 219–225.
17. SHAPIRO, R. A., B. T. WAKIMOTO, E. M. SUBERS & N. M. NATHANSON. 1989. Proc. Natl. Acad. Sci. USA **86**: 9039–9043.
18. HULME, E. C., C. A. M. CURTIS, M. WHEATLEY, A. AITKEN & A. C. HARRIS. 1989. Trends Pharmacol. Sci. **10** (Suppl.): 22–25.
19. STRADER, C. D., I. S. SIGAL & R. A. F. DIXON. 1989. Trends Pharmacol. Sci. **10** (Suppl.): 26–30.
20. FUKUDA, K., T. KUBO, I. AKIBA, A. MAEDA, M. MISHINA & S. NUMA. 1987. Nature **327**: 623–625.
21. BUJO, H., J. NAKAI, T. KUBO, K. FUKUDA, I. AKIBA, A. MAEDA, M. MISHINA & S. NUMA. 1988. FEBS Lett. **240**: 95–100.
22. HAMMER, R., C. P. BERRIE, N. J. M. BIRDSALL, A. S. V. BURGEN & E. C. HULME. 1980. Nature **283**: 90–92.
23. HAMMER, R., E. GIRALDO, G. B. SCHIAVI, E. MONFERINI & H. LADINSKY. 1986. Life Sci. **38**: 1653–1662.
24. BIRDSALL, N. J. M., E. C. HULME, M. KEEN, E. K. PEDDER, D. POYNER, J. M. STOCKTON & M. WHEATLEY. 1986. Biochem. Soc. Symp. **52**: 23–32.
25. LADINSKY, H., E. GIRALDO, E. MONFERINI, G. B. SCHIAVI, M. A. VIGANÒ, L. D. CONTI, R. MICHELETTI & R. HAMMER. 1988. Trends Phamacol. Sci. **9** (Suppl.): 44–48.
26. MAEDA, A., T. KUBO, M. MISHINA & S. NUMA. 1988. FEBS Lett. **239**: 339–342.
27. PFAFFINGER, P. J., J. M. MARTIN, D. D. HUNTER, N. M. NATHANSON & B. HILLE. 1985. Nature **317**: 536–538.
28. NIERENBERG, M., S. WILSON, H. HIGASHIDA, A. ROTTER, K. KRUEGER, N. BUSIS, R. RAY, J. G. KENIMER & M. ADLER. 1983. Science **222**: 794–799.
29. FUKUDA, K., H. HIGASHIDA, T. KUBO, A. MAEDA, I. AKIBA, H. BUJO, M. MISHINA & S. NUMA. 1988. Nature **335**: 355–358.
30. NEHER, E., A. MARTY, K. FUKUDA, T. KUBO & S. NUMA. 1988. FEBS Lett. **240**: 88–94.
31. BARISH, M. E. 1983. J. Physiol. **342**: 309–325.
32. BROWN, D. A. & P. R. ADAMS. 1980. Nature **283**: 673–676.
33. HIGASHIDA, H. & D. A. BROWN. 1986. Nature **323**: 333–335.
34. PERALTA, E., A. ASHKENAZI, J. WINSLOW, J. RAMACHANDRAN & D. CAPON. 1988. Nature **334**: 434–437.
35. LAI, J., L. MEI, W. R. ROESKE, F.-Z. CHUNG, H. I. YAMAMURA & J. C. VENTER. 1988. Life Sci. **42**: 2489–2502.
36. ASHKENAZI, A., J. W. WINSLOW, E. G. PERALTA, G. L. PETERSON, M. I. SCHIMERLIK, D. J. CAPON & J. RAMACHANDRAN. 1988. Science **238**: 672–675.

37. Kubo, T., H. Bujo, I. Akiba, J. Nakai, M. Mishina & S. Numa. 1988. FEBS Lett. 241: 119–125.
38. Strader, C. D., R. A. F. Dixon, A. H. Cheung, M. R. Candelore, A. D. Blake & I. S. Sigal. 1987. J. Biol. Chem. 262: 16439–16443.
39. Kobilka, A., T. Kobilka, T. Daniel, J. Regan, M. Caron & R. Lefkowitz. 1988. Science 240: 1310–1316.
40. Franke, R. R., T. P. Sakmar, D. D. Oprian & H. G. Khorana. 1988. J. Biol. Chem. 263: 2119–2122.
41. Shapiro, R. A. & N. M. Nathanson. 1989. Biochemistry 28: 8946–8950.
42. Wess, J., M. R. Brann & T. I. Bonner. 1989. FEBS Lett. 258: 133–136.
43. Cottechia, S., S. Exum, M. Caron & R. Lefkowitz. 1990. Proc. Natl. Acad. Sci. USA 87: 2896–2990.
44. Lechleiter, J., R. Hellmiss, K. Duerson, D. Ennulat, N. David, D. Clapham & E. Peralta. 1990. EMBO J. 9: 4381–4390.

Studies of the Pharmacology, Localization, and Structure of Muscarinic Acetylcholine Receptors[a]

MARK R. BRANN, HANS B. JØRGENSEN,
ETHAN S. BURSTEIN, TRACY A. SPALDING,
JOHN ELLIS, S. V. PENELOPE JONES,
AND DAVID HILL-EUBANKS

Molecular Neuropharmacology Section
Department of Psychiatry
and
Vermont Comprehensive Cancer Center
University of Vermont
Burlington, Vermont 05405

INTRODUCTION

The first direct evidence for multiple muscarinic receptor subtypes came from a pharmacological comparison of the receptors expressed by brain and heart.[1-3] These receptors were then purified to apparent homogeneity from porcine brain[4] and heart.[5] Sufficient peptide sequence was obtained from these preparations to allow the cloning of the m1[6] and m2[7,8] muscarinic receptor subtypes from these tissues. Using homology cloning, the human and rat forms of these receptors as well as three additional subtypes (m3–m5) were identified.[9,10] The sequences of these receptor subtypes were subsequently confirmed in the human[11,12] and defined in several other species.[13,14] Comparison of the amino acid sequences of the five muscarinic receptor subtypes indicates that they are derived from a highly conserved gene family. Maximum sequence homology among the members occurs in the seven hydrophobic regions that are predicted to be transmembrane (TM) domains. These are the regions where muscarinic receptors have the most sequence homology with other receptors that mediate signal transduction by coupling with G-proteins.[15]

Prior to their molecular cloning, muscarinic receptors were divided into two or at most three subtypes based on pharmacological differences. Thus, pharmacological approaches were inadequate to determine reliably either the tissue-specific expression

[a] Some of the described work was supported by Public Health Services Grants (PHS R01 AG05214, PHS R29 NS29634 and NSF BNS-9111629), by Lilly Research Labs, Schering Plough Pharmaceuticals, Boehringer Ingelheim, and Allergan Pharmaceuticals.

of each of the subtypes or their physiological function.[2,3] Based on the sequences of the receptor cDNAs it has been possible to prepare subtype-selective antibody and cDNA probes to determine the tissue-specific expression of each of the receptor subtypes. In this paper we discuss data collected using these approaches and their physiological implications. Availability of receptor clones has also inspired many studies attempting to define functional regions within the receptor subtypes. We discuss the current status of our understanding of the structure/function relationships of the ligand binding and G-protein coupling domains of the muscarine acetylcholine receptor subtypes.

PHARMACOLOGICAL PROPERTIES

Muscarinic receptors in different tissues can be distinguished pharmacologically. Three tissues widely used to evaluate the pharmacology of the distinct receptor subtypes are the vas deferens, heart, and ileum: M1 receptors enhance neurogenic contraction in vas deferens; M2 receptors slow the heart, and M3 receptors contract ileal smooth muscle. Using data from several of the most selective muscarinic compounds, we correlated the pharmacologies of these physiologically defined subtypes (M1–M3) with that of the genetically defined subtypes (m1–m5) as determined using radioligand binding assays. Strong correlations are observed between the m2 and M2 receptors, and these receptors are the easiest to differentiate from the other subtypes. The pharmacology of the M3 receptor correlates the best with m3, but caution should be exercised as even the most discriminating compounds have a modest selectivity (<10 fold). Unfortunately, even when a relatively large number of compounds are considered, it is impossible to unequivocally assign the M1 receptor of vas deferens to a genetically defined subtype.[3,16] We recently completed a study of the pharmacology of the cloned receptors using functional assays in living transfected mammalian cells. If one assumes that M1 = m1, M2 = m2, and M3 = m3, then a precise 1:1 relationship is observed for the pharmacology of cloned and endogenously expressed receptor subtypes (FIG. 1). These data suggest that the small deviations that were previously observed between the pharmacologies of cloned receptors and tissue preparations were due to artifacts associated with binding assays that were performed using nonphysiological conditions.

ANATOMICAL LOCALIZATION

As indicated above, pharmacological data alone are inadequate to evaluate the anatomical distribution of the muscarinic receptor subtypes. Fortunately, the molecular cloning of the receptor subtypes has provided new tools to evaluate their tissue-specific expression. For example, using cloned DNA or oligodeoxynucleotides it is possible to measure the distribution of the mRNA that encodes each of the receptor subtypes. Both approaches have been used to map the distribution of muscarine receptor mRNAs.[9–11, 17–21] Similarly, the encoded receptor proteins can be measured by use of subtype-selective antibodies. One approach is to prepare synthetic peptides based on the predicted sequences of the receptor proteins. These peptides have been conjugated

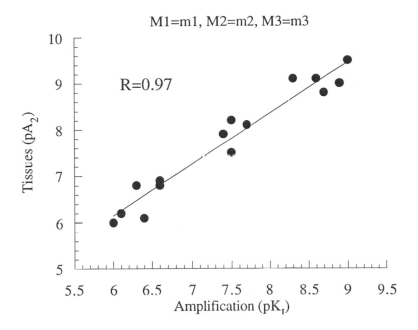

FIGURE 1. Correlation of the affinities of antagonists for muscarinic receptor subtypes expressed by tissues and transfected cells. Data for the peripheral tissue preparations are pA_2 values for the M1 vas deferens, M2 heart, and M3 illeum.[59] Data for the cloned receptors are pKI values determined by antagonism of carbachol-induced amplification responses measured in living mammalian cells transfected with each of the receptor subtypes (Jørgensen and Brann, unpublished observations).

to carrier protein and used as antigens.[22,23] Another approach is to express fragments of the cloned receptors as fusions with antigenic proteins in *E. coli*, and use these proteins as antigens.[24-26]

Overall, the above approaches have indicated that the muscarinic receptors are differentially distributed in peripheral tissues and have provided evidence for the molecular identities of the pharmacologically defined receptor subtypes. In addition to inhibitory M1 receptors, vas deferens has M2 receptors that enhance neurogenic contractions.[27] Both m1 and m2 receptor proteins are present in vas deferens.[28] Many studies have indicated a homogeneous population of M2 muscarinic receptors in heart,[2,3] and only m2 mRNA[18,19] and protein[28] have been detected in heart. Similarly, the majority of ileal receptors have an M2 pharmacology,[29] and m2 mRNA[19] and protein[28] are the predominant molecular species in ileum. As indicated above, ileum also has a functionally defined M3 subtype. Small amounts of m3 mRNA[19] has been detected in ileum, but no m3 protein.[28] M1 and M3 muscarinic receptors are present in submaxillary glands,[30] and high levels of both m1 and m3 mRNAs[19] and proteins[28] have also been observed. Sympathetic ganglia have both M1 and M2 receptor subtypes,[31] and both m1 and m2 receptor proteins are present.[28] Binding studies have indicated that muscarinic receptors in peripheral lung and NG108-15 cells have a unique "M1-like" pharmacology, which been termed the "M4" receptor.[32] m4 mRNA is the predominant subtype in both lung[32] and NG108-15 cells.[11] Both m2 and m4 proteins are present in peripheral lung.[28]

All five of the receptor subtype mRNAs and proteins have been detected in the brain.[17,18,20,21,25] m1 and m4 mRNA and protein are widely expressed in cerebral cortex, basal ganglia (including caudate-putamen), and hippocampus. Thus, these receptors are likely to play major roles as postsynaptic muscarinic receptors in various cognitive and motor functions and are likely to be major contributors to the M1 responses that have been measured in these brain regions. m4 mRNA and protein are the most abundant subtype in caudate-putamen, possibly explaining the anomolous M1-like "M4" pharmacology of binding sites in this brain region.[33] m2 receptor and protein predominates in the brainstem and cholinergic cells of the basal forebrain and caudate-putamen. The distribution of m2 versus m4 receptors may account for differences in the pharmacology of cAMP inhibition by muscarinic receptors among brain regions.[34] These data are also consistent with receptor autoradiography of M2 binding sites.[35-37] Pharmacological studies have indicated M2 receptors inhibit acetylcholine release, and the finding of m2 mRNA in cholinergic cells[21] and m2 protein in cholinergic neurons[25] establishes this subtype as a major presynaptic muscarinic receptor. It should also be noted that both m2 mRNA and protein are widely expressed by noncholinergic cells in various brain regions including the cerebral cortex.[20,21,25] m3 mRNA is present within the cerebral cortex, hippocampus, and thalamus, but not in basal ganglia.[17,18,20,21]

m1 mRNA is expressed by the majority of medium-sized neurons of the caudate-putamen and m4 is expressed by ~ 50% of these neurons. Within the caudate-putamen, the m4 receptor is co-expressed with dopamine receptors, implicating a direct interaction with dopaminergic neurotransmission and the control of dopamine-mediated psychomotor function. The m5 receptor is expressed by the dopaminergic neurons within the substantia nigra pars compacta, leading to the suggestion that this receptor may be the muscarinic receptor that mediates direct stimulation of dopamine release by acetylcholine.[21] A composite of data from immunological and mRNA studies is presented in TABLE 1.

Overall, because of the complex expression patterns of muscarinic receptors within the brain and the paucity of cellular information concerning the behavioral function of the neuronal systems involved, it is difficult to unequivocally relate individual receptor subtypes with the individual behavioral effects of muscarinic drugs. For example, the antiparkinsonian site of action of drugs such as trihexiphenidyl is likely to be within the basal ganglia. Since this drug has high affinity for both m1 and m4 receptors and both of these receptors are expressed by the basal ganglia, either or both of these receptors could be the relevant site of action.

The subtype-selective targeting of cholinergic agonists for treatment of Alzheimer's disease is similarly problematic, because brain regions involved in cognitive function express all five of the muscarinic receptor subtypes. Studies in animal models using muscarinic antagonists have tended to discount M2/m2 receptors, and considering their presynaptic location one would expect that M2/m2 receptor stimulation should be avoided. On the other hand, "M1" selective agonists have not proven to be more effective in clinical trials than acetylcholinesterase inhibitors. In this regard, it should be noted that biochemical experiments suggest that these "M1" agonists are, in fact, weak partial agonists with selectivity for m2/m4 receptors. Thus the receptor subtype(s) that should be targeted in the treatment of Alzheimer's disease remains to be established.

TABLE 1. Distribution of Muscarinic Receptor Subtypes within the Brain and among Peripheral Tissues

	m1	m2	m3	m4	m5
Brain					
Cerebral cortex	+ + + +	+ +	+ +	+ +	
Striatum	+ + + +	+ +		+ + + +	
Thalamus		+ + +	+ + +	+ + +	
Brainstem and cerebellum		+ + + +	+		
Hippocampus	+ + + +	+ + +	+ +	+ + +	+ +
Olfactory tubercle	+ + + +	+ +		I I I I	
Substantia Nigra (DA)					+ + +
Basal forebrain (ACH)		+ + + +	+ +		
Peripheral Tissues					
Sympathetic ganglia	+ +	+ + + +			
Vas deferens	+ + +	+ +			
Submaxillary gland	+ +	+ +	+ + +		
Atrium		+ + + +			
Peripheral lung	+	+ + +		+ + +	
Uterus		+ + +		+ +	
Ileum		+ + +	+	+	

Data are collected from immunoprecipitations[25,26,28], in situ hybridization histochemistry,[17,18,20,21] and immunocytochemistry.[25] DA refers to dopamine-containing neurons, ACH refers to acetylcholine-containing neurons.

Because of their discrete patterns of expression, the m4 and m5 receptors represent compelling therapeutic targets. Within the periphery, expression of the m4 receptor is most prominent in the lung. Since M1-selective drugs are useful in the treatment of asthma, an m4-selective antagonist may avoid m1-mediated side effects (e.g., in sympathetic ganglia). Similarly, m5 has a very limited distribution within the brain. If the m5 receptor is the one responsible for enhanced release of dopamine, then this receptor may be a useful target for therapeutic modulation of dopaminergic tone (e.g. in Tourette's syndrome and schizophrenia).

STRUCTURAL FEATURES

The N-terminal regions of all the muscarinic receptor subtypes have sites for N-linked glycosylation. Removal of these sites by point mutations does not influence ligand binding or levels of receptor in the membrane.[38] Similarly, replacement of the entire N-terminal region of the m1 receptor (spanning the sites of N-linked glycosylation) with various unrelated sequences has little effect on ligand binding and levels of receptor in the membrane (Jørgensen, Hill-Eubanks, and Brann, unpublished observations).

Muscarinic receptor subtypes can be differentiated based on their selectivities for G-proteins and functional responses. For example, the m1, m3, and m5 receptors selectively couple with a pertussis toxin (PTX) insensitive G-protein to stimulate phos pholipase C, while m2 and m4 selectively couple with a PTX-sensitive G-protein that inhibits adenylyl cyclase. The m2 and m4 receptors also weakly stimulate phospholipase C via a PTX-sensitive G-protein. These receptors can also be classified based on which ion channels they modulate. Briefly m1, m3, and m5 open Ca^{2+}-dependent potassium channels and inhibit the m-current via PTX-insensitive G-proteins, while m2 and m4 open non-specific cation conductances and inwardly rectifying potassium conductances and inhibit calcium conductances via PTX-sensitive G-proteins.[2,3,39]

Chimeric muscarinic receptors have been constructed in which individual epitopes have been exchanged between subtypes that differ in their functional selectivities. Chimeras between m1 and m2 receptors were constructed and expressed in oocytes where selective coupling to electrophysiological responses were examined. These studies demonstrated that the third cytoplasmic loop (i3) was sufficient to define functional selectivity for ion channels, while the C-terminus of the receptor did not qualitatively influence electrophysiological responses. In contrast to the importance of the i3 loop in defining functional selectivity, this region had no effect on the selectivities of these receptors for ligands.[40]

Chimeras between m2 and m3 receptors were constructed and expressed in mammalian cells where selective coupling to G-proteins and biochemical responses were examined. These studies confirmed a critical role of the i3 loop in defining functional responses, and the lack of influence of the C-terminus. These studies also demonstrated that the N-terminal region of the i3 loop, proximal to TM5, was critical in defining coupling selectivity.[41-43] The latter results are consistent with data from experiments that employed deletion mutants. These experiments have shown that only regions proximal to the TM domains are involved in coupling to G-proteins.[44] A critical role for the N-terminus of the i3 loop has recently been extended to electrophysiological[45] and calcium responses[46] by expressing very similar m2/m3 chimeras in oocytes. In studies of chimeric beta-adrenergic/muscarinic receptors, it was observed that both the N-terminal region of i3 and the i2 loop must be exchanged to reverse the functional phenotype.[47]

Muscarinic receptors can be classified based on their differential sensitivities to various ligands. In fact, the first compelling evidence for the existence of multiple muscarinic receptor subtypes came from the demonstration that muscarinic receptors expressed by brain have higher affinity for the antagonist pirenzepine than those expressed by heart.[1] Analysis of the affinity profiles of many muscarinic antagonists indicates that they can be divided into families based on their selectivities among the subtypes. For example, trihexyphenidyl, pirenzepine, and derivatives such as UH-AH 37 have higher affinity for the m1 and m4 receptor than for the other subtypes. On the other hand, himbacine, methoctramine, and derivatives of AF-DX 116 have much higher affinity for m2 and m4 than for m5 receptors[16]; a similar profile of relative selectivity for the individual subtypes has been observed for the allosteric antagonist gallamine.[48] While much less is known about agonist interaction with the receptors, many agonists have higher potency and efficacy at the m2 and m4 receptors.[3]

To investigate which epitopes within the receptor contribute to the subtype selectivity of ligands, the binding properties of several chimeric m2/m3 and m2/m5 receptors have been investigated. As indicated earlier, exchange of the i3 loop between m2 and m3 receptors does not influence antagonist affinity, but does reverse the relative affinity for some agonists.[43] Analysis of the series of chimeric m2/m5 and m2/m3 receptors has demonstrated that multiple regions contribute the subtype selectivity of several antagonist ligands, and these ligands can be classified according to which regions contribute to binding selectivity.[42,49] For example, himbacine and AQ-RA 741 have a very similar binding profile among the chimeras, suggesting that they recognize similar structural epitopes. The high affinity of both of these drugs for m2 receptors is highly dependent on the N- and C-terminal regions. On the other hand, the higher affinity of UH-AH 37 for m5 receptors is largely defined by differences within the TM6 and/or third outer loop, a region that does not influence the binding of the former compounds. This same region defines the subtype selectivity of the allosteric antagonist gallamine, in spite of the fact that gallamine and UH-AH 37 have divergent selectivities among the wild-type subtypes.[50]

Amino Acids Involved in Ligand Binding

Labeling with covalent ligands and site-directed mutagenesis have been used to identify amino acids within the muscarinic receptors that contribute to ligand binding. [³H]propylbenzilylcholine mustard (PBCM) covalently attaches to muscarinic receptors through its ammonium headgroup. By sequencing fragments of purified muscarinic receptors that have been labeled with this ligand, its primary attachment site has been shown to be an aspartic acid in TM3.[51] Using similar methods, a muscarinic agonist has been shown to attach to the same amino acid.[52] A critical role of this amino acid in ligand binding has been confirmed using site-directed mutagenesis, as substitutions at this position disrupt binding.[53] The observation that an ammonium headgroup is a critical feature of most muscarinic ligands[54] suggests that formation of an ionic bond between the ammonium headgroup and the aspartic acid of TM3 may be a general feature of ligand binding to muscarinic receptors.

Examination of the TM domains of the five muscarinic receptors indicates a series of threonine and tyrosine residues which are identical for the five muscarinic receptor subtypes, but are not conserved in other G-protein coupled receptor sequences. The hypothesis that these amino acids may contribute to the binding of muscarinic ligands was tested by substituting phenylalanine and alanine for the tyrosine and threonine residues, respectively. The working hypothesis was that one or more of the hydroxyl groups contributed by one or more of these amino acids would form a hydrogen bond with the ester group of acetylcholine. A tyrosine and a threonine located in TM5 and 6, respectively (the TM's that bound the i3 loop), had the most dramatic effects on decreasing agonist binding and activation of PI metabolism. None of these substitutions had any effect on antagonist binding.[55] However, too many of the substitutions selectively reduced agonist affinity, for the results to be interpreted as being due solely to direct interactions between the ligand and the individual residues (e.g. hydrogen bonding to the ester region). Molecular models of a muscarinic acetylcholine receptor that are based on the structure of bacteriorhodopsin,[56] suggest that the major role of these multiple hydroxyl-containing amino acids may be to stabilize the aspartate of TM3, and that it is this stabilization that influences agonist binding.[57]

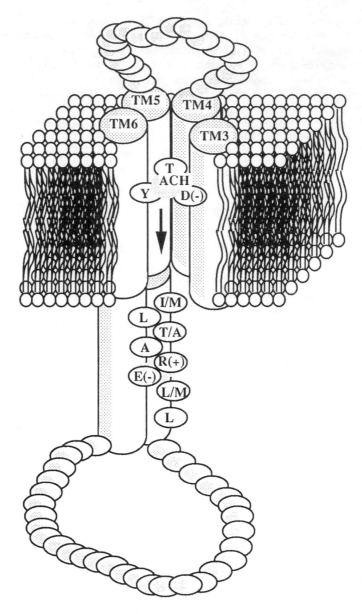

FIGURE 2. Model of agonist activation of a muscarinic receptor. The amino acids that strongly influence agonist binding are marked: Asp (D) of TM3, Thr (T) of TM5, and Tyr (Y) of TM6. Agonist binding leads to a conformational change in the i3 loop. The N-terminal region of the i3 loop is a continuation of the TM5 alpha helix. Conserved residues that contribute to G-protein coupling are indicated. The C-terminal region of the i3 loop was also involved in G-protein coupling. The critical residues are shown. The functionally critical residues of the N- and C-terminal regions of the i3 loop are predicted to face each other. Agonist binding (acetylcholine, ACH) may alter the conformation of TM5 relative to TM6 and consequently lead to a conformation of the i3 loop that activates G-proteins.

Amino Acids Involved in G-protein Coupling

In order to identify specific residues within the i3 loop that contribute to G-protein coupling, we subjected the 20 amino acids adjacent to the TM 5 to random-saturation mutagenesis (a region that defines G-protein coupling selectivity, see above). Analysis of the pattern of amino acids that are resistant to mutation would lie on a functional face of this helix (Hill-Eubanks, Jørgensen and Brann, unpublished observations). These findings are consistent with computational predictions based on primary sequence. Computational models predict the presence of a helix within analogous regions (N-terminal region of the i3 loop) of several G-protein coupled receptors.[58] Based on these data, we propose the model of agonist activation of muscarinic receptors that is drawn in FIGURE 2.

ACKNOWLEDGMENTS

M.R.B. would like to thank former members of his laboratory at the NINDS, J. Wess, E. Novotny, D. Gdula, D. Weiner, and A. Levey, who are responsible for many of the previously published results.

REFERENCES

1. HAMMER, R., C. P. BERRIE, N. J. M. BIRDSALL, A. S. V. BURGEN, & E. C. HULME. 1980. Pirenzepine distinguishes between different subclasses of muscarinic receptors. Nature **283**: 90–92.
2. HULME, E. C., N. J. M. BIRDSALL & N. J. BUCKLEY. 1990. Muscarinic receptor subtypes. Annu. Rev. Pharmacol. Toxicol. **30**: 633–673.
3. JONES, S. V. P., A. I. LEVEY, D. M. WEINER, J. ELLIS, E. NOVOTNY, S.-H. YU, F. DORJE, J. WESS & M. R. BRANN. 1992. Muscarinic acetylcholine receptors. *In* Molecular Biology of G-Protein-Coupled Receptors. M. R. Brann, Ed.: 170–197. Birkhauser. Boston.
4. HAGA. K. & T. HAGA. 1983. Affinity chromatography of the muscarinic acetylcholine receptor. J. Biol. Chem. **258**: 13575–13579.
5. PETERSON, G. L., G. S. HERRON, M. YAMAKI, D. S. FULLERTON & M. J. SCHIMERLIK. 1984. Purification of the muscarinic acetylcholine receptor from porcine atria. Proc. Natl. Acad. Sci. USA **81**: 4993–4997.
6. KUBO, T., K. FUKUDA, A. MIKAMI, A. MAEDA, H. TAKAHASHI, T. MISHINA, K. HAGA, A. HAGA, A. ICHIYAMA, K. KANAGAWA, M. KOJIMA, H. MATSUO, T. HIROSE & S. NUMA. 1986. Cloning, sequencing and expression of complementary DNA encoding the muscarinic acetylcholine receptor. Nature **323**: 411–416.
7. KUBO, T., A. MAEDA, K. SUGIMOTO, I. AKIBA, A. MIKAMI, T. TAKAHASI, K. HAGA, A. HAGA, A. ICHIYAMA, K. KANAGAWA, H. MATSUO, T. HIROSE & S. NUMA. 1986. Primary structure of porcine cardiac muscarinic acetylcholine receptor deduced from the cDNA sequence. FEBS Lett. **209**: 367–372.
8. PERALTA, E. G., A. ASHKENAZI, J. W. WINSLOW, J. RAMACHANDRAN & D. J. CAPON. 1987. Primary structure and biochemical properties of an M2 muscarinic receptor. Science **236**: 600–605.
9. BONNER, T. I., N. J. BUCKLEY, A. C. YOUNG & M. R. BRANN. 1987. Identification of a family of muscarinic acetylcholine receptor genes. Science **237**: 527–532.
10. BONNER, T. I., A. YOUNG, M. R. BRANN & N. J. BUCKLEY. 1988. Cloning and expression of the human and rat m5 muscarinic receptor genes. Neuron **1**: 403–410.

11. PERALTA, E. G., J. W. WINSLOW, G. L. PETERSON, D. H. SMITH, A. ASHKENAZI, J. RAMACHAN-DRAN, M. I. SCHIMERLIK & D. J. CAPON. 1987. Distinct primary structures, ligand-binding properties and tissue-specific expression of four human muscarinic acetylcholine receptors. EMBO J. 6: 3923–3929.

12. LIAO, C-F., A. P. N. THEMMEN, R. JOHO, C. BARBERIS, M. BIRNBAUMER & L. BIRNBAUMER. 1989. Molecular cloning and expression of a fifth muscarinic acetylcholine receptor. J. Biol. Chem. 264: 7328–7337.

13. SHAPIRO, R. A., B. T. WAKIMOR, E. M. SUBERS & N. M. NATHANSON. 1989. Characterization and functional expression in mammalian cells of genomic and cDNA clones encoding a Drosophila muscarinic acetylcholine receptor. Proc. Natl. Acad. Sci. USA 86: 9039–9043.

14. TIETJE, K. M., P. S. GOLDMAN & N. M. NATHANSON. 1990. Cloning and functional analysis of a gene encoding a novel muscarinic acetylcholine receptor expressed in chick heart and brain. J. Biol. Chem. 265: 2828–2834.

15. BRANN, M. R., Ed. 1992. In Molecular Biology of G-Protein-Coupled Receptors. Birkhauser. Boston.

16. DORJE, F., J. WESS, G. LAMBRECHT, E. MUTSCHLER & M. R. BRANN. 1991. Antagonist binding studies at five cloned human muscarinic receptor subtypes. J. Pharmacol. Exp. Ther. 256: 727–733.

17. BRANN, M. R., N. J. BUCKLEY & T. I. BONNER. 1988. The striatum and cerebral cortex express different muscarinic receptor mRNAs. FEBS Lett. 230: 90–94.

18. BUCKLEY, N. J., T. I. BONNER & M. R. BRANN. 1988. Localization of a family of muscarinic receptor mRNAs in rat brain. J. Neurosci. 8: 4646–4652.

19. MAEDA, A., T. KUBO, M. MISHINA & S. NUMA. 1988. Tissue distribution of mRNAs encoding muscarinic acetylcholine receptor subtypes. FEBS Lett. 239: 339–342.

20. WEINER, D. M. & M. R. BRANN. 1989. Distribution of m1–m5 muscarinic acetylcholine receptor mRNAs in rat brain. Trends Pharmacol. Sci. (Suppl.) 4: 115.

21. WEINER, D. M., A. LEVEY & M. R. BRANN. 1990. Expression of muscarinic acetylcholine and dopamine receptor mRNAs within the basal ganglia. Proc. Natl. Acad. Sci. USA 87: 7050–7054.

22. LUTHIN, G. R., J. HARKNESS, R. P. ARTYMYSHYN & B. B. WOLFE. 1988. Antibodies to a synthetic peptide can be used to distinguish between muscarinic acetylcholine receptor binding sites in brain and heart. Mol. Pharmacol. 34: 327–333.

23. LEVEY, A. I., W. SIMONDS, A. SPIEGEL & M. R. BRANN. 1989. Characterization of muscarinic receptor subtype specific antibodies. Soc. Neurosci. Abstr. 15: 64.

24. LEVEY, A. I., T. M. STORMANN & M. R. BRANN. 1990. Bacterial expression of human muscarinic receptor fusion proteins and generation of subtype-specific antisera. FEBS Lett. 275: 65–69.

25. LEVEY, A. I., C. KITT, W. SIMONDS, D. PRICE & M. R. BRANN. 1991. Identification and localization of muscarinic receptor subtype proteins in rat brain. J. Neurosci. 11: 3218–3226.

26. WALL, S. J., R. P. YASUDA, F. HORY, S. FLAGG, B. M. MARTIN, E. I. GINNS & B. B. WOLFE. 1991. Production of antisera selective for m1 muscarinic receptors using fusion proteins: Distribution of m1 receptors in rat brain. Mol. Pharm. 39: 643–649.

27. ELTZE, M., G. GMELIN, J. WESS, C. STROHMANN, R. TACKE, E. MUTSCHLER & G. LAMBRECHT. 1988. Muscarinic M1 and M2 receptors mediating opposite effects on neuromuscular transmission in rabbit vas deferens. Eur. J. Pharmacol. 151: 205–221.

28. DORJE, F., A. LEVEY & M. R. BRANN. 1991. Immunological detection of muscarinic receptor subtype proteins (m1–m5) in rabbit peripheral tissues. Mol. Pharmacol. 40: 459–462.

29. CANDELL, L. M., S. H. YUN, L. L. TRAN & J. EHLERT. 1990. Differential coupling of subtypes of the muscarinic receptor to adenylate cyclase and phosphoinositide hydrolysis in longitudinal muscle of the rat ileum. Mol. Pharmacol. 38: 689–697.

30. MEI, L., W. R. ROESKE, K. T. IZUTSU & H. I. YAMAMURA. 1990. Characterization of muscarinic acetylcholine receptors in human labial salivary glands. Eur. J. Pharmacol. 176: 367–370.

31. NEWBERRY, N. R. & T. PRIESTLY. 1987. Pharmacological differences between two muscarinic responses of the rat superior cervical ganglion in vitro. Br. J. Pharmacol. **92**: 817–826.

32. LAZARENO, S., N. J. BUCKLEY & F. ROBERTS. 1990. Characterization of muscarinic M4 binding sites in rabbit lung, chicken heart, and NG108-15 cells. Mol. Pharmacol. **38**: 805–815.

33. WAELBROECK, M., M. TASTENOY, J. CAMUS & J. CHRISTOPHE. 1990. Binding of selective antagonists to four muscarinic receptors (M1-M4) in rat forebrain. Mol. Pharmacol. **38**: 267–273.

34. MCKINNEY, M., D. ANDERSON, C. FORRAY & E. E. EL-FAKAHANY. 1989. Characterization of the striatal M2 muscarinic receptor mediating inhibition of cyclic AMP using selective antagonists: A comparison with the brainstem M2 receptor. J. Pharmacol. Exp. Ther. **250**: 565–572.

35. CORTES, R. & J. M. PALACIOS. 1986. Muscarinic cholinergic receptor subtypes in the rat brain. I. Quantitative autoradiographic studies. Brain Res. **362**: 227–238

36. MASH, D. C., D. D. FLYNN & L. T. POTTER. 1985. Loss of M2 muscarine receptors in the cerebral cortex in Alzheimer's disease and experimental cholinergic denervation. Science **228**: 1115–1117.

37. MASH, D. C. & L. T. POTTER. 1985. Autoradiographic localization of M1 and M2 muscarine receptors in the rat brain. Neuroscience **19**: 551–564

38. VAN KOPPEN, C. J. & N. M. NATHANSON. 1990. Site-directed mutagenesis of the m2 muscarinic acetylcholine receptor. J. Biol. Chem. **265**: 20887–20892.

39. JONES, S. V. P. 1993. Muscarinic receptor subtypes: Modulation of ion channels. Life Sci. **52**: 457–464.

40. KUBO, T., H. BUJO, I. AKIBA, J. NAKAI, M. MISHINA & S. NUMA. 1988. Location of a region of the muscarinic acetylcholine receptor involved in selective effector coupling. FEBS Lett. **241**: 119–125.

41. WESS, J., M. R. BRANN & T. I. BONNER. 1989. Identification of a small intracellular region of the muscarinic m3 receptor as a determinant of selective coupling to PI turnover. FEBS Lett. **258**: 133–136.

42. WESS, J., T. I. BONNER & M. R. BRANN. 1990. Chimeric m2/m3 muscarinic receptors: Role of carboxyl terminal receptor domains in selectivity of ligand binding and coupling to phosphoinositide hydrolysis. Mol. Pharm. **38**: 872–877.

43. WESS, J., T. I. BONNER, F. DORJE & M. R. BRANN. 1990. Delineation of muscarinic receptor domains conferring selectivity of coupling to G proteins and second messengers. Mol. Pharm. **38**: 517–523.

44. SHAPIRO, R. A. & N. M. NATHANSON. 1989. Deletion analysis of the mouse m1 muscarinic acetylcholine receptor: Effects on phosphoinositide metabolism and down-regulation. Biochemistry **28**: 8946–8950.

45. LECHLEITER, J., S. GIRARD, D. CLAPMAN & E. PERALTA. 1991. Subcultural patterns of calcium release determined by G protein-specific residues of muscarinic receptors. Nature **350**: 505–508.

46. LECHLEITER, J., R. HELLMISS, K. DUERSON, D. ENNULAT, N. D. CLAPHAM & E. PERALTA. 1991. Distinct sequence elements control the specificity of G protein activation by muscarinic acetylcholine receptor subtypes. EMBO J. **9**: 4381–4390.

47. WONG, S. K. F., E. M. PARKER & E. M. ROSS. 1990. Chimeric muscarinic cholinergic: β-adrenergic receptors that activate Gs in response to muscarinic agonists. J. Biol. Chem. **265**: 6219–6224.

48. ELLIS, J., J. HUYLER & M. R. BRANN. 1991. Allosteric regulation of cloned m1-m5 muscarinic receptor subtypes. Biochem. Pharmacol. **42**: 1927–1932.

49. WESS, J., D. GDULA & M. R. BRANN. 1992. Chimeric m2/m5 muscarinic receptors: Identification of receptor domains conferring antagonist binding selectivity. Mol. Pharm. **41**: 369–374.

50. ELLIS, J., M. SEIDENBERG & M. R. BRANN. 1993. Use of chimeric muscarinic receptors to investigate epitopes involved in allosteric interactions. Mol. Pharm. In press.

51. CURTIS, C. A. M., M. WHEATLEY, S. BASAL, N. J. M. BIRDSALL, P. EVELEIGH, E. K. PEDDER, D. POYNER & E. C. HULME. 1989. Propylbenzilylcholine mustard labels an acidic residue in transmembrane helix 3 of the muscarinic receptor. J. Biol. Chem. **264**: 489–495.

52. BIRDSALL, N. J. M., T. A. SPALDING, J. E. T. CORRE, C. A. M. CURTIS & E. C. HULME. 1993. Studies on muscarinic receptors using nitrogen mustards. Life Sci. **52:** 561.
53. FRASER, C. M., C. D. WANG, D. A. ROBINSON, J. D. GOCAYNE & J. C. VENTER. 1989. Mol. Pharm. **36:** 840–847.
54. SCHULMAN, J. M., M. L. SABIO & R. L. DISCH. 1983. J. Med. Chem. **26:** 817–823.
55. WESS, J., D. GDULA & M. R. BRANN. 1991. Site-directed mutagenesis of the m3 muscarinic receptor: Identification of a series of threonine and tyrosine residues involved in agonist but not antagonist binding. EMBO J. **10:** 3729–3734.
56. HENDERSON, R., J. M. BALDWIN, T. A. CESKA, F. ZEMLIN, E. BECKMANN & K. H. DOWNING. 1990. Model for the structure of bacteriorhodopsin based on high-resolution electron cryomicroscopy. Mol. Biol. **213:** 899–493.
57. BRANN, M. R., V. J. KLIMKOWSKI & J. ELLIS. 1993. Structure/function relationship of muscarinic acetylcholine receptors. Life Sci. **52:** 405–512.
58. STRADER, C. D. & R. A. F. DIXON. 1992. Genetic analysis of the β-adrenergic receptor. *In* Molecular Biology of G-Protein-Coupled Receptors. M. R. Brann, Ed.: 62–75. Birkhauser. Boston.
59. ELTZE, M., E. MUTSCHLER, U. MOSER, T. FRIEBE, GUBITZX, R. TACKE & G. LAMBRECHT. 1993. Characterization of the vasodilatory muscarinic receptor in perfused rat kidney by the use of agonist and antagonists. Life Sci. **52:** 579.

Coupling of Muscarinic Receptor Subtypes to Ion Channels: Experiments on Neuroblastoma Hybrid Cells[a]

D. A. BROWN,[b] H. HIGASHIDA,[c] M. NODA,[c]
N. ISHIZAKA,[c] M. HASHII,[c] N. HOSHI,[c]
S. YOKOYAMA,[c] K. FUKUDA,[d] M. KATAYAMA,[d]
T. NUKADA,[d,e] K. KAMEYAMA,[e] J. ROBBINS,[b]
S. J. MARSH,[b] AND A. A. SELYANKO[b]

[b]Department of Pharmacology
University College London
London, WC1E6BT, United Kingdom

[c]Department of Biophysics
Neuroinformation Research Institute
Kanazawa University School of Medicine
Kanazawa 920, Japan

[d]Departments of Medical Chemistry and Molecular Genetics
Kyoto University Faculty of Medicine
Kyoto 606, Japan

[e]Department of Neurochemistry
Brain Research Institute
University of Tokyo
Faculty of Medicine
Tokyo 113, Japan

INTRODUCTION

Activation of muscarinic acetylcholine receptors (mAChRs) induces a great variety of ionic conductance changes in different cell types. These include opening or closing of K channels, opening of cation channels, closing of Ca channels, and opening or closing of Cl channels.[1] This variation in response arises from two principal factors:

[a] This work was supported by grants from the Japanese Ministry of Education, Science and Culture and from the United Kingdom Medical Research Council.

237

different cells possess different types of ion channels and different cells also express different genetic variants (subtypes) of mAChR, which may not couple with equal facility to any given ion channel.

One approach to studying the coupling preferences of different subtypes is to express the individual genes in different cells equipped with appropriate ion channels and then determine which receptor most effectively opens or closes the ion channel in question. While this does not necessarily duplicate precisely what happens in a normal cell expressing multiple receptor subtypes and multiple channel types, it is useful in laying down some of the ground rules for coupling preferences, upon which further tests on normal cells can be based.

Thus far, five genetic subtypes of mAChR, denoted m1 through m5, have been cloned and expressed[2-4] and membrane ionic responses to individual subtypes recorded in oocytes[5,6] and in several cell lines: neuroblastoma hybrid cells,[7-10] fibroblasts,[11] epithelial cells,[12] secretory cells,[13] and insect cells.[14]

In this paper we evaluate the information so far yielded about coupling preferences and possible transduction mechanisms from experiments on two neural cell lines (NG108-15 neuroblastoma × glioma hybrid cells and NL308 neuroblastoma × fibroblast hybrids) that have been stably transfected with DNA for m1, m2, m3, and m4 mAChRs, and consider to what extent this might be representative of equivalent coupling processes in mammalian nerve cells.

NG108-15 NEUROBLASTOMA × GLIOMA HYBRID CELLS

When differentiated (by cyclic AMP or a cyclic AMP–generating system), these cells extend neuritic processes, become electrically excitable, and express a variety of neuron-related ionic currents.[15-18] They also express a low level of endogenous mAChRs corresponding to the genetic m4[7] or pharmacological M_4[19,20] subtype: amounts of endogenous receptor expression are around 100 fmoles/mg protein[7] or $5-10 \times 10^3$ receptors per cell.[10]

In the present experiments, cells were transfected with cDNAs or genomic DNA for pig m1, m2, or m3 receptors or rat m3 and m4 receptors, to express additional receptors to a level between 3 and 10 times that of the endogenous mAChR.[7] The identity of these additional receptors was verified at the mRNA level using specific probes[7] and also at the receptor level using specific displacement of [^{3}H]N-methylscopolamine (NMS) binding.[10]

The principal results obtained when these transformed cells were challenged with ACh or other mAChR stimulants[7-10] are summarized in FIGURE 1. Both biochemical and electrophysiological effects fell into two broad categories: (1) Cells transformed to express m1 or m3 mAChRs responded with increased inositol phosphate production, increased intracellular Ca and activation of a Ca-dependent K current ($I_{K(Ca)}$), and inhibition of a voltage-gated K current $I_{K(M)}$. All three effects were resistant to Pertussis toxin (PTX). None of these effects were seen in non-transformed cells or in cells transformed to express m2 or additional m4 receptors.

(2) Non-transformed cells (expressing endogenous m4 receptors) responded with a reduced PGE$_1$-stimulated cyclic AMP production[21] and inhibition of a specific ω-conotoxin–sensitive component[22] of voltage-gated Ca-current $I_{Ca(N)}$ (see below).

These responses were significantly greater in cells expressing additional m2 or m4 receptors, but not in those expressing additional m1 or m3 receptors. Inhibition of $I_{Ca(N)}$ was totally prevented by PTX.

The three membrane current responses—activation of $I_{K(Ca)}$, inhibition of $I_{K(M)}$, and inhibition of $I_{Ca(N)}$—and possible transduction pathways leading to their inhibition, will now be considered in more detail.

Activation of Ca-dependent K Current ($I_{K(Ca)}$)

This may be most simply attributed to the generation of Ins-1,4,5-P_3 and subsequent release of Ca ions from intracellular stores, for three reasons:

(1) As indicated in FIGURE 1, only those receptors (m1 and m3) whose stimulation led to an increased inositol phosphate production activated $I_{K(Ca)}$.

(2) As shown in FIGURE 2 (see also Neher et al.[8]), the outward current was accompanied by a rise in intracellular [Ca]. (In fact, the outward current slightly *preceded* the recorded rise in somatic [Ca], by about 0.6 sec on average,[23] but this may be attributed to differences between submembrane and recorded mean somatic levels). Further, when the rise in [Ca] was suppressed by intracellular BAPTA (20 mM) or extracellular BAPTA/AM (100 μM), then so was the outward current.[23]

(3) Similar changes in intracellular [Ca] and membrane current were observed after intracellular application of Ins-1,4,5-P_3.[23,24] When the effects of intracellularly applied Ins-1,4,5-P_3 were prevented [e.g., by adding 1 mM heparin to the pipette or in the continued presence of a high concentration (100 μM) of Ins-1,4,5-P_3] then so was the response to external ACh.[23]

Overall, the effects of ACh on m1- and m3-transformed NG108-15 cells closely resembles that of non-transformed cells to bradykinin (BK) reported previously.[25-27] Indeed, considerable cross-desensitization between ACh- and BK-induced outward currents can be observed at the level of the Ins-1,4,5-P_3/Ca-release mechanism, resulting from depletion of internal Ca stores[27]: this strongly suggests a common pathway. Further, effects of both ACh and BK are mediated by a PTX-insensitive G-protein.[7] This may be G_q/G_{11} since BK-induced inositol phosphate production in NG108-15 cells is attenuated by antibodies to G_q/G_{11},[29] and there is a preliminary report[30] to the effect that the BK-induced outward current may be reduced by these same antibodies.

Inhibition of the Voltage-dependent M Current ($I_{K(M)}$)

$I_{K(M)}$ is a non-inactivating time- and voltage-dependent K current activated between rest potential and 0 mV. Its properties in NG108-15 cells, recently described in some detail,[31] closely resemble those of the equivalent current originally reported in sympathetic neurons.[32] As in sympathetic ganglion cells, inhibition of $I_{K(M)}$ in NG108-15 cells induces a net inward current at steady-state (FIG. 3, a) and, in unclamped cells, a depolarization and increased excitability (FIG. 3, b). Inhibition of $I_{K(M)}$ accounts for the second phase of inward current following the initial outward (Ca-activated) current on application of ACh to m1- or m3-transformed cells (see FIG. 2 and Fukuda et al.[7]).

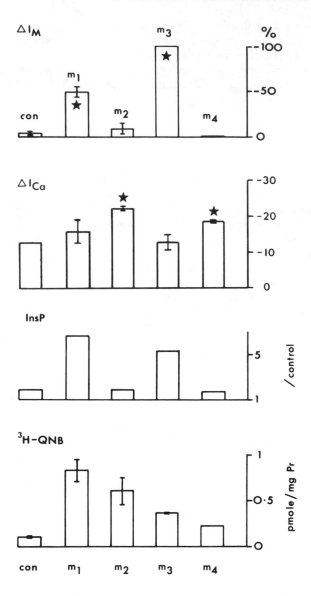

FIGURE 1. Summary of effects of mAChR stimulation in control and transformed NG108-15 cells. Histograms show (from bottom to top): specific [³H]-(–)quinuclidinyl benzilate (QNB) binding, fmoles/mg protein[7]; total inositol phosphate production (as fraction of resting production) induced by 10-min incubation with 1 mM carbachol[7]; % depression of peak Ca current amplitude by 1 mM ACh[9]; and % inhibition of the voltage-gated K current $I_{K(M)}$ (= I_M).[10] Bars show s.e.m. ★ = Significant difference from control (non-transfected:con) cells.

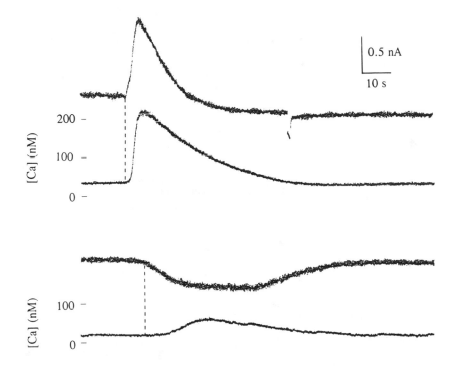

FIGURE 2. Changes in membrane current (upper records) and intracellular [Ca] (lower records) during applications of 100 μM ACh to an m1-transformed NG108-15 cells. Records were obtained using patch-electrodes containing 100 μM Indo-1 plus 100 μM BAPTA (see Robbins et al.[31] for experimental details). (*Upper trace*) ACh induced an initial outward current resulting from activation of $I_{K(Ca)}$. (*Lower trace*) Obtained after several applications of ACh, the initial outward current had dissipated leaving only an inward current resulting from inhibition of $I_{K(M)}$.

The "Messenger" for $I_{K(M)}$ Inhibition

There is clearly a close relation between the ability of receptors to inhibit $I_{K(M)}$ and their effectiveness in activating phospholipase C, in the sense that both are affected by the same type of receptor (m1 or m3 mAChR or BK). However, the precise link between these two responses (if any) is not yet clear. Some of the key experiments leading to this uncertainty are summarized in FIGURE 4.

Thus, neither Ca nor Ins-1,4,5-P_3 are likely to be the "messenger" for several reasons. Inhibition of $I_{K(M)}$ persists when the release of Ca and/or activation of $I_{K(Ca)}$ is suppressed or attenuated by such procedures as repeated application of ACh (which "desensitizes" the Ca release process) or buffering internal Ca with 20 mM BAPTA (FIG. 4). Even when Ca release is recorded, inhibition of $I_{K(M)}$ may *precede* the rise in Ca by several seconds, as in FIGURE 2. Internal application of Ins-1,4,5-P_3 at concentrations up to 100 μM does not inhibit $I_{K(M)}$ nor does it prevent the subsequent inhibition of $I_{K(M)}$ by ACh (FIG. 4), even though it does inhibit the activation of $I_{K(Ca)}$ by ACh.

As an alternative, we have previously suggested that diacylglycerols arising in

a

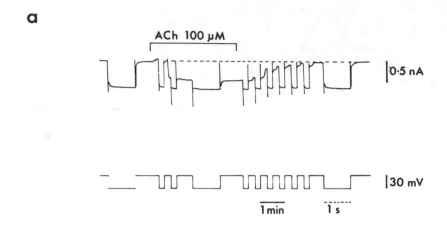

b

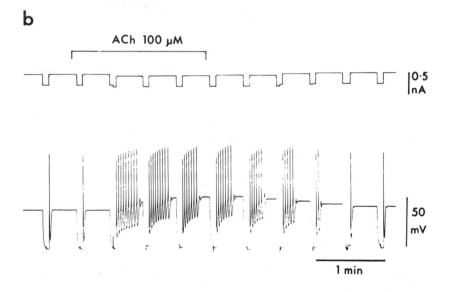

FIGURE 3. Responses of m1-transformed NG108-15 cells to 100 μM ACh recorded under (a) voltage-clamp and (b) current clamp. (a) The cell was held at − 30 mV and commanded to − 60 mV for 1 sec each 30 sec. ACh induced an inward current at − 30 mV due to inhibition of $I_{K(M)}$: this is reflected by the reduced amplitude of the current transients during each 1 sec hyperpolarizing step. (b) ACh produced a depolarization at the resting potential of − 50 mV and induced repetitive spike discharges following each hyperpolarizing step. (Note that DC current was applied during the application of ACh in b to limit the depolarization.)

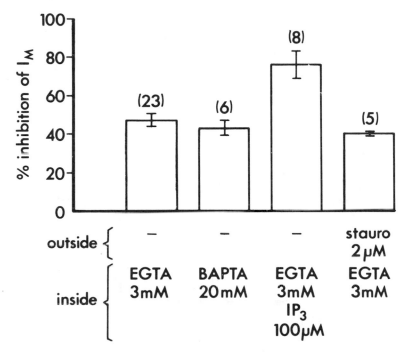

FIGURE 4. Partial summary of some tests on $I_{K(M)}$ inhibition by ACh in m1-transformed NG108-15 cells.[23] Each block shows the mean % inhibition of $I_{K(M)}$ produced by 100 μM ACh in patch-clamped cells. (Bars show s.e.m.; number of cells in brackets). Patch electrodes normally contained 3 mM EGTA. In the second block, this was replaced with 20 mM BAPTA. In the third block, 100 μM Ins-1,4,5-P₃ was added to the pipette solution. In the fourth block, 2 μM staurosporin was added to the bathing solution. Note that none of these procedures reduced the inhibition of $I_{K(M)}$, although both 20 mM intra-pipette BAPTA and 100 μM intra-pipette Ins-1,4,5-P₃ suppressed the initial activation of $I_{K(Ca)}$ by ACh.

parallel with inositol phosphates from phospolipase C stimulation might be responsible for $I_{K(M)}$ inhibition, on the basis that exogenous diacylglycerols or other activators of protein kinase C (PKC), such as phorbol dibutyrate, could also inhibit $I_{K(M)}$.[24,25,33] However, recent observations do not support this pathway as an obligatory step in $I_{K(M)}$ inhibition since inhibition of PKC with staurosporin (FIG. 4) or down-regulation by pretreatment with phorbol dibutyrate did not significantly impair the effect of ACh.[23]

A key question then is whether a cytosolic or diffusible messenger is really required at all or whether instead K(M)-channel closure might be mediated by some more local interaction between the activated G-protein and the channel. We have attempted to test this by recording single K(M) channels using cell-attached patch electrodes and then stimulating receptors outside the patch, in the manner previously applied to sympathetic neuron K(M) channels.[34] Channels with an appropriate time- and voltage-dependence could be detected in a proportion of such patches, showing sustained activity on depolarizing some 30 mV from rest potential (FIG. 5). Extra-patch applica-

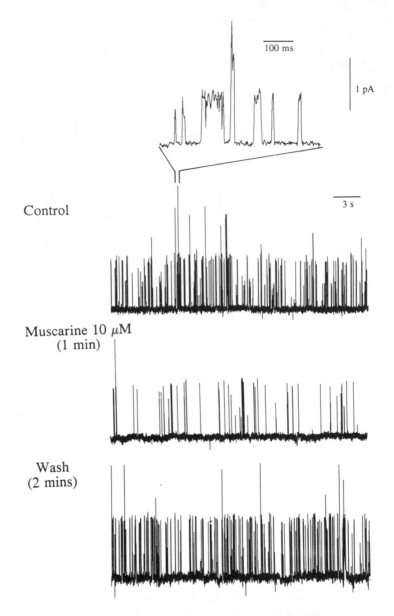

FIGURE 5. Effects of extra-patch application of 10 μM muscarine on single K(M)-channel activity recorded with a cell-attached patch electrode from an m1-transformed NG108-15 cell. See Selyanko *et al.*[34] for methodology. The patch-pipette was filled with Krebs' solution containing 2.5 mM K, with 100 nM apamin and 10 nM charybdotoxin (to block Ca-activated channels) and 500 nM tetrodotoxin (to block Na channels). The patch was held at 10 mV depolarized to rest potential. The bathing fluid contained 25 mM K (to prevent any depolarizing effect of the muscarine[34]) and 0 [Ca] with raised [Mg], to prevent Ca entry. Records show (from top to bottom): 30 sec control

tion of muscarine then clearly inhibited channel activity. This indicates that some remote signaling between extra-patch receptors and intra-patch channels can indeed occur, though whether this is an absolute requirement for channel closure is not yet known. Further experiments on isolated patches may clarify this.

Inhibition of the Voltage-gated Ca Current ($I_{Ca(N)}$)

ACh can reduce the amplitude of the composite voltage-activated Ca current in non-transfected cells through activation of the endogenous m4/M_4 receptors.[9,20] This complicates the assessment of the relative efficiencies of the different exogenous receptors in inhibiting I_{Ca}. However, the inhibitory effect of ACh was clearly and consistently increased in cells transformed to express exogenous m2 receptors, but was not significantly increased in cells expressing m1 or m3 receptors.[9] From this we conclude that both m2 and m4 mAChRs can inhibit I_{Ca}, though their relative effectiveness is difficult to ascertain.

Identity of the Inhibitable Current

The component of Ca current inhibitable by ACh (or noradrenaline) can be clearly identified as that component of high-threshold current inhibitable by ω-conotoxin.[22] Although kinetically different from that current originally designated the "N" current,[35] we may therefore refer to the current as $I_{Ca(N)}$ for convenience. This designation is important because this component of Ca current probably corresponds to that in nerve terminals responsible for the charge of Ca necessary to release transmitter.[36] Hence, the effects seen in NG108-15 cells may be relevant to feedback inhibition of transmitter release.

Nature of the G Protein Link

Inhibition of I_{Ca} by both exogenous m2 and endogenous m4 receptors is totally prevented by PTX. Three principal species of PTX-sensitive G-protein alpha subunits can be identified in NG108-15 cells: α_{i2}, α_{i3}, and α_o.[37-39] To test which might be most effective in coupling endogenous m4 mAChRs to Ca channels, we overexpressed α_{i2}, α_{i3}, and α_{oA}, plus α_{i1} (which is normally undetectable in NG108-15 cells) by DNA transfection.[40,41] Transfected cells showed about twofold higher levels of immu-

---

activity; activity 1 min after adding 10 μM muscarine to the solution bathing the cell; and activity 2 min after removing muscarine. The top record shows an expanded segment of control activity. Channels were identified as K(M) channels by their voltage dependence and by the time dependence of averaged currents during step depolarizations.[34] (A. A. Selyanko & J. Robbins, unpublished results.)

noreactive peptide in each case, indicating that each exogenous α subunit was expressed to about the same level.[41] The effects of this overexpression are summarized in FIGURE 6. Cells overexpressing either α_{i3} or α_{oA}, but not those overexpressing α_{i1} or α_{i2}, showed a significant increase in inhibition of I_{Ca} by ACh. This accords with previous experiments suggesting that the inhibition of I_{Ca} in NG108-15 cells by opiates[42] and noradrenaline[43] is mediated by α_o in preference to α_{i1} or α_{i2}.

Several recent studies have indicated that there is appreciable specificity in the choice of G-protein linking different receptors to Ca currents, with the following pairings: in rat dorsal root ganglion cells, neuropeptide Y selects α_o in preference to α_i whereas bradykinin receptors can couple to both α_o and $\alpha_{i1,2,3}$,[44] in GH3 cells, muscarinic receptors couple to α_{oA} whereas somatostatin receptors couple to α_{oB};[45] and in NG108-15 cells, opiate and noradrenaline α2 receptors couple to α_{oA} whereas somatostatin receptors do not.[46] Thus, the present data provide a further example, by suggesting that muscarinic m4 receptors can couple to both α_{oA} and α_{i3}, but not to α_{i2} or α_{i1}.

A second point of interest was that cells expressing exogenous α_{i1} showed a significantly reduced inositol phosphate response to bradykinin (FIG. 6). Since this effect was not prevented by PTX (which ADP-ribosylates α/β/γ complexes), it indicates that activation of phospholipase C by BK (and possibly by m1 or m3 mAChRs) might be affected by excess free α_{i1} subunits, which may be cytosolically located.[47]

NL308 NEUROBLASTOMA × FIBROBLAST HYBRID CELLS

Our reason for testing mAChR ion channel coupling in these cells is that Chalazonitis *et al.*[48] reported that they were hyperpolarized by ACh, through atropine-sensitive (i.e., muscarinic) receptors, and that this hyperpolarization was due to an increased K conductance. Since the endogenous mAChR in NL308 cells as judged from RNA blot hybridizations appears to be the same as that in NG108-15 cells (i.e., m4), and since no such hyperpolarization occurs in NG108-15 cells, this suggested that NL308 cells might exhibit a different coupling pathway between mAChRs and K-channels than that in NG108-15 cells.

Accordingly, NL308 cells were transfected with DNA for m1, m2, m3, and m4 receptors as described previously[7] and clones expressing significantly increased [3H]QNB binding selected for further study.[49] Some of the principal results are summarized in FIGURE 7. The following points should be noted.

As in NG108-15 cells, only cells transformed to express exogenous m1 or m3 receptors showed an increased inositol phosphate production when challenged with a choline ester. This was not due solely to different receptor numbers since the number of expressed m2 and m3 receptors were not significantly different, but stimulation of m2-transformed cells did not detectably increase inositol phosphate production.

Unlike NG108-15 cells, an outward K current could be generated by applying ACh to cells from all four subclones, i.e., to cells expressing all four mAChRs. The mean amplitude of this outward current (in responding cells) varied approximately in proportion to the receptor density in each subclone.

However, the responses of m2- or m4-transformed cells differed from those of m1- or m3-transformed cells in three significant respects. First, m2/m4-transformed

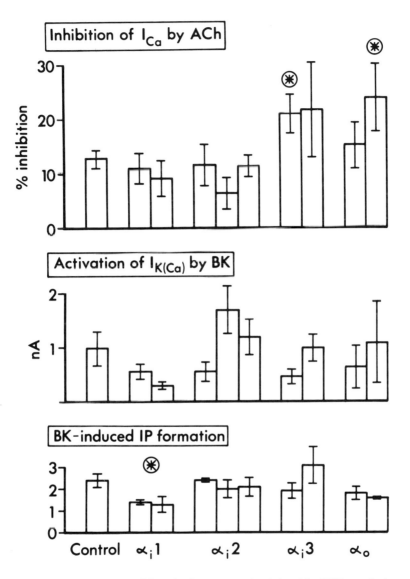

FIGURE 6. Effects of G-protein alpha subunit overexpression induced by DNA transfection on responses of NG108-15 cells to stimulation of endogenous m4 mAChRs and bradykinin receptors.[40,41] Results are shown for non-transfected cells and for two clones transfected with DNA for α_{i1}, three clones with α_{i2}, two clones with α_{i3}, and two clones with α_o. Histograms show (from bottom to top): Ins-P_3 formation by 10 μM bradykinin (BK) for 10 sec (expressed as a ratio of that formed in control cells not exposed to BK); mean amplitude (nA) of the initial outward current $I_{K(Ca)}$ produced by focal application of 10 μM BK in 3 μl ejection volume; the mean % inhibition of the high-threshold Ca current produced by focal application of 1 mM ACh in 3 μl ejection volume. Each block shows mean data for each clone; bars give s.e.m.. * Significant difference from non-transfected controls ($p<0.05$).

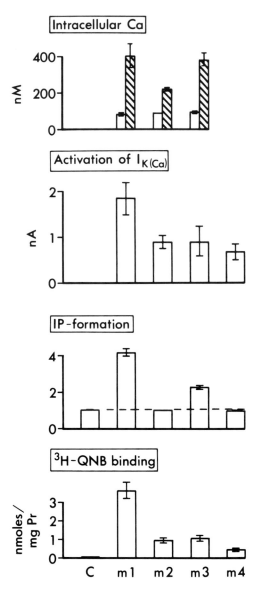

FIGURE 7. Effects of transfecting DNA for m1, m2, m3, and m4 mAChRs on NL 308 neuro-blastoma × fibroblast hybrid cells.[49] Histograms show (from bottom to top): saturable [³H]QNB binding (fmoles/mg protein; 1 nM (−)-[³H]QNB, 43.9 Ci/mmol); inositol phosphate production by 1 mM carbachol (determined as in Fukuda *et al.*[7]); maximum amplitude of initial outward current $I_{K(Ca)}$ produced by focal application of ACh; and maximum ACh-induced increases in intracellular [Ca] measured with Fura-2/AM (open bars, before ACh; hatched bars, after ACh). Each block gives mean response. Bars show s.e.m.

cells were much less sensitive to ACh than m1/m3-transformed cells, by a factor of about 100. Second, outward currents in m2/m4-transformed cells tended to show oscillations (see FIG. 9), whereas those in m1/m3-transformed cells were "smooth," like those in NG108-15 cells (cf. FIG. 2). Third, the responses of m2 or m4-transformed cells were completely annulled by pretreatment with PTX, whereas those of m1 or m3-transformed cells were not significantly changed after PTX treatment, implying two totally different G-protein systems.

Nature of the K Current

Because of the differences between the effects of stimulating m1/m3 receptors on the one hand and m2/m4 receptors on the other, we wondered which species of K current was activated and whether both receptor pairs activated the same species of current. The answer to the second question seems to be yes, and the species of K current activated appears to be a Ca-activated current similar to that in NG108-15 cells, for the following reasons.[19] (1) Responses to both m1/m3 and m2/m4 stimulation were blocked by 20 nM charybdotoxin or 1 mM TEA, and both showed a characteristic voltage-dependent block by 1 mM Ba. They were unaffected by 1 mM 4-amino-pyridine, 1 μM MCDP, or 0.4 μM apamin. (2) Responses were affected by buffering external Ca with 10 μM BAPTA-AM. (3) Application of ACh to both m1/m3 and m2/m4 mAChR-transformed cells induced rises in intracellular Ca, albeit of somewhat different form (see below). Very similar effects occur on applying BK to these cells.[50]

K Channels

Using cell-attached patch-electrodes, the opening of single channels of three types, with average conductances 10–14, 34–40, and 75–86 pS, could be recorded within the patch electrode following application of ACh to the membrane outside the patch in both m1- and m2-transformed cells (FIG. 8). This implies a remote coupling from the receptor to the channel, as would be expected were the receptors to activate the channel through the release of intracellular Ca. Although we have not yet established that these channels are indeed Ca-activated, they somewhat resemble those recorded from NG108-15 cells following extra-patch application of BK, which can also be activated by intracellular Ca or Ins-1,4,5-P_3 injections.[51,52]

Intracellular Ca

Changes in intracellular [Ca] induced by ACh were recorded using Fura-2/AM. Significant elevations in intracellular [Ca] were recorded in m1, m2, and m3-transformed cells (FIG. 7). To determine the temporal relationship between the rise in Ca and the outward current, membrane currents were recorded using patch-pipettes containing Indo-1. Although in some cells (most notably m1- or m3-transformed cells) a close temporal relationship between dI and d[Ca] similar to that illustrated in FIGURE 2 could be seen, in many others (especially m2-transformed cells) the two responses

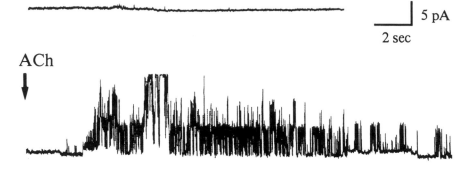

FIGURE 8. Continuous recording of single channel currents activated by ACh in an m2-transformed NL308 cell (clone NLPM2-301). Records show channel activity recorded with a cell-attached pipette containing 140 mM KCl and 5 mM HEPES (pH 7.4 with KOH), held at + 30 mV. ACh (3 μl, 100 μM) was applied to the membrane outside the patch via the bathing fluid at the time indicated by the arrow.

showed rather striking temporal dissociations. Thus, in the cell illustrated in FIGURE 9, the initial outward current following the first application of ACh clearly preceded any detectable rise in somatic [Ca] and also showed a multiphasic nature. With repeated applications of ACh, the multiphasic nature of the current became increasingly obvious, while the Ca signal became progressively smaller and more dissociated from the current response.

One possible explanation for the dissociation of current and Ca signal is that much of the current is generated in neuritic processes, in response to local changes in [Ca] not registered at the soma. Thus, membrane current responses devoid of somatic Ca signals could readily be evoked by focal application of ACh to neurites. The spatial distribution of the Ca changes was further studied using Fura-2/AM imaging (FIG. 10). Two points emerged: Ca changes in the soma frequently showed oscillations and Ca changes in the soma and neurites were often out of phase. (Although the strongest signals were obtained in the soma, this only refers to bulk cytoplasmic Ca, not the submembrane Ca changes near the K channels.)

Interpretation

We interpret these observations to indicate that both m1/m3 mAChR and m2/m4 mAChR may induce increases in intracellular Ca and thereby activate Ca-dependent K channels, but via different mechanisms. As in NG108-15 cells, m1/m3 receptors activate phospholipase C through a PTX-insensitive G-protein (probably of the G_q/G_{11} family). In contrast, the effects of m2/m4 receptors are mediated through a different, PTX-sensitive, G-protein. This may also couple to PLC, perhaps through

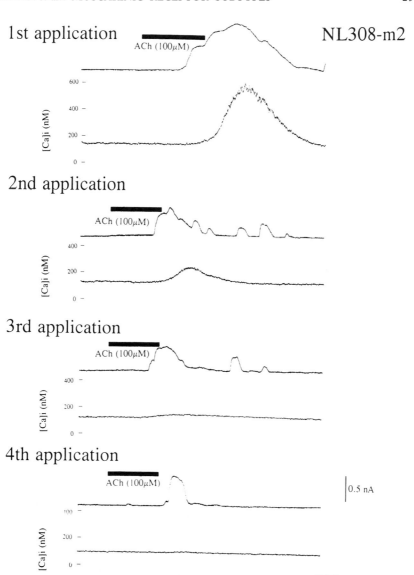

FIGURE 9. Effects of repeated applications of ACh (at about 10-min intervals) on membrane current (upper records) and intracellular [Ca] (lower records) in an m2-transformed NL308 cell recorded with a patch-pipette containing 100 μM Indo-1 plus 100 μM BAPTA.[31] ACh (100 μM) was applied via the bathing fluid for the duration indicated by the bars. Note the dissociation between the outward current ($I_{K(Ca)}$) responses and changes in intracellular [Ca].

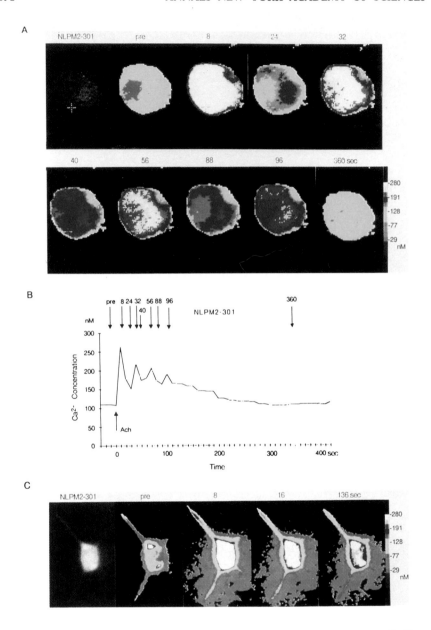

FIGURE 10. Digital image analysis of intracellular [Ca] in m2-transformed NL308 cells (clone NLPM2-301) following loading with Fura-2/AM. (**A**) The spatial distribution of [Ca] in a single cell observed before (pre) and at various times after application of 100 μM ACh. (**B**) The time-course of average Ca concentration changes in the same cell. Arrows refer to the images in **A**. Note the oscillations in [Ca]. (**C**) [Ca] rises in another cell with processes following application of ACh. Note the rise of [Ca] in both soma and processes.

release of βγ subunits from the activated G-proteins,[53] or it may affect internal Ca through a different pathway. Finally, there is a tendency for the two pathways—m1/m3 and m2/m4—to be spatially separated, with the latter directed more toward processes than soma: local, asynchronous activation of Ca rises at discrete foci could account for the oscillatory current responses. However, this distinction is not absolute, nor do we know whether it arises because of local distribution of expressed receptors, Ca-release pathways, or K channels: tests on cells co-transfected with DNA for both m2 and m3 receptors may help to clarify this.

This cross-talk between m2 and m3 receptors onto Ca-activated ion channels in NL308 cells has not been reported in other neural cell lines such as NG108-15 cells, nor in fibroblast or epithelial cell lines.[11,12] However, it does resemble the cross-talk onto Ca-activated Cl channels in oocytes described by Lechleiter et al.,[6] in terms of differential agonist sensitivity and G-protein involvement, and may also involve a comparable spatial difference in the initial Ca signal.

DISCUSSION

Responses to Stimulating m1 and m3 Receptors

Activation of Ca-dependent Ion Channels

m1 and m3 mAChR can clearly activate Ca-dependent channels in a variety of cells. This is a direct result of their primary coupling to phospholipase C and consequential formation of Ins-1,4,5-P_3 and release of Ca ions. The principal variable determining the final response then becomes the species of Ca-dependent channel finally activated. In oocytes, this is a Cl channel,[5] whereas in neuroblastoma cells, fibroblasts, and epithelial cells the initial current is carried by K channels.[7,11,12] These channels correspond to the small (5–15 pS) and/or medium (35–85 pS) conductance K(Ca) channels, rather than the large (>200 pS) "BK" channels.[51,52,54] However, their pharmacological properties appear to differ from one cell type to another, and, even in the same cell, may not be uniform.[52] Further, a component of current in fibroblasts is also carried by Ca-activated Cl channels[54] while in NL308 cells we have detected a component apparently carried by Ca-dependent non-specific channels.[49]

Notwithstanding the widespread occurrence of this linkage in expression cells, it is difficult to pinpoint the precise role of such ionic responses in primary cells. Thus, while it is clear that activation of K(Ca) and Cl(Ca) channels is an important component of the response of secretory cells to mAChR stimulation, there are very few clearly established examples of such responses in primary mammalian neurons, in spite of the widespread occurrence of the appropriate receptors and channels. The reason for this is quite unknown.

Inhibition of K Currents

In contrast to the activation of $I_{K(Ca)}$, inhibition of $I_{K(M)}$ in NG108-15 cells does represent a major component of muscarinic excitation in neurons.[55] Further, and in accordance with the results obtained with expressed receptors in NG108-15 cells, inhibition of $I_{K(M)}$ in sympathetic neurons can clearly be ascribed to activation of pharmacologically defined M1 receptors.[56,57] (In contrast, in cortex[58] and hippocampus,[59] inhibition of $I_{K(M)}$ has been attributed to activation of M2 receptors. Our data suggest that these are more likely to be m3 (M3) receptors—a conclusion more in keeping with the observations of Pitler and Alger[60]). NG108-15 cells therefore provide a reasonably faithful model for this type of ionic channel coupling.

However, the ability of mAChR stimulation to close K channels is not restricted to K(M) channels. Thus, in the brain and in the enteric nervous system, another prime target is a class of apamin-resistant K(Ca) channels.[1] This effect has not been reported so far in any cells or cell lines following activation of cloned receptors, presumably because the appropriate channels are absent.

Responses to Stimulating m2 or m4 Receptors

Inhibition of Ca Currents

Experiments on NG108-15 cells have suggested that the "N-type" (ω-conotoxin sensitive) Ca current can be inhibited either through endogenous m4 receptors or through exogenous m2 receptors.[9] Equivalent effects have been identified in two types of primary mammalian neuron: in rat sympathetic neurons, inhibition of $I_{Ca(N)}$ is mediated by M4 receptors[57] whereas in rat cholinergic forebrain neurons inhibition of an equivalent high-voltage activated Ca current is inhibited via M2 receptors.[61] In both cases the inhibition is prevented by Pertussis toxin, suggesting a comparable G-protein link to that in NG108-15 cells, though the specific G-protein involved remains to be identified. The principal biochemical response to m2/m4 receptor activation is an inhibition of adenylate cyclase. However, from experiments on rat sympathetic neurons[62] and on the equivalent inhibition produced by catecholamines,[63] it seems likely that the activated G-protein interacts in a rather localized ("membrane-delimited") manner with the Ca channels, rather than through any such intracellular biochemical pathway. One obvious physiological consequence of $I_{Ca(N)}$ inhibition might be the reduction of transmitter release.[36,63] Effects in transformed NG108-15 cells might therefore provide a fair representation of the widespread auto-inhibitory effect of ACh on its own release, most especially in the central nervous system.

Activation of K Currents

The activation of Ca-dependent K currents in NL308 cells and of Ca-dependent Cl currents in oocytes[6] appear to represent an interesting form of "cross-talk" between m2/m4 receptors on the one hand and m1/m3 on the other, whereby two sets of receptors affect the same ion channels but through quite different G-proteins. This cannot be ascribed to "overexpression" of m2 or m4 receptors as such, since the densities of m2 and m3 receptors in both the present experiments and in those of Lechleiter *et al.*[6] were very comparable. At present, however, it is difficult to place this into a physiological perspective, since activation of these channels through m2 or m4 receptors requires much higher agonist concentrations than those needed to produce an equivalent effect via m1 or m3 receptors; and no equivalent effect in primary cells has so far been identified.

A much more widespread effect associated with pharmacologically defined M2 receptors is activation of a Ca-independent, inwardly rectifying K current. This is seen in both nerve cells[1] and cardiac cells.[64] No such effect has been detected in any of the neuroblastoma, fibroblast, or epithelial cell lines so far used for studying cloned receptors, presumably because the appropriate channels are absent. More relevant, perhaps, is the recently reported enhancement of an endogenous inwardly rectifying current in transformed cells of a secretory cell line[13]: how far this replicates the effects in primary cells remains to be determined.

CONCLUSIONS

In spite of the indirect nature of the coupling between mAChRs and ion channels, some qualitatively rather distinct patterns of interaction have been deduced from studies on transformed cells, which appear reasonably representative of some of the major pathways operating in primary cells. However, there is also evidence for points of cross-talk between the lines of communication. A major requirement for the future is to define the quantitative parameters determining coupling preferences between the different receptor subtypes and ion channel species and thereby establish the precise limits to these coupling preferences and the extent to which they may be subject to physiological modification.

ACKNOWLEDGMENTS

We are particularly grateful to the late Professor S. Numa, with whom the work was started.

REFERENCES

1. NORTH, R. A. 1989. Muscarinic cholinergic receptor regulation of ion channels. *In* The Muscarinic Receptors. J. H. Brown, Ed.: 341–346. Humana Press. New Jersey.
2. BONNER, T. I. 1989. The molecular basis of muscarinic receptor diversity. Trends Neurosci. **12**: 148–151.
3. FUKUDA, K., T. KUBO, A. MAENA, I. AKIBA, J. BUJO, J. NAKAI, M. MISHINA, H. HIGASHIDA, E. NEHER, A. MARTY & S. NUMA. 1989. Selective effector coupling of muscarinic acetylcholine receptor subtypes. Trends Pharmacol. Sci. (Suppl.)**IV**: 4–10.
4. HULME, E. C., N. J. M. BIRDSALL & N. J. BUCKLEY. 1990. Muscarinic receptor subtypes. Ann. Rev. Pharmacol. Toxicol. **30**: 633–673.
5. FUKUDA, K., T. KUBO, I. AKIBA, A. MAEDA, M. MISHINA & S. NUMA. 1987. Molecular distinctions between muscarinic acetylcholine receptor subtypes. Nature **327**: 623–625.
6. LECHLEITER, J., S. GIRARD, D. E. CLAPHAM & E. PERALTA. 1991. Subcellular patterns of calcium release determined by G-protein specific residues of muscarinic receptors. Nature **350**: 505–508
7. FUKUDA, K., H. HIGASHIDA, T. KUBO, A. MAEDA, I. AKIBA, II. BUJO, M. MISHINA & S. NUMA.

1988. Selective coupling with K^+ currents of muscarinic acetylcholine receptor subtypes in NG108-15 cells. Nature **335**: 355–358.

8. NEHER, E., A. MARTY, K. FUKUDA, T. KUBO & S. NUMA. 1988. Intracellular calcium release mediated by two muscarinic receptor subtypes. FEBS Lett. **240**: 88–94.
9. HIGASHIDA, H., M. HASHII, K. FUKUDA, M. P. CAULFIELD, S. NUMA & D. A. BROWN. 1990. Selective coupling of different muscarinic acetylcholine receptors to neuronal calcium currents in DNA-transfected cells. Proc. R. Soc. Lond. **242**: 68–74.
10. ROBBINS, J., M. P. CAULFIELD, H. HIGASHIDA & D. A. BROWN. 1991. Genotypic m3-muscarinic receptors preferentially inhibit M-currents in DNA-transfected NG108-15 neuroblastoma × glioma hybrid cells. Eur. J. Neurosci. **3**: 820–824.
11. JONES, S. V. P., J. L. BARKER, N. J. BUCKLEY, T. I. BONNER, R. M. COLLINS & M. R. BRANN. 1988. Cloned muscarinic receptor subtypes expressed in A9L cells differ in their coupling to electrical responses. Mol. Pharmacol. **34**: 421–426.
12. JONES, S. V. P., C. J. HEILMAN & M. R. BRANN. 1991. Functional responses of cloned muscarinic receptors expressed in CHO-K1 cells. Mol. Pharmacol. **40**: 242–247.
13. JONES, S. V. P. 1991. Effects of muscarinic receptor subtypes on an inward potassium conductance and on exocytosis. Neurosci. Abstr. **17**: 67.
14. VASUDEVAN, S., L. PREMKUMAR, S. STOWE, P. W. GAGE, H. REILANDER & C-H. CHUNG. 1992. Muscarinic acetylcholine receptor produced in recombinant bacalovirus infected *Sf9* insect cells couples with endogenous G-proteins to activate ion channels. FEBS Lett. **311**: 7–11.
15. HAMPRECHT, B. 1977. Structural, electrophysiological and pharmacological properties of neuro-blastoma-glioma cell hybrids in cell culture. Int. Rev. Cytol. **49**: 99–170.
16. BROWN, D. A. & H. HIGASHIDA. 1988. Voltage- and calcium-activated potassium currents in mouse neuroblastoma × rat glioma hybrid cells. J. Physiol. **397**: 149–165.
17. BROWN, D. A., R. J. DOCHERTY & I. MCFADZEAN. 1989. Calcium channels in vertebrate neurons. Experiments on a neuroblastoma hybrid model. Ann. N.Y. Acad. Sci. **560**: 358–372.
18. DOCHERTY, R. J., J. ROBBINS & D. A. BROWN. 1991. NG 108-15 neuroblastoma × glioma hybrid cell line as a model neuronal system. *In* Cellular Neurobiology: a Practical Approach. J. Chad & H. V. Wheal, Eds.: 257–321. IRL Press. Oxford.
19. LAZARENO, S., N. J. BUCKLEY & F. F. ROBERTS. 1990. Characterisation of muscarinic M_4 binding sites in rabbit lung, chicken heart and NG 108-15 cells. Mol. Pharmacol. **38**: 805–815.
20. CAULFIELD, M. P. & D. A. BROWN. 1991. Pharmacology of the putative M_4 muscarinic receptor mediating Ca-current inhibition in neuroblastoma × glioma hybrid (NG 108-15) cells. Br. J. Pharmacol. **104**: 39–44.
21. KUROSE, H., T. KATADA, T. AMANO & M. UI. 1983. Specific uncoupling by islet-activating protein, pertussis toxin, of negative signal transduction via α-adrenergic, cholinergic and opiate receptors in neuroblastoma × glioma hybrid cells. J. Biol. Chem. **258**: 4870–4875.
22. CAULFIELD, M. P., J. ROBBINS & D. A. BROWN. 1992. Neurotransmitters inhibit the ω-conotoxin-sensitive component of Ca current in neuroblastoma × glioma hybrid (NG 108-15) cells, not the nifedipine-sensitive component. Pflug. Arch. **420**: 486–492.
23. ROBBINS, J., S. J. MARSH & D. A. BROWN. 1993. On the mechanism of M-current inhibition by muscarinic m1 receptors in DNA-transfected NG108-15 cells. J. Physiol. **469**: 153–178.
24. BROWN, D. A. & H. HIGASHIDA. 1988. Inositol 1,4,5-trisphosphate and diacylglycerol mimic bradykinin effects on mouse neuroblastoma × rat glioma hybrid cells. J. Physiol. **397**: 185–207.
25. HIGASHIDA, H. & D. A. BROWN. 1986. Two polyphosphoinositide metabolites control two K^+ currents in a neuronal cell. Nature **323**: 333–335.
26. BROWN, D. A. & H. HIGASHIDA. 1988. Membrane current responses of NG108-15 mouse neuroblastoma × rat glioma hybrid cells to bradykinin. J. Physiol. **397**: 167–184.
27. KIMURA, K. & H. HIGASHIDA. 1992. Dissection of bradykinin-evoked responses by buffering intracellular Ca^{2+} in neuroblastma × glioma hybrid NG108-15 cells. Neurosci. Res. **15**: 213–220.
28. ROBBINS, J. 1993. Agonist-induced inhibition of IP_3-activated $I_{K(Ca)}$ in NG108-15 neuroblastoma cells. Pflug. Arch. **422**: 364–370.

29. GUTOWSKI, S., A. SMRCKA, L. NOWAK, Q. WU, M. SIMON & P. C. STERNWEISS. 1991. Antibodies to the α_q subfamily of guanine nucleotide-binding regulatory protein α-subunits attenuate activation of phosphatidylinositol 4,5-bisphosphate hydrolysis by hormones. J. Biol. Chem. 266: 20616–20624.

30. BELARDETTI, F., S. GUTOWSKI & P. STERNWEISS. 1992. An anti-Gα_q antibody blocks the PLC-dependent bradykinin response in NG108-15 cells. Soc. Neurosci. Abstr. 18: 335.21.

31. ROBBINS, J., J. TROUSLARD, S. J. MARSH & D. A. BROWN. 1992. Kinetic and pharmacological properties of the M-current in rodent neuroblastoma × glioma hybrid cells. J. Physiol. 451: 159–185.

32. ADAMS, P. R., D. A. BROWN & A. CONSTANTI. 1982. M-currents and other potassium currents in bullfrog sympathetic neurones. J. Physiol. 330: 537–572.

33. SCHAFER, S., P. BEHE & H. MEVES. 1991. Inhibition of the M-current in NG108-15 neuroblastoma × glioma hybrid cells. Pflug. Arch. 418: 581–591.

34. SELYANKO, A. A., C. E. STANSFELD & D. A. BROWN. 1992. Closure of potassium M-channels by muscarinic acetylcholine-receptor stimulants requires a diffusible messenger. Proc. R. Soc. Lond. B 250: 119–125.

35. FOX, A. P., M. NOWYCKY & R. W. TSIEN. 1987. Kinetic and pharmacological properties distinguishing three types of calcium currents in chick sensory neurones. J. Physiol. 394: 149–172.

36. HIRNING, L. D., A. P. FOX, E. W. MCLESKEY, B. M. OLIVERA, S. A. THAYER & R. J. MILLER. 1988. Dominant role of the N-type Ca^{2+} channels in evoked release of norepinephrine from sympathetic neurons. Science 239: 57–61.

37. MILLIGAN, G., P. GIERSCHIK, A. M. SPIEGEL & W. A. KLEE. 1986. The GTP-binding regulatory proteins of neuroblastoma × glioma, NG108-15 and glioma, C6, cells. FEBS Lett. 195: 225–230.

38. MULLANEY, I., A. I. MAGEE, C. G. UNSON & G. MILLIGAN. 1988. Differential regulation of amounts of the guanine nucleotide binding proteins G$_i$ and G$_o$ in neuroblastoma × glioma hybrid cells in response to dibutyryl cyclic AMP. Biochem. J. 256: 649–656.

39. ASANO, T., R. MORISHITA, M. SANO & K. KATO. 1989. The GTP-binding proteins G$_o$ and G$_{i2}$ of neural cloned cells and their changes during differentiation. J. Neurochem. 53: 1195–1198.

40. NUKADA, T., M. MISHINA & S. NUMA. 1987. Functional expression of adenylate cyclase-stimulating G-protein. FEBS Lett. 211: 5–9.

41. NUKADA, T., M. HASHII, K. KAMEYAMA & H. HIGASHIDA. 1990. Effects on bradykinin- and acetylcholine-induced signal transduction of overexpression of the α-subunit of GTP binding protein in NG108-15 cells. Bull. Jap. Neurochem. Soc. 29: 382–383.

42. HESCHELER, J., W. ROSENTHAL, W. TRAUTWEIN & G. SCHULTZ. 1987. The GTP-binding protein, G$_o$, regulates neuronal calcium channels. Nature 325: 445–447.

43. MCFADZEAN, I., I. MULLANEY, D. A. BROWN & G. MILLIGAN. 1989. Antibodies to the GTP-binding protein G$_o$ antagonize noradrenaline-induced calcium current inhibition in NG108-15 hybrid cells. Neuron 3: 177–182.

44. EWALD, D. A., I-H. PANG, P. C. STERNWEISS & R. J. MILLER. 1989. Differential G protein-mediated coupling of neurotransmitter receptors to Ca^{2+} channels in rat dorsal root ganglion neurons in vitro. Neuron 2: 1185–1193.

45. KLEUSS, C., J. HESCHELER, C. EWEL, W. ROSENTHAL, G. SCHULTZ & B. WITTIG. 1991. Assignment of G-protein subtypes to specific receptors inducing inhibition of calcium currents. Nature 353: 43–48.

46. TAUSSIG, R., S. SANCHEZ, M. RIFO, A. G. GILMAN & F. BELARDETTI. 1992. Inhibition of the ω-conotoxin-sensitive calcium current by distinct G proteins. Neuron 8: 799–809.

47. ROTROSEN, D., J. I. GALLIN, A. M. SPIEGEL & H. L. MALECH. 1988. Subcellular localization of G$_i$ in human neutrophils. J. Biol. Chem. 263: 10958–10964.

48. CHALAZONITIS, A., J. D. MINNA & M. NIRENBERG. 1977. Expression and properties of acetylcholine receptors in several clones of mouse neuroblastoma × L cell somatic hybrids. Exptl. Cell Res. 105: 269–280.

49. NODA, M., K. KATAYAMA, D. A. BROWN, J. ROBBINS, S. J. MARSH, N. ISHIZAKA, K. FUKUDA,

N. Hoshi, S. Yokoyama & H. Higashida. 1993. Coupling of m2 and m4 muscarinic acetylcholine receptor subtypes to Ca^{2+}-dependent K^+-channels in transformed NL308 neuroblastoma × fibroblast hybrid cells. Proc. R. Soc. Lond. B **251:** 215–224.

50. Higashida, H., Y. Okano, N. Hoshi, Y. Yada, S. Yokoyama, T. Asaga, T. Fu & Y. Nozawa. 1990. Bradykinin induces inositol 1,4,5-trisphosphate-dependent hyperpolarization in K^+ M-current deficient hybrid NL308 cells: comparison with NG108-15 neuroblastoma × glioma hybrid cells. Glia **3:** 1–12.

51. Higashida, H. & D. A. Brown. 1988. Ca-dependent K channels in neuroblastoma hybrid cells activated by intracellular inositol trisphosphate or extracellular bradykinin. FEBS Lett. **238:** 395–400.

52. Robbins, J., I. McFadzean & D. A. Brown. 1992. Effects of bradykinin on ion conductances in NG108-15 neuroblastoma × glioma hybrid cells recorded with patch-clamp electrodes. *In* Recent Progress on Kinins. G. Bonner, H. Fritz, Th. Unger, A. Roscher & K. Luppertz, Eds. Part 1:98–107. Birkhauser Verlag. Basel.

53. Camps, M., C. Hou, D. Siridopoulos, J. B. Stock, K. H. Jakobs & P. Gierschik. 1992. Stimulation of phospholipase C by guanine-nucleotide binding protein subunits. Eur. J. Biochem. **206:** 821–831.

54. Jones, S. V. P., J. L. Barker, M. Goodman & M. R. Brann. 1990. Inositol trisphosphate mediates cloned muscarinic receptor-activated conductances in transfected mouse fibroblast A9L cells. J. Physiol. **421:** 499–519.

55. Brown, D. A. 1988. M-currents. *In* Ion Channels. T. Narahashi, Ed. **1:** 55–99.

56. Marrion, N. V., T. G. Smart, S. J. Marsh & D. A. Brown. 1989. Muscarinic suppression of the M-current in the rat sympathetic ganglion is mediated by receptors of the M_1-subtype. Br. J. Pharmacol. **98:** 557–573.

57. Bernheim, L., A. Mathie & B. Hille. 1992. Characterization of muscarinic receptor subtypes inhibiting Ca^{2+} current and M current in rat sympathetic neurons. Proc. Natl. Acad. Sci. USA **89:** 9544–9548.

58. Constanti, A. & J. A. Sim. 1987. Muscarinic receptors mediating suppression of the M-current in guinea-pig olfactory cortex neurones may be of the M_2-subtype. Br. J. Pharmacol. **90:** 3–5.

59. Dutar, P. & R. A. Nicoll. 1988. Classification of muscarinic responses in hippocampus in terms of receptor subtypes and second messenger systems: electrophysiological studies in vitro. J. Neurosci. **8:** 4214–4224.

60. Pitler, T. A. & B. E. Alger. 1990. Activation of the pharmacologically-defined M_3 muscarinic receptor depolarizes hippocampal pyramidal cells. Brain Res. **534:** 257–262.

61. Allen, T. G. J. & D. A. Brown. 1993. M_2 muscarinic receptor-mediated inhibition of a high voltage-activated calcium current in rat magnocellular cholinergic basal forebrain neurones. J. Physiol. **466:** 173–189.

62. Mathie, A., L. Bernheim & B. Hille. 1992. Inhibition of N- and L-type calcium channels by muscarinic receptor activation in rat sympathetic neurons. Neuron **8:** 907–914.

63. Lipscombe, D., S. Kongsamut & R. W. Tsien. 1989. α-Adrenergic inhibition of sympathetic neurotransmitter release mediated by modulation of N-type calcium channel gating. Nature **340:** 639–642.

64. Noma, A. 1987. Chemical-receptor dependent potassium channels in cardiac muscle. *In* Electrophysiology of Single Cardiac Cells. D. Noble & T. Powell, Eds.: 223–246. Academic Press. New York.

Cardiac Chloride Channels: Incremental Regulation by Phosphorylation/ Dephosphorylation[a]

DAVID C. GADSBY, TZYH-CHANG HWANG,
MINORU HORIE, GEORG NAGEL,
AND [b]ANGUS C. NAIRN

Laboratory of Cardiac/Membrane Physiology
[b]Laboratory of Cellular and Molecular Neuroscience
1230 York Avenue
The Rockefeller University
New York, New York 10021

INTRODUCTION

It has recently become clear (for reviews[1-3]) that cardiac Cl⁻ channels regulated by protein kinase A (PKA)[4-6] are closely similar, if not identical, to epithelial cystic fibrosis transmembrane conductance regulator (CFTR) channels.[7,8] Thus, in intact, mammalian, ventricular myocytes catecholamines activate a macroscopic Cl⁻ conductance through the classical β-adrenoceptor-G_s-cAMP-PKA pathway,[9,10] and both whole-cell and single-channel currents through these PKA-regulated cardiac Cl⁻ channels resemble those flowing in epithelial CFTR Cl⁻ channels in all biophysical and biochemical properties so far examined.[1-3] For example, in addition to the characteristic small ohmic conductance of single cardiac Cl⁻ channels in excised membrane patches exposed to symmetrical [Cl⁻] solutions,[11,12] native cardiac Cl⁻ channels share with recombinant human epithelial CFTR Cl⁻ channels an absolute requirement for hydrolyzable nucleoside triphosphates to open the channels after their phosphorylation by PKA.[11,13] Moreover, Northern blot analysis has demonstrated that CFTR mRNA exists in guinea pig, rabbit, and human heart.[11,14] The deduced amino acid sequence of rabbit cardiac CFTR cDNA corresponding to the first of the two nucleotide binding domains shows 98% homology[14] with human CFTR cloned from sweat duct cells,[15] including conservation of the phenylalanine at position 508, the deletion of which accounts for ~70% of the cystic fibrosis diagnosed in the caucasian population. Although the complete sequence of cardiac CFTR is not yet known, preliminary

[a] The original research described here and the preparation of this manuscript were supported by grants from the National Institutes of Health (HL-14899 and HL-49907), the New York Heart Association, and the Cystic Fibrosis Foundation.

259

results using polymerase chain reaction techniques suggest that, in comparison to epithelial CFTR,[15] cardiac CFTR lacks 30 amino acids from the predicted first cytoplasmic loop, implying that it is an alternatively spliced variant.[16]

While the primary, disease-causing defect in patients with cystic fibrosis is localized to epithelial cells lining the respiratory and gastrointestinal tracts, technical difficulties tend to hamper studies of the regulation of CFTR Cl^- channels in those cells.[17] These include the relatively low density of CFTR channels in native epithelia[18,19] in comparison to the densities of other Cl^- channels with larger unitary conductances, some showing outward rectification, that can obscure the current signals from CFTR channels,[20,21] and a relative unsuitability of epithelial cells for the isolation and manipulation necessary for whole-cell recording.[22] On the other hand, the normal role of CFTR in the heart and the cardiological consequences of its possible dysfunction in cystic fibrosis patients remain to be clarified.[1-3] But cardiac myocytes do afford the opportunity for semiquantitative, functional studies of the natural mechanisms of regulation of CFTR Cl^- channels in their native environment.[22] For instance, wide-tipped pipettes equipped with a pipette perfusion device[23] can provide rapid diffusional access of small molecules to the interior of a cardiac myocyte at specified times during an experiment. During such thorough intracellular dialysis with solutions incorporating ATP and GTP, but no cAMP, no whole-cell CFTR Cl^- conductance can be detected until the cell is stimulated, either by agonists like isoproterenol or forskolin that increase the cellular [cAMP] level or by direct introduction of cAMP into the cell.[4,9] Under those experimental conditions, the cellular regulatory mechanisms that control gating of CFTR channels can be probed by the introduction of various chemical agents, including nucleotides and inhibitors of protein kinases and phosphatases.[9,24,25] Complementary recordings of unitary Cl^- channel currents can be obtained using giant, excised, inside-out patches of myocyte membrane, which permit relatively unhindered access of kinases and their inhibitors and nucleotides to the cytoplasmic surface.[11,26]

In contrast to the activation pathway of the PKA-regulated cardiac Cl^- channels, the general features of which have been outlined,[4,9-11,27,28] the deactivation pathway is poorly understood, although it presumably involves dephosphorylation by cellular protein phosphatases. By recording macroscopic Cl^- current flowing through excised inside-out patches from NIH 3T3 fibroblasts stably expressing recombinant CFTR, Berger *et al.*[29] were able to show that channels phosphorylated by PKA could be dephosphorylated by purified phosphatase 2A but not by phosphatase 1 or 2B. Which phosphatase (or phosphatases) dephosphorylates the channel *in situ*, in the intact cell, remains to be determined. The presence in CFTR molecules of multiple consensus sequences for PKA phosphorylation[15] and the finding that, upon stimulation, PKA phosphorylates four or five serines in the regulatory domain of CFTR[30,31] lend weight to this question. To investigate the dephosphorylation of cardiac CFTR Cl^- channels, we have introduced into the cell okadaic acid and microcystin, both potent inhibitors of phosphatases 1 and 2A. Maximally effective concentrations of okadaic acid and/or microcystin were found to enhance the Cl^- conductance activated by isoproterenol or forskolin and to slow, and render incomplete, its deactivation following washout of the agonist. The results suggest that phosphatase 1 and/or 2A is absolutely required for full dephosphorylation, but that some other phosphatase can also partially dephosphorylate the channel, and they indicate incremental regulation of cardiac CFTR Cl^- channels by PKA-mediated phosphorylation at functionally distinguishable sites.[32]

METHODS

The materials and methods were essentially as previously described.[9,11]

Whole-Cell Experiments

Single ventricular myocytes were isolated by collagenase digestion of guinea pig hearts.[9] Normal Tyrode's solution contained (in mM): 145 NaCl, 5.4 KCl, 1.8 CaCl$_2$, 0.5 MgCl$_2$, 5 HEPES (pH 7.4 with NaOH), and 5.5 glucose. The modified Tyrode's solution for superfusion of myocytes contained (in mM): 145 NaCl, 1 CdCl$_2$, 1.5 MgCl$_2$, 5 HEPES (pH 7.4 with NaOH), and 5.5 glucose. All superfusion solutions were prewarmed to 36°C. The standard pipette solution for intracellular dialysis contained (in mM): 85 aspartic acid, 5 pyruvic acid, 10 EGTA, 20 TEACl, 5 Tris$_2$-creatine phosphate, 10 MgATP, 0.1 Tris$_{2.5}$-GTP, 2 MgCl$_2$, 5.5 glucose, and 10 HEPES (pH 7.4 with CsOH): free [Ca^{2+}] and [Mg^{2+}] were estimated to be <1 nM and ~1 mM, respectively.[33,34] In 109 mM Cl$^-$ pipette solution, aspartate was replaced by Cl$^-$.

Whole-cell currents were recorded via wide-tipped, low resistance (0.5–2 MΩ) borosilicate pipettes (Mercer Glass Works, Inc., NY), fitted with an intrapipette perfusion device.[23] A gigaohm seal was obtained with gentle suction (~ – 20 cm H$_2$O), and the membrane then ruptured by more vigorous suction controlled with a 10 ml syringe. Once the cell interior was equilibrated (~ 3 min) with the pipette solution, the cell was exposed to modified Tyrode's solution and the holding potential set at 0 mV to inactivate Na$^+$- and Ca^{2+}-channel currents. K$^+$-channel currents were minimized by omitting K$^+$ from intra- and extracellular solutions, and including 20 mM TEA$^+$ in pipette solutions. Omission of K$^+$ also prevented currents generated by the Na$^+$/K$^+$ pump.[35] Na$^+$/Ca^{2+} exchange current was prevented by the omission of internal Na$^+$ and of internal and external Ca^{2+}.[36]

Two 3 M KCl half cells connected the clamp amplifier to the pipette interior and to the chamber, to minimize liquid junction potentials. Currents were elicited by 80-msec voltage pulses to potentials from + 100 mV to – 100 mV in 20 mV increments. Current and voltage signals were filtered at 2 kHz, digitized on-line at 8 kHz, and stored in an IBM PC-AT computer for analysis with ASYST software (Keithley Instruments, Inc., Taunton, MA). Steady-state current-voltage (I-V) relationships were plotted from the current levels averaged over the final 12.5 msec of each pulse.

Giant-Patch Experiments

Ventricular myocytes were stored in high [K$^+$], low [Ca^{2+}] medium for 8–24 h so that large sarcolemmal blebs would develop. Wide tipped (12–20 μm diameter) borosilicate glass pipettes of ~ 100 kΩ resistance, made hydrophobic at the tips by coating them with a mineral oil-parafilm mixture, were sealed to the blebs with light suction. After a gigaohm seal was established, the membrane patch was excised and transferred to a continuously perfused, temperature controlled (25°C) chamber in which solutions could be exchanged in ~ 1 sec[37] by switching manual (Hamilton,

Reno, NV) or computer-controlled electric (General Valve, NJ) valves. Pipette and bath solutions, and the 0 mV holding potential, were designed to inhibit currents through Ca^{2+}, Na^+, and K^+ channels, and the Na^+/Ca^{2+} exchanger. The pipette solution contained (in mM): 145 NMG-Cl, 5 CsCl, 2 $BaCl_2$, 2.3 $MgCl_2$, 0.5 $CdCl_2$, 10 HEPES (pH 7.4 with NMG), and the bath solution contained (in mM): 140 NMG-aspartate, 10 EGTA, 10 HEPES, 2 $MgCl_2$, 20 TEA-OH (pH 7.4 with Tris). Na^+/K^+ pump current, activated by a brief exposure to bath Na^+ plus MgATP, provided a convenient indicator of inside-out patch (rather than vesicle) formation. PKA catalytic subunit was prepared as described[38] and dialyzed into (in mM) 10 EGTA, 10 HEPES, 20 TEA-Cl, 2 $MgCl_2$, 85 aspartic acid (pH 7.4 with ~120 CsOH), to a protein concentration of 0.7 mg/ml. Patch current was recorded with a LIST EPC-7 amplifier, stored on video tape, and then filtered at 7 Hz and digitized at 20 Hz off-line, and analyzed with programs written in ASYST.

RESULTS AND DISCUSSION

FIGURE 1 shows that activation of β-adrenoceptors by isoproterenol in guinea pig ventricular myocytes elicits a substantial Cl^- conductance. The isoproterenol-induced current varied approximately linearly with membrane potential and reversed sign near 0 mV, close to the Cl^- ion equilibrium potential, when the internal and external $[Cl^-]$ were almost equal (FIG. 1, B and D, b–a); but, after the internal $[Cl^-]$ had been lowered towards 0 mM to generate a steep inwardly directed gradient of Cl^- ion concentration, the inward current practically disappeared, leaving a strong outwardly rectifying conductance (FIG. 1, B and D, d–c). In contrast to this $[Cl^-]$ sensitivity of the isoproterenol-induced current, membrane current in the absence of isoproterenol was practically unaffected by the same drastic reduction of intracellular $[Cl^-]$ (FIG. 1, C, a, c, e), demonstrating the lack of a measurable background Cl^- conductance in these dialyzed myocytes. As implied by the linear steady-state I-V relationship for the isoproterenol-induced current (FIG. 1, D, b–a), this Cl^- conductance shows little or no time dependence, the current reaching its new amplitude almost instantaneously when the membrane potential is stepped from one level to another (FIG. 1, B).

Activation of the covert Cl^- current requires PKA-dependent phosphorylation because it can be elicited by exposure to isoproterenol,[4,9] histamine,[9,39] or forskolin[5,9] (all of which raise cellular cAMP levels), or by direct intracellular application of cAMP or PKA catalytic subunit,[4] and because intracellular application of a specific peptide inhibitor of PKA (PKI, 5-24-amide[40]) can abolish, or prevent, Cl^- current activation by isoproterenol,[9] forskolin[24] (FIG. 2), or intracellular cAMP.[27] Complete abolition of the forskolin-induced Cl^- conductance by PKI (FIG. 2, c–a) confirms that it must have been activated exclusively via PKA. The prompt deactivation of the Cl^- conductance upon removal of isoproterenol (FIG. 1, A) or introduction of PKI (FIG. 2) must therefore reflect dephosphorylation, its speed indicating that cellular phosphatases must be continuously and highly active.

PKA-dependent activation of this cardiac Cl^- conductance and its time and voltage independence are properties that suggest similarity of the underlying cardiac Cl^- channels to CFTR.[8] These, and other, similarities were confirmed by measuring unitary Cl^- channel currents in excised giant patches.[37] Single-channel conductance was ohmic

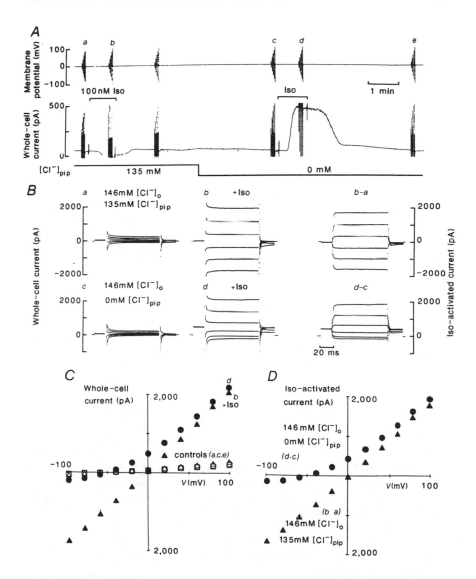

FIGURE 1. A time- and voltage-independent Cl⁻ conductance activated by isoproterenol (Iso). (**A**) Chart recording of membrane potential (top) and whole-cell current. (**B**) Superimposed records of whole-cell currents for pulses to ± 100, ± 60 and ± 20 mV, from the 0 mV holding potential, in the presence or absence of isoproterenol under conditions of approximately symmetrical or asymmetrical Cl⁻ concentrations. (**C** and **D**) Steady-state I-V relationships of whole cell current or isoprenaline-activated difference current as indicated. (From Bahinski *et al.*[4] Reprinted with permission.)

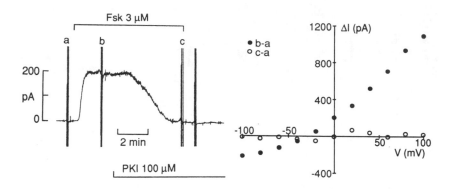

FIGURE 2. Abolition of forskolin (Fsk)-induced Cl⁻ conductance by PKA inhibitor peptide, PKI (5-24-amide). (*Left*) Chart record of current changes at 0 mV showing complete abolition of the Cl⁻ current elicited by forskolin by concomitant intracellular dialysis with 100 μM PKI. (*Right*) Steady-state difference I-V relationships obtained by subtraction as indicated. (From Hwang et al.[24] Reprinted with permission.)

and ~ 12 pS with approximately symmetrical 150 mM Cl⁻ solutions, channel gating was relatively slow, and channel open probability was roughly independent of membrane potential between − 60 and + 60 mV.[26] Channel activation required PKA-mediated phosphorylation because currents were not activated by MgATP alone, nor by PKA catalytic subunit in the absence of MgATP, nor by PKA plus MgATP in the presence of PKI (FIG. 3, A); but withdrawal of PKI then led to the appearance of unitary Cl⁻-channel currents, which could survive for many minutes (≥ 15 min) after washing off the PKA catalytic subunit, even in the presence of PKI to prevent further kinase activity (FIG. 3, A). This persistence suggests that dephosphorylation proceeds slowly in the excised patches, presumably because phosphatases have been largely washed away. FIGURE 3(B) shows that these cardiac Cl⁻ channels also share the hallmark of CFTR, namely the absolute requirement of high micromolar levels of ATP (or other hydrolyzable nucleoside triphosphate) for the opening of PKA-phosphorylated channels.[11,13] Given this close resemblance, examination of the phosphorylation and dephosphorylation of cardiac CFTR *in situ* ought to yield insight into the regulation of epithelial CFTR.

The phosphorylation and dephosphorylation of the Cl⁻ channels (or of closely associated regulatory proteins) implied by FIGURES 1 to 3 can be described by the following simple scheme:

$$D \underset{\beta}{\overset{\alpha}{\rightleftharpoons}} P \qquad \text{(Scheme 1)}$$

where D and P represent dephosphorylated (deactivated) and phosphorylated (activated) states of the Cl⁻ channels, and α and β represent pseudo-first (or first) order rate constants for PKA-dependent phosphorylation and for dephosphorylation. If this scheme holds, then the steady-state amplitude of the Cl⁻ conductance, proportional

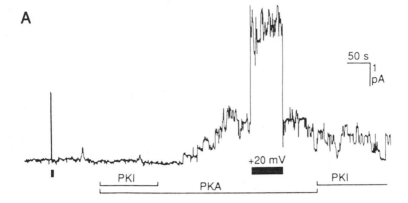

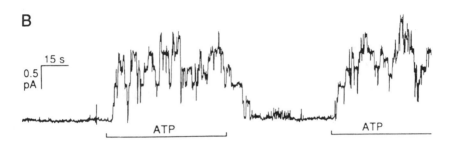

FIGURE 3. Phosphorylated Cl⁻ channels in excised inside-out patches need ATP to open. **(A)** 100 μM PKI peptide prevented channel activation by 100 nM PKA plus 500 μM MgATP, but failed to close the phosphorylated channels after withdrawal of PKA. **(B)** Reversible activation of phosphorylated Cl⁻ channels by ATP. The current record begins ~ 30 sec after the withdrawal of PKA and ATP had closed all channels; re-exposure to 500 μM MgATP reopened the channels. (From Nagel et al.[11] Reprinted with permission.)

to the fractional occupancy of the phosphorylated state, P, reflects both α and β through α/(α + β). But the rate of deactivation of the conductance on sudden withdrawal of agonist or introduction of PKI should reflect only β, assuming that α then falls to zero[24]; moreover, as long as β ≠ 0, that deactivation should proceed to completion. The scheme predicts, then, that inhibition of phosphatases contributing to β should increase the steady-state level of forskolin-induced protein phosphorylation, and hence Cl⁻ conductance, and decrease the rate of conductance decline upon removal of forskolin.

To test this, we introduced into the pipette okadaic acid, a potent inhibitor of protein phosphatases 1 and 2A,[41-43] two of the four major mammalian cytosolic serine and threonine phosphatases.[44] As FIGURE 4 illustrates, introduction of 10 μM okadaic acid both enhanced the Cl⁻ current activated by 1 μM forskolin (FIG. 4, A), without altering its reversal potential (FIG. 4, B), and slowed the deactivation of the Cl⁻ conductance on washing out the forskolin (FIG. 4, A). Also, the concentration dependence of okadaic acid's effect on the steady level of forskolin-induced Cl⁻ conductance

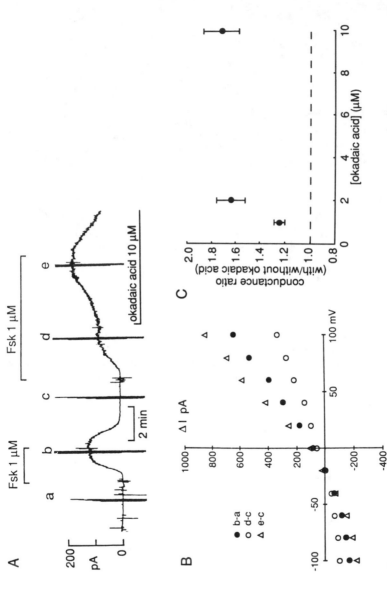

FIGURE 4. Okadaic acid enhances the forskolin-induced Cl⁻ conductance and slows its deactivation on washout of forskolin. (A) Chart record of current trace at 0 mV showing enhancement of forskolin-induced Cl⁻ current by intrapipette application of 10 μM okadaic acid. (B) Steady-state difference I-V relationships from A. (C) Concentration dependence of okadaic acid effect on Cl⁻ conductance. Mean (± sem) ratios of the forskolin-induced Cl⁻ conductance with and without internal okadaic acid (e.g., |e−c| / |d−c| from A and B) are plotted against pipette okadaic acid concentration. (From Hwang *et al.*[24] Reprinted with permission.)

(FIG. 4, C) suggests that 10 μM okadaic acid was practically a maximally effective concentration. These two effects of okadaic acid are consistent with a reduction of β (Scheme 1), but not to zero, because then all Cl⁻ channels would have remained trapped in the phosphorylated, activated state after washout of agonist. On the other hand, if additional, okadaic acid–insensitive phosphatases contribute to β, then decay of the Cl⁻ conductance after washing out forskolin is expected to be slowed but complete.

However, FIGURE 5 shows that, in the presence of 10 μM okadaic acid, deactivation of the enhanced forskolin-induced Cl⁻ conductance was slowed but *incomplete*: the residual sustained component of Cl⁻ conductance displayed the same reversal potential (FIG. 5, D), the same shape current-voltage relationship (FIG. 5, D), the same time independence (FIG. 5, B), and the same [Cl⁻] sensitivity (FIG. 5, A and E) as the full forskolin-induced conductance. Moreover, because it was insensitive to PKI (FIG. 5, A and D), it could not have resulted from persisting PKA activity but must have reflected permanently phosphorylated Cl⁻ channels.

To ensure that the slow conductance deactivation seen in the presence of 10 μM okadaic acid (FIGS. 4 and 5) did not result from incomplete inhibition of types 1 and 2A phosphatases, due to rapid diffusion of the lipophilic okadaic acid out through the cell membrane, we examined the effects of intrapipette microcystin, a hydrophilic, membrane-impermeant inhibitor of phosphatases 1 and 2A.[45] FIGURE 6 (A and B) demonstrates that, as found for okadaic acid, introduction of 5 μM microcystin maximally enhanced the forskolin-induced Cl⁻ conductance (10 μM microcystin had no further effect; FIG. 6, B: e–d, △) and yet deactivation of the Cl⁻ conductance on withdrawal of forskolin was still not abolished. Even 10 μM microcystin and 10 μM okadaic acid applied together failed to prevent a slowed, though incomplete, deactivation of the Cl⁻ conductance (FIG. 6, D). Our conclusions from these results are that full dephosphorylation of the Cl⁻ channels requires a type 1 and/or 2A phosphatase, and that a phosphatase other than type 1 or 2A can partially dephosphorylate PKA-phosphorylated Cl⁻ channels. The simplest scheme (Scheme 2) consistent with these properties includes at least two phosphorylated states:

$$D \underset{\beta}{\overset{\alpha}{\rightleftharpoons}} P_1 \underset{\beta'}{\overset{\alpha'}{\rightleftharpoons}} P_1P_2 \qquad \text{(Scheme 2)}$$

where dephosphorylation of state P_1, but not of state P_1P_2, requires phosphatase 1 and/or 2A. With sufficient okadaic acid to reduce β to zero, only states P_1 and P_1P_2 should be occupied in the presence of forskolin, but only state P_1 should be occupied after removal of forskolin. The reduced Cl⁻ conductance after washout of forskolin argues that state P_1P_2 is associated with a higher whole-cell Cl⁻ conductance than state P_1: the simplest explanation for such an effect would be a higher open probability of Cl⁻ channels in state P_1P_2 than in state P_1.[24] The rate of conductance decline following withdrawal of forskolin should reflect the magnitude of the okadaic acid–insensitive phosphatase, β′.

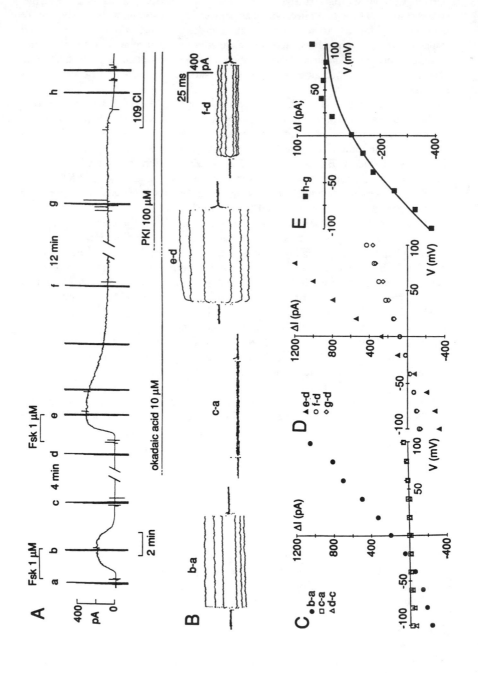

We can rule out an alternative scheme (Scheme 3) with two phosphorylated states each accessible from the dephosphorylated state, e.g.,

$$P_1 \underset{\beta}{\overset{\alpha}{\rightleftharpoons}} D \underset{\beta'}{\overset{\alpha'}{\rightleftharpoons}} P_2 \qquad \text{(Scheme 3)}$$

where, again, β is assumed to be okadaic acid–sensitive, because, if okadaic acid reduces β to zero, P_1 should then become an absorbing state. So, according to Scheme 3, in okadaic acid, the initial forskolin-induced increase in Cl^- conductance (resulting from population of both P_1 and P_2) should be followed by a slower increase, or decrease (depending on whether the conductance contribution of P_1, or P_2, is the greater), to an eventual steady level reflecting occupancy of only P_1, and that level would then be unaffected by forskolin removal. FIGURES 5 and 6 show that does not happen. Indeed, in the presence of okadaic acid, repeated brief stimulations of PKA in a given myocyte result in the same residual level of Cl^- conductance after each withdrawal of agonist, and the same conductance increment upon agonist reapplication[24]; this result is also incompatible with Scheme 3, which predicts the residual Cl^- conductance to progressively grow as the residual occupancy of P_1 grows at the expense of the non-conducting state D.

An alternative to Scheme 2 that cannot be ruled out on the basis of the whole-cell current data illustrated here is that there are two populations of Cl^- channels with practically identical properties except that dephosphorylation of one population, but not the other, requires an okadaic acid–sensitive phosphatase(s). The different whole-cell Cl^- conductances could then be accounted for by different single-channel conductances, and/or different channel open probabilities, and/or different numbers of each channel type per cell. However, different single-channel conductances seem an unlikely explanation since amplitude histograms of unitary Cl^- currents in giant excised patches[11] (Nagel, Hwang & Gadsby, unpublished results) or in outside-out patches excised from myocytes with raised intracellular [cAMP][12] are dominated by a single amplitude. On the other hand, preliminary results indicate that the open probability of a single Cl^- channel can jump, in a phosphorylation-dependent manner, between two levels that differ by a factor of roughly three (Nagel, Hwang & Gadsby, unpublished results; but cf. Ehara and Matsuura[17]). This supports the implication of Scheme 2 that a single population of Cl^- channels, displaying two distinct conducting modes depending on differential phosphorylation via PKA, can explain all the results shown here.

As already discussed, PKA-regulated cardiac Cl^- channels are practically indistin-

---

FIGURE 5. Okadaic acid prevents complete deactivation of the forskolin-activated Cl^- conductance. (A) Whole-cell current at 0 mV showing enhancement of forskolin-induced Cl^- conductance by 10 µM pipette okadaic acid, and persistence of a Cl^--sensitive conductance component after removal of forskolin. (B) Superimposed sample traces of difference currents from A, determined by subtraction of digitized records of currents elicited by voltage pulses to ±20, ±40, and ±60 mV, as indicated. (C–E) Steady-state whole-cell difference I-V relationships as indicated. The smooth curve in E shows a non-linear least-squares fit of the constant field equation to the data, yielding an estimated Cl^- permeability coefficient, P_{Cl}, of 4.5×10^{-8} cm/sec. (From Hwang *et al.*[24] Reprinted with permission.)

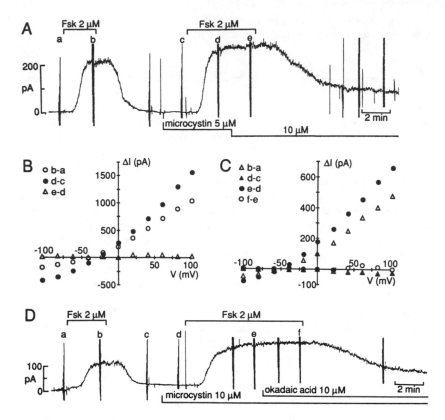

FIGURE 6. Concentration dependence of microcystin action and occlusion of okadaic acid effect by microcystin. (A) Chart record of current at 0 mV showing enhancement of forskolin-induced Cl⁻ current by 5 μM microcystin, with no further effect of 10 μM microcystin. (B) Difference I-V relationships showing forskolin-activated Cl⁻ conductance in the absence and presence of 5 μM microcystin, and lack of effect of doubling the [microcystin]. (D) Chart record showing occlusion of the influence of 10 μM okadaic acid by 10 μM microcystin. (C) Difference I-V relationships from D. (From Hwang et al.[24] Reprinted with permission.)

guishable from CFTR in all functional and biochemical characteristics so far examined.[11,14] Since it is known that, during activation of CFTR, PKA phosphorylates four or five serines in CFTR's cytoplasmic regulatory domain both *in vitro* and *in vivo*,[30,31] this provides a reasonable biochemical basis to support our conclusion that there are at least two functionally distinct phospho-forms of the same population of PKA-regulated cardiac Cl⁻ channels. This contrasts with the earlier conclusion, derived from qualitative functional analysis of CFTR mutants (with 1 to 4 serine/alanine point mutations) expressed in HeLa cells, that phosphorylation at a *single* serine might suffice to activate a Cl⁻ channel[30] and, hence, that the multiple phosphorylations are degenerate. However, a very recent study of expressed CFTR channels with a variable number of mutated serine residues also suggested that macroscopic Cl⁻ conductance might increase in proportion to phosphorylation at multiple sites.[32]

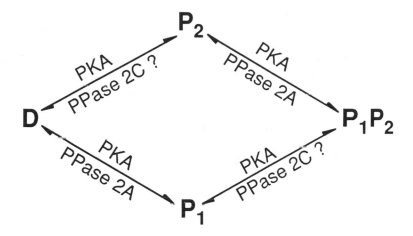

FIGURE 7. Proposed random, sequential, phosphorylation scheme for the CFTR Cl⁻ channel. PKA activates the inactive channel by phosphorylating it at the P_1 site, while additional phosphorylation at the P_2 site, to yield P_1P_2, is suggested to increase the open probability of the active channel (active channels require MgATP to open). Preliminary results[46] imply that channel molecules with only the P_2 site phosphorylated are nonconducting.

Can we identify the okadaic acid–insensitive (β') and –sensitive (β) phosphatases in Scheme 2? Mammalian protein-serine/threonine phosphatases have been classified as types 1, 2A, 2B, and 2C.[44] Under our experimental conditions (nominally Ca^{2+}-free pipette solutions containing 10 mM EGTA and ~ 1 mM free Mg^{2+}), in the presence of okadaic acid, only type 2C phosphatases are likely to be active since type 2B phosphatases are Ca^{2+}/calmodulin-dependent and show an absolute requirement for Ca^{2+}.[44] The okadaic acid–sensitive phosphatase is probably 2A, because preliminary tests reveal that introduction of phosphorylated inhibitor 1 neither enhances the for-skolin-induced Cl⁻ conductance nor slows its deactivation. Similarly, Berger *et al.*[29] recently found that PKA-activated macroscopic Cl⁻ current in excised patches containing expressed CFTR channels was diminished by direct application of purified type 2A but not type 1 phosphatase.

The sequential-phosphorylation model in Scheme 2 is the simplest that can explain our data, but it arbitrarily imposes the constraint of ordered phosphorylation. In the absence of further information, a scheme incorporating random phosphorylation of the sites represented by P_1 and P_2 (FIG. 7; cf. Scheme 2) should be considered the most likely. Further information to support Scheme 2 and/or FIGURE 7 will require determination, at the single-channel level, of the gating properties of the various postulated phosphorylated states. Preliminary whole-cell tests indicate that, in the presence of okadaic acid, intracellular introduction of the relatively non-specific phosphatase inhibitor, orthovanadate, further enhances forskolin-induced Cl⁻ conductance which then shows no sign of deactivation on washout of forskolin, consistent with both Scheme 2 and FIGURE 7.[46]

It has recently become clear that ATP or some other hydrolyzable nucleoside triphosphate is required to open the PKA-phosphorylated Cl⁻ channel[11,13] (but cf.

Quinton and Reddy[47]), and that CFTR's two nucleotide binding folds interact differently with ADP.[48] Given that there are at least two functioning Cl⁻ channel phospho–forms (Scheme 2 and Fig. 7), it will be interesting to learn how nucleoside triphosphates interact with each phospho-form, in other words to understand how the different PKA-phosphorylated states influence channel gating by ATP and analogs.

ACKNOWLEDGMENTS

We thank Dr. Y. Tsukitani for generously providing okadaic acid, and Peter Hoff for invaluable technical assistance.

REFERENCES

1. HUME, J. R. & R. D. HARVEY. 1992. Chloride conductance pathways in heart. Am. J. Physiol. **261:** C399–C412.
2. ACKERMAN, M. J. & D.E. CLAPHAM. 1993. Cardiac chloride channels. Trends Cardiovasc. Med. **3:** 23–28.
3. HWANG, T.-C. & D. C. GADSBY. 1993. Chloride channels in mammalian heart cells. Curr. Top. Membr. Transp. (In press.)
4. BAHINSKI, A., A. C. NAIRN, P. GREENGARD & D. C. GADSBY. 1989. Chloride conductance regulated by cyclic AMP-dependent protein kinase in cardiac myocytes. Nature **340:** 718–721.
5. HARVEY, R. D. & J. R. HUME. 1989. Autonomic regulation of a chloride current in heart. Science **244:** 983–985.
6. MATSUOKA, S., T. EHARA & A. NOMA. 1990. Chloride-sensitive nature of the adrenaline-induced current in guinea-pig cardiac myocytes. J. Physiol. **425:** 579–598.
7. COLLINS, F. S. 1992. Cystic fibrosis: molecular biology and therapeutic implications. Science **256:** 774–779.
8. WELSH, M. J., M. P. ANDERSON, D. P. RICH, H. A. BERGER, G. M. DENNING, L. S. OSTEDGAARD, D. N. SHEPPARD, S. H. CHENG, R. J. GREGORY & A. E. SMITH. 1992. Cystic fibrosis transmembrane conductance regulator: a chloride channel with novel regulation. Neuron **8:** 821–829.
9. Hwang, T.-C., M. Horie, A. C. Nairn & D. C. GADSBY. 1992. Role of GTP-binding proteins in the regulation of cardiac chloride conductance. J. Gen. Physiol. **99:** 465–489.
10. TAREEN, F. M., K. ONO, A. NOMA & T. EHARA. 1991. β-adrenergic and muscarinic regulation of the chloride current in guinea-pig ventricular cells. J. Physiol. **440:** 225–241.
11. NAGEL, G. A., T.-C. HWANG, K. L. NASTIUK, A. C. NAIRN & D. C. GADSBY. 1992. The protein kinase A-regulated Cl channel resembles CFTR (Cystic Fibrosis Transmembrane conductance Regulator). Nature **360:** 81–84.
12. EHARA, T. & H. MATSUURA. 1993. Single channel study of the cyclic AMP-regulated chloride current in guinea-pig ventricular myocytes. J. Physiol. **464:** 307–320.
13. ANDERSON, M. P., H. A. BERGER, D. R. RICH, R. J. GREGORY, A. E. SMITH & M. J. WELSH. 1991. Nucleoside triphosphates are required to open the CFTR chloride channel. Cell **67:** 775–784.
14. LEVESQUE, P. C., P. J. HART, J. R. HUME, J. L. KENYON & B. HOROWITZ. 1992. Expression of cystic fibrosis transmembrane regulator Cl⁻ channels in heart. Circ. Res. **71:** 1002–1007.
15. RIORDAN, J. R., J. M. ROMMENS, B-S. KEREM, N. ALON, R. ROZMAHEL, Z. GRZELCZAK, J. ZIELENSKI, S. LOK, N. PLAVSIC, J.-L. CHOU, M. L. DRUMM, M. C. IANNUZZI, F. S. COLLINS & L.-C. TSUI. 1989. Identification of the cystic fibrosis gene: Cloning and characterization of complementary DNA. Science **245:** 1066–1073.

16. HOROWITZ, B., S. S. TSUNG, P. HART, P. C. LEVESQUE & J. R. HUME. 1993. Alternative splicing of CFTR Cl⁻ channels in heart. Am. J. Physiol. 264: H2214–H2220.

17. FISCHER, H., K.-M. KREUSEL, B. ILLEK, T. E. MACHEN, U. HEGEL & W. CLAUSS. 1992. The outwardly rectifying Cl⁻ channel is not involved in cAMP-mediated Cl⁻ secretion in HT-29 cells: evidence for a very-low-conductance Cl channel. Pflügers Arch. 422: 159–167.

18. HAYSLETT, J. P., H. GOGELEIN, K. KUNZELMANN & R. GREGER. 1987. Characteristics of apical chloride channels in human colon cells (HT₂₉). Pflügers Arch. 410: 487–494.

19. KROUSE, M. E., G. HAGIWARA, J. CHEN, N. J. LEWISTON & J. J. WINE. 1989. Ion channels in normal human and cystic fibrosis sweat gland cells. Am. J. Physiol. 257: C129–C140.

20. SHOEMAKER, R. L., R. A. FRIZZELL, T. M. DWYER & J. M. FARLEY. 1986. Single chloride channel currents from canine tracheal epithelial cells. Biochim. Biophys. Acta 858: 235–242.

21. FULLER, C. M. & D. J. BENOS. 1992. CFTR! Am. J. Physiol. 263: C267–C286.

22. KAPLAN, J. H. 1993. Molecular biology of carrier proteins. Cell 72: 13–18.

23. SOEJIMA, M. & A. NOMA. 1984. Mode of regulation of the ACh-sensitive K-channel by the muscarinic receptor in rabbit atrial cells. Pflügers Archiv. 400: 424–431.

24. HWANG, T.-C., M. HORIE & D. C. GADSBY. 1993. Functionally distinct phospho-forms underlie incremental activation of PKA-regulated Cl⁻ conductance in mammalian heart. J. Gen. Physiol. 101: 629–650.

25. HORIE, M., T.-C. HWANG & D. C. GADSBY. 1992. Pipette GTP is essential for receptor-mediated regulation of Cl⁻ current in dialyzed myocytes from guinea-pig ventricle. J. Physiol. 455: 235–246.

26. NAGEL, G. A., T.-C. HWANG, A. C. NAIRN & D.C. GADSBY. 1992. Regulation of PKA-activated Cl conductance in guinea pig ventricular myocytes: single-channel studies. J. Gen. Physiol. 100: 70a.

27. BAHINSKI, A., D. C. GADSBY, P. GREENGARD & A. C. NAIRN. 1989. Chloride conductance regulated by protein kinase A in isolated guinea-pig ventricular myocytes. J. Physiol. 418: 32P.

28. HARVEY, R. D., C. D. CLARK & J. R. HUME. 1990. Chloride current in mammalian cardiac myocytes. J. Gen. Physiol. 95: 1077–1102.

29. BERGER, H. A., S. M. TRAVIS & M. J. WELSH. 1993. Regulation of the cystic fibrosis transmembrane conductance regulator Cl⁻ channel by specific protein kinases and protein phosphatases. J. Biol. Chem. 268: 2037–2047.

30. CHENG, S. H., D. P. RICH, J. MARSHALL, R. J. GREGORY, M. J. WELSH & A. L. SMITH. 1991. Phosphorylation of R domain by cAMP-dependent protein kinase regulates the CFTR chloride channel. Cell 66: 1027–1036.

31. PICCIOTTO, M., J. COHN, G. BERTUZZI, P. GREENGARD & A. NAIRN. 1992. Phosphorylation of the cystic fibrosis transmembrane conductance regulator. J. Biol. Chem. 267: 12742–12752.

32. CHANG, X.-B., J. A. TABCHARANI, Y.-X. HOU, T. J. JENSEN, N. KARTNER, N. ALON, J. W. HANRAHAN & J R RIORDAN. 1993. Protein kinase A (PKA) still activates CFTR chloride channel after mutagenesis of all 10 PKA concensus phosphorylation sites. J. Biol. Chem. 268: 11304–11311.

33. FABIATO, A. & F. FABIATO. 1979. Calculator programs for computing the composition of the solutions containing multiple metals and ligands used for experiments in skinned muscle cells. J. Physiol. (Paris). 75: 463–505.

34. TSIEN, R. Y. & T. J. RINK. 1980. Neutral carrier ion-selective microelectrodes for measurement of intracellular free calcium. Biochim. Biophys. Acta 599: 623–638.

35. GADSBY, D. & M. NAKAO. 1989. Steady-state current-voltage relationship of the Na/K pump in guinea-pig ventricular myocytes. J. Gen. Physiol. 94: 511–537.

36. KIMURA, J., A. NOMA & H. IRISAWA. 1986. Na-Ca exchange current in mammalian heart cells. Nature 319: 596–597.

37. HILGEMANN, D. W. 1990. Regulation of cardiac Na⁺-Ca⁺ exchange in giant excised sarcolemmal membrane patches. Nature 344: 242–245.

38. KACZMAREK, L. K., K. R. JENNINGS, F. STRUMWASSER, A. C. NAIRN, U. WALTER, F. D. WILSON & P. GREENGARD. 1980. Microinjection of catalytic subunit of cyclic AMP-dependent protein

kinase enhances calcium action potentials of bag cell neurons in cell culture. Proc. Natl. Acad. Sci. USA **77**: 7487–7491.

39. HARVEY, R. D. & J. R. HUME. 1990. Histamine activates the chloride current in cardiac ventricular myocytes. J. Cardiovasc. Electrophysiol. **1**: 309–317.

40. CHENG, H. C., B. E. KEMP, R. B. PEARSON, A. J. SMITH, L. MICONI, S. M. VAN PATTEN & D. A. WALSH. 1986. A potent synthetic peptide inhibitor of the cAMP-dependent protein kinase. J. Biol. Chem. **261**: 989–992.

41. TAKAI, A., C. BIALOJAN, M. TROSCHKA & J. C. RUEGG. 1987. Smooth muscle myosin phosphatase inhibition and force enhancement by black sponge toxin. FEBS Lett. **217**: 81–84.

42. BIALOJAN, C. & A. TAKAI. 1988. Inhibitory effect of a marine-sponge toxin, okadaic acid, on protein phosphatases. Biochem. J. **256**: 283–290.

43. HESCHELER, J., G. MIESKES, J. C. RUEGG, A. TAKAI & W. TRAUTWEIN. 1988. Effects of a protein phosphatase inhibitor, okadaic acid, on membrane currents of isolated guinea-pig cardiac myocytes. Pflügers Arch. **412**: 248–252.

44. COHEN, P. 1989. The structure and regulation of protein phosphatases. Ann. Rev. Biochem. **58**: 453–508.

45. HONKANEN, R. E., J. ZWILLER, R. E. MOORE, S. DAILY, B. S. KHATRA, M. DUKELOW & A. L. BOYNTON. 1990. Characterization of microcystin-LR, a potent inhibitor of type 1 and type 2A protein phosphatases. J. Biol. Chem. **265**: 19401–19404.

46. HWANG, T.-C., G. NAGEL, A. C. NAIRN & D. C. GADSBY. 1993. Dephosphorylation of cardiac CFTR Cl⁻ channels requires multiple protein phosphatases. Biophys. J. **64**: A343.

47. QUINTON, P. M. & M. M. REDDY. 1992. Control of CFTR chloride conductance by ATP levels through non-hydrolytic binding. Nature **360**: 79–81.

48. ANDERSON, M. P. & M. J. WELSH. 1992. Regulation by ATP and ADP of CFTR chloride channels that contain mutant nucleotide-binding domains. Science **257**: 1701–1704.

Inhibition of the Cystic Fibrosis Transmembrane Conductance Regulator By ATP-Sensitive K⁺ Channel Regulators[a]

DAVID N. SHEPPARD AND MICHAEL J. WELSH[b]

Howard Hughes Medical Institute
Departments of Internal Medicine and Physiology and Biophysics
University of Iowa College of Medicine
Iowa City, Iowa 52242

INTRODUCTION

Chloride channels located in the apical membrane of airway epithelia play a key role in regulating the quantity and composition of respiratory tract fluid. Cystic fibrosis (CF),[1] a common lethal genetic disease in Caucasians, is characterized by defective cAMP-stimulated Cl⁻ secretion by airway epithelia. This defect is caused by the loss of apical membrane Cl⁻ channels activated by an increase in cellular levels of cAMP.[2]

Genetic studies identified and cloned the single gene that is mutated in CF chromosomes.[3,4] This gene encodes a protein called the cystic fibrosis transmembrane conductance regulator (CFTR). Amino acid sequence analysis and comparison with other proteins suggested that CFTR consists of two repeats of a unit containing a membrane-spanning domain, composed of six putative transmembrane segments and a nucleotide-binding domain (NBD), containing Walker A and Walker B motifs[5] (FIG. 1). The two repeats are separated by a large polar segment called the R domain, which contains multiple consensus sequences for phosphorylation by cAMP-dependent protein kinase (PKA). The predicted topology of CFTR, with the exception of the R domain, resembles that of a family of proteins called the traffic ATPases[6] or the ATP-binding cassette (ABC) transporters.[7] Most members of this family are ATP-dependent transporters, including bacterial periplasmic permeases, the yeast STE6 gene product, and P-glycoprotein, which is responsible for multidrug resistance by cancer cells.

[a] Work from authors' laboratory was supported in part by the Howard Hughes Medical Institute, the National Heart, Lung and Blood Institute, and the National Cystic Fibrosis Foundation.

[b] Address correspondence to: Michael J. Welsh, M.D., Howard Hughes Medical Institute, Department of Internal Medicine, 500 EMRB, University of Iowa College of Medicine, Iowa City, IA 52242.

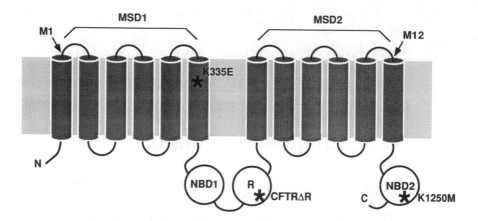

FIGURE 1. Model showing the proposed domain structure of CFTR. MSD refers to the membrane-spanning domains, NBD refers to the nucleotide-binding domains, and R refers to the R domain. The location of CFTR mutants K335E, K1250M, and CFTRΔR are indicated. The membrane is represented by the shaded area.

EVIDENCE THAT CFTR IS A Cl⁻ CHANNEL

Recent work has demonstrated that CFTR is a Cl⁻ channel. That conclusion is based on several observations. First, expression of recombinant CFTR in a wide variety of epithelial and nonepithelial cells generated cAMP-activated Cl⁻ channels.[8-12] Second, the properties of cAMP-regulated Cl⁻ channels generated by recombinant CFTR were the same as those of cAMP-regulated Cl⁻ channels in the apical membrane of secretory epithelia, where the CF defect is located.[13] Third, mutation of specific basic residues within the first membrane-spanning domain to acidic residues altered the anion selectivity of CFTR Cl⁻ channels.[14] Fourth, when purified recombinant CFTR was incorporated into planar lipid bilayers it displayed regulated Cl⁻ channel activity.[15]

REGULATION OF CFTR Cl⁻ CHANNELS

The CFTR Cl⁻ channel is regulated by cAMP-dependent phosphorylation and by intracellular nucleotides. In experiments using excised inside-out membrane patches, addition of the catalytic subunit of PKA to the cytosolic side of the patch activated CFTR Cl⁻ channels.[16,17] The R domain is the site of PKA-dependent regulation because four serine residues in the R domain are phosphorylated *in vivo* when cellular levels of cAMP increase.[18] Moreover, expression of CFTR in which part of the R domain has been deleted produces Cl⁻ channels that are constitutively active, that is, channel activation does not require an increase in cellular cAMP levels.[19] Intracellular nucleotides, such as ATP, also regulate the CFTR Cl⁻ channel: once phosphorylated by PKA, cytosolic ATP is required to maintain channel activity.[20] ATP activates the channel by a mechanism independent of both PKA and the R domain; however, ATP regulation is Mg^{2+}-dependent and competitively inhibited by ADP.[21] The NBDs were shown to be the site of nucleotide regulation because site-directed mutations in both NBDs modulated regulation.[21]

COMPARISON OF CFTR Cl⁻ CHANNELS AND ATP-SENSITIVE K⁺ CHANNELS

Intracellular ATP also regulates a class of K⁺ channels (ATP-sensitive K⁺ channels; K-ATP channels).[22-24] The modulation of K-ATP channels in excitable cells by intracellular ATP couples cellular metabolism and electrical activity; this process is believed to be important in preserving cellular homeostasis. It has been speculated that CFTR Cl⁻ channels fulfill a similar homeostatic mechanism in epithelia.[25] Regulation of CFTR Cl⁻ channels by intracellular ATP may represent a strategy by which secretory epithelial cells balance the rate of transepithelial Cl⁻ secretion with cellular ATP levels, thereby controlling cell volume and ionic composition.[20,25]

Interestingly, K-ATP channels share some functional properties with CFTR Cl⁻ channels. Both CFTR and K-ATP channels can be regulated by PKA-dependent phosphorylation.[16,17,26] In addition, both channels are relatively insensitive to changes in membrane voltage and to the intracellular free Ca²⁺ concentration.[13,23] The most notable similarity between these channels is regulation by intracellular ATP. The effect of ATP, however, differs for the two types of channel: ATP activates CFTR Cl⁻ channels, whereas it inhibits K-ATP channels[20,22,23]; for both channels the effect of ATP is competitively inhibited by ADP.[21,27] Another difference is that non-hydrolyzable analogs of ATP inhibit K-ATP channels, whereas hydrolyzable nucleotides are required to regulate CFTR.[20,27]

K-ATP channels have a distinct pharmacology: they are inhibited by sulphonylureas, such as tolbutamide and glibenclamide, a group of hypoglycemia-inducing drugs used to treat diabetes mellitus.[28] K-ATP channels are also activated by a novel class of drugs known as K⁺ channel openers.[29,30] These agents include cromakalim, a potent smooth muscle relaxant with hypotensive and bronchodilatory activity *in vivo*, as well as diazoxide, an antihypertensive drug also used to treat some pancreatic carcinomas.[31]

PHARMACOLOGY OF CFTR Cl⁻ CHANNELS

In contrast to K-ATP channels, the pharmacology of CFTR Cl⁻ channels is poorly defined. Millimolar concentrations of extracellular diphenylamine-2-carboxylate (DPC) are required to achieve significant inhibition of CFTR Cl⁻ currents[10,13,32]; the arylaminobenzoate NPPB (100–200 μM) applied extracellularly is reported to be less efficient than DPC.[32] Extracellular DIDS, a stilbene-disulphonic acid derivative that blocks several types of epithelial Cl⁻ channels does not affect CFTR Cl⁻ currents at concentrations of 500 μM.[11,13] Similarly, the Cl⁻ channel blockers Zn²⁺ and the indanyloxyacetic acid derivative IAA-94 are ineffective inhibitors when applied extracellularly at concentrations of 100 and 40 μM, respectively.[13]

The search for modulators of CFTR Cl⁻ channels is important in at least two respects. First, no high affinity inhibitors of CFTR Cl⁻ channels have been identified

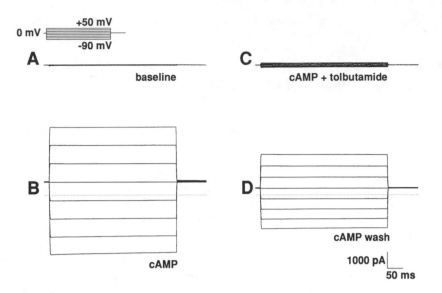

FIGURE 2. Tolbutamide inhibits CFTR Cl⁻ currents. Whole-cell currents were recorded from an NIH 3T3 fibroblast stably expressing wild-type CFTR. The inset shows that the holding voltage was 0 mV and voltage was stepped from +50 mV to −90 mV in 20 mV decrements. Dotted line represents the zero current level. (A) Baseline conditions. (B) Currents recorded 2 min after adding 10 μM forskolin and 100 μM IBMX (cAMP). (C) Currents recorded 3 min after adding 500 μM tolbutamide. (D) Recovery of currents, 15 min after washing the drug from the bath. The patch-pipette contained 120 mM NMDG, 43 mM Cl⁻ and 1 mM MgATP; the bath solution contained 140 mM NaCl. (From Sheppard and Welsh.[34] Reprinted with permission of the Rockefeller University Press.)

yet; such inhibitors might be useful as agents for distinguishing CFTR Cl⁻ channels and as probes of the mechanism of permeation. Second, novel pharmacological activators of CFTR Cl⁻ channels might provide an important therapy for the defective Cl⁻ secretion across CF epithelia. Because of the similarities in regulation by ATP, we have examined the effects of sulphonylureas and K⁺ channel openers on CFTR Cl⁻ currents, measured using the whole-cell configuration of the patch-clamp technique.[33,34]

EFFECT OF SULPHONYLUREAS AND K⁺ CHANNEL OPENERS ON CFTR Cl⁻ CURRENTS

The effect of the sulphonylurea drugs, tolbutamide and glibenclamide, was tested following CFTR Cl⁻ current activation with cAMP agonists. Both tolbutamide (500 μM) and glibenclamide (25 μM) inhibited CFTR Cl⁻ currents (FIGS. 2 and 3). Inhibition showed little voltage dependence, developed slowly, and was reversible with tolbutamide (FIG. 2) but not with glibenclamide (FIG. 3). These characteristics are similar to the effect of sulphonylureas on K-ATP channels.[35,36] Sulphonylurea inhibition

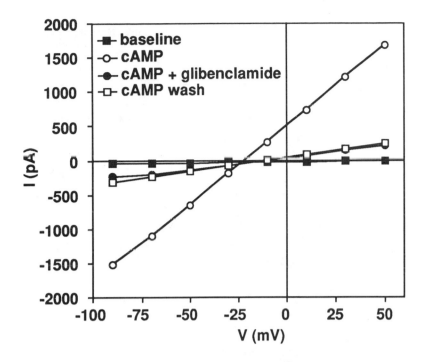

FIGURE 3. Glibenclamide inhibits CFTR Cl⁻ currents. Steady-state I-V relationships for CFTR Cl⁻ currents recorded under basal conditions (■), following cell stimulation with cAMP agonists (○), 3 min after addition of 25 µM glibenclamide to the bath (●), and after removing glibenclamide (□) are shown. Currents were recorded from an NIH 3T3 fibroblast expressing CFTR as described in Fig. 2 (From Sheppard and Welsh.[34] Reprinted with permission of the Rockefeller University Press.)

was concentration dependent; half-maximal inhibition occurred at about 150 µM tolbutamide and 20 µM glibenclamide, respectively. This effect was weaker than their inhibition of K-ATP channels in pancreatic β cells,[37] but not for K-ATP channels in cardiac myocytes.[35]

In contrast to their effect on K-ATP channels, the K⁺ channel openers diazoxide, BRL 38227 (lemakalim, the biologically active enantiomer of cromakalim), and minoxidil sulphate did not stimulate CFTR Cl⁻ currents, either under baseline or cAMP-stimulated conditions. Instead, these agents inhibited CFTR Cl⁻ currents. FIGURE 4 shows the effect of minoxidil sulphate on CFTR Cl⁻ currents; similar results were obtained with BRL 38227 and diazoxide. As was observed for the sulphonylureas, inhibition showed little voltage dependence, developed slowly, and was poorly reversible. Inhibition by K⁺ channel openers was concentration dependent; half-maximal inhibition occurred at about 40 µM minoxidil sulphate, 50 µM BRL 38227, and 250 µM diazoxide. This effect was weaker than their stimulation of K-ATP channels in vascular smooth muscle.[30]

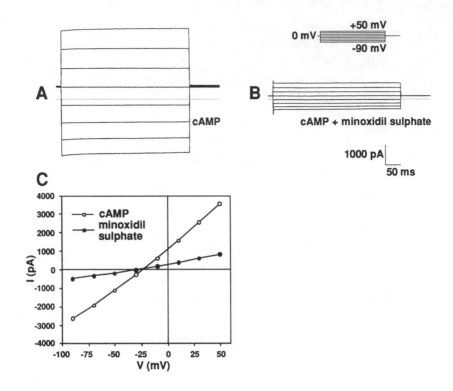

FIGURE 4. Minoxidil sulphate inhibits CFTR Cl⁻ currents. Traces are from an NIH 3T3 fibroblast stably expressing CFTR. Under basal conditions, membrane currents measured $< \pm 100$ pA at $+50$ and -90 mV. (**A**) Currents recorded 2 min after adding cAMP agonists. (**B**) Currents recorded 3 min after adding 100 μM minoxidil sulphate. (**C**) Corresponding I-V relationship is shown. Currents were measured as described in FIG. 2. (From Sheppard and Welsh.[34] Reprinted with permission of the Rockefeller University Press.)

EFFECT OF GLIBENCLAMIDE ON CFTR MUTANTS

We were interested to learn how K-ATP channel regulators interact with CFTR. We therefore examined the effect of glibenclamide, the most potent inhibitor we had identified, on Cl⁻ currents generated by several CFTR mutants. We studied CFTR-containing mutations that affect each of the three types of domains of CFTR: K335E, which contains a mutation in the sixth transmembrane segment; K1250M, which contains a mutation in the second NBD; and CFTRΔR, where part of the R domain has been deleted (amino acids 708–835). We thought it possible that if glibenclamide specifically interacts with one of these domains, its inhibitory properties might be altered.

K335E forms Cl⁻ channels that are similar to wild-type CFTR, except that the anion selectivity sequence of wild-type CFTR (Br⁻ > Cl⁻ > I⁻) is altered such that I⁻ > Br⁻ > Cl⁻.[14] However, this mutation had no effect on glibenclamide inhibition compared with wild-type CFTR. The mutation K1250M, located in the Walker A

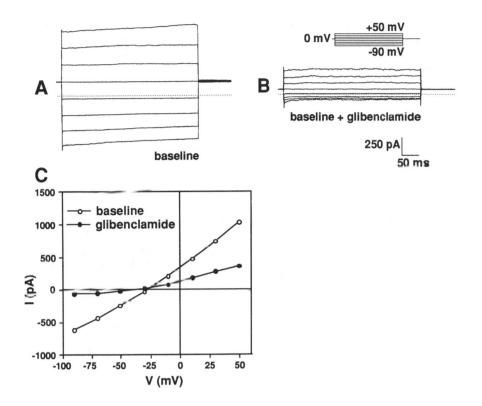

FIGURE 5. Glibenclamide inhibits CFTRΔR. Traces are from a C127 cell stably expressing CFTRΔR. (A) CFTRΔR currents recorded in the absence of cAMP agonist. (B) Currents recorded 3 min after adding 100 μM glibenclamide. (C) Corresponding I-V relationship is shown. Currents were measured as described in Fig. 2. (From Sheppard and Welsh.[34] Reprinted with permission of the Rockefeller University Press.)

motif of NBD 2 impairs the regulation of CFTR Cl⁻ channels by intracellular nucleotides[21]; MgATP is less potent at activating K1250M compared with wild-type CFTR and the competitive inhibition of CFTR Cl⁻ channels by intracellular ADP is abolished.[21] Despite these changes in regulation, inhibition of K1250M Cl⁻ currents by glibenclamide was similar to that observed with wild-type CFTR. This data suggests that residues K335 and K1250 do not form a critical part of the glibenclamide interaction site. Nevertheless, other residues within the membrane-spanning domains and nucleotide-binding domains may contribute to the interaction.

CFTRΔR forms Cl⁻ channels that are active, independent of cAMP-dependent phosphorylation.[19] FIGURE 5 shows a family of CFTRΔR currents recorded in the absence of cAMP agonists. Although CFTRΔR Cl⁻ channels are constitutively active, they possess the same biophysical properties as wild-type CFTR Cl⁻ channels. Deletion of part of the R domain did not prevent inhibition by glibenclamide. However, inhibition showed significant voltage dependence, with greater potency at hyperpolarizing voltages. This result suggests that the R domain in some way affects the response to glibenclamide such that deletion of part of the R domain partially relieves the block of CFTR Cl$^-$ currents by glibenclamide at depolarized voltages. Nevertheless, the data also indicate that the major part of the R domain is not required for glibenclamide interaction.

IMPLICATIONS FOR DISEASE

The discovery of novel pharmacological activators of CFTR Cl$^-$ channels might provide a new therapeutic strategy for treating CF patients. Although K-ATP channel regulators did not activate CFTR, the observation that CFTR was inhibited by these agents suggests an interaction; it is possible that related compounds might prove to be valuable activators of CFTR. A similar relationship between channel activators and inhibitors exists with the dihydropyridine-sensitive Ca^{2+} channel agonists and antagonists.[38]

It is, however, interesting to consider the possibility that these or related agents might be of value in another disease that probably involves CFTR Cl$^-$ channels. CFTR is located within the apical membrane of Cl$^-$ secreting intestinal epithelial cells.[39,40] Chloride efflux through CFTR Cl$^-$ channels probably contributes to the watery diarrhea caused by microbial toxins such as cholera toxin and heat-stable *E. coli* enterotoxin.[41] The development of therapeutically active blockers of CFTR Cl$^-$ channels might therefore provide a treatment for some forms of diarrhea. The potency of glibenclamide inhibition of CFTR Cl$^-$ channels suggests that it may be of value in the design and synthesis of such drugs.

ACKNOWLEDGMENTS

We thank our laboratory colleages and collaborators for their advice and critical comments and Theresa Mayhew for typing the manuscript.

REFERENCES

1. BOAT, T. F., M. J. WELSH & A. L. BEAUDET. 1989. Cystic Fibrosis. *In* The Metabolic Basis of Inherited Disease. C. R. Scriver, A. L. Beaudet, W. S. Sly & D. Valle, Eds.: 2649–2680. McGraw-Hill, Inc. New York.
2. QUINTON, P. M. 1990. Cystic fibrosis: a disease in electrolyte transport. FASEB J. 4: 2709–2717.
3. RIORDAN, J. R., J. M. ROMMENS, B-S. KEREM, N. ALON, R. ROZMAHEL, Z. GRZELCZAK, J. ZIELENSKI, S. LOK, N. PLAVSIC, J-L. CHOU, M. L. DRUMM, M. C. IANNUZZI, F. S. COLLINS & L-C. TSUI. 1989. Identification of the cystic fibrosis gene: cloning and characterization of complementary DNA. Science 245: 1066–1073.
4. KEREM, B-S., J. M. ROMMENS, J. A. BUCHANAN, D. MARKIEWICZ, T. K. COX, A. CHAKRAVARTI, M. BUCHWALD & L-C. TSUI. 1989. Identification of the cystic fibrosis gene: genetic analysis. Science 245: 1073–1080.
5. WALKER, J. E., M. SARASTE, M. J. RUNSWICK & N. J. GAY. 1982. Distantly related sequences in the α- and β-subunits of ATP synthase, myosin, kinases and other ATP-requiring enzymes and a common nucleotide binding fold. EMBO J. 1: 945–951.

6. AMES, G. F., C. S. MIMURA & V. SHYAMALA. 1990. Bacterial periplasmic permeases belong to a family of transport proteins operating from *Escherichia coli* to human: Traffic ATPases. FEMS. Microbiol. Rev. **75:** 429–446.

7. HYDE, S. C., P. EMSLEY, M. J. HARTSHORN, M. M. MIMMACK, U. GILEADI, S. R. PEARCE, M. P. GALLAGHER, D. R. GILL, R. E. HUBBARD & C. F. HIGGINS. 1990. Structural model of ATP-binding proteins associated with cystic fibrosis, multidrug resistance and bacterial transport. Nature **346:** 362–365.

8. RICH, D. P., M. P. ANDERSON, R. J. GREGORY, S. H. CHENG, S. PAUL, D. M. JEFFERSON, J. D. McCANN, K. W. KLINGER, A. E. SMITH & M. J. WELSH. 1990. Expression of cystic fibrosis transmembrane conductance regulator corrects defective chloride channel regulation in cystic fibrosis airway epithelial cells. Nature **347:** 358–363.

9. DRUMM, M. L., H. A. POPE, W. H. CLIFF, J. M. ROMMENS, S. A. MARVIN, L-C. TSUI, F. S. COLLINS, R. A. FRIZZELL & J. M. WILSON. 1990. Correction of the cystic fibrosis defect in vitro by retrovirus-mediated gene transfer. Cell **62:** 1227–1233.

10. ANDERSON, M. P., D. P. RICH, R. J. GREGORY, A. E. SMITH & M. J. WELSH. 1991. Generation of cAMP-activated chloride currents by expression of CFTR. Science **251:** 679–682.

11. KARTNER, N., J. W. HANRAHAN, T. J. JENSEN, A. L. NAISMITH, S. SUN, C. A. ACKERLEY, E. F. REYES, L-C. TSUI, J. M. ROMMENS, C. E. BEAR & J. R. RIORDAN. 1991. Expression of the cystic fibrosis gene in non-epithelial invertebrate cells produces a regulated anion conductance. Cell **64:** 681–691.

12. BEAR, C. E., F. DUGUAY, A. L. NAISMITH, N. KARTNER, J. W. HANRAHAN & J. R. RIORDAN. 1991. Cl⁻ channel activity in *Xenopus* oocytes expressing the cystic fibrosis gene. J. Biol. Chem. **266:** 19142–19145.

13. ANDERSON, M. P., D. N. SHEPPARD, H. A. BERGER & M. J. WELSH. 1992. Chloride channels in the apical membrane of normal and cystic fibrosis airway and intestinal epithelia. Am. J. Physiol. **263:** L1–L14.

14. ANDERSON, M. P., R. J. GREGORY, S. THOMPSON, D. W. SOUZA, S. PAUL, R. C. MULLIGAN, A. E. SMITH & M. J. WELSH. 1991. Demonstration that CFTR is a chloride channel by alteration of its anion selectivity. Science **253:** 202–205.

15. BEAR, C. E., C. LI, N. KARTNER, R. J. BRIDGES, T. J. JENSEN, M. RAMJEFSINGH & J. R. RIORDAN. 1992. Purification and functional reconstitution of the cystic fibrosis transmembrane conductance regulator (CFTR). Cell **68:** 809–818.

16. BERGER, H. A., M. P. ANDERSON, R. J. GREGORY, S. THOMPSON, P. W. HOWARD, R. A. MAURER, R. MULLIGAN, A. E. SMITH & M. J. WELSH. 1991. Identification and regulation of the cystic fibrosis transmembrane conductance regulator-generated chloride channel. J. Clin. Invest. **88:** 1422–1431.

17. TABCHARANI, J. A., X-B. CHANG, J. R. RIORDAN & J. W. HANRAHAN. 1991. Phosphorylation-regulated Cl⁻ channel in CHO cells stably expressing the cystic fibrosis gene. Nature **352:** 628–631.

18. CHENG, S. H., D. P. RICH, J. MARSHALL, R. J. GREGORY, M. J. WELSH & A. E. SMITH. 1991. Phosphorylation of the R domain by cAMP-dependent protein kinase regulates the CFTR chloride channel. Cell **66:** 1027–1036.

19. RICH, D. P., R. J. GREGORY, M. P. ANDERSON, P. MANAVALAN, A. E. SMITH & M. J. WELSH. 1991. Effect of deleting the R domain on CFTR-generated chloride channels. Science **253:** 205–207.

20. ANDERSON, M. P., H. A. BERGER, D. P. RICH, R. J. GREGORY, A. E. SMITH & M. J. WELSH. 1991. Nucleoside triphosphates are required to open the CFTR chloride channel. Cell **67:** 775–784.

21. ANDERSON, M. P. & M. J. WELSH. 1992. Regulation by ATP and ADP of CFTR chloride channels that contain mutant nucleotide-binding domains. Science **257:** 1701–1704.

22. NOMA, A. 1983. ATP-regulated K⁺ channels in cardiac muscle. Nature **305:** 147–148.

23. COOK, D. L. & C. N. HALES. 1984. Intracellular ATP directly blocks K⁺ channels in pancreatic B-cells. Nature **311:** 271–273.

24. Ashcroft, S. J. & F. M. Ashcroft. 1990. Properties and functions of ATP-sensitive K-channels. Cell Signal. **2:** 197–214.
25. Quinton, P. M. 1990. Cystic fibrosis. Righting the wrong protein. Nature **347:** 226.
26. Ribalet, B., S. Ciani & G. T. Eddlestone. 1989. ATP mediates both activation and inhibition of K(ATP) channel activity via cAMP-dependent protein kinase in insulin-secreting cell lines. J. Gen. Physiol. **94:** 693–717.
27. Dunne, M. J., J. A. West-Jordan, R. J. Abraham, R. H. Edwards & O. H. Petersen. 1988. The gating of nucleotide-sensitive K^+ channels in insulin-secreting cells can be modulated by changes in the ratio ATP^{4-}/ADP^{3-} and by nonhydrolyzable derivatives of both ATP and ADP. J. Membr. Biol. **104:** 165–177.
28. Sturgess, N. C., M. L. Ashford, D. L. Cook & C. N. Hales. 1985. The sulphonylurea receptor may be an ATP-sensitive potassium channel. Lancet **2:** 474–475.
29. Edwards, G. & A. H. Weston. 1990. Structure-activity relationships of K^+ channel openers. Trends Pharmacol. Sci. **11:** 417–422.
30. Standen, N. B., J. M. Quayle, N. W. Davies, J. E. Brayden, Y. Huang & M. T. Nelson. 1989. Hyperpolarizing vasodilators activate ATP-sensitive K^+ channels in arterial smooth muscle. Science **245:** 177–180.
31. Dunne, M. J. & O. H. Petersen. 1991. Potassium selective ion channels in insulin-secreting cells: physiology, pharmacology and their role in stimulus-secretion coupling. Biochim. Biophys. Acta **1071:** 67–82.
32. McCarty, N. A., B. N. Cohen, M. W. Quick, J. R. Riordan, N. Davidson & H. A. Lester. 1992. Diphenylamine-2-carboxylate (DPC) blocks the CFTR Cl^- channel from the cytoplasmic side. Biophys. J. **61:** A10 (Abstr.).
33. Hamill, O. P., A. Marty, E. Neher, B. Sakmann & F. J. Sigworth. 1981. Improved patch-clamp techniques for high-resolution current recording from cells and cell-free membrane patches. Pfluegers Arch. **391:** 85–100.
34. Sheppard, D. N. & M. J. Welsh. 1992. Effect of ATP-sensitive K^+ channel regulators on CFTR chloride currents. J. Gen. Physiol. **100:** 573–591.
35. Belles, B., J. Hescheler & G. Trube. 1987. Changes of membrane currents in cardiac cells induced by long whole-cell recordings and tolbutamide. Pfluegers Arch. **409:** 582–588.
36. Gillis, K. D., W. M. Gee, A. Hammoud, M. L. McDaniel, L. C. Falke & S. Misler. 1989. Effects of sulfonamides on a metabolite-regulated ATP_i-sensitive K^+ channel in rat pancreatic B-cells. Am. J. Physiol. **257:** C1119–C1127.
37. Zunkler, B. J., S. Lenzen, K. Manner, U. Panten & G. Trube. 1988. Concentration-dependent effects of tolbutamide, meglitinide, glipizide, glibenclamide and diazoxide on ATP-regulated K^+ currents in pancreatic B-cells. Naunyn Schmiedeberg's. Arch. Pharmacol. **337:** 225–230.
38. Hess, P., J. B. Lansman & R. W. Tsien. 1984. Different modes of Ca channel gating behaviour favoured by dihydropyridine Ca agonists and antagonists. Nature **311:** 538–544.
39. Crawford, I., P. C. Maloney, P. L. Zeitlin, W. B. Guggino, S. C. Hyde, H. Turley, K. C. Gatter, A. Harris & C. F. Higgins. 1991. Immunocytochemical localization of the cystic fibrosis gene product CFTR. Proc. Natl. Acad. Sci. U.S.A. **88:** 9262–9266.
40. Denning, G. M., L. S. Ostedgaard, S. H. Cheng, A. E. Smith & M. J. Welsh. 1992. Localization of cystic fibrosis transmembrane conductance regulator in chloride secretory epithelia. J. Clin. Invest. **89:** 339–349.
41. Fondacaro, J. D. 1986. Intestinal ion transport and diarrheal disease. Am. J. Physiol. **250:** G1–G8.

The ClC Family of Voltage-Gated Chloride Channels: Structure and Function[a]

THOMAS J. JENTSCH, MICHAEL PUSCH,
ANNETT REHFELDT,[b] AND KLAUS STEINMEYER

Centre for Molecular Neurobiology
Hamburg University
Martinistrasse 52
D-20246 Hamburg, Germany

Because of the impact of molecular biological techniques, remarkable progress has been made in recent years in the molecular characterization of voltage-gated chloride channels. Even though our current view of this field is still very limited, already three different structural classes of voltage-gated chloride channels have emerged: the ClC-family of chloride channels, discovered by expression cloning of the *Torpedo* electric organ Cl channel,[1] and with more than three members identified to date[1-3]; the I_{Cln} chloride channel,[4] identified by expression cloning starting from a kidney epithelial cell line; and phospholemman, the major heart plasma membrane target for protein phosphorylation, which unexpectedly leads to chloride channel activity when expressed in oocytes.[5] This article will focus exclusively on the ClC family of chloride channels identified and characterized in our laboratory.

ClC-0, THE TORPEDO ELECTRIC ORGAN CHLORIDE CHANNEL

ClC-0 was the first voltage-gated chloride channel to be cloned.[1] The electric organ of *Torpedo* was chosen as a source of mRNA for expression cloning because the pioneering studies of Miller and colleagues[6-10] have shown that it is a very abundant source for voltage-gated chloride channels, and because the electroplax chloride channel can be easily expressed in *Xenopus* oocytes.[11] Furthermore, extensive biophysical data are available on the reconstituted channel.[6-10] A hybrid-depletion cloning strategy[12] was chosen because the active mRNA size was in the 9–10 kb range.[1] Expression from the single cloned cDNA gave currents typical for the *Torpedo* electric organ

[a] Projects in the lab are supported, in part, by the Bundesministerium für Forschung und Technologie, the Deutsche Forschungsgemeinschaft, the U.S. Cystic Fibrosis Foundation, and the U.S. Muscular Dystrophy Association.
[b] Dedicated to Annett Rehfeldt who died in April 1993.

chloride channel (as studied in reconstitution), including the peculiar "double-barreled" characteristics when studied at the single-channel level.[13] In this model, postulated by Miller and colleagues,[8] the channel is composed of two identical (functional) "proto-channels," which can open and close (in a voltage-dependent manner) independently of each other, but which can also be closed by a common gate. The gate operating on the single "protochannel" is fast and opened by depolarization, whereas the common gate is slow (in the second range) and open by hyperpolarization. All these features, at first described with reconstituted native channels, can also be observed when ClC-0 is expressed from the cloned cDNA, suggesting that there are no other essential subunits. While it seems plausible that a *functional* homodimer should be assembled from more than one monomer of the ClC-0 protein, no conclusion on the stoichiometry of this putative homomultimeric channel is possible to date.

ClC-0 is a protein of 805 amino acids with a calculated molecular weight of 89 kD.[1] When analyzed by SDS-PAGE, it runs at roughly 75 kD. This difference may be due to the difficulties in estimating molecular masses by gel electrophoresis (especially with hydrophobic proteins). There is no sequence homology of ClC chloride channels to any other known proteins, including CFTR[14] (the chloride channel affected in cystic fibrosis), the I_{Cln} chloride channel,[4] and phospholemman.[5] Hydropathy analysis suggests the presence of several hydrophobic domains that may span the lipid bilayer, and we originally[1] suggested 12 to 13 transmembrane spans, named D1 through D13. Recent experiments with ClC-2 (described below) seem to exclude D13 as a transmembrane domain, leading to a tentative model of 12 membrane spans. The cDNA predicts no cleavable N-terminal signal peptide, suggesting that both the amino- and the carboxy-terminus are on the cytoplasmic side of the membrane.

The *Torpedo* channel co-purifies with the Na^+/K^+-ATPase when purified by sucrose gradients, suggesting its presence in the non-innervated surface of the electrocyte.[7] These cells are huge multinucleated polarized cells with a very high density of nicotinic acetylcholine receptors on the innervated plasma membrane. It has been previously suggested that a chloride channel at the opposite, non-innervated surface of the cell serves to clamp its voltage to the chloride equilibrium potential.[7] During discharge of the organ, this would ensure a low transcellular electrical resistance and the generation of a transcellular voltage in the range of 100 mV. These voltages would then add up to more than 100 V due to the arrangement of electrocytes in large stacks (like batteries connected in series). Using antibodies to bacterial fusion proteins and peptides derived from ClC-0, we could now confirm this localization by immunofluorescence studies (C. Ortland and T. J. Jentsch, unpublished results).

ClC-1, THE MAJOR CHLORIDE CHANNEL FROM SKELETAL MUSCLE

Because the *Torpedo* electric organ is ontogenetically derived from skeletal muscle, we suspected that mammalian skeletal muscle expresses a homologue of ClC-0. By homology screening we indeed identified a muscle chloride channel termed ClC-1 (initially from rat). Overall identity to ClC-0 is about 55%, with highest similarity in putative transmembrane regions. It is a larger protein than ClC-0 due to extensions at the N- and C-terminus. Expression is highest in differentiated skeletal muscle, with much lower mRNA levels in heart and a smooth muscle cell line (A10). Developmental Northern analysis revealed a dramatic increase in mRNA levels in rat skeletal muscle during the first few weeks after birth.[2] This exactly parallels the observed[15] increase of chloride conductance in rodent muscle during that time. Expression in oocytes[2] led to 9-anthracene-carboxylic acid (9-AC)–sensitive chloride currents, which showed inward rectification in the positive voltage range and deactivated more slowly at voltages more negative than – 100 mV (Fig. 1). All these observations are fully compatible with previous data on macroscopic muscle chloride conductance,[16] strongly suggesting that ClC-1 is the major skeletal muscle chloride channel.

This raised the important question whether genetic alterations in ClC-1 could be the cause of some forms of myotonia. Myotonia (muscle stiffness) is a symptom of several human diseases.[17] There are two purely myotonic diseases (with symptoms restricted to skeletal muscles) in humans, autosomal recessive generalized myotonia (GM, Becker's myotonia[18]) and autosomal dominant myotonia congenita (MC, or Thomsen's disease[19]). In addition, there are animal models for recessive myotonia (ADR mice)[20–22] and dominant myotonia (myotonic goats[23]). Many biophysical and biochemical changes have been found in myotonic muscle. These include reduction of macroscopic chloride conductance,[22–24] but also alterations in sodium channel kinetics[24,25] and changes in parvalbumin levels.[26] Myotonia can be elicited in vitro by application of 9-AC,[17] suggesting that a large reduction of chloride conductance is in principle sufficient to cause myotonia. The mechanism is quite simple: in skeletal muscle, chloride accounts for some 70 to 80% of resting membrane conductance[16] and thus assumes a role similar to the one potassium conductance plays in most other cells, namely stabilization of resting voltage. If this conductance is blocked or genetically absent, the action potential will repolarize more slowly. This gives sodium channels enough time to recover from inactivation before the membrane is fully repolarized, leading to a train of action potentials after a single stimulus. A direct consequence of these "myotonic runs" is the impairment of muscle relaxation seen in myotonia.

Using myotonic ADR mice as a model system at first, we could show that the ClC-1 chloride channel is indeed genetically destroyed in this strain.[27] In that particular case the mechanism of inactivation is the insertion of a mouse transposon into an intron, which interrupts D9 coding sequence. On the mRNA level, this leads to several different splice variants in which exonic ClC-1 sequence is followed by transposon sequence. None of these mRNAs is able to encode functional channels. In addition to ADR mice, several other independent myotonic mouse strains exist.[20] Since their mutations are allelic, other mutations in ClC-1 (e.g., point mutations) will be responsible for myotonia in these cases.

Turning to the human diseases, a tight linkage of the CLC-1 locus to both recessive Becker's myotonia and dominant Thomsen's disease was found.[28] In several Becker families a point mutation in ClC-1 was identified. Its predicted consequence[28] is a Phe to Cys exchange in putative transmembrane span D8. Thus, in humans both recessive and dominant myotonias are due to mutations in the ClC-1 channel. This has interesting consequences for the structure of the channel: a recessive disease can be simply explained by a total loss of function, in which the normal allele leads to a level of chloride

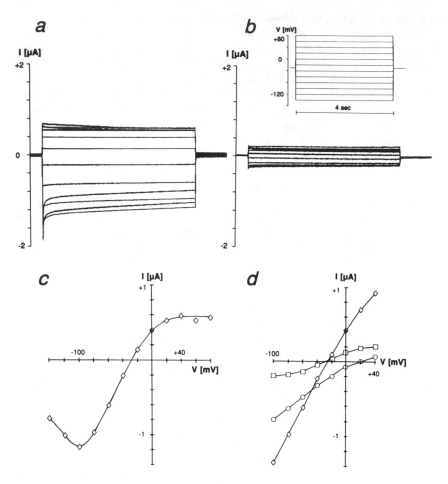

FIGURE 1. Electrophysiological properties of ClC-1 expressed in *Xenopus* oocytes. (a and b) Current traces from two-electrode voltage-clamp experiments. The clamp program is shown in the inset above b. (b) 0.1 mM 9-AC has been added to the bath for more than 15 min. (c) Quasi–steady-state currents taken from (a) at the end of the 4-sec test pulse. Note the inward rectification in the positive voltage range, and the deactivation at negative voltages, which leads to a current maximum at ≈ – 100 mV. (d) Current-voltage relationship of a different ClC-1 injected oocyte. Measurements were done in normal saline containing 103 mM Cl (◇); low-chloride saline (7 mM Cl⁻) (O); and 15 min after application of 0.1 mM 9-AC in normal saline (☐). (From Steinmeyer *et al.*[2] Reprinted with permission.)

channel expression sufficient to prevent myotonia; to explain a dominant form of inheritance, however, one needs to postulate a (homo)multimeric structure of the channel; dominant mutations will lead to non-functional subunits, which can still associate with the normal ones and will lead to their inactivation. A homomultimeric structure of ClC chloride channels is of course in line with the remarks made above for ClC-0.

ClC-2, A UBIQUITOUS CHLORIDE CHANNEL INVOLVED IN VOLUME REGULATION

ClC-2 was identified by homology screening with ClC-1 and was originally cloned from rat heart and brain.[3] However, Northern blot analysis and cloning from the T84 intestinal epithelial cell line demonstrated its presence in every tissue and cell line examined. This included typical fibroblastic, epithelial, and neuronal tissues and cells, and suggested an important (probably housekeeping) function for every cell. The electrophysiological properties, when studied in the *Xenopus* oocyte expression system,[3] were intriguing: ClC-2 is closed under resting conditions, but can be activated by very strong hyperpolarization (in excess of − 100 mV). These large negative voltages will never be reached *in vivo*. Once activated by hyperpolarization, ClC-2 has a nearly linear, slightly inward-rectifying current-voltage relationship and a Cl over I selectivity.

Subsequent studies[29] revealed that ClC-2, when expressed in *Xenopus* oocytes, can be activated reversibly by extracellular hypotonicity (Fig. 2). This is probably due to cell swelling, since the slow time-course of activation is compatible with the low water permeability of *Xenopus* oocytes.[30] On the other hand, hypertonicity did not change the activation by strong hyperpolarization. When activated by hypotonicity, ClC-2 is much less activated by hyperpolarization, but still displays some inward-rectification.

Replacement of the ClC-2 amino-terminus with corresponding segments of ClC-0 or ClC-1 yielded constitutively open channels that were no longer responsive to medium tonicity.[29] An extensive study[29] using site-directed mutagenesis identified two distinct regions in the amino-terminus: an "essential" region of about 15 amino acids where deletions lead to the "open" phenotype as seen with the chimeric channels described above; and, further downstream, a "modulating" region, where deletions lead to an "intermediate" phenotype. The latter mutants resemble channels partially opened by hypotonicity, and, in contrast to wild-type ClC-2, can now be closed by hypertonicity. The structural requirements on the "essential" region are relaxed, since several charge deletions had no effect. However, some other point mutations (either generating new charges or increasing local hydrophobicity) also resulted in the "open" phenotype.

The effect of the N-terminal inactivating domain is largely position-independent[29]: firstly, it can be separated from the channel backbone by "spacer" peptides without loss of function; secondly, and most important, it confers volume and voltage sensitivity when transplanted to other regions of the channel in N-terminal deletion mutants. It was also functional when inserted after D13, virtually excluding this (highly conserved) domain as a membrane-spanning region. Thus these transplantation studies can also be used as a test for topological models.

These results suggest a model reminiscent of the "ball-and-chain" model[31-33] for cation channels: An N-terminal, position-independent inactivating particle ("ball") binds to a putative "receptor" on the channel and causes its closure. In contrast to K-channels, the ClC-2 channel is inactivated at rest, and the "ball" is released by hyperpolarization or cell swelling. We hypothesized that this causes a change in the affinity of the receptor.[29] Also in contrast to the ball-and-chain model, deactivation times are not determined by the chain length. Thus another step (maybe conformational changes in the protein after the binding of the "ball") must be rate-limiting.

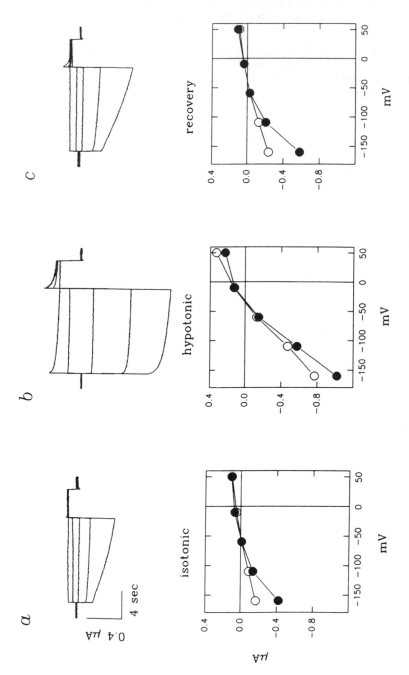

FIGURE 2. Activation of ClC-2 chloride channel expressed in *Xenopus* oocytes by voltage and cell volume. Top, voltage clamp traces, bottom, current-voltage relationships; (○) currents present immediately after the test pulse, (●) currents at the end of the 4-sec test pulse. Tonicity was reduced by a factor of two for the hypotonic challenge, and oocytes were incubated for more than 10 min in the respective salines. (From Gründer *et al.*[29] Reprinted with permission.)

TABLE 1. Cloned ClC Chloride Channels

	Tissue Distribution	Function	Voltage Dependence	Pore	Structure
ClC-0	*Torpedo* electric organ, skeletal muscle, brain	Stabilization of V in electric organ and muscle	Slow gate opened by hyperpolarization, fast gate opened by depolarization	Linear 10 pS, double-barrel Cl>Br, I block	12 hydrophobic transmembrane domains (TMD) (putative)
ClC-1	Mammals, skeletal muscle (smooth muscle, heart)	Stabilization of V in skeletal muscle Defect: myotonia	Deactivates with hyperpolarization inward rectifier in positive V range	?? pS Cl>Br>I	12 TMDs
ClC-2	Mammals, ubiquitous	Cell volume regulation (regulatory volume decrease)	Slowly activates with hyperpolarization closed at resting V linear once activated	?? pS Cl≥Br>I	12 TMDs

COMMON FEATURES OF ClC CHANNELS

Work in the past few years has revealed a family of voltage-gated chloride channels with diverse functions and tissue distribution (TABLE 1). At this time, three members of this gene family have been published, but other members are forthcoming in our group and undoubtedly also in other labs. All ClC chloride channels described so far have very similar topological properties as suggested by hydropathy analysis. The best guess at this time for the number of transmembrane regions is twelve, but much work needs to be done to experimentally confirm or refute this assumption. The same is true for channel subunit stoichiometry: both the "double-barreled" structure of the *Torpedo* channel ClC-0 and the finding of dominantly inherited myotonia with mutations in ClC-1 suggest a homomultimeric channel, but the number of subunits is unknown. As to electrophysiology, the voltage-dependencies of ClC-0, ClC-1, and ClC-2 are very different. No common, high-affinity inhibitor of ClC-channels is known to date. A common feature, however, seems to be ion selectivity, as all three channels conduct chloride better than iodide. Unfortunately, single-channel data are only available[13] for ClC-0. In the future more members of this gene family will emerge, and site-directed mutagenesis will yield additional insights into the structure-function relationship of this important channel family.

ACKNOWLEDGMENTS

We wish to thank all our co-workers for their enthusiastic work.

REFERENCES

1. JENTSCH, T. J., K. STEINMEYER & G. SCHWARZ. 1990. Nature **348**: 510–514.
2. STEINMEYER, K., C. ORTLAND & T. J. JENTSCH. 1991. Nature **354**: 301–304.
3. THIEMANN, A., S. GRÜNDER, M. PUSCH & T. J. JENTSCH. 1992. Nature **356**: 57–60.
4. PAULMICHL, M., Y. LI, K. WICKMAN, M. ACKERMAN, E. PERALTA & D. CLAPHAM. 1992. Nature **356**: 238–241.
5. MOORMAN, J. R., C. J. PALMER, J. E. JOHN, M. E. DURIEUX & L. R. JONES. 1992. J. Biol. Chem. **267**: 14551–14554.
6. WHITE, M. M. & C. MILLER. 1979. J. Biol. Chem. **254**: 10161–10166.
7. WHITE, M. M. & C. MILLER. 1981. Biophys. J. **35**: 455–462.
8. MILLER, C. 1982. Phil. Trans. R. Soc. Lond. B **299**: 401–411.
9. MILLER, C. & M. M. WHITE. 1984. Proc. Natl. Acad. Sci. USA **81**: 2772–2775.
10. RICHARD, E. A. & C. MILLER. 1990. Science **247**: 1208–1210.
11. SUMIKAWA, K., I. PARKER, T. AMANAO & R. MILEDI. 1984. EMBO J. **3**: 2291–2294.
12. LÜBBERT, H., B. J. HOFFMAN, T. P. SNUTCH, T. VAN DYKE, A. J. LEVINE, P. R. HARTIG, H. A. LESTER & N. DAVIDSON. 1987. Proc. Natl. Acad. Sci. USA **84**: 4332–4336.
13. BAUER, C. K., K. STEINMEYER, J. R. SCHWARZ & T. J. JENTSCH. 1991. Proc. Natl. Acad. Sci. USA **88**: 11052–11056.
14. RIORDAN, J. R., J.M. ROMMENS, B.-S. KEREM, N. ALON, R. ROZMAHEL, Z. GRZELCZAK, J. ZIELENSKI, S. LOK, N. PLAVSIK, J.-L. CHOU, M. L. DRUMM, M. C. IANUZZI, F. S. COLLINS & L.-C. TSUI. 1989. Science **245**: 1066–1073.
15. CONTE CAMERINO, D., A. DE LUCA, M. MAMBRINI & G. VRBOVÀ. 1989. Pflügers Arch. **413**: 568–570.

16. BRETAG, A. H. 1987. Physiol. Rev. **67**: 618–724.
17. RÜDEL, R. & F. LEHMANN-HORN. 1985. Physiol. Rev. **65**: 310–356.
18. BECKER, P. E. 1977. Myotonia Congenita and Syndromes Associated with Myotonia. Thieme. Stuttgart.
19. THOMSEN, J. 1876. Arch. Psychiatr. Nervenkrankh. **6**: 702–718.
20. RÜDEL, R. 1990. Trends Neur. Sci. **13**: 1–3.
21. REINIGHAUS, J., E.-M. FÜCHTBAUER, K. BERTRAM & H. JOCKUSCH. 1988. Muscle Nerve **11**: 433–439.
22. MEHRKE, G., H. BRINKMEIER & H. JOCKUSCH. 1988. Muscle Nerve **11**: 440–446.
23. LIPICKY, R. J. & S. H. BRYANT. 1966. J. Gen. Physiol. **50**: 89–111.
24. FRANKE, C., P. A. IAIZZO, H. HATT, W. SPITTELMEISTER, K. RICKER & F. LEHMANN-HORN. 1991. Muscle Nerve **14**: 762–770.
25. IAIZZO, P. A., C. FRANKE, W. SPITTELMEISTER, K. RICKER, R. RÜDEL & F. LEHMANN-HORN. 1991. Neuromusc. Disord. **1**: 47–53.
26. STUHLFAUTH, I., J. REININGHAUS, H. JOCKUSCH & C. W. HEIZMANN. 1984. Proc. Natl. Acad. Sci. USA **81**: 4814–4818.
27. STEINMEYER, K., R. KLOCKE, C. ORTLAND, M. GRONEMEIER, H. JOCKUSCH, S. GRÜNDER & T. J. JENTSCH. 1991. Nature **354**: 304–308.
28. KOCH, M. C., K. STEINMEYER, C. LORENZ, K. RICKER, F. WOLF, B. ZOLL, F. LEHMANN-HORN, K. H. GRZESCHIK & T. J. JENTSCH. 1992. Science **257**: 797–800.
29. GRÜNDER, S., A. THIEMANN, M. PUSCH & T. J. JENTSCH. 1992. Nature **360**: 759–762.
30. ZHANG, R. & A. S. VERKMAN. 1991. Am. J. Physiol. **260**: C26–C34.
31. ARMSTRONG, C. M. & F. BEZANILLA. 1977. J. Gen. Physiol. **70**: 567–590.
32. HOSHI, T., W. N. ZAGOTTA & R. W. ALDRICH. 1990. Science **250**: 533–538.
33. ZAGOTTA, W. N., T. HOSHI & R. W. ALDRICH. 1990. Science **250**: 568–571.

The Role of the Calcium Release Channel of Skeletal Muscle Sarcoplasmic Reticulum in Malignant Hyperthermia[a]

DAVID H. MacLENNAN
AND S. R. WAYNE CHEN

Banting and Best Department of Medical Research
University of Toronto
Charles H. Best Institute
112 College Street
Toronto, Ontario, M5G 1L6, Canada

An underlying rationale for carrying out basic research in biological systems has been the prospect of understanding the molecular basis for diseases and abnormalities afflicting both human and animal species. As our knowledge of the proteins of the sarcoplasmic and endoplasmic reticulum has advanced,[1,2] researchers have been able to implicate some of these proteins in specific abnormalities. For example, abnormalities in Ca^{2+} release channel function have been postulated to be the underlying cause of an inherited abnormality, malignant hyperthermia.

Malignant hyperthermia (MH) in humans is an abnormal response to the administration of potent inhalational anesthetics and depolarizing muscle relaxants.[3,4] Major features of the syndrome are rapidly elevated temperature, cellular ion imbalance, and skeletal muscle contracture. Ventricular fibrillation, pulmonary edema or coagulopathy, and obstructive renal failure are some of the physiological problems that can result in neurological or kidney damage or death. Fortunately, through careful monitoring of the potential onset of the syndrome and immediate intervention, including administration of the antidote, dantrolene, morbidity and mortality can usually be circumvented. In swine, the same course of events can be triggered, mainly in animals homozygous for the defect, by various forms of stress, the abnormality being referred to as the porcine stress syndrome.

Since cytoplasmic Ca^{2+} is essential for muscle contracture, and since Ca^{2+} can also trigger metabolic events leading to heat production, it was reasonable to assume that elevated intracellular Ca^{2+} is a causal factor in the MH syndrome. Studies of Ca^{2+} release from sarcoplasmic reticulum isolated from an MH-susceptible (MHS) human[5]

[a] This work was supported by grants to D.H.M. from the Medical Research Council of Canada and the Muscular Dystrophy Association of Canada. S.R.W.C. was a postdoctoral fellow of the Medical Research Council of Canada

and MHS swine[6,7] showed alterations in sensitivity of Ca^{2+}-induced Ca^{2+} release and of the rate of Ca^{2+} release, while studies of other properties of the protein revealed differences in ryanodine binding[8] and in tryptic digestion patterns.[9]

Research in our laboratory over the past decade has been concerned with the cloning of cDNAs encoding the various isoforms of sarcoplasmic or endoplasmic reticulum proteins,[2] expressing them in functional form in heterologous systems,[10] and examining the effects of mutagenesis of individual amino acids on function.[10,11] These studies have led us to an interest in naturally occurring human and animal mutations leading to genetic abnormality or disease.[11,12]

RYANODINE RECEPTOR MUTATIONS IN MALIGNANT HYPERTHERMIA

Our studies of the genetic basis for MH were initiated concurrently with our cloning of cDNAs encoding the Ca^{2+} release channel (ryanodine receptor) of human skeletal muscle sarcoplasmic reticulum.[13] We located the gene (the *RYR1* gene) on chromosome 19q13.1,[14] thereby gaining access to a series of previously defined polymorphic markers in this region that could be used in linkage analysis. We then identified a series of restriction-fragment length polymorphisms within the *RYR1* gene.[15,16] In a linkage study involving nine families, including 23 meioses, we found that the MH phenotype segregated with chromosome 19q markers, including our markers in the *RYR1* gene.[15] Cosegregation of MH with the *RYR1* markers, resulting in a lod score of 4.2 (the log of the odds favoring linkage) at a linkage distance of zero centimorgans, indicated that at least some forms of human MH are likely to be caused by mutations in the *RYR1* gene.

The finding of linkage of MH to the *RYR1* gene provided the rationale for searching for specific mutations in *RYR1* that might be causal of MH. Accordingly, studies were initiated to compare the *RYR1* cDNA sequence from MH normal (MHN) Yorkshire swine with MHS Pietrain swine. A single point mutation, the replacement of C1843 with T, leading to the replacement of Arg^{615} with Cys, was found in this comparative study.[17] The substitution of T for C1843 in the nucleotide sequence leads to the loss of a *Hin*PI restriction endonuclease site and to the gain of a *Hgi*A1,[17] thereby providing the basis for our development of a diagnostic assay for the presence of the mutation in genomic DNA. We found that the polymerase chain reaction (PCR)[18] can be used to amplify a 659 bp segment of genomic DNA that contains a constant, "built in control" *Hgi*A1 digestion site, together with the variant *Hgi*A1 site.[19] Complete cutting at the variant site defines a homozygous MH animal; no cutting defines a normal animal; and 50% cutting indicates a heterozygote (FIG. 1).[19]

The presence of the C1843 to T mutation was found to correlate with MH in five major breeds of lean, heavily muscled swine. Haplotyping within the gene suggested that the mutant gene had been inherited from a common ancestor in these five breeds.[17] Since the gene appears to add a few percent to lean dressed carcass weight,[20,21] it would appear that it has been propagated in lean, heavily muscled breeds of swine throughout the world to gain the benefit of increased meat production.

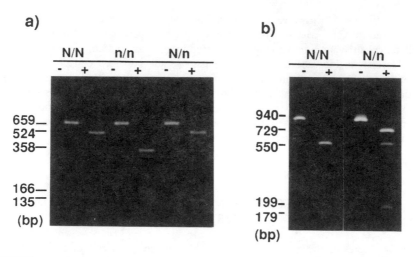

FIGURE 1. (a) Detection of the pig C1843 to T mutation by PCR amplification and subsequent digestion of the amplified product with *HgiAI*.[19] A 659 bp PCR product was amplified from genomic DNA from *N/N* Yorkshire, *n/n* Pietrain, *N/n* Yorkshire × Pietrain and *n/n* British Landrace pigs. In each case, the first lane (−) represents the PCR-amplified product, while the second lane (+) represents the same product after digestion with *HgiAI*. Since the mutation of C1843 to T creates a *HgiAI* site,[17] digestion of the *N/N* genotype with *HgiAI* generates 524 and 135 bp fragments from the constant *HgiAI* site, while digestion of the *n/n* genotype generates 358, 166, and 135 bp fragments through a combination of digestion of the constant and variant *HgiAI* sites. Fragments of 524, 358, 166, and 135 bp are generated in a *N/n* genotype from full digestion of the constant *HgiAI* site, full digestion of one allele at the variant *HgiAI* site, but no digestion of the other allele at the variant *HgiAI* site. (b) Detection of the human C1840 to T mutation by PCR amplification (−) and subsequent digestion (+) of the amplified product with *RsaI*.[19] A 922 bp PCR product was amplified from genomic DNA of normal (N/N) and susceptible (N/n) individuals. The mutation of C1840 to T deletes a *RsaI* site.[23] *RsaI* digestion of the N/N product generated 550, 199, and 179 bp fragments, while digestion of the N/n product generated 729, 550, 199, and 179 bp fragments. (From Ostu *et al.*[19] Reprinted with permission.)

Invariant cosegregation of MH and the C1843 to T mutation in swine was demonstrated in an extensive linkage study.[22] To establish linkage, crosses were made between homozygotes and heterozygotes from lines with defined haplotypes for the flanking markers, glucose phosphate isomerase (*GPI*) and 6-phosphogluconate dehydrogenase (*PGD*). For each animal, halothane challenge testing and *GPI* and *PGD* haplotyping were carried out to establish inheritance of one or two copies of the MH gene, while the presence of C1843 or T1843 in each of the two alleles was determined in our diagnostic test. In the study of 376 animals, including 338 informative meioses, complete linkage was observed, leading to a lod score of 102 at a linkage distance of zero centimorgans.

Using our diagnostic test, it has become possible to identify all swine carrying the MH mutation and, through appropriate breeding procedures, to eliminate this abnormality from swine populations. Elimination of the gene will clearly be of economic benefit to pork producers. However, the gene was propagated because of its contribution to production of lean meat. Accordingly, it may be reintroduced into certain breeding programs so that heterozygous slaughter animals will be generated that are less susceptible to stress-induced death, but carry one copy of the MH gene to promote muscle hypertrophy.

As a result of our success in identifying the porcine MH mutation, we searched for the corresponding mutation in humans.[23] The mutation of Arg^{614} to Cys has been found in about 1 to 2% of human MH families[23,24] where it segregates with MH (FIG. 1). The finding that the mutation is so closely linked to MH in swine, together with the finding that it reappears and is also linked to MH in a second species, is powerful evidence that the C1843 to T mutation is causal of porcine MH and at least some forms of human MH.

What causes other forms of human MH? In our analysis of a second human MH family, the substitution of Arg for Gly^{248} was tentatively identified as a causal mutation for human MH, but close linkage has not been established.[16] Current research in our laboratory is aimed at discovering additional causal mutations in the *RYR1* gene, either by sequencing of cDNAs or using amplification of each of the 100 or more single *RYR1* exons in genomic DNA from some 15 chromosome 19–linked MH families, followed by detection of potential mutations by single-strand conformational polymorphism (SSCP) analysis[25] and sequencing. It is anticipated that several new mutations will emerge from such analyses, not only from our laboratory, but from studies worldwide.

It is becoming increasingly clear, however, that all cases of familial MH are not linked to *RYR1*[26-28] and estimates of *RYR1*-linked MH range as low as about 30%.[28] As yet, no second MH gene has been clearly identified, but the adult sodium channel gene is a candidate.[29] Since the critical goal of MH research is to identify MHS individuals prior to anesthesia, research will no doubt continue at a high level to identify causal mutations and to develop diagnostic tests for them.

STRUCTURE/FUNCTION RELATIONSHIPS IN THE RYANODINE RECEPTOR

As indicated earlier, a major interest in our laboratory is to understand structure/function relationships in sarcoplasmic reticulum proteins. Thus the finding that the Arg^{615} to Cys mutation[17] can cause MH leads to speculation as to how this residue is involved in the function of the Ca^{2+} release channel. Since the Arg^{615} to Cys mutation alters the sensitivity of Ca^{2+}-induced Ca^{2+} release, it is possible that Arg^{615} is directly or indirectly involved in the binding of modulators of the Ca^{2+} release channel and that its alteration leads to hypersensitive channel gating. This postulate is supported by analogy to a homologous Ca^{2+} release channel, the inositol triphosphate (IP_3) receptor.[30] IP_3 binds to the NH_2-terminal part of the IP_3 receptor in a region homologous to that of the corresponding sequence in the ryanodine receptor.[31] There is 31% sequence identity between the 39 amino acids surrounding Arg^{615}. This homologous region is, therefore, a candidate site for ligand binding in both receptors.

Such speculation has led us to try to identify which other portions of the ryanodine receptor are, in fact, involved in ligand binding and in regulation of Ca^{2+} release through the binding of regulatory ligands in the Ca^{2+} release channel. The Ca^{2+} release channel is a homotetrameric complex constructed from a 565 kD subunit.[1]

Transmembrane sequences are located in the COOH-terminal fifth of each subunit[13,32] and the remainder of the subunit is cytoplasmic, bridging the gap between the sarcoplasmic reticulum and the transverse tubule. Transmembrane sequences in the tetramer may combine to form the transmembrane portion of the Ca^{2+} release channel and cytoplasmic sequences from each subunit appear to interact to enclose four extended channels that radiate from the central transmembrane channel and exit in peripheral vestibules.[33] Single channel measurements in planar bilayers have shown that Ca^{2+} release is mediated by a ligand-gated channel with a conductance greater than 100 pS in 50 mM Ca^{2+}.[34,35] Although it is not clear what signals open the channel in the muscle cell, Ca^{2+} and ATP act synergistically to open the channel in isolated vesicles, and Mg^{2+} and calmodulin inhibit channel opening.[1] Dantrolene, the clinical antidote for MH reactions,[3] inhibits halothane-induced[6] and Ca^{2+}-induced[36] Ca^{2+} release from sarcoplasmic reticulum preparations. In single Ca^{2+} release channels studied in planar lipid bilayers,[37] dantrolene at 5–25 µM first activates and then inactivates the channel.

cDNA encoding the rabbit skeletal muscle ryanodine receptor has been cloned and functionally expressed in Chinese hamster ovary cells[32,38] in Dr. Shosaku Numa's laboratory. Whole cell measurements showed that caffeine or ryanodine could induce Ca^{2+} release from intracellular organelles of the transformed cells, but not from non-transformed cells.[38] We have also functionally expressed rabbit *RYR1* cDNA in COS-1 cells. The partially purified ryanodine receptor expressed in COS-1 cells was characterized by single channel recordings in planar lipid bilayers.[39] These channels were responsive to pharmacological and physiological ligands that modulate native ryanodine receptors of sarcoplasmic reticulum, indicating that these ligand-binding sites are encoded within the primary structure of the ryanodine receptor.[39]

Analysis of the deduced amino acid sequences of ryanodine receptors has led to predictions of the location of transmembrane sequences and regulatory regions. We proposed that a modulator-binding region could be located within residues 2619–3016 in the cardiac ryanodine receptor, since predicted ATP- and calmodulin-binding sites and a phosphorylation site were located in that sequence.[40] In support of this prediction, a unique calmodulin kinase phosphorylation site was identified in the cardiac ryanodine receptor at residue 2809 and phosphorylation appeared to increase channel opening time.[41] Dr. Numa's group[32] postulated that a modulator-binding region could be located close to transmembrane segment M_1 in their model for the skeletal muscle ryanodine receptor. They identified potential Ca^{2+}, ATP, and calmodulin-binding sites in the sequence between residues 4253 and 4499. In support of these predictions, Fill *et al.*[42] demonstrated that polyclonal antibodies reacting against epitopes in amino acid sequences 4445–4586 or 4760–4877 could induce abnormal gating of the Ca^{2+} release channel by decreasing open probability and stabilizing subconductance states without blocking the conduction pathway.

In recent studies[43] we have used combined biochemical, immunological, and electrophysiological approaches to study Ca^{2+} binding and regulatory sites of the ryanodine receptor. We used $^{45}Ca^{2+}$ overlay and ruthenium red overlay methods with trpE fusion proteins to identify and localize possible Ca^{2+}-binding sites in segments of the ryanodine receptor. We began by constructing 14 trpE fusion proteins of the skeletal ryanodine receptor covering about 90% of the receptor (FIG. 2). Our overlay studies showed that a fusion protein containing residues 4014 to 4765 (FP13) is a major Ca^{2+}-binding fusion protein. The strong Ca^{2+}-binding domain of FP13 was then localized in subfragments FP13b, containing residues 4246 to 4377, and in FP13C, containing residues 4364 to 4529. By expressing even shorter sequences, we eventually demonstrated that Ca^{2+} could bind to short fragments consisting of residues 4246–4267 (13b$_1$), 4382–4417 (13c$_1$), and 4478–4512 (13c$_2$) each of which Dr. Numa had previously predicted to contain a Ca^{2+} binding site (FIG. 2).

We then made polyclonal antibodies against these sequences and determined whether they affected Ca^{2+} release in planar bilayers. We observed that the anti-13c$_2$ antibody activated the Ca^{2+} release channel by increasing both open probability and opening time without affecting channel conductance. The antibody-activated channel was still subject to modulation by Ca^{2+}, Mg^{2+}, ATP, ryanodine, and ruthenium red.

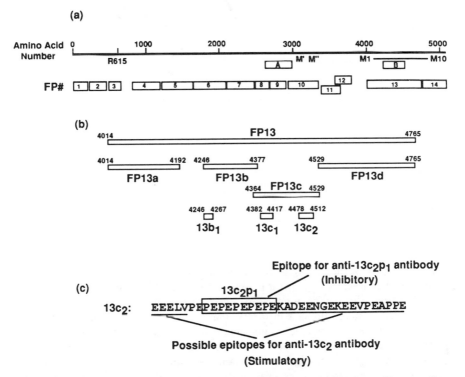

FIGURE 2. Mapping of Ca^{2+} binding sequences by analysis of expressed fragments of the ryanodine receptor. (a) Upper line: illustrates features of the linear sequence of the ryanodine receptor. The protein is 5,035 amino acids in length with predicted transmembrane sequences (M', M", M1–M10) in the COOH-terminal fifth of the molecule. R^{615} refers to the porcine MH mutation. Box A[13] and box B[32] refer to predicted regulatory regions. The lower line illustrates those fragments of the molecule that were expressed as fusion proteins.[43] (b) The boundaries of fusion protein 13 and subfragments FP13a, F13b, FP13c, FP13d, 13b$_1$, 13c$_1$, and 13c$_2$, are defined by their first and last amino acid residues. (c) The sequence of fragment 13c$_2$. The PE repeat sequence, the epitope for the anti-13c$_2$p$_1$ antibody, which inhibited channel activation, is boxed. The remainder of the sequence constitutes the epitope for the anti-13c$_2$ antibody, which activated the channel by raising its sensitivity to Ca^{2+} induced Ca^{2+} release, as illustrated in Figure 3. (From Chen et al.[44] Reprinted with permission.)

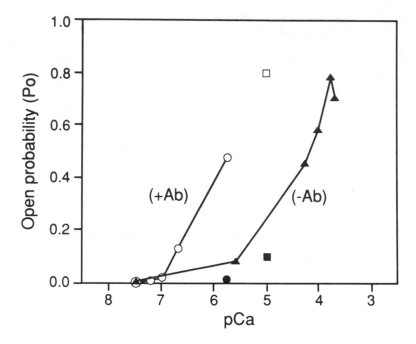

FIGURE 3. Effect of *cis* Ca^{2+} on channel open probability (Po) in the presence and absence of site-directed anti-13c$_2$ antibody. Single channel recordings were made at $+40$ mV in symmetrical 250 mM NaCl and 50 mM Tris/100 mM HEPES, pH 7.4. Data represented by circles, squares, and triangles are from different experiments.[43] Open symbols denote open probabilities in the presence of anti-13c$_2$ antibody and solid symbols indicate open probabilities in the absence of the antibody. (From Chen *et al.*[43] Reprinted with permission.)

The main effect of the antibody was to increase the Ca^{2+} sensitivity of the channel so that the channel could open in the presence of a much lower free Ca^{2+} concentration (FIG. 3). Thus, the specific modulatory effect of this antibody, directed against a probable Ca^{2+}-binding domain in the skeletal muscle ryanodine receptor, strongly suggested that the 13c$_2$ sequence, is involved in the Ca^{2+}-induced Ca^{2+} release mechanism. Since Ca^{2+} was able to activate the anti-13c$_2$ antibody-Ca^{2+} release channel complex, it was unlikely that the anti-13c$_2$ antibody bound directly to the Ca^{2+} activation site, but might bind to a site adjacent to it (FIG. 2). Alternatively, Ca^{2+} activation of the channel may occur through sites other than the Ca^{2+}-binding site in 13c$_2$.

We were intrigued by a proline-glutamate (PE) repeat sequence, the predicted high affinity Ca^{2+} binding site within the 13c$_2$ sequence.[32] To see whether this sequence was the major epitope for the anti-13c$_2$ antibody, which had been purified from an anti-13c$_2$ rabbit anti-serum using a GST-13c$_2$ antigen affinity column, a PE repeat of 10 amino acids (13c$_2$p$_1$ in FIG. 2) was synthesized and the immunocrossreactivity of the anti-13c$_2$ antibody with the synthetic peptide was measured. We found that the anti-13c$_2$p$_1$ antibody was at such a low level (0.5%) in the purified anti-13c$_2$ antibody that the stimulatory effect of the anti-13c$_2$ antibody on single channel activity was

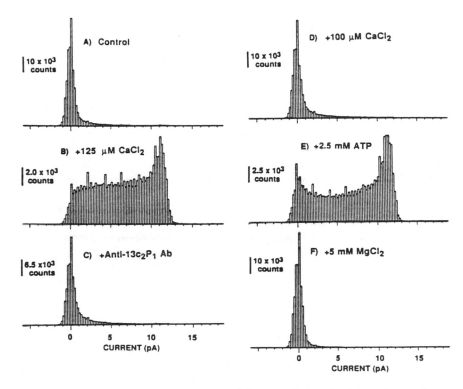

FIGURE 4. Effect of anti-13c₂p₁ antibody on Ca²⁺-activated channel activities. Sucrose density gradient–purified skeletal muscle ryanodine receptors were incorporated into planar lipid bilayers and single channel currents were recorded at +40 mV in symmetrical 250 mM NaCl and 50 mM Tris/ 100 mM HEPES buffer, pH 7.4. Recordings were filtered at 750 Hz and digitized at 5 KHz. Sequential additions (*cis*) of 125 µM CaCl₂ (**B**), 0.4 µg of anti-13c₂p₁ antibody (**C**), 100 µM CaCl₂ (**D**), 2.5 mM ATP (**E**), and 5 mM MgCl₂ (**F**) were made to the control channel (**A**) which was recorded in the presence of 4 µM free Ca²⁺ (0.2 mM EGTA plus 0.2 mM CaCl₂).[44] Amplitude histograms are shown for each condition. (From Chen *et al.*[44] Reproduced with permission.)

unlikely to have been due to the anti-13c₂p₁ antibody. Further evidence that the anti-13c₂ antibody was not acting through the PE repeat sequence (13c₂p₁) came from an examination of the effect on Ca²⁺ sensitivity of the channel of the anti-13c₂p₁ antibody, purified from the anti-13c₂ rabbit antiserum using the PE repeat peptide bound to a column. We found that the anti-13c₂p₁ antibody inhibited Ca²⁺-induced channel activity, the opposite effect to that found with the anti-13c₂ antibody. The anti-13c₂p₁ antibody reduced the open probability by 20-fold and the opening time constant by threefold when the channel had been activated with 125 µM CaCl₂ (FIG. 4). Further addition of 100 µM CaCl₂ could not reactivate the channel.

The inhibition by the anti-13c₂p₁ antibody was specific for the Ca²⁺ activation pathway and had little effect on the ATP activation pathway. When the Ca²⁺-activated channel was inhibited by the anti-13c₂p₁ antibody, the addition of 2.5 mM ATP could reactivate it (FIG. 4). The observation that the anti-13c₂p₁ antibody reacted with the Ca^{2+} activation pathway, but appeared to leave the ATP activation pathway intact, suggests that Ca^{2+} and ATP can act independently, even though they undoubtedly interact synergistically within the cell. The antibody might provide a useful tool for further investigation of the Ca^{2+} and ATP activation pathways.

Further study with caffeine revealed that the antibody could inactivate caffeine-induced Ca^{2+} release and that this activation could not be reversed by further addition of Ca^{2+}, but could be reversed by the addition of ATP. These results also suggest that the antibody specifically inactivated the Ca^{2+} activation pathway, since caffeine is believed to affect the Ca^{2+} release channel via the Ca^{2+} activation mechanism.[45] The facts that the inhibitory antibody did not alter the unitary conductance of single Ca^{2+}-release channels and that ATP could reactivate the antibody-inhibited channel suggest that the antibody was not blocking the Ca^{2+} release pore or interfering non-specifically with conformational changes that would gate the channel.

These data suggest that an inhibitory antibody reacted with the PE repeat of sequence $13c_2$, while an activating antibody reacted elsewhere in the 35-residue $13c_2$ fragment, either against the sequence EEELV at the N-terminus or against part of the sequence KADEENGEKEEVPEAPPE at the C-terminus of peptide $13c_2$ (FIG. 2). The simplest explanation for these findings is that the PE repeat is a part of the actual Ca^{2+} binding and activation site and the anti-$13c_2p_1$ antibody can inhibit this interaction, thereby inhibiting Ca^{2+}-induced Ca^{2+} release. By contrast, the interaction of anti-$13c_2$ at an adjacent site could enhance the interaction of Ca^{2+} with the Ca^{2+} binding site (possibly the PE repeat), increasing the Ca^{2+} sensitivity for channel activation by more than an order of magnitude (FIG. 3). Alternatively, the PE repeat sequence may have formed part of a structure involved in conformational changes induced specifically by Ca^{2+} binding near the PE repeat or at a distant site. Antibodies could inhibit or enhance such conformational changes by binding to the PE repeat or adjacent to it, resulting in inhibition or activation of Ca^{2+}-induced Ca^{2+} release.

SUMMARY

In this short review, we have described studies that have identified Arg^{615} in the Ca^{2+} release channel as a residue that influences channel sensitivity to Ca^{2+} induced Ca^{2+} release,[7] rate of Ca^{2+} release,[8] and channel closing.[46] We have also described studies that confirm Dr. Numa's predictions[32] that residues 4246–4267, 4382–4417, and 4478–4512 contain Ca^{2+} binding sites. The site between residues 4483 and 4494 (the PE repeat sequence) may be a key binding site for Ca^{2+} activation of the channel. Other residues in the sequence 4478–4512 may also contribute to activation of the channel. Thus our studies have contributed to basic knowledge of regulation of Ca^{2+} release function. They have also provided practical benefits in defining a disease gene, in development of a diagnostic test for porcine MH that is of economic benefit, and in laying the foundation for human MH diagnostic tests that may prevent anesthesia-induced morbidity and mortality.

ACKNOWLEDGMENTS

We gratefully acknowledge the contributions of numerous colleagues and collaborators in the studies described in this paper.

REFERENCES

1. FLEISCHER, S. & M. INUI. 1989. Annu. Rev. Biophys. Biophys. Chem. 18: 333–364.
2. LYTTON, J. & D. H. MacLENNAN. 1992. In The Heart and Cardiovascular System. H. A. Fozzard, E. Haber, R. B. Jennings, A. M. Katz & H. E. Morgan, Eds. 2: 1203–1222. Raven Press. New York.
3. BRITT, B. A. 1991. In Thermoregulation: Pathology, Pharmacology and Therapy. E. Schonbaum & P. Lomax, Eds.: 179–292. Pergamon Press Inc. New York.
4. O'BRIEN, P. J. 1987. Vet. Res. Comm. 11: 527–559.
5. ENDO, M., S. YAGI, T. ISHIZUKA, K. HORIUTI, Y. KOGA & K. AMAHA. 1983. Biomed. Res. 4: 83–92.
6. OHNISHI, S. T., S. TAYLOR & G. A. GRONERT. 1983. FEBS Lett. 161: 103–107.
7. O'BRIEN, P. J. 1986. Can. J. Vet. Res. 50: 318–328.
8. MICKELSON, J. R., E. M. GALLANT, L. A. LITTERER, K. M. JOHNSON, W. E. REMPEL & C. F. LOUIS. 1988. J. Biol. Chem. 263: 9310–9315.
9. KNUDSON, C. M., J. R. MICKELSON, C. F. LOUIS & K. P. CAMPBELL. 1990. J. Biol. Chem. 265: 2421–2424.
10. MARUYAMA, K., D. M. CLARKE, J. FUJII, T. W. LOO & D. H. MacLENNAN. 1989. Cell Motil. Cytoskel. 14: 2634.
11. MacLENNAN, D. H. 1990. Biophys. J. 58: 1355–1365.
12. MacLENNAN, D. H. & M. S. PHILLIPS. 1991. Science 256: 789–794.
13. ZORZATO, F., J. FUJII, K. OTSU, M. PHILLIPS, N. M. GREEN, F. A. LAI, G. MEISSNER & D. H. MacLENNAN. 1990. J. Biol. Chem. 265: 2244–2256.
14. MACKENZIE, A. E., R. G. KORNELUK, F. ZORZATO, J. FUJII, M. PHILLIPS, D. ILES, B. WIERINGA, S. LE BLOND, J. BAILLY, H. F. WILLARD, C. DUFF, R. G. WORTON & D. H. MacLENNAN. 1990. Am. J. Hum. Genet. 46: 1082–1089.
15. MacLENNAN, D. H., C. DUFF, F. ZORZATO, J. FUJII, M. PHILLIPS, R. G. KORNELUK, W. FRODIS, B. A. BRITT & R. G. WORTON. 1990. Nature 343: 559–561.
16. GILLARD, E. F., K. OTSU, J. FUJII, C. L. DUFF, S. DE LEON, V. K. KHANNA, B. A. BRITT, R. G. WORTON & D. H. MacLENNAN. 1992. Genomics 13: 1247–1254.
17. FUJII, J., K. OTSU, F. ZORZATO, S. DE LEON, V. K. KHANNA, J. WEILER, P. J. O'BRIEN & D. H. MacLENNAN. 1991. Science 253: 448–451.
18. SAIKI, R. K., D. H. GELFORD, S. STOFFEL, S. J. SCHARF, R. HIGUCHI, G. T. HORN, K. B. MULLIS & H. A. ERLICH. 1988. Science 239: 487–494.
19. OTSU, K., M. S. PHILLIPS, V. K. KHANNA, S. DE LEON & D. H. MacLENNAN. 1992. Genomics 13: 835–837.
20. WEBB, A. J. & S. P. SIMPSON. 1986. Anim. Prod. 43: 493–503.
21. SIMPSON, S. P. & A. J. WEBB. 1989. Anim. Prod. 49: 503–509.
22. OTSU, K., V. K. KHANNA, A. L. ARCHIBALD & D. H. MacLENNAN. 1991. Genomics 11: 744–755.
23. GILLARD, E. F., K. OTSU, J. FUJII, V. K. KHANNA, S. DE LEON, J. DERDEMEZI, B. A. BRITT, C. L. DUFF, R. G. WORTON & D. H. MacLENNAN. 1991. Genomics 11: 751–755.
24. HOGAN, K., F. COUCH, P. A. POWERS & R. C. GREGG. 1992. Anesth. Analg. 75: 441–448.
25. ORITA, M., H. IWAHANA, H. KANAZAWA, K. HAYOSHI & T. SEKIYA. 1989. Proc. Natl. Acad. Sci. USA 86: 2766–2770.
26. LEVITT, R. C., N. NOURI, A. E. JEDLICKA, V. A. McKUSICK, A.R. MARKS, J. G. SHUTACK, J. E. FLETCHER, H. ROSENBERG & D. A. MEYERS. 1991. Genomics 11: 543–547.
27. DEUFEL, T., A. GOLLA, D. ILES, A. MEINDL, T. MEITINGER, D. SCHINDELHAUER, A. DEVRIES, D. PONGRATZ, D. H. MacLENNAN, K. J. JOHNSON & F. LEHMANN-HORN. 1992. Am. J. Hum. Genet. 50: 1151–1161.
28. LEVITT, R. C., A. OLCKERS, S. MEYERS, J. E. FLETCHER, H. ROSENBERG, H. ISAACS & D. A. MEYERS. 1992. Genomics 14: 562–566.
29. OLCKERS, A., D. A. MEYERS, S. MEYERS, E. W. TAYLOR, J. E. FLETCHER, H. ROSENBERG, H. ISAACS & R. C. LEVITT. 1992. Genomics 14: 829–831.

30. Furuichi, T., S. Yoshikawa, A. Miyawaki, K. Wada, N. Maeda & K. Mikoshiba. 1989. Nature 342: 32–38.
31. Mignery, G. A. & T. C. Sudhof. 1990. EMBO J. 9: 3893–3898.
32. Takeshima, H., S. Nishimura, T. Matsumoto, H. Ishida, K. Kangawa, N. Minamino, H. Matsuo, M. Ueda, M. Hanaoka, T. Hirose & S. Numa. 1989. Nature 339: 439–445.
33. Wagenknecht, T., R. Grassucci, J. Frank, A. Saito, M. Inui & S. Fleischer. 1989. Nature 338: 167–170.
34. Smith, J. S., R. Coronado & G. Meissner. 1985. Nature 316: 446–449.
35. Smith, J. S., T. Imagawa, J. Ma, M. Fill, K. P. Campbell & R. Coronado. 1988. J. Gen. Physiol. 92: 1–26.
36. Ohta, T., S. Ito & A. Ohga. 1990. Eur. J. Pharmacol. 178: 11–19.
37. Nelson, T. E. & M. Lin. 1993. Biophys. J. 64: A380.
38. Penner, R., E. Neher, H. Takeshima, S. Nishimura & S. Numa. 1989. FEBS Lett. 259: 217–221.
39. Chen, S. R. W., D. M. Vaughan, J. A. Airey, R. Coronado & D. H. MacLennan. 1993. Biochemistry 32: 3743–3753.
40. Otsu, K., H. F. Willard, V. K. Khanna, F. Zorzato, N. M. Green & D. H. MacLennan. 1990. J. Biol. Chem. 265: 13472–13483.
41. Witcher, D. R., R. J. Kovacs, H. Schulman, D. C. Cefali & L. R. Jones. 1991. J. Biol. Chem. 266: 11144–11152.
42. Fill, M., R. Mejia-Alvarez, F. Zorzato, P. Volpe & E. Stefani. 1991. Biochem. J. 273: 449–457.
43. Chen, S. R. W., L. Zhang & D. H. MacLennan. 1992. J. Biol. Chem. 267: 23318–23326.
44. Chen, S. R. W., L. Zhang & D. H. MacLennan. 1993. J. Biol. Chem. 268: 13414–13421.
45. Rousseau, E. & G. Meissner. 1989. Am. J. Physiol. 256: H328–H333.
46. Fill, M., R. Coronado, J. R. Mickelson, J. Vilven, J. Ma, B. A. Jacobson & C. F. Louis. 1990. Biophys. J. 50: 471–475.

Ion Channel Mutations in Periodic Paralysis and Related Myotonic Diseases[a]

ROBERT H. BROWN, JR.

Harvard Medical School
Day Neuromuscular Research Laboratory, CNY-6
Department of Neurology
Massachusetts General Hospital-East
Building 149
13th Street
Charlestown, Massachusetts 02112

INTRODUCTION

In the last four years, a series of reports have implicated mutations in voltage-sensitive sodium and chloride channels in the pathogenesis of hyperkalemic periodic paralysis, myotonia congenita, and related diseases (TABLE 1). This review outlines the clinical features of these disorders, selected physiological studies of muscle from affected patients, the respective ion channel mutations, and an approach to understanding the biophysical basis of the diseases.

CLINICAL FEATURES

The central problem in the periodic and paramyotonia congenita is recurring attacks of weakness (TABLE 2). As the names imply, hypokalemic and hyperkalemic periodic paralysis are characterized by, respectively, low and high levels of serum potassium during paralytic spells.[1-3] Both disorders selectively target skeletal muscle; the spells of weakness are not accompanied by sensory symptoms or disturbance of cardiac function or mentation. Attacks last anywhere from minutes to hours and can be triggered by rest after intense exertion. During attacks of either type of periodic paralysis, affected muscles are depolarized and electrically inactive. In both disorders, chronic therapy with carbonic anhydrase inhibitors such as acetazolamide may prevent attacks.[4]

[a] R.H.B. receives generous support from the C.B. Day Investment Company, the Muscular Dystrophy Association, and National Institutes of Health grant 5R01-AR41025-02.

TABLE 1. Classification of Periodic Paralysis and Myotonias

Disease	Gene	Chromosome
Non-dystrophic		
Periodic paralysis		
Hyperkalemic	Na channel	17
Hypokalemic (rarely)	?	
Paramyotonia congenita	Na channel	17
Pure		
K-sensitive		
Myotonia congenita		
Autosomal dominant		
Typical (Thomsen's)	Cl channel	7
K-sensitive	Na channel	17
Autosomal recessive	Cl channel	7
(Becker's generalized)		
Schwartz-Jampel	?	
Dystrophic		
Myotonic dystrophy	protein kinase	19q

In hyperkalemic paralysis, onset is in early childhood, while the hypokalemic form does not begin until puberty. Hyperkalemic paralysis attacks are initiated by fasting and may be terminated by carbohydrate intake. Reciprocally, in hypokalemic paralysis, excessive carbohydrates may trigger attacks.[1,3]

Patients with hyperkalemic paralysis typically demonstrate severe muscle rigidity because of excessive electrical excitation of the muscle membrane. This feature, myotonia, is rare in hypokalemic paralysis. Myotonia is usually exacerbated by exposure to cold. Myotonia is the predominant symptom in the disorders myotonia congenita and paramyotonia congenita. The stiffness in myotonia congenita is not associated with paralytic episodes; these may be encountered in some patients with paramyotonia congenita after cooling of muscles without abnormalities in serum potassium levels. In most of the myotonias, repetitive contractions reduce muscle stiffness, while in paramyotonia, repeated contractions paradoxically accentuate the stiffness (hence *para-myotonia*).[5]

Because the periodic paralyses, myotonia congenita, and paramyotonia congenita are not associated with muscle deterioration, they are described as "non-dystrophic." Nonetheless, in these illnesses, frequent paralytic crises over time can provoke a slowly progressive, irreversible proximal weakness. By contrast, the other major myotonic muscle disorder, myotonic dystrophy, is characterized by progressive, disabling, distal muscle degeneration. Unlike the non-dystrophic myotonias, myotonic dystrophy is a multi-system disease with numerous characteristic features including frontal balding, cataracts, cardiac conduction defects, and insulin-resistance in addition to myotonia.[6]

The inheritance pattern in both forms of periodic paralysis, paramyotonia congenita and myotonic dystrophy, is autosomal dominant. Myotonia congenita may be either dominant (Thomsen's disease) or recessive (Becker's type).[1]

TABLE 2. Clinical Features of Periodic Paralysis, Paramyotonia Congenita, and Myotonia Congenita

	Periodic Paralyses		Paramyotonia Congenita	Myotonia Congenita
	Hypokalemic	Hyperkalemic		
Recurrent weakness	Yes	Yes	Rarely	No
Onset	Puberty	Infancy	Infancy	Infancy
Attack duration	Hours to days	Minutes to days	Minutes to days	Minutes to days
Interictal interval	Hours to days	Minutes to days	Minutes to days	Minutes to days
Myotonia	No	Yes	Yes	Yes
Triggers	Cold	Cold	Cold	Cold
	Rest after exercise	Rest after exercise	Rest after exercise	Rest after exercise
	Carbohydrates	Fasting	Fasting	Fasting
	Potassium	Carbohydrates	Warming	
Ameliorates	Exercise	Exercise	Exercise	Exercise
Therapy	Acetazolamide	Acetazolamide	Acetazolamide	Acetazolamide
		Mexiletine	Mexiletine	Mexiletine

MOLECULAR BASIS OF PERIODIC PARALYSIS
AND RELATED MYOTONIAS

Hyperkalemic Paralysis and Paramyotonia Congenita

Extensive physiological data implicate a potassium-induced abnormality in sodium conductance in the pathogenesis of paralytic episodes in hyperkalemic paralysis.[5,7–12] Direct measurements of membrane potentials in hyperkalemic paralytic muscle revealed that potassium triggered depolarization in excess of the potential shift predicted by the Nernst equation; this effect was completely blocked by tetrodotoxin (TTX), a specific blocker of sodium channels.[7,9,11] Cannon grew cultured myotubes from an individual with hyperkalemic periodic paralysis (mutation Met-1592-Val, Fig. 1) and recorded the activity of single sodium channels using patch clamp techniques.[12] At 3.5 mM of extracellular potassium, the latency, duration, and conductance of sodium

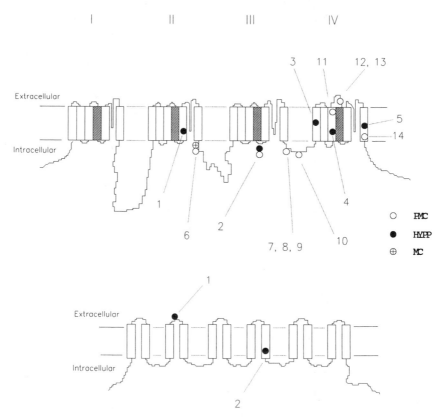

FIGURE 1. Schematic representation of voltage-dependent skeletal muscle sodium (**A**) and chloride (**B**) channels. Positions and phenotypes of mutations are as indicated. Details for each mutation are in TABLE 1.

channel openings were normal in the diseased muscle; by contrast, at 10 mM of extracellular potassium the channel behavior in affected myotubes was distinctly abnormal, with prolonged or repetitive openings that were abolished by TTX. These data strongly implicate potassium as a primary factor triggering an aberrant sodium channel gating mode and persistent sodium current in channels containing this particular mutation.

Fontaine and others explored this hypothesis by testing for genetic linkage analysis between the skeletal muscle sodium channel and hyperkalemic paralysis.[13] Sodium channels in skeletal muscle are heterodimers composed of a large alpha subunit (hSKMI, 260 kD) and a smaller beta subunit (36 kD) that is not essential for channel opening but may be necessary for correct channel gating.[14] The messenger RNA encoded by the hSKMI gene is specific to skeletal muscle and is not expressed in brain, heart, uterus, liver, or spleen; in humans, neuronal sodium channels are likely to be the products of separate genes, as has been demonstrated in other mammals. As illustrated in FIGURE 1(A), the hSKMI alpha subunit is a polypeptide of 1,836 amino acids; hydrophobicity analysis predicts that the channel has characteristic intracytoplasmic, extracellular, and membrane-spanning domains. Four major membrane-spanning domains, believed to have arisen by gene duplication, are each composed of six transmembrane segments that have been highly conserved during evolution. Conserved regions appear important for channel gating, ion selectivity, and as receptor sites for a variety of drugs and neurotoxins.

Fontaine used the complementary DNA for hSKMI as a probe for channel polymorphisms to map this skeletal muscle sodium channel gene to the long arm of human chromosome 17 and document tight genetic linkage of hSKMI to a large pedigree with hyperkalemic paralysis.[13] Subsequent studies have confirmed and extended this analysis to show linkage of both myotonic and nonmyotonic forms of hyperkalemic disease and of paramyotonia congenita to the same sodium channel locus on chromosome 17q.[15-18] Analogous genetic studies of myotonia congenita indicate linkage to at least two loci.[19] Although some individuals with features resembling myotonia congenita have mutations in the skeletal muscle sodium channel,[20] the defect in most patients clearly resides in the gene for a skeletal muscle chloride channel, as discussed below.[21-23]

In the last three years, a rapidly increasing family of missense mutations in the skeletal muscle sodium channel have been reported in hyperkalemic periodic paralysis, paramyotonia congenita, and some forms of myotonia congenita (TABLE 3). The mutations causing well-defined hyperkalemic periodic paralysis are located toward the cytoplasmic ends of the membrane spanning regions in domains II and IV.[24-26] Mutations causing paramyotonia congenita phenotype are located either within the III–IV cytoplasmic loop or toward the extracellular ends of S3 and S4 alpha helices in the channel domain.[27-29] One mutation causes a mixed hyperkalemic paralysis–paramyotonia phenotype, varying between affected individuals within a family; this is located toward the inner or cytoplasmic membrane surface.[20] Another mutation produces marked, chronic muscle stiffness suggestive of myotonia congenita in some but not all family members. This is located in a cytoplasmic loop associated with transmembrane segment S6 in domain II.[20] (These mutations have recently been the subject of excellent reviews.)[30,31]

TABLE 3. Sodium Channel Alpha Subunit Mutations and Polymorphisms in Hyperkalemic Periodic Paralysis, Paramyotonia Congenita, and Myotonia Congenita[a] and Chloride Channel Mutations in Myotonia Congenita

	Disease	Position	Domain	Exon	Genotype	Author	Reference
Sodium Channel Alpha Subunit Mutations							
1. Thr → Met	HYPP	704	II, S53	13	C2188T	Ptacek et al.	24
2. Ala → Thr	HYPP-MC	1156	III, S4–5	19	G3466A	McClatchey et al.	20
3. Met → Val	HYPP	1360	IV, S1	23	A4078G	Lehmann-Horn et al.	31
4. Phe → Leu	HYPP[b]	1419	IV, S3		C4257G	Rudolf et al.	26
5. Met → Val	HYPP	1592	IV, S6	24	A4774G	Rojas et al.	25
6. Ser → Phe	PMC–MC	804	II, S6	14	C2411T	McClatchey et al.	20
7. Gly → Val	PMC	1306	III-IV	22	G3917T	McClatchey et al.	28
8. Gly → Glu	PMC[c]	1306	III-IV	22	G3917A	Lerche et al.	31
9. Gly → Ala	PMC[d]	1306	III-IV	22	G3917C	Lerche et al.	31
10. Thr → Met	PMC	1313	III-IV	22	C3938T	McClatchey et al.	28
11. Leu → Arg	PMC	1433	IV, S3	24	T4298G	Ptacek et al.	29
12. Arg → His	PMC	1448	IV, S4	24	C4343A	Ptacek et al.	27
13. Arg → Cys	PMC	1448	IV, S4	24	C4342T	Ptacek et al.	20
14. Val → Met	PMC	1589	IV, S6	24	G4765A	Heine et al.	31
Sodium Channel Alpha Subunit Polymorphisms							
1. Met / Val		203	I, S3		607A/G	George et al.	49
2. Glu / Lys		860	II-III		2578G/A	George et al.	49
3. Glu / Asp		860	II-III		2580G/T	Wang et al.	50
4. Asp / His		932	II-III		2794G/C	Ptacek et al.	29
5. Asp / Asn		1376	IV, S1–S2		4126G/A	Rojas et al.	25
6. Glu / Gly		1606	IV, S6		4817A/G	Wang et al.	51
Chloride Channel Mutations							
1. Gly → Glu	Thomsen's	180	D3–D4		G180A	George et al.	38
2. Phe → Cys	Becker's	412	D8		T1238G	Koch et al.	23

Abbreviations: hyperkalemic periodic paralysis (HYPP), paramyotonia congenita (PMC), and myotonia congenita (MC).
[a] Modified from Lehmann-Horn et al.[31]
[b] Equine.
[c] "Myotonia permanens."
[d] "Myotonia fluctuans."

Myotonia Congenita

Linkage studies in myotonia congenita have shown genetic heterogeneity among affected families.[20-22] Though some pedigrees with this clinical picture are linked to the sodium channel gene, the originally described dominant form (Thomsen's disease) and the recessive generalized form (Becker's type) are not linked to chromosome 17.[21,22] Investigations of myotonic goat muscle showed a relative loss of permeability to chloride and suggested a possible chloride channel defect.[32-35] Steinmeyer and colleagues reported that the myotonic mouse mutant adr/adr lacks chloride conductance because of a genetic defect in a voltage-sensitive chloride channel.[36,37] Koch and associates reported that families with both Thomsen's and Becker's myotonia are genetically linked to the homologous human chloride channel gene on chromosome 7q; moreover, a specific mutation (substituting cysteine for phenylalanine in the D8 membrane segment) was identified in a patient with Becker's generalized myotonia[23] (FIG. 1, B). Another mutation, a glycine to glutamic acid substitution between the D3 and D4 segments on the extracellular channel face, was found by George and colleagues to cosegregate with the disease in three unrelated Thomsen's disease pedigrees.[38] Though the mutations for these two myotonic disorders are found in the same gene, in one case the phenotype is recessive (Becker's), while in the other it is dominant (Thomsen's). The recessive form, which is clinically more severe, is postulated to result from a loss of functional channel protein, while the dominant disease could occur if the defective protein somehow inactivates chloride channel subunits encoded by the normal gene (a "dominant negative" mutation), possibly by multimer formation.

Myotonic Dystrophy

Myotonic dystrophy is a progressive disorder characterized by moderately severe muscle stiffness, wasting of distal limb and selected face and neck muscles, and other systemic features. The defective gene for myotonic dystrophy on the long arm of chromosome 19 encodes a protein kinase, myotonin.[39,40] In myotonic dystrophy patients, the kinase gene contains an enlarged nucleotide triplet repeat (CTG) in the 3' end of the gene. The function of the protein and the effect of this mutation in myotonic dystrophy remain undefined. Conceivably, the altered kinase could affect phosphorylation of the sodium channel and thus impair channel gating, similar to the abnormality described above in hyperkalemic paralysis.

FUNCTIONAL SIGNIFICANCE OF SODIUM CHANNEL MUTATIONS

At least three regions of the sodium channel appear important for the inactivation process and thus may be perturbed in periodic paralysis and paramyotonia congenita.[13,42] (1) The intracellular loop between domains III and IV (FIG. 1) is thought to act as an inactivation gate that swings into the inner vestibule of the open channel to effect normal fast inactivation. Alteration of this region by intracellular proteases, site-specific antibodies, and site-directed mutagenesis disrupts inactivation.[42,43] How

TABLE 4. Disorders of Ion Channels and Related Receptors

Disease	Channel or Receptor
Human	
Inherited	
Cystic Fibrosis	Chloride channel (CFTR[a])
Hyperkalemic periodic paralysis	Sodium channel
Paramyotonia congenita	"
Myotonia congenita (K-sensitive)	"
Myotonia congenita	Chloride channel (voltage-sensitive)
Becker's generalized (recessive)	"
Thomsen's (dominant)	"
Malignant hyperthermia	Ryanodine receptor
Slow channel syndrome	Acetylcholine receptor
Sporadic	
Myasthenia gravis	Acetylcholine receptor
Lambert-Eaton myasthenic syndrome	Calcium channel
Animal	
Hyperkalemic periodic paralysis	Sodium channel
Myotonic goat	Chloride channel (Voltage-sensitive)
Myotonic adr/adr mouse	"
Dystrophic dys/dys mouse	Calcium channel (DHP[b] receptor)
Porcine malignant hyperthermia	Ryanodine receptor

[a] Cystic fibrosis transmembrane conductance regulator.
[b] Dihydropyridine receptor.

do the mutations within this loop, associated with paramyotonia congenita, confer unusual cold sensitivity upon channel inactivation? One may speculate that the III–IV loop mutation, Gly-1306-Val, disrupts a putative Gly-Gly hinge region upon which the inactivation particle may pivot,[44] and could thereby render normal motion of the loop (and hence inactivation itself) more critically temperature dependent. (2) Residues lining the inner channel vestibule may serve as an inactivation gate receptor; in the Shaker potassium channel, the analogous region has been shown to be critical for normal inactivation. It is interesting that the four hyperkalemic paralysis mutations are in putative membrane alpha helices near the cytoplasmic membrane surface (FIG. 1). These mutations might partially destabilize the interaction between the inactivation gate and its receptor. How elevated extracellular potassium triggers this effect is unresolved at this time. (3) At least one extracellular site, when bound by peptide toxins, impairs inactivation.[45]

Two additional mechanistic points are noted. First, as shown by Cannon, if only a small fraction of mutant channels is in the non-inactivating mode, the muscle cell will be depolarized and unable to contract. That experimental observation is in excellent agreement with theoretical computations using a Hodgkin-Huxley model. We have developed such a model using published voltage clamp data for skeletal muscle; this demonstrates that as few as 3–6% of sodium channels in a non-inactivating state produce depolarization and loss of excitability.[46]

Second, this abnormal sodium channel behavior is necessary but not sufficient to generate sustained myotonia. Cannon has shown using experimental data in rats[47] and the aforementioned computer model[46] that, even with a fraction of non-inactivating sodium channels, myotonic discharges persist after the end of a current stimulus only if there is activity-driven accumulation of potassium accumulates in the transverse-tubules. That is, both membrane hyperexcitability (the sodium channel mutation) and the distinctive anatomic structure of the transverse-tubules are required for myotonia.

CONCLUSIONS

The last four years have witnessed remarkable progress in delineating the molecular pathogenesis of neuromuscular disorders characterized by disordered membrane excitability. The only major disease in the group as yet not explained by dysfunction of a specific molecule is hypokalemic periodic paralysis.[48] These disorders appear generally to fall within a growing category of human diseases arising from dysfunction of ion channels and receptors (TABLE 4). It is likely that this family of diseases will continue to expand rapidly and eventually encompass other inherited disturbances of excitable membranes, such as the familial epilepsies.

REFERENCES

1. ENGEL, A. G. 1986. The periodic paralyses. *In* Myology. A. G. Engel & B. Q. Banker, Eds.: 1843–1870. McGraw Hill. New York.
2. RIGGS, J. E. 1988. The periodic paralyses. Neurol. Clin. **6**: 485–498.
3. BARCHI, R. L. 1988. The myotonic syndromes. Neurol. Clin. **6**: 473–483.
4. GRIGGS, R. C., W. K. ENGEL & J. S. RESNICK. 1970. Acetazolamide treatment of periodic paralysis. Ann. Intern. Med. **73**: 39–48.
5. RUDEL, R. & F. LEHMANN-HORN. 1985. Membrane changes in cells from myotonia patients. Physiol. Rev. **65**: 310–356.
6. ROSES, A. D. 1993. Myotonic Dystrophy. *In* The Molecular and Genetic Basis of Neurologic Disease. R. N. Rosenberg, S. B. Prusiner, S. DiMauro, R. L. Barchi & L. M. Kunkel, Eds.: 633–646. Butterworth-Heinemann.
7. CREUTZFELD, O. D., B. C. ABBOTT, W. M. FOWLER & C. M. PEARSON. 1963. Muscle membrane potentials in episodica adynamia. Electroencephalogr. Clin. Neurophysiol. **15**: 508–515.
8. RUDEL, R. 1986. The pathophysiological basis of the myotonias and the periodic paralyses. *In* Myology. A. G. Engel & B. Q. Banker, Eds.: 1297–1311.
9. LEHMANN-HORN, F., R. RUDEL, K. RICKER, H. LORKOVIC, R. DENGLER & C. HOPF. 1983. Two cases of adynamia episodica hereditaria: in vitro investigations of muscle cell membrane contraction parameters. Muscle Nerve **6**: 113–121.
10. LEHMANN-HORN, F. & R. RUDEL. 1987. Membrane defects in paramyotonia congenita (Eulenberg). Muscle Nerve **10**: 633–641.
11. RICKER, K., I. M. CAMACHO, P. GRAFE, F. LEHMANN-HORN & R. RUDEL. 1989. Adynamia episodica hereditaria: what causes the weakness? Muscle Nerve **12**: 883–891.

12. CANNON, S. C., R. H. BROWN, & D. P. COREY. 1991. A sodium channel defect in hyperkalemic periodic paralysis: potassium induced failure of inactivation. Neuron 6: 619–626.

13. FONTAINE, B., T. S. KHURANA, E. P. HOFFMAN, G. BRUNS, J. L. HAINES, J. A. TROFATTER, M. P. HANSON, D. McKENNA-YASEK, J. GUSELLA & R. H. BROWN, JR. 1990. Hyperkalemic periodic paralysis and the adult muscle sodium channel alpha-subunit gene. Science 250: 1000–1002.

14. CATTERALL, W. 1988. Structure and function of voltage-sensitive ion channels. Science 242: 50–61.

15. PTACEK, L. J., F. TYLER & J. S. TIMMER. 1991. Analysis in a large hyperkalemic periodic paralysis pedigree supports tight linkage to a sodium channel locus. Am. J. Hum. Genet. 49: 378–382.

16. KOCH, M. C., K. RICKER, M. OTTO, T. GRIMM, E. P. HOFFMAN, R. RUDEL, L. BENDER, B. ZOLL, P. S. HARPER & F. LEHMANN-HORN. 1991. Confirmation of linkage of hyperkalemic periodic paralysis to chromosome 17. J. Med. Genet. 28: 583–586.

17. EBERS, G. C., A. L. GEORGE, R. L. BARCHI, S. S. TING-PASSADOR, R. G. KALLEN, G. M. LATHROP, J. S. BECKMAN, A. F. HAHN, W. F. BROWN, R. D. CAMPBELL & A. J. HUDSON. 1991. Paramyotonia congenita and hyperkalemic paralysis are linked to the adult muscle sodium channel gene. Ann. Neurol. 30: 810–816.

18. PTACEK, L. J., J. S. TRIMMER, W. S. AGNEW, J. W. ROBERTS, J. H. PETAJAN & M. LEPPERT. 1991. Paramyotonia congenita and hyperkalemic periodic paralysis map to the same sodium channel gene. Am. J. Hum. Genet. 49: 851–854.

19. PTACEK, L. J., P. McMANNIS, H. KWIECINSKI & M. LEPPERT. 1992. Genetic heterogeneity in patients with the temperature-sensitive paramyotonia congenita phenotype. Ann. Neurol. 32: 250(A).

20. McCLATCHEY, A. I., D. McKENNA-YASEK, D. CROS, H. G. WORTHEN, R. W. KUNCL, S. M. DiSILVA, D. R. CORNBLATH, J. F. GUSELLA & R. H. BROWN, JR. 1992. Novel mutations in families with unusual and variable disorders of the skeletal muscle sodium channel. Nature Genetics 2: 148–152.

21. ABDALLA, J. A., W. L. CASLEY, H. K. COUSIN, A. J. HUDSON, E. G. MURPHY, F. C. CORNELIS, L. HASHIMOTO & G. C. EBERS. 1991. Linkage of Thomsen disease to the T-cell receptor beta (TCRB) locus on chromosome 7q35. Am. J. Hum. Genet. 51: 579–584.

22. ABDALLA, J. A., W. L. CASLEY, A. J. HUDSON, E. G. MURPHY, H. K. COUSIN, H. A. ARMSTRONG & G. C. EBERS. 1991. Linkage analysis of candidate loci in autosomal dominant myotonia congenita. Neurol. 42: 1561–1564.

23. KOCH, M. C., K. STEINMEYER, C. LORENZ, K. RICKER, F. WOLF, M. OTTO, B. ZOLL, F. LEHMANN-HORN, K.-H. GRZESCHIK & T. J. JENTSCH. 1992. The skeletal muscle chloride channel in dominant and recessive myotonia. Science 257: 797–800.

24. PTACEK, L. J., A. L. GEORGE, R. C. GRIGGS, R. TAWIL, R. G. KALLEN, R. L. BARCHI, M. ROBERTSON & M. F. LEPPERT. Identification of a mutation in the gene causing hyperkalemic periodic paralysis. Cell 67: 1021–1027.

25. ROJAS, C. V., J. WANG, L. S. SCHWARTZ, E. P. HOFFMAN, B. R. POWELL & R. H. BROWN, JR. 1991. A met-to-val mutation in the skeletal muscle Na channel alpha-subunit in hyperkalemic periodic paralysis. Nature 354: 387–389.

26. RUDOLPH, J. A., S. J. SPIER, G. BYRNS, C. V. ROJAS, D. BERNOCO & E. P. HOFFMAN. 1992. Periodic paralysis in quarter horses: a sodium channel mutation disseminated by selective breeding. Nature Genetics 2: 144–147.

27. PTACEK, L. J., A. L. GEORGE, JR., R. L. BARCHI, R. C. GRIGGS, J. E. RIGGS, M. ROBERTSON & M. F. LEPPERT. 1992. Mutations in an S4 segment of the adult skeletal muscle sodium channel cause paramyotonia congenita. Neuron 8: 891–897.

28. McCLATCHEY, A. I., P. VAN DEN BERGH, M. P. PERICAK-VANCE, W. RASKIND, C. VERELLEN, D. McKENNA-YASEK, K. RAO, J. L. HAINES, T. BIRD, R. H. BROWN, JR. & J. F. GUSELLA. 1992. Temperature-sensitive mutations in the III–IV cytoplasmic loop region of the skeletal muscle sodium channel gene in paramyotonia congenita. Cell 68: 769–774.

29. PTACEK, L. J., L. GOUW, H. KWIECINSKI, P. McMANIS, J. R. MENDELL, R. J. BAROHN, A. L.

GEORGE, JR., R. L. BARCHI, M. ROBERTSON & M. F. LEPPERT. 1993. Sodium channel mutations in paramyotonia congenita and hyperkalemic periodic paralysis. Ann. Neurol. 33(3): 300–307.

30. PTACEK, L. J., K. J. JOHNSON & R. C. GRIGGS. 1993. Genetics and physiology of the myotonic muscle disorders. New Engl. J. Med. 328: 482–489.

31. LEHMANN-HORN, F., R. RUDEL & K. RICKER. 1993. Non-dystrophic myotonias and periodic paralyses. Neuromuscular Disorders 3: 161–169.

32. BRYANT, S. H. & A. MORALES-AGUILERA. 1971. Chloride conductance in normal and myotonic muscle fibres and the action of monocarboxylic aromatic acids. J. Physiol. (Lond.) 219: 367–383.

33. BRYANT, S. H 1979 Myotonia in the goat. Ann. N.Y. Acad. Sci. 317: 314–325.

34. ADRIAN, R. H. & S. H. BRYANT. 1974. On the repetitive discharge in myotonic muscle fibers. J. Physiol. (Lond.) 240: 505–515.

35. ADRIAN, R. H. & M. W. MARSHALL. 1976. Action potential reconstruction in normal and myotonic muscle fibers. J. Physiol. (Lond.) 258: 125–143.

36. STEINMEYER, K., C. ORTLAND & T. J. JENTSCH. 1991. Primary structure and functional expression of a developmentally regulated skeletal muscle chloride channel. Nature 354: 301–304.

37. STEINMEYER, K., R. KLOCKE, C. ORTLAND, M. GRONEMEIER, H. JOCKUSCH, S. GRUNDER & T. JENTSCH. 1991. Inactivation of muscle chloride channel by transposon insertion in myotonic mice. Nature 354: 304–308.

38. GEORGE, A. L., JR., M. A. CRACKOWER, J. A. ABDALLA, A. J. HUDSON & G. C. EBERS. 1993. Molecular basis of Thomsen's disease (autosomal denotement myotonia congenita). Nature Genetics 3: 305–310.

39. ROSES, A. D., M. A. PERICAK-VANCE, D. A. ROSS, L. YAMAOTA & R. J. BARTLETT. 1986. RFLPs at the D19S19 locus of human chromosome 19 linked to myotonic dystrophy. Nucl. Acids Res. 14: 5569.

40. BROOK, J. D., M. E. MCCURRACH, H. G. HARLEY, A. J. BUCKLER, D. CHURCH, H. ABURATANI, K. HUNTER, V. P. STANTON, J.-P. THIRLON, T. HUDSON, R. SOHN, B. ZEMELMAN, R. G. SNELL, S. A. RUNDLE, S. CROW, J. DAVIES, P. SHELBOURNE, J. BUXTON, C. JONES, V. JUVONEN, K. JOHNSON, P. S. HARPER, D. J. SHAW & D. E. HOUSMAN. 1992. Molecular basis of myotonic dystrophy: expansion of a trinucleotide (CTG) repeat at the 3′ end of a transcript encoding a protein kinase family member. Cell 69: 385–395.

41. STUHMER, W., F. CONTI, H. SUZUKI, X. D. WONG, M. NODA, N. YAHAGI, H. KUBO & S. NUMA. 1989. Structural parts involved in activation and inactivation of the sodium channel. Nature 339: 597–603.

42. ARMSTRONG, C. M., F. BEZANILLA & E. ROJAS. 1973. Destruction of sodium inactivation in squid axons perfused with pronase. J. Gen. Physiol. 62: 375–391.

43. VASSILEV, P. M., T. SCHEUER & W. A. CATTERALL. 1989. Inhibition of inactivation of single sodium channels by a site directed antibody. Proc. Natl. Acad. Sci. USA 86: 8147–8151.

44. WEST, J. W., D. E. PATTON, T. SCHEUER, Y. WANG, A. L. GOLDIN & W. A. CATTERALL. 1992. A cluster of amino acid residues required for fast Na+ channel inactivation. Proc. Natl. Acad. Sci. USA 89: 10910–10914.

45. TREJEDOR, E. J. & W. A. CATTERALL. 1988. Site of inactivation of alpha-scorpion toxin derivatives in domain I of the sodium channel alpha subunit. Proc. Natl. Acad. Sci. USA 85: 8742–8746.

46. CANNON, S. C. & D. P. COREY. 1993. Loss of sodium channel inactivation by anemone toxin (ATXII) mimics the myotonic state in hyperkalemic periodic paralysis. J. Physiol. (Lond.) (In press.)

47. CANNON, S. C., R. H. BROWN, JR. & D. P. COREY. 1993. Theoretical reconstruction of myotonia and paralysis caused by incomplete inactivation of sodium channels. Biophys. J. 65: 270–288.

48. FONTAINE, B., J. TROFFATER, G. A. ROULEAU, J. L. HAINES, J. F. GUSELLA & R. H. BROWN, JR. 1992. Different gene loci for hyperkalemic and hypokalemic periodic paralysis. Neuromusc. Dis. 1: 235–238.

49. GEORGE, A. L., JR., J. KOMISAROF, R. G. KALLEN & R. L. BARCHI. 1992. Primary structure of the adult human skeletal muscle voltage-dependent channel. Ann. Neurol. **31**: 131–137.
50. WANG, J. Z., J. ZHOU, S. M. TODOROVIC, W. G. FEERO, F. BARANY, R. CONWIT, I. HAUSMANOWA-PETRUSEWICZ, A. FIDZIANSKA, K. ARAHATA, H. B. WESSEL & E. P. HOFFMAN. 1993. Molecular genetic and genetic correlations in sodium channelopathies: lack of founder effect and evidence for a second gene. Am. J. Hum. Genet. **52**: 1074–1084.
51. WANG, J. Z., C. V. ROJAS, J. ZHOU, L. S. SCHWARTZ, H. NICHOLAS & E. P. HOFFMAN. 1992. Sequence and genomic structure of the human adult skeletal muscle sodium channel alpha-subunit gene on 17q. Biochem. Biophys. Res. Commun. **182**: 794–801.

The Role of Channel Formation in the Mechanism of Action of Tumor Necrosis Factors[a]

BRUCE L. KAGAN, TAJIB MIRZABEKOV,
DAVID MUNOZ, RAE LYNN BALDWIN,[b]
AND BERNADINE WISNIESKI[b]

West Los Angeles VA Medical Center
Departments of Psychiatry and [b]Microbiology and Molecular Genetics
UCLA School of Medicine
Los Angeles, California

INTRODUCTION

Tumor necrosis factor (TNF) and lymphotoxin (LT) are polypeptide cytokines with pleiotropic effects. They are important in host defense against bacteria, parasites, and viruses, and have inflammatory and metabolic effects. TNF and LT are also cytotoxins that are selectively toxic for malignant cells. The mechanism of action by which TNF and LT exert these diverse effects remains obscure despite a great deal of effort in this area.

TNF has been implicated in the pathophysiology of wasting (cachexia) and hypertriglyceridemia. Indeed, it was independently discovered as the hormone (cachetin) responsible for these effects.[1] TNF plays a key role in regulating the immune response[2] and has been suggested to contribute to a number of rheumatic disorders.[3]

More recently TNF and LT have been implicated in the pathophysiology of AIDS.[4] (1) AIDS patients often exhibit extreme cachexia and wasting. (2) Lahdevirta *et al.*[5] have reported that all AIDS patients, 50% of patients with AIDS-related complex (ARC), and some patients with lymphadenopathy syndrome have elevated serum levels of TNF (100–200 pg/ml). (3) It has been reported that monocytes from HIV-infected patients produce high levels of TNF and lymphocytes from HIV-infected patients produce high levels of an LT-like substance.[6,7] (4) TNF has been reported to activate HIV replication in HIV-infected cells.[8–11] (5) Purified LT can kill chronically HIV-infected MOLT-4 cells.[4] (6) AIDS patients are subject to frequent opportunistic infections and these infectious agents are potent stimulators of TNF production by mono-

[a] This work was supported by the Department of Veterans Affairs and National Institute of Mental Health grant MH43433 (B.L.K.) and by National Institutes of Health grant GM22240 (B.J.W.). R.L.B. had a National Institutes of Health Atherosclerosis Predoctoral Training Grant Award (2T32 HLO7386).

317

cytes and macrophages. This may explain why frequent infection appears to speed the progression of AIDS.

In addition to the general role we have suggested for TNF in AIDS pathophysiology, a more specific role may be played by TNF in the central nervous system (CNS) effects of AIDS. (1) TNF is produced by microglia and astrocytes in the CNS.[12] (2) TNF can kill oligodendrocytes and cause demyelination.[13] (3) In AIDS patients, high levels of TNF correlate with encephalopathy.[14]

Multiple lines of evidence support the hypothesis that damage to the neurons producing CNS disease is not a direct effect of HIV infection, but an induced effect of viral proteins or cytokines induced by HIV.[15] (1) Multinucleated giant cells formed by HIV-induced fusion of macrophages or microglia can be found in close proximity to hyperplastic cells of the cerebral microvasculature. This suggests blood-brain–barrier alterations including increased vascular permeability, vasculitis, or necrosis.[16] These areas are also deficient in myelin.[17] (2) Cortical neuronal loss has been described in AIDS patients brains,[18] and neuronal cell culture studies indicate that the gp120 envelope protein of HIV can be neurotoxic *in vitro* possibly by increasing intracellular free calcium.[19,20] Thus viral envelope proteins can be toxic without a direct neuronal infection. (3) The rapid and relatively complete responsiveness of some AIDS dementia complex (ADC) patients treated with zidovudine suggests that the initial damage to the CNS is reversible.[21] This suggests that diffusible factors induced by HIV may be responsible for initial ADC pathology. (4) HIV-1 can trigger the production by monocytes of cytokines, including TNF, by binding to CD4.[22] CD4-positive T cells can be induced to produce LT by a similar mechanism.[23] TNF and other cytokines (IL-1, IL-6) augment HIV replication in T cells and macrophages.[23] Thus a positive feedback loop may generate increasingly large amounts of TNF. (5) TNF has been implicated in the pathogenesis of other neurologic diseases such as multiple sclerosis[24] or cerebral malaria.[25] The failure to find consistently elevated levels of TNF in the cerebrospinal fluid (CSF) of AIDS patients may reflect the importance of *local* tissue levels of TNF rather than systemic levels. Because TNF can be produced locally by macrophages and microglia in the brain and may often exert its effects through direct cell-cell contact,[1] CSF levels may not correspond to local tissue levels of TNF. (6) In disease and trauma states, TNF has been shown to cause white matter lesions, astrogliosis, and vascular changes.[24,25] Taken together these data strongly suggest that TNF (and other cytokines) may mediate the bulk of CNS pathology seen in AIDS.[15]

TNF is coded as a prohormone (26 kD) that can appear as a transmembrane cell surface protein that can be proteolytically clipped to the mature 17 kD form (157 amino acids).[26] This monomer non-covalently aggregates to a trimer, which is believed to be the physiologically active species.[27] The amino acid sequence of TNF is highly conserved amongst mammalian species (See Aggarwal[28] for a review of TNF structure). Human TNF has no carbohydrate and has a pI of 5.3. A single disulfide bridge exists between cysteines 69 and 101. The protein contains almost no α-helix and about 60% β-sheet by circular dichroism. This is confirmed by the crystal structure, which shows anti-parallel β-sheet and a novel face-to-edge packing of these sheets.[29,30] The sheets are organized in a "jelly-roll" motif characteristic of viral coat proteins. Substantial similarity is observed between TNF and capsid proteins from satellite tobacco necrosis viruses (STNV) and foot and mouth disease virus (FMDV). The three-dimensional structure is also remarkable for a central "channel" that extends more than halfway down the threefold axis of symmetry of the trimer. TNF binds to most cells through a single class of high affinity receptors (0.1–1.0 nM), although a second class of lower affinity receptors is sometimes observed. TNF receptors are homologous to NGF receptors, β (or B) cell antigen, and Shope fibroma virus antigen.[31] Both TNF and LT bind to the same receptor.[28] The primary sequences of TNF and LT show about 28% identity and 52% homology.[28] A remarkable feature of the TNF protein is its ability to renature and recover partial activity after treatment with urea, SDS, or guanidinium.[28]

The ability of TNF to depolarize muscle cells suggested that the plasma membrane might be a site of action of TNF.[32] TNF's effects on oligodendrocyte necrosis and periaxonal swelling were also consistent with a change in cell membrane permeability.[13] Furthermore, in model membrane studies, TNF exhibited a pH-dependent ability to insert into liposomes and to allow efflux of calcein.[33,34] These studies, coupled with the structural studies showing a potential channel down the threefold axis of TNF,[29,30] led us to test the effects of TNF and LT on planar phospholipid bilayer membranes. We report here that both TNF and LT can form pH-dependent ion channels, and we suggest that these channels may play a role in cytokine-mediated cytotoxicity.

METHODS

Solvent-free lipid bilayers were formed by the union of two monolayers across an aperture separating two aqueous phases. We have described these techniques in detail elsewhere.[35] The formation of such bilayers was carried out in a Teflon® chamber with two compartments separated by a thin (30 μm) Teflon® film with an aperture diameter of 50–200 μm. Each of the compartments was connected by plastic tubing to syringes filled with aqueous salt solutions. Initially, the level of solutions in both compartments was raised up to a level just below the aperture. Then, a 1% solution of lipid in hexane was carefully spread at the surfaces of the aqueous phases of both compartments. A small amount of squalene (usually 20 μl of a 1% solution in pentane) was spread at the partition between compartments. After solvent evaporation (15–20 min), the bilayer was formed by the gentle raising of the solution surfaces at both compartments to a level above the aperture. Formation of the bilayer was verified by monitoring its electrical characteristics (capacitance and conductance).

To monitor membrane formation, a triangular wave or square pulse of 10–20 mV and 100 Hz frequency was used. At the time when the levels of solutions are below the aperture level, the capacitance response was very small. After the raising of the solution levels above the aperture, the capacitance increases when monolayers are opposed. Union of two monolayers into a bilayer is indicated by a sharp increase in capacitance response. To measure conductance, a DC voltage of 100 mV is applied to the bilayer. Suitable membranes in these conditions showed a stable, voltage-independent current of less than 1 pA for at least 10 minutes.

Preparation of TNF-containing Liposomes

TNF-containing proteoliposomes were prepared by incubation of liposomes with recombinant human TNF at a low pH. Liposomes were prepared as follows. Ten milligrams of lipid (200 μl of purified soybean phospholipids in hexane at a concentration of 50 mg/ml) was added to each vial and then dried under a stream of nitrogen. One milliliter of salt solution (100 mM NaCl, 10 mM DGA-Tris, pH 4.5) was then added to each vial. The lipid-salt solution mixture was sonicated 15 min by pulse-sonication. To the resulting liposome mixture, 1 μg of TNF was added, and then liposomes were incubated at 37°C for 1 hour. A control sample of liposomes was incubated in the same conditions but without protein addition. After 1 h of incubation at the low pH, both liposome samples were adjusted to pH 7.5 by addition of Tris from a 200 mM stock solution.

Preparation of Membranes from Proteoliposomes

Formation of planar lipid membranes from proteoliposomes was carried out using the method developed by Schindler.[36] Lipid monolayers were formed on the air/salt solution interface using liposomes instead of a lipid/hexane mixture. Monolayers were formed from the disruption of liposomes on the air/solution interfaces. Proteoliposomes were added to the cis compartment and protein-free liposomes were added to the trans compartment. After 30 min, monolayers were formed. Planar membranes were then prepared as described above by a gentle raising of surface levels of both solutions above the hole in the partition between the two compartments.

Lipids

Soybean phospholipids (Avanti) or azolectin (Sigma, L-α-phosphatidylcholine II-S) were used in the experiments. These lipids were purified in two steps from divalent cations and proteolipids by the method of Labarca et al.,[37] and from neutral and oxidized lipids by the method of Kagawa et al.[38] Purified lipids were dissolved in hexane to a concentration of 50 mg/ml and stored at −20°C. To prevent oxidation, the second step of lipid purification was re-applied every two weeks.

Electrodes

The measurement of the electrical parameters of lipid bilayers requires electrical connection of the membrane to the recording equipment. Silver-silver chloride electrodes with or without agar bridges (used for measurements in conditions of asymmetric solutions) were routinely used. Electrode asymmetry was always less than 1 mV.

Recording Equipment

A Keithley 427 current amplifier was routinely used for measuring membrane current. A signal generator and oscilloscope were employed to monitor membrane capacitance (usually only at the stage of bilayer formation). Typical capacitances were of the order of 0.8 μF/cm². The source of DC voltage was a battery with voltage divider or a standard signal generator. For single-channel experiments requiring low noise and high resolution, we employed a commercially available voltage clamp amplifier (Axopatch 1C, Axon Instruments, Sunnyvale, CA) with suitable head stage (CV-3B). For data acquisition, a digital tape recorder and video cassette recorder allowed recording of large amounts of data. A storage oscilloscope was used for monitoring membrane capacitance and single-channel recordings.

RESULTS

Five to fifteen minutes after addition of TNF to one side of a lipid membrane, the conductance increased by discrete steps. FIGURE 1 shows the current response of a TNF-treated membrane to voltage. In the absence of TNF the membrane conductance is ohmic and equal to 5–10 pS, and membranes are stable to voltages of +140 mV to −140 mV. The side to which TNF is added is taken as ground; hence, voltages correspond to the "cytoplasmic" voltage. The conductance induced by TNF is due to formation of ion-permeable channels. Observed single-channel conductances are heterogeneous, but can be grouped into two main classes, one centered at ~5 to 10 pS, and a second, larger class ranging from ~200 to 2,000 pS. The most frequently observed event is 5 pS at a sodium chloride concentration of 100 mM. Although channels can form at pH 7.2, channel formation is dramatically enhanced by lowering the pH of the aqueous phase containing TNF.[39]

FIGURE 2 shows current fluctuations due to the presence of TNF in membranes formed from proteoliposomes. These records demonstrate that TNF can readily form channels at pH 7.2 and that this method of reconstitution can reliably produce single channels for recording. Since the incorporation of TNF into liposomes at acidic pH has been demonstrated by photolabeling,[33] there can be little doubt that these channels are due to the presence of TNF in the membrane. Their properties (voltage dependence, kinetics, ionic selectivity) are quite similar to TNF channels incorporated from aqueous solution. Since membranes are more stable at pH 7.2, this enhances our ability to record channels successfully.

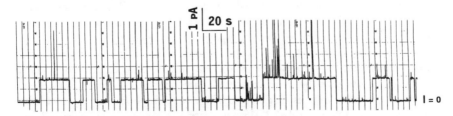

FIGURE 1. Membrane current fluctuations induced by TNF. A planar lipid membrane was formed from soybean phospholipids. Salt solutions contained: 100 mM NaCl, 2 mM MgCl, 10 mM dimethyl glutaric acid-NaOH (pH 4.5) in the *cis* side and 100 mM NaCl, 2 mM MgCl, 10 mM Tris-HCl (pH 7.5) in the *trans* side. T = 22°C. Addition of 500–700 ng/ml of TNF to the membrane resulted in the formation of channels with conductance approximately 5–7 pS. More rarely, current transitions (channels) with conductances 0.1–1 nS were observed (data not shown).

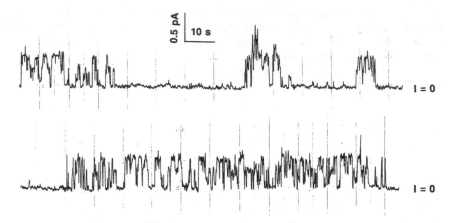

FIGURE 2. Ion channel conductance transitions of a lipid bilayer formed from TNF-containing proteoliposomes. For formation of the planar membrane on a 150 μm hole in the Teflon® film, two lipid monolayers were used: a monolayer formed from TNF-containing azolectin liposomes on one side and a monolayer formed from pure azolectin liposomes on the other side. Membrane voltage was +40 mV. Symmetric salt solutions contained 100 mM NaCl, 10 mM DGA-Tris, pH 7.5.

We recently proposed that the observation of channels with a variety of single-channel conductance amplitudes might be a result of TNF-trimer aggregation in the membrane.[39] Therefore we tried to find the conditions under which TNF might be disaggregated.

To accomplish this we added TNF together with a low concentration of the nonionic detergents Triton X-100 or octylglucopyranoside, or used high salt concentrations in the aqueous solutions. Addition of 0.1–5.0 μg/ml of detergents (at this concentration the detergents do not change the electrical properties of lipid bilayers) did not change the channel-forming activity and amplitude distribution of observed TNF channels. Incorporation of TNF into membranes at 1 M NaCl also revealed a wide distribution of amplitudes with peaks at 10, 50, 80, 160, 320, 360, 500 pS and 2.2 nS approximately (FIG. 3). Whether these peaks represent multiple conductance states of a channel or multiple molecular species is uncertain.

Addition of lymphotoxin (LT or TNFβ) to the lipid bilayer also results in formation of ion channels. FIGURE 4 shows the macroscopic currents induced by LT in a planar lipid bilayer. Note that the conductance increases at (trans) positive voltages and decreases at negative voltages. These recordings were made in conditions of a pH gradient (LT side: 4.5/*trans* side 7.2). In symmetric pH conditions, the voltage dependence is much less apparent. FIGURE 5 shows current fluctuations due to single channels of LT. Note that the predominant size of 6 ± 1 pS is very close to the single-channel conductance of TNF under these conditions. As with TNF, larger conductance steps can also be seen at later times after addition, suggesting a possible aggregation of LT in the membrane.[39]

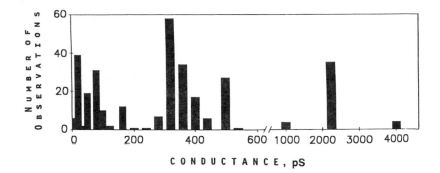

FIGURE 3. Amplitude distribution histogram of ion channels formed by TNF in azolectin membranes at high salt concentrations. Solutions with asymmetric pH contained 1 M NaCl, 2 mM MgCl$_2$, 10 mM DGA-NaOH, pH 4.5 in the *cis* side and 1 M NaCl, 2 mM MgCl$_2$, 10 mM Tris-HCl, pH 7.5 in the *trans* side. TNF was added to the *cis* side to a concentration of 500–700 ng/ml. Data from six experiments were used. The number of events is 324.

DISCUSSION

The following data support our conclusion that channel formation is intrinsic to TNF and LT. First, in the absence of added TNF and LT, no channel activity is observed. Second, three separate, highly purified preparations of TNF and LT (recombinant human TNF from Genentech, recombinant human LT from Genentech, and natural human TNF from Calbiochem) produced similar channel activity. Third, boiling TNF for 5 min, which eliminates biological activity, also eliminates channel-forming activity. Fourth, channel activity is greatly enhanced by low pH. This is consistent with reports showing that membrane insertion[33] and permeabilization[34,40,41] increase at low pH. Although our results seem to contrast with those of Young,[42] who found no effect of TNF on lipid membranes, we should point out that his experiments were not performed in the presence of a pH gradient, a requirement for optimal TNF channel activity that is highly reminiscent of findings with diphtheria toxin.[43] Furthermore, his membranes contained cholesterol, which renders membranes more rigid and may inhibit protein insertion.

Despite intensive study, the cellular mechanisms of action of TNF remain unclear. No enzymatic activity has been identified, and some evidence suggests TNF must be internalized to act. Other evidence suggests that cell surface TNF receptors can mediate the action of TNF. Although TNF receptors are necessary for cells to be sensitive to TNF, there is no direct correlation between the number of receptors and the sensitivity.[1] After binding to the receptor, TNF is internalized and degraded.

Several lines of evidence implicate a role for membrane damage in TNF action. (*1*) A degradation product of TNF secreted into the medium is lytic for liposomes.[44] (*2*) TNF can release calcium from liposomes of pH < 5.[34] (*3*) TNF can induce lysis of internal membranes.[45] (*4*) Acidic phospholipids can increase the ability of TNF to lyse liposomes.[41] (*5*) TNF can increase the calcein permeability of negatively charged liposomes at pH 5–6 and even neutral liposomes release calcium when treated with

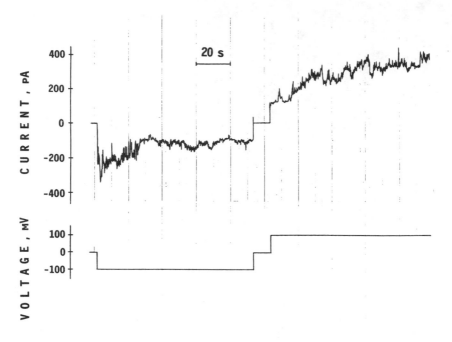

FIGURE 4. Potential dependence of the lymphotoxin-induced steady-state current. Lymphotoxin was added to a final concentration of 2 µg/ml. The figure demonstrates that at negative membrane potential (– 100 mV) lymphotoxin-induced membrane current decreases. After application of positive voltage (+ 100 mV) current again increases to the original level.

TNF at pH 4.5.[40] This correlates well with the increased hydrophobicity of TNF as measured by ANS fluorescence.[39,40] Other groups have also observed that TNF can insert into liposomes without causing efflux, and that this insertion is pH dependent.[33] (6) A TNF mutant with poor toxicity to L929 cells also fails to cause calcium efflux from liposomes.[40] It has also been observed that TNF leads to internucleosomal DNA cleavage and that inhibitors of ADP-ribosylation block TNF toxicity.[46,47] These investigators noticed the similarities between the cell killing induced by TNF and by diphtheria toxin (DT), which has been found to form channels in lipid bilayers,[43] and in target cells. DT and TNF both lyse cells in a time- and concentration-dependent manner, bind to cell surface receptors, and are then endocytosed. This leads to an apoptotic cell death with early DNA fragmentation, and the toxicity of both proteins can be blocked by ADP-ribosylation inhibitors.[1]

Although the above data demonstrate convincingly that TNF and LT can form ion-permeable channels in lipid membranes in a pH-dependent manner, the question of whether these channels are relevant to the action of TNF *in vivo* remains. Addition of TNF to human U937 histiocytic lymphoma cells rapidly increases $^{22}Na^+$ uptake by approximately 100–300%, in the presence or absence of ouabain. The simplest explanation for this enhanced Na^+ uptake is a direct permeabilization of the target cell membrane by TNF.[39]

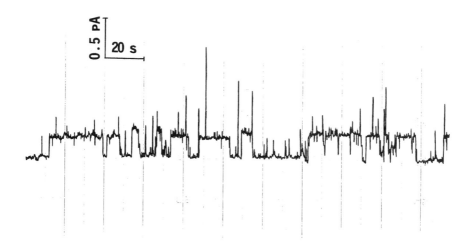

FIGURE 5. Current fluctuations of a planar lipid bilayer induced by one-sided addition of lymphotoxin (TNFβ or LT). The membrane was bathed by salt solutions with asymmetric pH (*cis*: 100 mM NaCl, 2 mM MgCl₂, 10 mM DGA-NaOH, pH 4.5; *trans*:100 mM NaCl, 2 mM MgCl₂, 10 mM Tris-HCl, pH 7.5). Holding potential was + 120 mV. LT was added to the solution with lower pH (*cis* side) to a final concentration of 600 ng/ml.

It is noteworthy that the three-dimensional structure of TNF shows striking homology to that of several viral coat proteins,[29] including the influenza hemagglutinin HA, which displays acid-dependent membrane fusogenic activity. This suggests a possible role for the "jelly roll" motif in facilitating acid-dependent membrane penetration. Our results raise the further possibility that acid-facilitated conformational changes and subsequent membrane penetration may allow the central "channel-like" region of the TNF trimer to assume an "open" state.

Channel formation by TNF would explain the rapid decrease in resting membrane potential in skeletal muscle observed by Tracey *et al.*,[32] the increased cellular Na⁺ and water levels seen in endotoxic shock,[48] and the myelin dilatation, oligodendrocyte necrosis, and periaxonal swelling seen by Selmaj and Raine.[13] Channel formation in an internal membrane might explain the inhibition of mitochondrial energy transfer caused by TNF.[49] The presence of TNF-specific receptors on the plasma membrane might compensate for the relatively low channel-forming activity of TNF seen with planar membranes at neutral pH (e.g., by facilitating membrane insertion). Localized acid pH effects are also possible both intracellularly (e.g., in endosomes) and extracellularly (e.g., near activated macrophages and osteoclasts).

The cytotoxic effects of TNF and LT could be explained quite simply by channel formation. The presence of these channels in the plasma membrane of target cells could induce efflux of vital intracellular ions such as K⁺ or Mg²⁺.[50,51] Alternatively, an influx of toxic elements such as Ca²⁺ could occur through an ion channel. Determination of the precise mechanism of killing must await further studies.

The pH dependence of channel formation may be relevant to the multiplicity of actions of TNF. Entry into an acidic endosome may induce TNF to form channels with altered voltage dependence and may have a different effect on the target cell than channel formation at the plasma membrane. Alternatively, channel formation may be related to entry of TNF into the cytosol as is the case for diphtheria toxin.[43] Finally, it must be considered that channel formation may represent an epiphenomenon that reflects the ability of TNF and LT to insert themselves into membrane environments under appropriate conditions.

ACKNOWLEDGMENTS

We gratefully acknowledge Genentech for supplies of TNF and LT, L. Greene, P. Marshall for expert technical assistance, and Dr. G. Eisenman and Dr. Y. Sokolov for valuable discussions.

REFERENCES

1. BEUTLER, B. & A. CERAMI. 1989. Annu. Rev. Immunol. 7: 625–655.
2. BONAVIDA, B. 1991. Biotherapy 3: 127–133.
3. MAURY, C. P. J. 1989. Scan. J. Rheumatol. 18: 3–5.
4. KOBAYASHI, N., Y. HAMAMOTO & N. YAMAMOTO. 1990. Virus Genes 4: 183–190.
5. LAHDEVIRTA, J., C. P. J. MAURY, A. M. TEPPO & H. REPO. 1988. Am. J. Med. 85: 289–291.
6. WRIGHT, S. C., A. JEWETT, R. MITSUYASU & B. BONAVIDA. 1988. J. Immunol. 141: 99–104.
7. LOMBARD, R. P., C. MODOUX, A. CRUCHAUD & J. M. DAYER. 1989. Clin. Immunol. Immunopathol. 50: 374–384.
8. MATSUYAMA, T., H. YOSHIYAMA, Y. HAMAMOTO, N. YAMAMOTO, G. SOMA, D. MIZUNO & N. KOBAYASHI. 1989. AIDS Res. Hum. Retroviruses 5: 139–146.
9. MATSUYAMA, T., Y. HAMAMOTO, G. SOMA, D. MIZUNO, N. YAMAMOTO & N. KOBAYASHI. 1988. J. Virol. 63: 2504–2509.
10. SUZUKI, M., N. YAMAMOTO, F. SHINOZAKI, K. SHIMADA, G. SOMA & N. KOBAYASHI. 1989. Lancet 1: 1206–1207.
11. FOLKS, T. M., K. A. CLOUSE, J. JUSTEMENT, A. RABSON, E. DUH, J. H. KEHRL & A. S. FAUCI. 1989. Proc. Natl. Acad. Sci. USA 86: 1365–1368.
12. RIGHI, M., L. MORI, G. DE LIBERO, M. SIRONI, A. BIONDI, A. MANTOVANI, S. D. DONINI & P. RICARDI-CASTAGNOLI. 1989. Eur. J. Immunol. 19: 1443–1448.
13. SELMAJ, K. W. & C. S. RAINE. 1988. Ann. Neurology 23: 339–346
14. MINTZ, M., R. RAPAPORT & J. M. OLESKE. 1989. Am. J. Dis. Child 143: 771–774.
15. MERRILL, J. E. & I. S. Y. CHEN. 1991. FASEB J. 5: 2391–2397.
16. SMITH, T. W., U. DEGIROLANI, D. HENIN, F. BOLGERT & J. J. HAUW. 1990. J. Neuropathol. Exp. Neurol. 49: 357–370.
17. DICKSON, D. W., A. L. BELMAN, Y. D. PARK, C. WILEY, D. S. HOROUPIAN, J. LLENA, K. KURE, W. D. LYMAN, R. MORECKI, S. MITSUDO & S. CHO. 1989. APMIS Suppl. 8: 40–47.
18. KETZLER, S., S. WEIS, H. HAIG & H. BUDKA. 1990. Acta Neuropathol. 80: 92–94.
19. BRENNEMAN, D. E., G. L. WESTBROOK, S. P. FITZGERALD, D. L. ENNIST, K. L. ELKINS, M. R. RUFF & C. B. PERT. 1988. Nature (London) 335: 639–642.
20. DREYER, E. B., P. K. KAISER, J. T. OFFERMANN & S. A. LIPTON. 1990. Science 248: 364–367.
21. PORTEGIES, P., J. DEGANS, J. M. A. LANGE, M. M. A. DEUX, H. SPEELMAN, M. BAKKER, S. A. DANNER & J. GOUDSMIT. 1989. Br. Med. J. 299: 819–821.
22. MERRILL, J. E., Y. KOYANAGI & I. S. Y. CHEN. 1989. J. Virol. 63: 4404–4408.
23. VYAKARNAM, A., J. MCKEATING, A. MEAGER & P. C. BEVERLY. 1990. AIDS 4: 21–27.

24. Hofman, F. M., D. R. Hinton, K. Johnson & J. E. Merrill. 1989. J. Exp. Med. 170: 607–612.
25. Grau, G. E., L. F. Fajardo, P. F. Pigue, B. Allet, P. H. Lambert & P. Vassalli. 1987. Science 237: 1210–1212.
26. Kriegler, M., C. Perez, K. DeFay, I. Albert & S. D. Lu. 1988. 53: 45–53.
27. Smith, R. A. & C. Baglioni. 1987. J. Biol. Chem. 262: 6951–6951.
28. Aggarwal, B. B. 1991. Biotherapy 3: 113–120.
29. Jones, E. Y., D. I. Stuart & N. P. C. Walker. 1989. Nature (London) 338: 225–228.
30. Eck, M. J. & S. R. Sprang. 1989. J. Biol. Chem. 264: 17595–17605.
31. Smith, C. A., T. Davis, D. Anderson, L. Solam, M. P. Beckmann, R. Jerzy, S. K. Dower, D. Cossman & R. G. Goodwin. 1990. Science 248: 1019–1023.
32. Tracey, K. J., S. F. Lowry, B. Beutler, A. Cerami, J. D. Albert & G. T. Shires. 1986. J. Exp. Med. 164: 1368–1373.
33. Baldwin, R. L., P. M. Chang, J. Bramhall, S. Graves, B. Bonavida & B. Wisnieski. 1988. J. Immol. 141: 2352–2357.
34. Yoshimura, T. & S. Sone. 1987. J. Biol. Chem. 262: 4597–4601.
35. Kagan, B. L. & Y. Sokolov. 1993. Methods Enzymol. (In press.)
36. Schindler, II. 1980. FEBS Lett. 122: 77–79.
37. Labarca, P., R. Coronado & C. Miller. 1980. J. Gen. Physiol. 76: 397–424.
38. Kagawa, Y., A. Kondrach & E. Racker. 1973. J. Biol. Chem. 248: 51–68.
39. Kagan, B. L., R. L. Baldwin, D. Munoz & B. J. Wisnieski. 1992. Science 255: 1427–1430.
40. Oku, N., R. Araki, H. Araki, S. Shibamoto, F. Ito, T. Nishihawa & M. Tsujimoto. 1987. J. Biochem. 102: 1303–1310.
41. Yoshimura, T. & S. Sone. 1990. Biochem. Int. 20: 697–705.
42. Young, J. D. 1988. Immunol. Lett. 19: 287–292.
43. Kagan, B. L., A. Finkelstein & M. Colombini. 1981. Proc. Natl. Acad. Sci. USA 778: 4950–4954.
44. Ohsawa, F. & S. Natori. 1988. J. Biochem. 103: 730–734.
45. Niitsu, Y., N. Watanabe, H. Sone, H. Neda, N. Yamauchi & I. Urushizaki. 1985. Jpn. J. Cancer Res. 76: 1193–1197.
46. Chang, M. P. & B. J. Wisnieski. 1990. Infect. Immun. 58: 2644–2650.
47. Schmid, D. S., R. Hornung, K. M. McGrath, N. Paul & N. H. Ruddle. 1987. Lymphokine Res. 6: 195–202.
48. Cunningham, Jr., J. N., N. W. Carter, F. C. Rector & D. W. Seldin. 1971. J. Clin. Invest. 50: 50–59.
49. Lancaster, Jr., J. R., S. M. Laster & L. R. Gooding. 1989. FEBS Lett. 248: 169–174.
50. Schein, S. J., B. L. Kagan & A. Finkelstein. 1978. Nature (London) 276: 154–163.
51. Kagan, B. L. 1983. Nature 302: 709–711.

Biophysical Aspect of Information Flow from Receptor to Channel

TOHRU YOSHIOKA, HIROKO INOUE,
KAZUHISA ICHIKAWA,[a] MANABU SAKAKIBARA,[b]
MOTOYUKI TSUDA,[c] AND HIDEO SUZUKI[d]

School of Human Sciences
Waseda University
2-579-15, Mikajima, Tokorozawa
Saitama 359

[a]*Fundamental Research Laboratory*
Fuji Xerox
Tokyo

[b]*School of Developmental Engineering*
Tokai University

[c]*School of Life Science*
Himeji Institute of Technology
Hyogo

[d]*School of Science and Engineering*
Waseda University
Tokyo, Japan

INTRODUCTION

Several years ago, molecular biology clarified the amino acid sequences of rhodopsin, transducin, phosphodiesterase, and ion channels involved in phototransduction of vertebrate photoreceptors.[1] According to the accumulated data, we can visualize how the received signal is transmitted from rhodopsin molecule to ion channels. The photosignal is initially received in the 11-*cis* retinal molecule, which binds basically to Lys-306 in the seventh helix of opsin. The helix VII is considered a principal region for photoreception along with helix II. The photoisomerization of retinal induces isomerization of X-pro, which is involved unusually in the helix structure. This isomerization will induce a large change in domain structure, which is composed of three cytosolic loops and some peptides involved in the C-terminal. This structural change will activate the G-protein. When the G-protein is associated with an activated rhodopsin, the binding constant of this protein to GDP is lowered and its affinity to GTP is increased. Thus

GTP-GDP exchange interaction occurs on the G-protein molecule. When GTP binds to $G_t\alpha$ the complex is divided in two. One part is GTP-$G_t\alpha$ and the other is $G_t\beta\gamma$, thought to remain in the membrane. GTP-$G_t\alpha$ becomes free from the membrane and can activate phosphodiesterase (PDE), which is composed of three subunits (P_α, P_β, and P_γ). It is estimated that an activated single rhodopsin can produce more than 500 GTP-$G_t\alpha$. The signal-carrying complex molecule GTP-$G_t\alpha$ will bind to P_γ (11 kD) of PDE and remove this smallest subunit from the others. By this process, PDE is activated and it hydrolyzes about 500 cGMPs to 5-GMPs. The hydrolyzation of cGMP will close ion channels and a hyperpolarized response is generated. FIGURE 1 illustrates this admittedly oversimplified model for information flow of vertebrate photoreceptor. This model does have some problems to be solved. One problem is how a large protein molecule such as $G_t\alpha$ could move freely in the cytosol and hit exactly a target molecule, $P\gamma$. This is very unlikely. Another problem is more serious because this model uses a negative second messenger, whose diffusion constant may be far less than that of a positive messenger. If we use a negative messenger system, we can not explain sufficiently the time resolution of photoresponse. Therefore there is still a possibility that the model will change again.

In the case of the photoreceptor system of invertebrate signal, the mechanism might be more complicated. In 1985, we found that phospholipase C, which hydrolyzes phosphatidylinositol (PI) and phosphatidylinositol 4,5-bisphosphate (PIP$_2$), is essential for photodetection in *Drosophila* eye when using a visual transduction mutant.[2,3] Alan Fein and his collaborators, however, found that inositol 1,4,5-trisphosphate (IP$_3$), which is a hydrolyzed product of PIP$_2$, is not involved in the phototransduction process.[4] Many researchers tried to solve this contradiction but failed. Furthermore, the response time of the invertebrate photoreceptor (for example, *Drosophila*) is estimated at less than several ten μsec (unpublished data). To make clear the complicated phenomena, we focused our attention on the characteristics of PIP$_2$ molecules and developed a new model to explain these contradictory results. The situation described above tells us that a totally different approach is needed to understand the actual mechanism of signal transmission in the invertebrate visual cell. Before proposing a new model of signal transduction in invertebrate photoreceptor, we would like to introduce "entropy," a fundamental concept to help consider information transmission from receptor to ion channel.

THE GENERAL THEORY FOR SIGNAL TRANSMISSION IN BIOLOGICAL SYSTEMS

It has been believed that any type of signal transmission can be explained by information theory, which was developed about 20 years ago. This established concept was successfully applied to genetics and neuroscience. In the case of neuroscience, the information theory was applied to the central nervous system (CNS) to analyze the information capacity of nerve fibers, information processing in CNS, and information transmission from endolympha to basilar membrane.[5] The concept can be expanded to determine the role of macromolecules in information transmission inside a cell. Here we would like to propose an information transmission model at the cellular level, focusing on the flow of entropy.

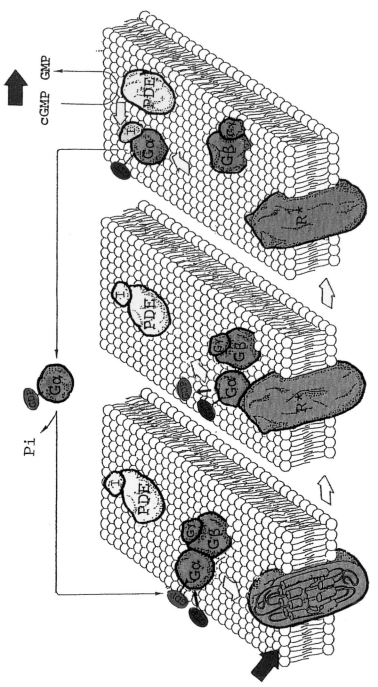

FIGURE 1. Visualized phototransduction mechanism for vertebrate photoreceptor. Gα, Gβ, and Gγ are subunits of transducin molecule. R; rhodopsin, I; γ subunit of phosphodiesterase.

The Relationships between Information and Entropy

In 1949, Shannon proposed a new formula describing relationships between the information quantity of a sentence (I) and the appearance probability of letters (P),[6]

$$I = -\log P. \tag{1}$$

This equation corresponds to the definition of entropy

$$S = -k\log P + \text{Const.} \tag{2}$$

where k is Boltzman's constant and P is probability of general definition. On the basis of the results accumulated by many researchers, the close correlation between entropy and information transmitted in biological systems can be understood by two theorems:

(*1*) According to the information theory and the Boltzman formula for entropy, a quantity of information of the molecular system in a physical unit represents an entropy reduction in the system caused by the receiving of the information.

(*2*) According to the second law of thermodynamics, entropy production in the process of receiving information always exceeds the entropy reduction mentioned above.

Thus we can say at present that information and entropy belong to the same category. Now we have established the theoretical base to use entropy instead of information. In the following section we will discuss entropy flow inside the cell.

Entropy Flow in the Cell

Firstly, we would like to point out that the cell has a metabolic system for entropy. Therefore, emitted entropy (S_{out}) produced at the time of signal reception always exceeds entropy absorbed inside of the cell (S_{int}). Therefore, the thermodynamical state of the receptor at the time of signal reception might be "negative entropy." Secondly, we would like to indicate that entropy is transmitted unidirectionally. In this case, the signal is propagated from one molecule to another one by conformational change of the molecule, just like a wave. Attention should be paid to the fact that entropy is not carried by moving molecules, but by the change of molecular state (FIG. 2).

METHOD OF SIGNAL TRANSMISSION BY ENTROPY CHANGE

As mentioned above, the received signal was transmitted from receptor to ion channel by the conformational change of signal molecules, as if it were a wave. Three different methods are remarkable for such entropy-type signal transmission: conformational, membranous, and concentration methods.

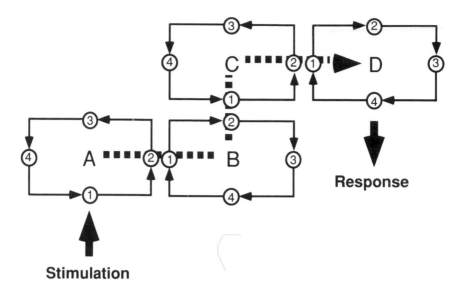

Stimulation

FIGURE 2. Signal flow from molecule A to molecule D. One, two, three, and four show different entropy states of the molecule, respectively. Stimulated signal is propagated from A→D.

Signal Transmission by Conformational Change in Protein Molecules

As is well known in thermodynamics, microscopic expression of entropy is given as

$$S = K\ln Z, \tag{3}$$

where K is Boltzmann constant and Z is the number of microscopic molecular states. If we consider a single protein molecule and compare solid and liquid states, we can show that entropy of protein in a solid state is greater than that in a liquid state:

$$\Delta S = S_{liquid} - S_{solid}$$
$$= R\ln(Z_{liq}/Z_{sol}) > 0, \tag{4}$$

where R is gas constant. When the protein molecule is dissolved and amino acid residues are moving freely in the solution, and if we assume each amino acid has 36 different configurations, the value of entropy of an amino acid is calculated as $S = R\ln 36 = 7.2$ e.u. (e.u. = cal/mol·deg). Actually in the native state, the amino acid may have only two different states, *cis* and *trans*, here the value of entropy might be $S = R\ln 2 = 1.38$ e.u. In the case of phosphorylation and dephosphorylation, we have considered two amino acid residues, serine and threonine. Then we get $S = R\ln 4 = 2.75$ e.u. Thus, protein can transmit entropy by conformational change.

The Entropy Transmission by Change in Membrane Structure

When phospholipase C hydrolyzes the acidic phospholipid molecule PIP_2 in the inner leaflet of the membrane, negative charges associated with phosphate group are removed from membrane surface in the form of IP_3 and drastic disturbance of membrane potential will occur.[7] This disturbance will be propagated on the membrane as if it were a spin wave in magnetic materials. This spin wave–like information wave packet may always have large decaying time constant, so that the disturbance can not propagate for a long distance.

The Entropy Transmission by Concentration Change

The second messenger system can also be expressed using entropy theory. When cAMP is used as a second messenger in a cell system, the concentration change might be twofold. The entropy change must be $S = -G/T = -(-RT\ln2)/T = R\ln2 = 1.38$ e.u., where G is free energy given as

$$G = RT\ln \text{[concentration of second messenger before stimulation/} \\ \text{concentration of second messenger after stimulation].} \qquad (5)$$

If Ca is used as a second messenger, the concentration change of Ca by the stimulation must be generally around tenfold. Then the entropy change will be $S = R\ln10 = 4.35$ e.u. Therefore, it is concluded that Ca is a more efficient second messenger than cyclic nucleotide.

ONE-WAY DIRECTIONAL INFORMATION TRANSMISSION FROM RECEPTOR TO CHANNEL

If a molecule is involved in information transmission, the molecule will change its thermodynamical state (FIG. 3). This is referred to as an entropy cycle. In this figure, the entropy change in process ① will be written as

$$\Delta S_1 = S_2 - S_1 = S(\chi_2, Z_1) - S(\chi_1, Z_1), \qquad (6)$$

when variable x or z changed, the entropy change by this process is expressed as

$$\Delta S_x = S_1 + \Delta S_3$$
$$\Delta S_z = S_2 + \Delta S_4$$

The order of each entropy is $\Delta S_x \leq \Delta S_z$. When $\Delta S_x \geq 0$, the entropy is transmitted freely in one direction.

If we accept the idea that a signal is propagated in the living cell in the form of entropy, heat production or temperature change must be associated with the transmembrane signaling process, because signal reception may cause an entropy change in the receptor system. Simultaneous changes in mechanical structure of signaling molecules

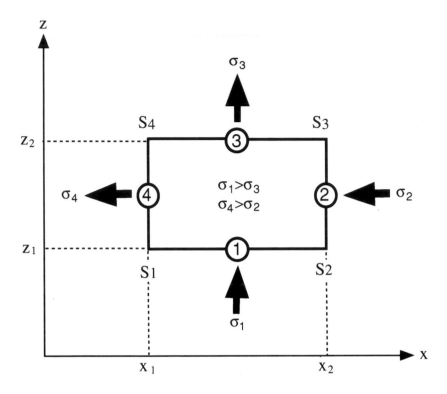

FIGURE 3. Cyclic change of entropy of signal-transducing molecule. S, entropy of the molecular system; σ, inlet or outlet of entropy; x and z variable describing molecular state.

and in concentration of the second messenger and membranous physical wave generation will be discussed in the following sections, independently.

Heat Production in Squid Retina in Response to Light

In 1985, Tasaki and Nakaye reported that exposure to a brief light pulse evoked a rapid increase in the temperature of the dark-adapted squid retina. The amount of heat generation was far greater than that associated with the stimulating pulse itself.[8] Using a pyroelectric heat detector made with polyvinylidene fluoride (PVDF) film, they observed the thermal response of a slice of the dark-adapted squid retina. With a brief pulse of light, a fairly rapid increase in temperature was observed approximately 20 msec after the onset of light pulse at 20°C. The rate of temperature increase reached maximum value about 90 msec after the onset. These thermal responses, however, disappeared when glucose was removed from the incubation medium or when an oxygen-utilization inhibitor, azide or cyanide, was added to it. These results suggest that ATP and/or phosphorylation is essential for heat production. The reason for the ATP requirement in the phototransduction process will be understood later by the consideration of the importance of the PI cycle in the retina. When PI turnover is blocked by the addition of any kind of inhibitor, octopus retina could not generate receptor potential.

Involvement of Phospholipase C for the Drosophila Photoresponse

Involvement of phospholipase C (PLC) in the *Drosophila* photoresponse was established in our laboratory using a visual mutant, *norpA*.[2,3,9] The PLC was localized in *Drosophila* retina and hydrolyzed PIP_2 as well as PI. The enzyme was associated with the plasma membrane and it was sensitive to pH and Ca. A degree of defect in PLC activity was found to be in parallel with the size of electroretinogram (ERG), which corresponds to receptor potential, especially in fly.[10] This parallelism was confirmed using a temperature-sensitive allele of *norpA* mutant, *K050*.[3] *K050* showed normal photosensitivity when kept at 18 °C, but the sensitivity was lost at 28 °C or more. If the involvement of PLC in phototransduction is correct, PLC activities in *K050* alleles kept at different temperature must show different enzyme activities. This prediction was completely verified.

Initially our findings were thought to support the idea that IP_3 might be a second messenger in the invertebrate photoreceptor. Actually Szuts *et al.* measured IP_3 formation on a rapid time scale in squid retina.[11] Before the findings on the concentration change in IP_3 by light flashes, Fein *et al.* reported that IP_3 injected into *Limulus* ventral eye photoreceptor cell, R-lobe, showed the production of a discrete wave of depolarization and a burst of waves, which have a similar waveform to the quantum bumps that are evoked by a single photon. These findings suggested that excitation and adaptation by injected IP_3 are mediated by a rise in intracellular Ca.[4] However, the idea that IP_3-induced intracellular Ca release causes visual excitation is problematic. The dilemma is that whereas EGTA blocks excitation by IP_3, it does not block the light response. Therefore, another as-yet-unidentified transmission system should be considered in the molecular mechanism of the invertebrate photoreceptor.

PIP₂ Breakdown May Cause Photoresponse in Hermissenda Photoreceptor B Cell

Considering the above description, the following two contradictory results may be true for phototransduction mechanism in invertebrate photoreceptors. (*1*) PIP_2-specific PLC is necessary for the production of photoreceptor potential. (*2*) IP_3, the hydrolyzed product of PIP_2, is not necessary for the generation of photoreceptor potential.

To construct a reasonable model that can explain the above description, we would like to demonstrate a novel idea: destruction of ionic equilibrium across membrane or the abrupt change in membrane potential. In general, phospholipid composition of the inner leaflet of the membrane is a mixture of PI, PS, PE, and PIP_2. Although the content of PIP_2 in the membrane is very low (less than 1%), the effect of IP_3 release from the membrane should be quite large, because IP_3 has five negative charges in neutral pH. If we neutralize these negative charges with positively charged material, nothing will happen by photostimulation. To confirm this idea we injected neomycin, which belongs to the aminoglycoside complex, and spermine, belonging to polyamine,

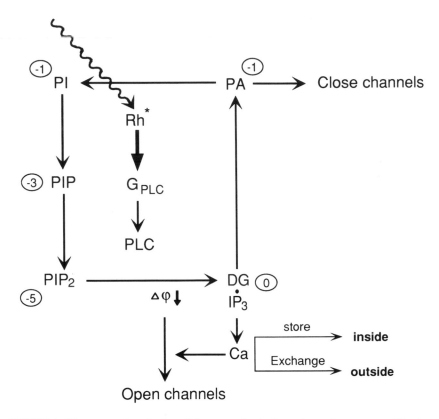

FIGURE 4. The proposed mechanism of photoreception by invertebrate photoreceptor. PA, phosphatidic acid; PI, phosphatidylinositol; PIP, phosphatidylinositol 4-monophosphate; PIP$_2$ phosphatidylinositol 4,5-bisphosphate; DG, diacylglycerol; IP$_3$, inositol 1,4,5-trisphosphate; $\Delta\varphi$, charge in membrane potential. Circled number represents valence of phospholipid.

into *Hermissenda* photoreceptor B cell.[12,13] For comparison, isobutylmethylxanthine (IBMX), which is known to suppress PI cycle,[14] and manoalide, which is an inhibitor to PLC activity, were also examined. When IBMX and neomycin was injected independently, photoresponse was almost suppressed and recovered gradually. The spermine-injected receptor showed partial suppression in photoresponse. When we examined manoalide, results were found to be inconsistent: occasionally the response was reduced largely, but sometimes the responses were increased, conversely. To explain these physiological data, we examined the effect of manoalide on PLC activity biochemically. It was found that the effect of manoalide and its concentration had an inverse relationship. The order of the effectiveness of the inhibitor was found to be neomycin > spermine > IBMX > manoalide. These results suggest that membrane potential should be changed totally by the release of a negative charge from the inner surface of the membrane. The voltage-sensitive ion channel can thus be opened by the reduction of membrane potential. The diagram of the mechanism is shown in FIGURE 4.

CONCLUSION

By introducing the physical concept of entropy, we proposed a novel model of phototransduction for invertebrate photoreception and demonstrated it experimentally. What is the advantage of the introduction of entropy? Firstly, we can point out that entropy can be propagated unidirectionally. Secondly, entropy is transmitted three different ways: conformational change of protein molecules, structural disturbance in membrane (phospholipid) molecules, and concentration change in second messengers. The conformational change in protein molecules is the fastest, while the second messenger system is the slowest (diffusion constant of second messenger is estimated as $D = 7 \times 10^{-6}$ cm^2/sec). We can select the most plausible model among these three models based on the response time.

The greatest advantage of the introduction of entropy is that we could establish the physiological significance of the measurement of heat production associated with the signal reception. According to the law of entropy, heat generation coupled with signal reception can be predicted theoretically. However, no one succeeded, except Tasaki and Nakaye, because of the poor time resolution and sensitivity of thermodetectors. Now we are ready to introduce the entropy concept for photoreception as well as signal reception in general.

It is likely that signal transduction thermodynamics will be a major step after molecular cloning of signaling molecules.

REFERENCES

1. Hara, T., Ed. 1988. Molecular Physiology of Retinal Protein. Yamada Science Foundation. Osaka.
2. Yoshioka, T., H. Inoue & Y. Hotta. 1985. J. Biochem. **97**: 1251–1254.
3. Inoue, H., T. Yoshioka & Y. Hotta. 1985. Biochem. Biophys. Res. Commun. **132**: 513–519.
4. Fein, A., R. Payne, S. W. Corson, M. J. Berridge & R. F. Irvine. 1984. Nature **311**: 157–160.
5. Halzmüller, W. 1981. Information in Biological System: The Role of Macromolecules. Cambridge University Press. Cambridge.
6. Shannon, C. E. & W. Weaver. 1949. The Mathematical Theory of Communication. University of Illinois.
7. Yoshioka, T. & H. Suzuki. 1989. Biosignal Transduction Mechanism: 12–13. Springer-Verlag. Tokyo.
8. Tasaki, T. & T. Nakaye. 1985. Science **227**: 654–655.
9. Toyoshima, S., N. Matsumoto, P. Wang, H. Inoue, T. Yoshioka, Y. Hotta & T. Oosawa. 1990. J. Biol. Chem. **265**: 14842–14848.
10. Hotta, Y. & S. Benzer. 1970. Proc. Natl. Acad. Sci. USA **67**: 1156–1163.
11. Szuts, E. Z., S. F. Wood, M. J. Reid & R. F. Irvine. 1986. Biochem. J. **240**: 929–932.
12. Sokabe, M., J. Hayase & K. Miyamoto. 1982. Proc. Jap. Acad. **58B**: 177–180.
13. Nomura, K., K. Haruse, K. Watanabe & M. Sokabe. 1990. J. Membr. Biol. **115**: 241–251.
14. Yoshioka, T., H. Inoue, M. Takagi, F. Hayashi & D. Amakawa. 1983. Biochim. Biophys. Acta **755**: 50–55.

A Sodium Channel Model

CHIKARA SATO AND GEN MATSUMOTO

Electrotechnical Laboratory
Tsukuba, Ibaraki 305, Japan

The complete amino acid sequence of an invertebrate sodium channel was previously determined by cloning and sequence analysis of the complementary DNA of the squid *Loligo bleekeri*.[1] This, together with other findings obtained for vertebrate sodium channels, has made it possible for us to elucidate the structural organization of the sodium channel with respect to its functions of voltage sensing, activation, ion selectivity, and inactivation.[2] This could be carried out partly because the sodium channel was well studied in the electric organ of the eel *Electrophorus electricus* and the rat brain (see papers in this book by Catterall, Noda and Imoto for review), but also because the squid sodium channel, as compared with vertebrate sodium channels, retains a rather short sequence length of 1,522 residues, corresponding to approximately three-fourths of the sequence length for rat sodium channels. In spite of their simplicity, amino acid sequences for the squid sodium channel closely resemble those for vertebrate sodium channels, especially for the segments of membrane-spanning portions S2, S3, S4, and of the linker between S5 and S6 (S5–6 region).[2] Furthermore, we adopted the assumption that all transmembrane segments S1–S6 form 3_{10}-helices, not α-helices. In the 3_{10}-helix S4 segments, the charged side chains are clustered largely on one side of the helix. Together with this assumption, the charged residue configuration in S1 to S6 segments for the respective domains led us to the conclusion that the octagonal structure illustrated in FIGURE 1 (a and b) correctly represents a sodium channel structure in which the S5–6 regions stably interact with the inner surface of the core pore formed by the S4 and S2 segments. In the resting membrane state, the tips of S5–6 remain at the positively charged sites of S4 nearest the positions where the negative charge residues of S2 are located (FIG. 1, c and d). In our sodium channel model, the S5–6 region serves the three principal functions of ion selectivity, voltage sensing, and activation. The S5–6 regions, negatively charged as a whole, are capable of sensing membrane potential, resulting in voltage-sensor function. The negatively charged sites, together with the size of the pore formed by the four S5–6 regions, enable selection of ions passing through the S5–6 pore.[2] When the membrane is depolarized, S5–6 can slide through the guiding pore formed by S4 and S2 to the cytoplasmic side (FIG. 2,c). This sliding is energetically favored by linearly aligned, positively charged sites of S4s (FIG. 1,f). In the resting state, the guiding pore is kept closed by the C-terminal segment from the cytoplasmic side (FIG 1,c–e; FIG. 2,a). When the tips of the S5–6 regions approach the cytoplasmic side upon depolarization, the C-terminal is repelled by the electrostatic interaction between the tips of the S5–6 pore and the C-terminal; the C-terminal is also directly repelled by the depolarized potential. In this way, the

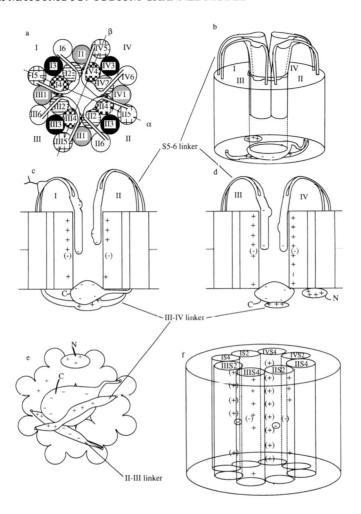

FIGURE 1. Schematic illustrations of our sodium channel model. The octagonal topology of trans-membrane segments S1–S6 and the S5–6 regions for the domains I–IV is shown from the extracellular side (a) and from the transmembrane side (b). Note that the domains form a circle in the octagonal structure in the sequence of I→IV→II→III instead of that of I→II→III→IV proposed by Noda *et al.* in 1984 (Nature 312: 121–127). Resting configurations of the S5–6 regions and C- and N-terminals are also illustrated as cross sections along line α (c) and line β (d), where segments (I and II) and (III and IV) are viewed, respectively. Resting configurations of the II–III and III–IV linkers, and N- and C-terminal regions are shown from the cytoplasmic side (e). Positive and negative charges of the S4 and S2 segments, respectively, are exposed on the inner surface of the guiding pore (see text) as shown in (f).

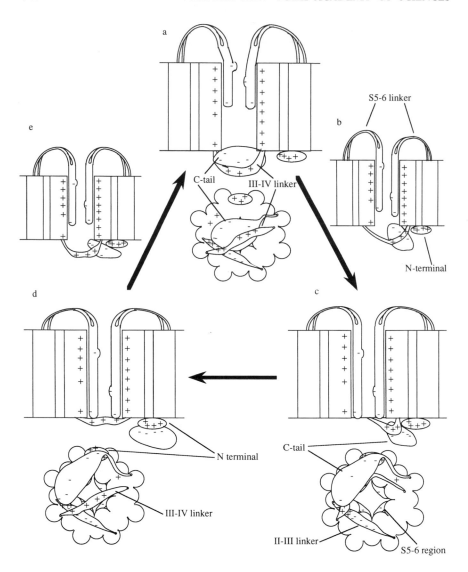

FIGURE 2. Schematic illustrations of the proposed tertiary structure model of the sodium channel corresponding to the resting (a), activated (c), and inactivated (d) states. The *upper* and *lower* pictures in (a), (c) and (d), represent the transmembrane and intracellular side views of the channel, respectively. The illustrations (b) and (e) represent transitional configurations between (a) and (c) and between (d) and (a), respectively.

sodium channel is activated (FIG. 2,c). The III–IV linker, which has been understood to play a crucial role in inactivation (see Catterall in this volume for review), covers the C-terminal from the cytoplasmic side in the resting state (FIG. 1,c and d; FIG. 2,a). Synchronously with the activation, the III–IV linker moves away from the channel pore (the S5–6 pore) together with the C-terminal (FIG. 2,c). However, the III–IV linker in this configuration (FIG. 2,c) is unstable, both electrically (since negative charges of the S5–6 tips are exposed on the cytoplasmic side, the tips attract the III–IV linker) as well as elastically, and the linker eventually reverts to its original configuration (FIG. 2,d). Thus, the sodium channel is inactivated.

The present model was developed mainly by determining the configuration with the least electrostatic energy, when we assume that all transmembrane segments form 3_{10}-helices. This model of ion selectivity, voltage sensing, and activation is consistent with other experiments employing site-directed mutagenesis, as reviewed by Noda and Imoto in this volume.

REFERENCES

1. SATO, C. & G. MATSUMOTO. 1992. Biochem. Biophys. Res. Commun. **186:** 61.
2. SATO, C. & G. MATSUMOTO. 1992. Biochem. Biophys. Res. Commun. **186:** 1158.

Identification of a Thr-to-Met Mutation in the Skeletal Muscle Sodium Channel Gene in Hyperkalemic Periodic Paralysis of a Japanese Family[a]

KIICHI ARAHATA,[b,c] JIANZHOU WANG,[d]
W. GREGORY FEERO,[d] HIROSHI HAYAKAWA,[e]
KOICHI HONDA,[f] HIDEO SUGITA,[b]
AND ERIC P. HOFFMAN[d]

[a,b]National Institute of Neuroscience
NCNP
Tokyo, Japan

[d]University of Pittsburgh
Pittsburgh, Pennsylvania

[e]Hitachi General Hospital
Ibaragi, Japan

[f]Yamagata-Ken Chuo Hospital
Yamagata, Japan

INTRODUCTION

Familial hyperkalemic periodic paralysis (HyperPP) is a dominant genetic disease characterized by transient attacks of muscular paralysis. Attacks can be precipitated by rest after exercise or potassium intake. Serum potassium level may elevate during the attacks. A candidate gene approach has been used to define the molecular basis for the human inherited condition. The disease has been found to be caused by point mutations in the adult muscle sodium channel gene on chromosome 17q. It is the

[a] This work was supported by grants from the NCNP of the Ministry of Health and Welfare (Japan; K.A.), the Ministry of Education, Science and Culture (Japan; K.A.), National Institute of Health (U.S.A.; E.P.H.), and the Muscular Dystrophy Association (U.S.A.; E.P.H.).

[c] Address correspondence to: Kiichi Arahata, M.D., Department of Neuromuscular Research, National Institute of Neuroscience, National Center of Neurology and Psychiatry, 4-1-1 Ogawa-higashi, Kodaira, Tokyo 187, Japan.

342

first genetic disease attributable to mutations of voltage-sensitive ion channels.[1,2] We examined two Japanese HyperPP families using a novel application of the ligase chain reaction (LCR) to detect the previously identified point mutations (Thr$_{704}$ to Met and Met$_{1592}$ to Val) in Caucasian HyperPP patients.

CLINICAL CASES

Patient 1

A 19-year-old female (FIG. 1, A) had the first episode of paralytic weakness at 3 years of age, characterized by generalized muscle weakness involving facial and pharyngeal muscles. Myotonia was not noted. Attacks were provoked at rest or sleep after exercises. Excessive sweating occurred prior to the episodes and serum potassium was elevated (TABLE 1). Acetazolamide treatment reduced the severity and frequency of the attack. The symptoms get worse before menstruation. Muscle biopsy showed myopathic changes with tubular aggregates. Her father was similarly affected since 12 years of age. He still feels slowness of movement after prolonged sitting on the floor. The patients had no thyroid dysfunction.

Patient 2

A 30-year-old man (FIG. 1, B). He noticed first paralytic attack at 16 years of age, characterized as generalized muscle weakness and dysesthesia in hands and feet. Serum potassium level was elevated during the attack (TABLE 1). He had grip myotonia and myotonic repetitive discharges. An attack is provoked by exposure to cold, rest after exercise, fasting, and oral administration of 5 g KCl. He was clinically normal between

(A) Family of Patient 1.

(B) Family of Patient 2.

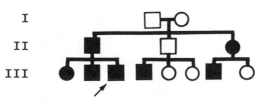

FIGURE 1. Pedigree of families.

TABLE 1. Clinical Data

Patient	Sex and Age (yr)	Duration of Symptoms	Attacks (per month)		Serum Potassium Levels (mEq/L)		Provocative Stimuli	Therapy	Sodium Channel Mutation
			Frequency	Duration	Ictal	Interictal			
1	F,19	16 yr	6	1–8 h	$4.5 \sim 5.7$	3.8	Menstruation, Morning	Acetazolamide	Thr_{704} to Met
2	M,30	14 yr	10	1 h	5.3	3.8	Morning, Cold, Exercise	Acetazolamide	Not Found

attacks. Acetazolamide treatment abolished the attack. Thyroid function was normal. Six other family members were similarly affected (FIG. 1, B).

DNA ISOLATION AND LCR ANALYSIS

DNA was isolated from whole blood. PCR and LCR (ligase chain reaction) primers flanking the target regions were synthesized from the cDNA sequence for the skeletal muscle sodium channel alpha subunit, and LCR analysis (20 cycles at 94°C for 1 min, then 65°C for 4 min) was done as described.[3]

RESULTS AND DISCUSSION

Patients in the first HyperPP family had the predicted Thr$_{704}$-to-Met mutation (C-T transition) of the skeletal muscle sodium channel alpha subunit, but no mutation was identified in the second Japanese family. Although both Japanese families had characteristic clinical features that can be observed in "adynamia episodica hereditadia" or primary HyperPP, Patient 2 in the second family had atypical findings with sensory abnormality during the attack, and the attack was provocative by cold exposure. Thus both genotype/phenotype heterogeneity exist in Japanese HyperPP families.

ACKNOWLEDGMENT

We are indebted to Ms. Kanako Goto for her expert technical assistance.

REFERENCES

1. Rojas, C. V., J. Z. Wang, L. S. Schwartz, E. P. Hoffman, B. R. Powell & R. H. Brown, Jr. 1991. A Met-to-Val mutation in the skeletal muscle Na+ channel alpha-subunit in hyperkalaemic periodic paralysis. Nature 354: 387–389.
2. Ptacek, L. J., A. L. George, R. C. Griggs, R. Tawil, R. G. Kallen, R. L. Barch, M. Robertson & M. F. Leppert. 1991. Identification of a mutation in the gene causing hyperkalaemic periodic paralysis. Cell 67: 1021–1027.
3. Feero, W. G., J. Wang, F. Barany, J. Zhou, S. M. Todorovic, R. Conwit, G. Galloway, I. Hausmanowa-Petrusewicz, A. Fidzianska, K. Arahata, H. B. Wessel, C. Wadelius, H. G. Marks, P. Hartlage, H. Hayakawa & E. P. Hoffman. 1993. Hyperkalemic periodic paralysis: Rapid molecular diagnosis and relationship of genotype to phenotype in 12 families. Neurology 43: 668–673.

Cloning and Characterization of Sodium Channel cDNA from Puffer Fish

MOHAMMED SHAHJAHAN, MAMORU YAMADA,
MANABU NAGAYA, MOTOHARU KAWAI,
AND ATSUSHI NAKAZAWA[a]

Department of Biochemistry
Yamaguchi University School of Medicine
Ube, Yamaguchi 755, Japan

Tetrodotoxin (TTX) selectively blocks sodium channel in excitable membranes. The puffer fish is thought to have TXX-resistant sodium channels, resulting in tolerance to the toxin in spite of its high accumulation in the tissues.[1] To clarify the exact nature of the sodium channel of puffer fish, we cloned the channel cDNA from *Fugu rubripes rubripes* (Tora Fugu in Japanese).

cDNA library was constructed in λgt10 vector using poly(A)$^+$ RNA isolated from the fish brain. Twelve partial cDNA clones were isolated, which were classified into three types (pfBNa1, pfBNa2, and pfBNa3) after compilation of the sequences. pfBNa2 directed the C-terminal 1717 residues of the channel peptide, while pfBNa1 encoded the 530-residue sequence corresponding to the N-terminal portion of the peptide. Overlapping sequences of these two peptides differed from each other in the connecting loop between putative membrane-spanning segments 5 and 6 of the repeat I among four homologous repeats in the channel peptide[1] (FIG. 1). pfBNa2 lacked a 14-residue

---

FIGURE 1. Comparison of amino acid sequences for the regions between putative membrane-spanning segments 5 and 6 of four internal repeats in various sodium channels. [I], repeat I; [II], repeat II; [III], repeat III; [IV] repeat IV. pfBNa1, puffer fish brain channel I; pfBNa2, puffer fish brain channel II; hHNa1, human heart TTX-insensitive channel[3]; rMNa2, rat heart and skeletal muscle TTX-insensitive channel[2]; rMNa1, rat skeletal muscle TTX-sensitive channel[3]; rBNa1, rat brain TTX-sensitive channel I[6]; rBNa2, rat brain TTX-sensitive channel II[6]; rBNa3, rat brain TTX-sensitive channel III[7]. Sets of identical residues are enclosed with solid lines. Gaps (– – –) have been inserted to achieve maximal homology. The membrane-spanning segments 5 and 6 in each repeat, I to IV are indicated below the alignments. Short segmental domains, SS1 and SS2, in each connecting loop between the segments 5 and 6 are also indicated. The 14-residue deletion found in the TTX-insensitive sodium channels is shown above the alignments.

[a] To whom correspondence should be addressed.

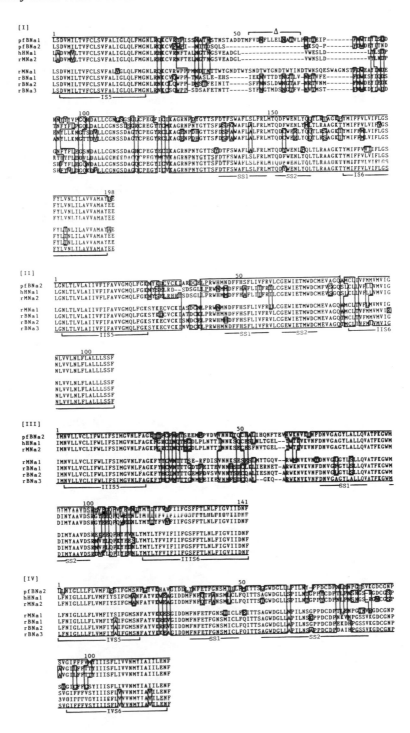

segment in the N-terminal part of the connecting loop, which is expected to extend outward from the membrane surface. The reported TTX-insensitive sodium channels of rat skeletal muscle[2] and human heart[3] also had deletions in the same position. The TTX-binding site was predicted from the site-directed mutagenesis studies[4,5] to lie in the seven-residue segment in the SS2 domain of the connecting loop in the repeat I (TQDCWER in the rat heart TTX-insensitive channel[2]). However, the sequence for the seven residues of pfBNa2 of puffer fish is TQDFWEN, which is the same as those of TTX-sensitive rat brain channels.[6] Therefore, it seems difficult to accept that the SS2 domain is responsible for the TTX resistance.

Northern blot analysis was performed with poly(A)$^+$ RNA isolated from the fish tissue using the cloned cDNA as a probe. The results indicated that the brain contained a mixture of mRNA with lengths ranging from 5,000–7,000 nucleotides. The skeletal muscle, the heart, and the intestine contained two distinct species of mRNA with about 6,000 and 1,500 nucleotides, respectively, while the liver contained only one species with about 6,000 nucleotides.

REFERENCES

1. TRIMMER, J. S. & W. S. AGNEW. 1989. Annu. Rev. Physiol. 51: 401–418.
2. ROGART, R. B., L. L. CRIBBS, L. K. MUGLIA, D. D. KEPHART & M. W. KAISER. 1989. Proc. Natl. Acad. Sci. USA 86: 8170–8174.
3. GELLENS, M. E., A. L. GEORGE, JR., L. CHEN, M. CHAHINE, R. HORN, R. L. BARCHI & R. G. KALLEN. 1992. Proc. Natl. Acad. Sci. USA 89: 554–558.
4. NODA, M., H. SUZUKI, S. NUMA & W. STUHMER. 1989. FEBS Lett. 229: 213–216.
5. SATIN, J., J. W. KYLE, M. CHEN, P. BELL, L. L. CRIBBS, H. FOZZARD & R. B. ROGART. 1992. Science 256: 1202–1205.
6. NODA, M., T. IKEDA, T. KAYANO, H. SUZUKI, H. TAKESHIMA, M. KURASAKI & S. NUMA. 1986. Nature 320: 188–192.
7. KAYANO, T., M. NODA, V. FLOCKERZI, H. TAKAHASHI & S. NUMA. 1988. FEBS Lett. 228: 187–200.

Identification of Ligand-Binding Sites that Form External Mouth of Ion Pore in Calcium and Sodium Channels

HITOSHI NAKAYAMA,[a] YASUMARU HATANAKA,
MOTOHIKO TAKI, EIICHI YOSHIDA,
AND YUICHI KANAOKA[b]

Faculty of Pharmaceutical Sciences
Hokkaido University
Sapporo 060, Japan

Since the primary structures of ion channels including sodium channels, calcium channels, and potassium channels, have been elucidated succeedingly,[1] it is intriguing to identify the structures relevant to their integral channel functions. Utilization of toxins or drugs that act specifically on the corresponding channels is one of the promising approaches to accomplish it.

We specifically photolabeled the sodium channel protein (250 kD) from eel electroplax using a bioactive tetrodotoxin (TTX) derivative possessing a (diazirino)trifluoroethylbenzoyl (DTB) group as a carbene precursor that reacts more efficiently to afford more stable photoproduct(s) than a nitrene-generating azidophenyl group,[2] as similarly shown in the 1,4-dihydropyridine (DHP) receptor–selective photoaffinity probe for calcium channels.[3] Photoaffinity-labeled regions were identified by probing labeled proteolytic fragments with several anti-peptide antibodies recognizing different segments of the sodium channel.[4] One fourth of the label occurred in tryptic fragments between Lys-1478 and Lys-1542 derived from the loop between segments S5 and S6 in repeat IV that had been proposed to be extracellular. One fifth of the labeling, however, was found in the tryptic fragments between Lys-1213 and Arg-1226 as well as Arg-1226 and Arg-1269. The latter fragment apparently contains the transmembrane segment S6 of repeat III. These data could imply that the TTX binding site is formed by close apposition of two discontinuous regions of the sodium channel sequence in repeat III and IV, as the sodium channel has only a single high affinity binding site for TTX. Similar results have been previously observed for the DHP binding site of skeletal muscle calcium channels by photoaffinity labeling[5] with the DHP analog (FIG. 1).

---

[a] Faculty of Pharmaceutical Sciences, Kumamoto University, Kumamoto 862, Japan.
[b] Present Address: Toyama Women's College, Toyama 930-01, Japan.

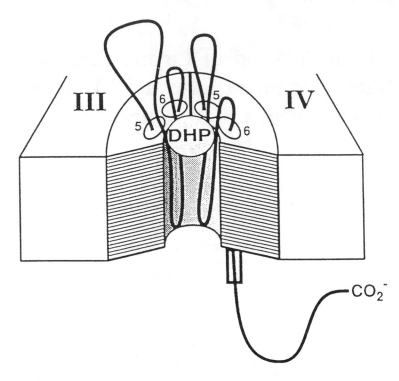

FIGURE 1. A model for dihydropyridine binding on the repeat III and IV of the L-type calcium channel. Note that the proposed *folding* of the segment between transmembrane helices S5 and S6 back into the membrane to form the lining of the transmembrane pore.[5] A similar *folding model* can be drawn for the sodium channel by the photoaffinity labeling results.

It is worthwhile to point out that photolabeled fragments with the TTX derivative were observed in the regions between S5 and S6 of repeat III and IV but not in the region of repeat I. We propose that C-11 of TTX, where the photoreactive DTB group is attached, is favored to orient the repeat III and IV, and the guanidinium group of TTX, which is apparently directed opposite to C-11, orients most likely to the negative charge cluster of the repeat I and/or II[4] (FIG. 2). This may be partly supported by the careful inspection of the recent results of site-directed mutagenesis[6] that mutation of Asp-1426 in the repeat III had no significant effects on the toxin sensitivity, whereas mutation of Lys-1422 to a negatively charged residue like those located at the equivalent position of repeat I and II, strongly reduced the sensitivity. Lys-1213 (Lys-1422 in the rat brain channel II) in repeat III, which is well conserved in all of the cloned sodium channels, is a likely participant in the interaction with the hemiacetal OH group (or more likely with its anionic form, $pK_a = 8.7$, after binding with sodium channels).

It is also worth pointing out that the two labeled regions in repeat III and IV, the connecting loops between S5 and S6 and the transmembrane segment S6, were identified as the binding site for both of TTX and DHP. Considering that TTX and DHP

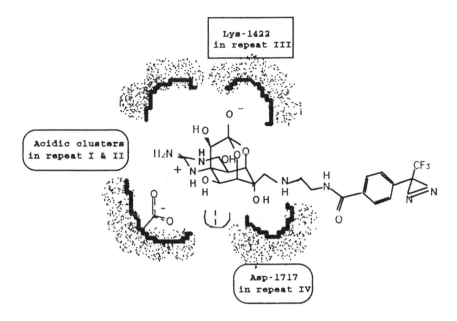

FIGURE 2. A proposed binding model for the photoreactive tetrodotoxin derivative on the sodium channel. Lys-1422 in repeat III is proposed to associate with the hemiacetal O^- of tetrodotoxin and stabilize the binding as well as acidic residues in repeat I, II, and IV.

reach their binding sites from the extracellular side to block the entry of sodium and calcium ions in the respective channel, the results strongly suggest that (*1*) the loop between S5 and S6 of the ion channels folds back into the membrane and contributes to the formation of the transmembrane ion pore and (*2*) the binding sites of two ligands must be located to the external mouth of the ion pore.

REFERENCES

1. CATTERALL, W. A. 1989. Science **242**: 50–61.
2. YOSHIDA, E., H. NAKAYAMA, Y, HATANAKA & Y. KANAOKA. 1990. Chem. Pharm. Bull. **38**: 982–987.
3. TAKI, M., H. NAKAYAMA & Y. KANAOKA. 1991. FEBS Lett. **283**: 259–262.
4. NAKAYAMA, H., Y. HATANAKA, E. YOSHIDA, K. OKA, M. TAKANOHASHI, Y. AMANO & Y. KANAOKA. 1992. Biochem. Biophys. Res. Commun. **184**: 900–907.
5. NAKAYAMA, H., M. TAKI, J. STRIESSNIG, H. GLOSSMANN, W. A. CATTERALL & Y. KANAOKA. 1991. Proc. Natl. Acad. Sci. USA **88**: 9203–9207.
6. TERLAU, H., S. H. HEINEMANN, W. STUEHMER, M. PUSCH, F. CONTI, K. IMOTO & S. NUMA. 1991. FEBS Lett. **293**: 93–96.

Patch-Clamp Study of Developmental Changes in Voltage-Dependent Ion Channels of Mouse Skeletal Muscle Fibers

TOHRU GONOI

Research Center for Pathogenic Fungi and Microbial Toxicoses
Chiba University
Chiba 260, Japan

INTRODUCTION

It is known that acetylcholine receptors of skeletal muscles undergo remarkable changes in their density as well as their properties and distribution in early postnatal days. However, similar postnatal changes have not been well understood for other types of ion channels. Our recent studies of postnatal changes in types of voltage-dependent Na, Ca, and inward-rectifier K channels of mouse skeletal muscle fibers are summarized in this paper. I also suggest a cellular mechanism that controls a part of these changes.

METHODS

Muscle fibers were isolated from M. flexor digitorum brevis of 0 to 30 day-old (P30) mice by collagenase digestion.[1-3] They were cultured in mixture of Dulbecco-Vogt's modified Eagle medium, 10% horse serum, and 5% newborn calf serum. The whole-cell patch-clamp method was used to record ionic currents from freshly isolated and cultured fibers. Compositions of extracellular recording solutions (pH 7.4) were as follows, with millimolar concentrations in parentheses. For recording Ba currents through Ca channels: $Ba(OH)_2$ (30), tetraethylammonium (TEA)-OH (90), TEA-Cl (10), methanesulfonate (140), glucose (30), and 3-(N-morpholino)propanesulfonic acid (MOPS, 10). Na currents: NaCl (5), TEA-Cl (145), KCl (5), $CaCl_2$ (1.5), $MgCl_2$ (1), glucose (5), and MOPS (5). K currents: KOH (20), tris(hydroxymethyl)-aminomethane (135), $Ca(OH)_2$ (1.5), HCl (8), methanesulfonate (150), and MOPS (10). Denervation of muscle fibers was performed by cutting the sciatic nerve under anesthesia.

RESULTS

Ca Channel Currents

Muscle fibers of newborn mice showed two distinct types of Ca channel currents, a low-threshold transient ($I_{Ca,transient}$) and high-threshold sustained currents ($I_{Ca,sustained}$). The mean specific amplitude of $I_{Ca,transient}$ at P1 was -2.7 A/Farad (F; membrane capacitance) at -30 mV test pulses. $I_{Ca,transient}$ decreased progressively in the postnatal days and became undetectable by P17 (FIG. 1). In contrast, the specific amplitude of $I_{Ca,sustained}$ at $+20$ mV test pulses increased fourfold from -6.9 A/F at P1 to -27.7 A/F at P30. Denervating muscle fibers at P8 or P17 did not interfere the disappearing process of $I_{Ca,transient}$. The increase of $I_{Ca,sustained}$ was suppressed by the denervation.

Na Currents

In muscle fibers of newborn mice geographutoxin II (GTX II) distinguished two different types of voltage-sensitive Na channels: GTX II-sensitive and -resistant channels, which corresponded to tetrodotoxin (TTX)-sensitive and -resistant channels, respectively.[4,5] The mean specific Na conductance (g_{Na}) for the total (GTX II-sensitive plus resistant) Na channels was 0.22 mS/µF at 5 mM $[Na]_0$ at P0. The total g_{Na}

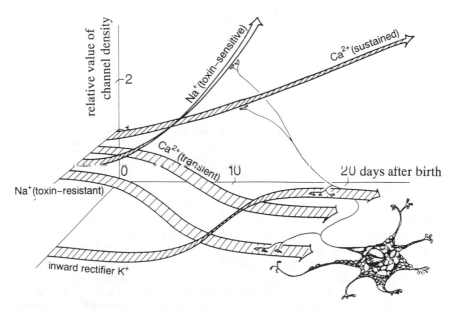

FIGURE 1. Changes in types of voltage-dependent ion channel densities in mouse skeletal muscle fibers during postnatal development. Nerve fibers are schematically drawn to indicate the processes that require innervation for the changes.

increased sixfold during the first 20 postnatal days. The specific g_{Na} for the GTX II-resistant channels was 0.15 mS/μF at P0, decreased progressively to become undetectable by P16.

In muscle fibers denervated at P4 or P12, the GTX II-resistant g_{Na} increased in the following days. The increase of the total g_{Na} was accelerated over the level of normal fibers in P4-denervated fibers, but it was suppressed in P12-denervated fibers.

Inward Rectifier K Currents

Inward rectifier K currents were hardly discernible in fibers acutely isolated from P1 mice. The steady-state component (I_{s-s}) and the slowly activated component (I_{rise}) of inwared rectifier currents became apparent between P8 and P16. The specific amplitudes of I_{s-s} and I_{rise} at a test-pulse of -100 mV increased to their respective plateau values of -68 and -15 μA/cm^2 at P20.

In fibers denervated at P4 the developmental increase of the specific I_{s-s} was suppressed to one tenth of that in normal fibers, and I_{rise} did not appear throughout the development. In muscle fibers denervated at P16 or P20, the specific amplitudes of I_{s-s} and I_{rise} decreased, reaching the levels of P4-denervated fibers in 2 to 4 days after denervation.

In fibers isolated from P1 mice and cultured in control culture medium, I_{s-s} and I_{rise} were hardly observable.[6] Within one day after addition of Ca^{2+} agonist, A23187, ionomycin, or ryanodine to a culture medium, significant increases of I_{s-s} (-106 μA/cm^2 at -100 mV for the case of A23187) and I_{rise} were observed. The inward rectifier current decreased to the level of control cultures within 11 h after a removal of A23187.

DISCUSSION AND CONCLUSIONS

In normal *in situ* development of mouse muscle fibers the channels of $I_{Ca,transient}$ and the toxin-resistant Na channels disappear, and inward rectifier K channels appear in the first few weeks after birth (FIG. 1). The channels of $I_{Ca,sustained}$ and the total Na channels increase during this period. The muscle fibers have to be innervated for the toxin-resistant Na channels to disappear and for the inward rectifier K channels to appear. Innervation is also required in mature fibers to keep these channel densities at their respective adult levels. In contrast, the channels of $I_{Ca,transient}$ disappear independent of innervation and do not re-appear even after denervation. I conclude that the mechanisms regulating densities of the voltage-dependent ion channels were heterogeneous among the different types of channels.

I propose the following hypothesis on the regulatory mechanisms and physiological roles of muscle inward rectifiers. The fibers with high muscular activity may have a relatively high cytosolic [Ca^{2+}], and more inward rectifiers may be induced in these fibers. The induced rectifiers help the re-uptake of K^+ ions, which have been extruded through repetitive membrane firing. Also, the induced rectifiers hyperpolarize the muscle membrane and reduce the probability of Ca^{2+} release from the sarcoplasmic reticulum and Ca^{2+} entry through voltage-gated Ca channels, thereby lowering cytosolic $[Ca^{2+}]$ of the active fibers to a preferable level.

REFERENCES

1. Gonoi, T. & S. Hasegawa. 1988. Post-natal disappearance of transient calcium channels in mouse skeletal muscle: Effects of denervation and culture. J. Physiol. (Lond.) **401**: 617–637.
2. Gonoi, T., Y. Hagihara, J. Kobayashi, H. Nakamura & Y. Ohizumi. 1989. Geographutoxin-sensitive and -insensitive sodium currents in mouse skeletal muscle developing *in situ*. J. Physiol. (Lond.) **414**: 159–177.
3. Gonoi, T. & S. Hasegawa. 1991. Postnatal induction and neural regulation of inward rectifiers in mouse skeletal muscle. Pflügers Arch. **418**: 601–607.
4. Gonoi, T., S. Sherman & W. A. Catterall. 1985. Voltage-clamp analysis of tetrodotoxin-sensitive and -insensitive sodium channels in rat muscle cells developing in vitro. J. Neurosci. **5**: 2559–2564.
5. Gonoi, T., Y. Ohizumi, H. Nakamura, J. Kobayashi & W. A. Catterall. 1987. The *Conus* toxin Geographutoxin II distinguishes two functional sodium channel subtypes in rat muscle cells developing *in vitro*. J. Neurosci. **7**: 1728–1731.
6. Gonoi, T. & S. Hasegawa. 1991. Induction of inward rectifiers in mouse skeletal muscle fibers in culture. Pflügers. Arch. **419**: 657–661.

L-Type Calcium Channel Regulates Depolarization-Induced Survival of Rat Superior Cervical Ganglion Cells *In Vitro*[a]

T. KOIKE, S. TANAKA,
AND A. TAKASHIMA[b]

Department of Natural Science
Saga Medical School
Nabeshima, Saga 849, Japan

[b]*Mitsubishi-Kasei Institute of Life Sciences*
Machida, Tokyo 194, Japan

Chronic depolarization with elevated K^+ has proved to support neuronal survival in a variety of cell types *in vitro* including sympathetic neurons.[1-3] The Ca^{2+} channel antagonists such as nimodipine and nifedipine effectively block the survival-promoting effect of elevated K^+ (>35 mM), while the Ca^{2+} channel agonists including Bay K 8644 and (+)-(S) 202 791 do not by themselves promote survival, but strongly augment the effect of high K^+, indicating that the activation of voltage-dependent L-type Ca^{2+} channels have a crucial role in this phenomenon.[2-4] Recently, measurements of intracellular free Ca^{2+} levels ($[Ca^{2+}]_i$) of neurons loaded with fura-2 have revealed a good correlation between cell survival and $[Ca^{2+}]_i$ of sympathetic or other neurons chronically exposed to various concentrations of extracellular K^+.[5,6] It has also been shown that elevated levels of $[Ca^{2+}]_i$ are associated with acquisition of trophic factor–independent survival of sympathetic and sensory neurons.[5,7-9] Since depolarization-mediated neuronal survival is dependent on Ca^{2+} influx through L-type Ca^{2+} channels, the functional state of this L-type Ca^{2+} channels may be critical for neuronal survival under depolarizing conditions. This study examines the responses of cultured sympathetic neurons to membrane depolarization as a function of treatment with NGF.

Dissociated sympathetic neurons were prepared from superior cervical ganglia of Wistar rats (P1) as described previously.[2,5,9] The anti-mitotic drug, fluorodeoxyuridine (20 μM) was added together with 20 μM uridine to the feeding medium for 5 days to kill non-neuronal cells. The neurons were fixed with 4% paraformaldehyde in Ca^{2+}-, Mg^{2+}-free phosphate-buffered saline, pH 7.2, and stained with anti-MAP 2

[a] This work was partly supported by Grants-in-aid from the Ministry of Education, Culture, and Science and a Grant from the Ministry of Health.

antibody. Sympathetic neurons were loaded with 4 μM fura-2 for 1 h and details of fluorescence imaging of fura-2 loaded neurons were described.[5,9]

Superior cervical ganglion cells dissociated from newborn rats and grown in the presence of NGF for 6 to 7 days were well hypertrophied with extensive neurites. Employing these cultured neurons, we reported [2,5,9] that membrane depolarization with elevated K^+ supported the survival of these neurons independent of NGF. However, as shown in Figure 1, elevated K^+ (40 mM) did not support the survival of sympathetic neurons grown for 8 h (designated as Day 0 in Fig. 1) or one day in the presence of NGF. At Day 3, these neurons became partially responsive to membrane depolarization by elevated K^+ (40 mM) (Fig. 1). As shown previously, sympathetic neurons treated with NGF for 5 to 7 days survived fully under depolarizing conditions. In order to have a correlation between responsiveness to high K^+ and intracellular calcium levels, we measured the time course of $[Ca^{2+}]_i$ changes of the neurons in response to high K^+. When sympathetic neurons, grown for 7 days in the presence of NGF, were loaded with fura-2, its fluorescence intensity was homogeneously distributed: no clear distinction between cytoplasmic and nuclear levels of free Ca^{2+} was observed in 80–90% of these neurons. Exposure to a high K^+ medium (40 mM) induced a rapid increase in $[Ca^{2+}]_i$ followed by sustained levels of $[Ca^{2+}]_i$ (Fig. 2, right) that could be maintained for up to 2 days; for example, the sustained level of $[Ca^{2+}]_i$ of the neurons

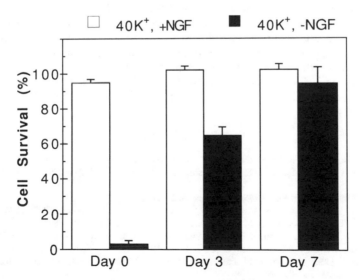

FIGURE 1. Enhancement by NGF of high K^+ (40 mM)-mediated neuronal survival. SCG (superior cervical ganglion) neurons were dissociated from newborn rats and grown for 8 h (designated as Day 0), 3 days (Day 3), and 7 days (Day 7) in the presence of 50 ng/ml NGF. The neurons were then exposed to a high K^+ medium (40 mM) containing NGF ($\square$) or anti-NGF ($\blacksquare$) for 36 h, and were examined for cell survival. Surviving neurons identified by anti-MAP 2 staining were counted up to 4,000 cells. Relative number of surviving neurons under depolarizing conditions are presented compared to the number of surviving neurons in the presence of NGF. The total cytoplasmic enzyme activities were compared in cases where cell counting was a difficult task, such as in the neurons grown for 7 days which tended to form aggregates. Mean ± SD ($n = 3$).

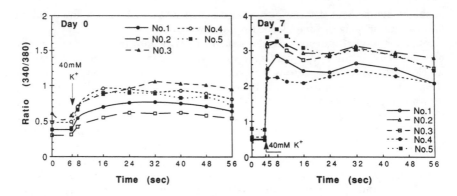

FIGURE 2. High K$^+$-induced increase in intracellular calcium levels ([Ca^{2+}]$_i$) of SCG neurons. SCG neurons grown in the presence of NGF for 8 h (designated as Day 0; left) and for 7 days (Day 7; right) were loaded with 4 μM fura-2 for 1 h. The neurons were then challenged to a high K$^+$ saline (40 mM), and the time course of changes in [Ca^{2+}]$_i$ levels of five individual neurons were recorded.

depolarized for 1 day was 247 ± 14 nM (n = 30), while the basal level of [Ca^{2+}]$_i$ was 93.0 ± 10.5 nM (n = 48). The rapid phase is mediated by a transient activation of N-type Ca^{2+} channels, while the sustained portion is due to L-type Ca^{2+} channels. When the neurons at Day 0 were exposed, however, to a high K$^+$ medium, no threshold response was observed followed by small plateau levels, which were soon diminished even under continuous exposure to a high K$^+$ medium (FIG. 2, left). These findings suggest that the capability of sympathetic neurons to respond to membrane depolarization is developmentally regulated and NGF enhances this response possibly through the upregulation of L-type Ca^{2+} channels (and also N-type channels).

REFERENCES

1. PHILLIPSON, O. T. & M. SANDLER. 1975. Brain Res. **90**: 278–281.
2. KOIKE, T., D. P. MARTIN & E. M. JOHNSON, JR. 1989. Proc. Natl. Acad. Sci. USA **86**: 6421–6425.
3. COLLINS, F. & J. D. LILI. 1989. Brain Res. **502**: 99–107.
4. GALLO, V., A. KINGSBURY, R. BALAS & O. S. JORGENSEN. 1987. J. Neurosci. **7**: 2203–2215.
5. KOIKE, T. & S. TANAKA. 1991. Proc. Natl. Acad. Sci. USA **88**: 3892–3896.
6. COLLINS, F., M. F. SCHMIDT, P. B. GUTHRIE *et al.* 1991. J. Neurosci. **11**: 2582–2587.
7. EICHLER, M. E., J. M. DUBINSKY & K. M. RICH. 1992. J. Neurochem. **58**: 263–269.
8. KOIKE, T. 1992. Neurotrophic Factors. Y. Tsukada & E. M. Shooter, Eds.: 83–96. Japan Scientific Societies Press. Tokyo.
9. TANAKA, S. & T. KOIKE. 1992. Biochim. Biophys. Acta (Mol. Cell Res.) **1175**: 114–122.

Dihydropyridine-Sensitive Calcium Current Mediates Neurotransmitter Release from Retinal Bipolar Cells

MASAO TACHIBANA, TAKASHI OKADA,
TOMOMI ARIMURA, AND KATSUNORI KOBAYASHI

Department of Psychology
Faculty of Letters
The University of Tokyo
7-3-1 Hongo, Bunkyo-ku
Tokyo 113, Japan

The investigation of the Ca^{2+}-neurotransmitter release coupling in the vertebrate central nervous system has been hindered largely due to technical difficulties arising from the small size of presynaptic terminals. In the goldfish retina, one type of bipolar cells has an extraordinarily large axon terminal (*ca.* 10 µm in diameter).[1] Using this preparation, we examined the properties of the presynaptic calcium current that mediates the release of neurotransmitter.

Bipolar cells with a large axon terminal were dissociated enzymatically from the goldfish retina,[2] and were voltage clamped by a patch pipette in the whole-cell recording configuration. A Ca^{2+} indicator, Fura-2, was introduced into the bipolar cell via patch pipette, and the spatiotemporal changes of intracellular free Ca^{2+} concentration ($[Ca^{2+}]_i$) and the calcium current (I_{Ca}) were monitored simultaneously.

A horizontal cell isolated from the catfish retina was voltage clamped in the whole-cell recording configuration, and used as a probe of glutamate, which is the leading candidate for the neurotransmitter of bipolar cells.[3]

I_{Ca} of bipolar cells was identified as the high-voltage activated, dihydropyridine-sensitive type (FIGS. 1 and 2). Neither the ω-conotoxin-sensitive type nor the low-voltage activated type was detected in this preparation.

Upon activation of I_{Ca} by a depolarizing pulse $[Ca^{2+}]_i$ increased rapidly at the axon terminal region, and then recovered slowly to the initial level after the cessation of the pulse (FIG. 1). On the other hand, $[Ca^{2+}]_i$ at the cell body did not change (FIG. 1), or increased slightly with a slower time course and reached a peak much later than the offset of depolarization. Nicardipine blocked both I_{Ca} and the increase of

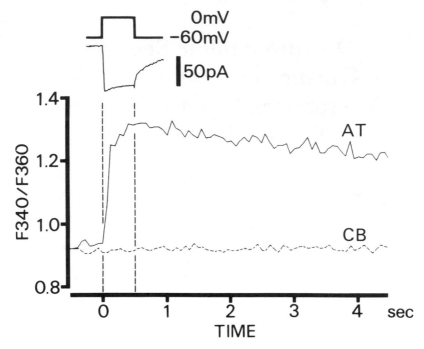

FIGURE 1. Simultaneous recordings of I_{Ca} and of temporal changes of $[Ca^{2+}]_i$ in a voltage-clamped bipolar cell. The membrane current evoked by a depolarizing pulse mostly consisted of I_{Ca} (the leakage current was not subtracted). Fura-2 (100 μM) was introduced into the cell via a patch pipette. Video images of Fura-2 fluorescence were obtained every 15 frames per sec by using a silicone-intensified target camera and its output was fed into an image processor (ARGUS-100/CA system, Hamamatsu Photonics). The main panel indicates the spatially averaged fluorescence ratios (F340/F360) in the axon terminal region (AT) and in the cell body region (CB).

$[Ca^{2+}]_i$. These results suggest that dihydropyridine-sensitive calcium channels are highly localized to the axon terminal region.

Activation of I_{Ca} of a bipolar cell evoked a neurotransmitter-induced current[3] in the horizontal cell closely apposed to the axon terminal of the bipolar cell (FIG. 2). I_{Ca} and the neurotransmitter-induced current were suppressed concomitantly by nifedipine and Co^{2+}, but not by ω-conotoxin. The larger the Ca^{2+} influx into bipolar cells, the larger the amplitude of the neurotransmitter-induced current.[3]

It is concluded that the neurotransmitter (probably glutamate) is released by the activation of dihydropyridine-sensitive calcium channels localized to the presynaptic terminals of bipolar cells.

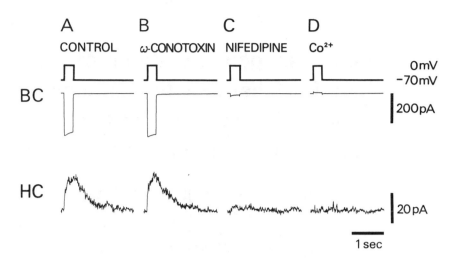

FIGURE 2. Effects of ω-conotoxin, nifedipine, and Co^{2+} on I_{Ca} of a goldfish bipolar cell (BC) and on the neurotransmitter-induced current recorded from a catfish horizontal cell (HC) closely apposed to the axon terminal of the BC. (**A**) A depolarizing pulse (*top trace*) applied to the BC evoked I_{Ca} in the BC (*middle trace*) and an outward current (the neurotransmitter-induced current) in the HC voltage-clamped at +40 mV (*bottom trace*). Both currents were not affected by ω-conotoxin (1 μM; **B**), but were suppressed by nifedipine (100 μM; **C**), and Co^{2+} (3.5 mM; **D**). ω-Conotoxin, nifedipine, or Co^{2+} caused little suppressive effect on the glutamate-evoked currents of HCs.

REFERENCES

1. ISHIDA, A. T., W. K. STELL & D. A. LIGHTFOOT. 1980. J. Comp. Neurol. **191:** 315–335.
2. KANEKO, A. & M. TACHIBANA. 1985. J. Physiol. **358:** 131–152.
3. TACHIBANA, M. & T. OKADA. 1991. J. Neurosci. **11:** 2199–2208.

Voltage-Dependent Block of L-Type Ca Channels by Caffeine in Smooth Myocytes of the Guinea Pig Urinary Bladder

M. YOSHINO, Y. MATSUFUJI AND H. YABU

Department of Physiology
Sapporo Medical University
Sapporo 060, Japan

Recent whole-cell patch clamp studies have shown a possible direct inhibitory action of caffeine on voltage-dependent Ca channel in smooth muscle cells.[1-2] However, those studies mainly measured Ba^{2+} currents passing through Ca channels and the more precise mechanism underlying the inhibitory action of caffeine remains unclear. In this study, we have examined the inhibitory action of caffeine on L-type Ca current in more detail using the whole-cell and cell-attached patch clamp techniques in smooth myocytes of the guinea pig urinary bladder.

The time course of inhibitory action of caffeine on Ca current (I_{Ca}) was examined by applying repetitive depolarizations at a frequency of 1 Hz. The inhibition of I_{Ca} induced by caffeine was found to occur in a biphasic manner when a low concentration of EGTA (0.05–0.5 mM) was included in the patch pipettes (FIG. 1,A). The initial transient decrease in I_{Ca} was eliminated by increasing EGTA concentration (20 mM) or adding both procaine (10 mM) and heparin (1 mg/ml) in the patch pipettes (FIG. 1,B), but the steady decrease in I_{Ca} was unaffected by those treatments. The result suggested that the initial transient decrease in I_{Ca} induced by caffeine was due to Ca-mediated inactivation of I_{Ca} caused by release of stored Ca and that the steady decrease in I_{Ca} was probably due to the direct blocking action of caffeine on I_{Ca}.

The degree of steady decrease in I_{Ca} by caffeine was found to be dose dependent with the K_d value of 20 mM. The steady-state activation curve remained unchanged by caffeine. The steady-state inactivation curve, however, was significantly shifted to the negative direction by caffeine in both 1.8 mM Ca^{2+} (FIG. 2) and Ba^{2+} solutions. This hyperpolarizing shift in the steady-state inactivation curve was not observed when we applied 8-bromo-cyclic AMP (100 μM) to the bath solution.

Cell-attached patch clamp recordings from smooth myocytes of the guinea pig urinary bladder were done with 100 mM Ba^{2+} as the charge carrier. Application of caffeine into the bath solution reduced the opening probability of L-type Ca channel

A

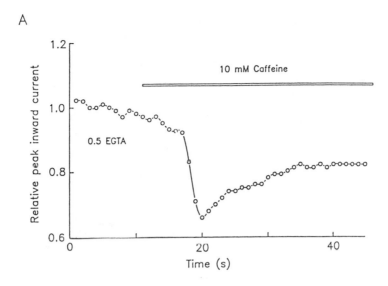

B

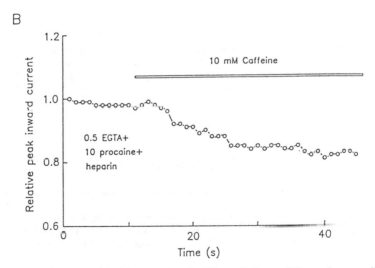

FIGURE 1. The time course of inhibitory action of caffeine on I_{Ca} in two different pipette conditions. I_{Ca} was obtained by depolarization (100 msec duration) to 10 mV from a holding potential of -60 mV at a frequency of 1 Hz. The relative peak I_{Ca} was plotted against time. Caffeine was applied for the periods indicated by the bars. 0.5 mM EGTA (**A**), and 0.5 mM EGTA with procaine (10 mM) and heparin (1 mg/ml) (**B**) were included in the patch pipettes.

current (slope conductance, 33 pS) without a significant change in single channel conductance. These results suggest that caffeine has a direct blocking action on I_{Ca}, presumably by preferential binding of caffeine to L-type Ca channel in the inactivated state similar to the action of 1,4-dihydropyridine Ca channel antagonists.[4]

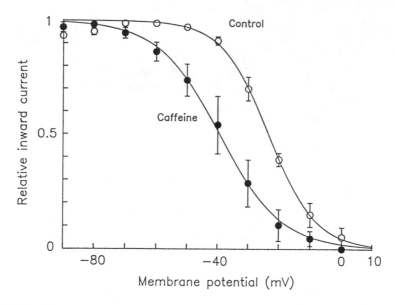

FIGURE 2. The effects of caffeine on steady-state inactivation curves for I_{Ca} in the absence (O) and presence (●) of 10 mM caffeine. The conventional two-pulse protocol was used: a 5-sec conditioning pulse to various potentials was following by a 100-msec test pulse to 10 mV. The maximum amplitude of the peak inward current evoked with no conditioning pulse was normalized to 1.0. Each point represents an averaged value obtained from 4–10 different cells. Smooth curves were fitted based on the Boltzmann distribution for the inactivation gating parameters as a function of the membrane potential. Control, $I = 1/[1 + \exp(V + 23.4)/7.5]$. Caffeine, $I = 1/[1 + \exp(V + 39)/10]$.

REFERENCES

1. MARTIN, C., C. DACQUET, C. MIRONNEAU & J. MIRONNEAU. 1989. Br. J. Pharmacol. **98:** 493–498.
2. HUGHES, A. D., S. HERING & T. B. BOLTON. 1990. Pflügers Arch. **416:** 462–466.
3. ZHOLOS, A. V., L. V. BAIDEN & M. F. SHUBA. 1991. Pflügers Arch. **419:** 267–273.
4. KURA, H., M. YOSHINO & H. YABU. 1992. J. Pharmacol. Exp. Ther. **261:** 724–729.

Cyclic AMP-Dependent Phosphorylation and Regulation of the Cardiac Dihydropyridine-Sensitive Ca Channel

AKIRA YOSHIDA,[a] MASAMI TAKAHASHI,
SEIICHIRO NISHIMURA,[b] HIROSHI TAKESHIMA,[b]
AND SHINICHIRO KOKUBUN[c]

Mitsubishi Kasei Institute of Life Sciences
Machida, Tokyo 194, Japan

[b]Kyoto University
Kyoto 606, Japan

[c]Nihon University School of Medicine
Tokyo 173, Japan

Stimulation of Ca^{2+} influx through the dihydropyridine (DHP)-sensitive Ca channel into cardiac muscle is essential for the isotropic effect caused by β-adrenergic drugs. cAMP-dependent protein kinase (PKA)-mediated phosphorylation participates in the stimulation of the Ca current, however, phosphoproteins involved in this regulation remain to be identified.[1,2] We investigated the phosphorylation of the α1 subunit in rabbit heart and CHO cells stably transfected with cDNA encoding the entire region of the α1 subunit (CCAR cells) and the effect of dibutyryl cAMP (db-cAMP) and forskolin on the Ca current in the cells.[3]

A polyclonal antibody, CR2, against a C-terminal segment of α1 subunit of rabbit cardiac L-type Ca channel immunoprecipitated more than 60% of [³H]PN200-110–labeled Ca channel solubilized from rabbit cardiac membranes. The CR2 recognized 200 kD and 250 kD polypeptides as major and minor antigens, respectively, in the rabbit cardiac membranes by immunoblotting. A 250 kD protein recognized by CR2 was expressed in the CCAR cells. cAMP-dependent protein kinase (PKA) phosphorylated the 250 kD but not the 200 kD protein *in vitro* (FIG. 1). The 250 kD protein in the CCAR cells was phosphorylated within 5 min after treatment with db-cAMP.

[a] New address: Advanced Research Center for Human Sciences, Waseda University, 2-579-15 Mikajima, Tokorozawa, Saitama 359, Japan.

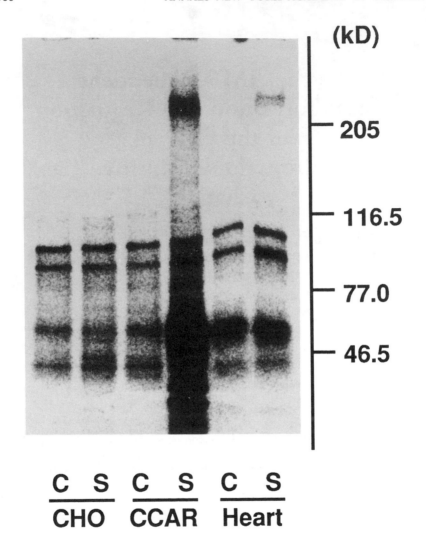

FIGURE 1. Phosphorylation of the cardiac α1 subunit from the rabbit heart and CCAR cell by PKA *in vitro*. The α1 subunit were solubilized from CHO cells, CCAR cells, and rabbit hearts, and immunoprecipitated with either control IgG (C) or CR2 (S). The immunoprecipitates were incubated with [γ-^{32}P]ATP and a catalytic subunit of PKA for 10 min at 30°C.

In order to determine whether PKA-mediated phosphorylation of the 250 kD form of α1 subunit could affect the Ca channel function, we investigated the electrophysiological properties of the CCAR cells. In CCAR cells, voltage-dependent Ba current was expressed and was completely inhibited by 10 μM nicardipine. The amplitude of dihydropyridine-sensitive Ba current observed in the cells increased significantly by treatment with db-cAMP or 8-bromo-cAMP (FIG. 2). We observed an

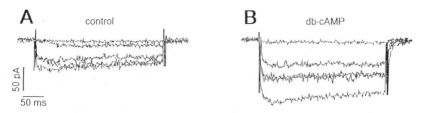

FIGURE 2. The effect of db-cAMP on I_{Ba} in CCAR cells. (**A**) Currents in the absence of db-cAMP recorded by depolarization to -40, -20, 0, $+20$, and $+40$ mV for 500 msec from a holding potential of -60 mV are superimposed. (**B**) Currents in the presence of db-cAMP (200 μM) recorded at same membrane potentials as those in **A** are superimposed.

increase in the amplitude of Ba current by db-cAMP and 8-bromo-cAMP in 8 and 14 experiments out of 19 and 25 in the CCAR cells, respectively. The amplitude of Ba current at $+20$ mV gradually increased after adding db-cAMP, and reached a steady level about 5 min thereafter. This was consistent with the time course of the phosphorylation of the α1 subunit. These results indicated that the Ca channel activity can be modulated by cAMP-dependent phosphorylation of the α1 subunit of cardiac Ca channel.

REFERENCES

1. CHANG, F. C. & M. M. HOSEY. 1988. J. Biol. Chem. **263**: 18929–18937.
2. YOSHIDA, A., M. TAKAHASHI, Y. FUJIMOTO, H. TAKISAWA & T. NAKAMURA. 1990. J. Biochem. **107**: 608–612.
3. YOSHIDA, A., M. TAKAHASHI, S. NISHIMURA, H. TAKESHIMA & S. KOKUBUN. 1992. FEBS Lett. **309**: 343–349.

Primary Structure and Tissue Distribution of a Novel Calcium Channel from Rabbit Brain[a]

TETSUHIRO NIIDOME[b,c] AND YASUO MORI[d]

Tsukuba Research Laboratories
Eisai Co., Ltd.
Tokodai, Tsukuba
Ibaraki 300-26, Japan

[d]*Departments of Medical Chemistry and Molecular Genetics*
Kyoto University Faculty of Medicine
Kyoto 606-01, Japan

Voltage-dependent calcium channels are essential for the regulation of a variety of cellular functions, including membrane excitability, muscle contraction, synaptic transmission, and other forms of secretion. At least four types of calcium channel (designated T-, L-, N-, and P-type calcium channels) have been distinguished by their electrophysiological and pharmacological properties. Recently, cDNA cloning studies have revealed the existence of multiple calcium channel gene products. Here we report the complete amino acid sequence of a novel brain calcium channel (designated BII) deduced from the cDNA sequence. The tissue distribution of BII mRNA, together with that of BI mRNA, has also been studied by northern blot analysis.[1]

FIGURE 1 shows the primary structure of the rabbit BII calcium channel deduced from the cDNA sequence; the open reading frame corresponding to the amino acid sequences of the skeletal muscle,[2] cardiac,[3] and BI[4] calcium channels was adopted and translation initiation site was assigned to the first ATG triplet that appears downstream of a nonsense codon. Two isoforms of BII calcium channel (BII-1 and BII-2; see FIG. 1 legend) differ from each other in the carboxy-terminal sequence beginning with amino acid residue 2,101. BII-1, and BII-2 are composed of 2,259 and 2,178 amino acid residues, respectively, with relative molecular masses of 254,249 and 245,595. Amino acid sequence comparison of the BII, BI-2, cardiac and skeletal muscle calcium channels reveals that the amino acid sequence of the BII calcium channel is more closely related to that of the BI-2 (59% amino acid identities between the BII-1/BI-2 and BII-2/BI-2 pairs) than to those of the other calcium channels (38, 40, 41, and 42% amino acid identities between the BII-1/cardiac, BII-2/cardiac, BII-1/skeletal,

[a] **GenBank Accession Number:** The nucleotide sequences of the rabbit BII cDNA will appear in the EMBL, GenBank and DDBJ nucleotide sequence data bases under the accession number X67855 (BII-1) and X67856 (BII-2).

[b] Address correspondence to: Tetsuhiro Niidome, Tsukuba Research Laboratories, Eisai Co., Ltd., 1-3 Tokodai 5-chome, Tsukuba-shi, Ibaraki 300-26, Japan.

BII-2/skeletal pairs, respectively). Significant sequence divergence among the four (or five) calcium channels is found in a region between repeat II and III, and in a carboxy-terminal region. RNA preparations from different rabbit tissues and from different regions of rabbit brain were subjected to northern blot analysis with a BII cDNA probe or a BI cDNA probe (FIG. 2). Two major hybridizable RNA species of ~ 10,500 and ~ 11,000 nucleotides were detected with a BII cDNA probe and a major hybridizable RNA species of ~ 9,400 nucleotides with a BI cDNA probe in the brain at high levels. The level of BII mRNA species was much higher in the cerebral cortex, hippocampus, and corpus striatum than in the olfactory bulb, mid brain, cerebellum, and medulla-pons, whereas the level of BI mRNA species was much higher in the cerebellum than in the other regions of brain.

Our results show that voltage-dependent calcium channels can be classified into at least two main subfamilies, according to the degrees of amino acid sequence homology between calcium channel pairs. One subfamily consists of the L-type calcium channels from skeletal muscle, heart, smooth muscle, pancreas, and brain, and the other subfamily consists of the BI calcium channel, which may represent P-type, and the BII calcium channel. The spatial distribution of BII mRNA in the brain is markedly different from that of BI mRNA. BII mRNA is abundant in the cerebral cortex, hippocampus, and corpus striatum, while BI mRNA is expressed predominantly in the cerebellum. Very recently, a partial cDNA encoding the class E calcium channel have been isolated from rat hippocampus.[5] The amino acid sequence of the class E calcium channel is most closely related to the class A and B calcium channels, which appear to represent P- and N-type, respectively. Northern blot analysis shows the class E calcium channel mRNA is found in different regions of the brain, its level being much higher in the hippocampus. These results suggest that the class E calcium channel is the rat counterpart of the BII calcium channel in the rabbit brain. What are the physiological roles for the BII calcium channel? Notably, CA3 and DRG neurons have a significant fraction of high threshold calcium channel current remaining in the presence of ω-Aga-IVA, ω-conotoxin, and dihydropyridines,[6] which are specific blockers of P-, N-, and L-type, respectively. The BII calcium channel may represent this unidentified fraction and be involved in long-term potentiation and/or delayed neuronal cell death.

REFERENCES

1. NIIDOME, T., M. S. KIM, T. FRIEDRICH & Y. MORI. 1992. FEBS Lett. 308: 7–13.
2. TANABE, T., H. TAKESHIMA, A. MIKAMI, V. FLOCKERZI, H. TAKAHASHI, K. KANGAWA, M. KOJIMA, H. MATSUO, T. HIROSE & S. NUMA. 1987. Nature 328: 313–318.
3. MIKAMI, A., K. IMOTO, T. TANABE, T. NIIDOME, Y. MORI, H. TAKESHIMA, S. NARUMIYA & S. NUMA. 1989. Nature 340: 230–233.
4. MORI, Y., T. FRIEDRICH, M. S. KIM, A. MIKAMI, J. NAKAI, P. RUTH, E. BOSSE, F. HOFMANN, V. FLOCKERZI, T. FURUICHI, K. MIKOSHIBA, K. IMOTO, T. TANABE & S. NUMA. 1991. Nature 350: 398–402.
5. SOONG, T. W. & T. P. SNUTCH. 1992. Soc. Neurosci. (Abstr.) 18: 1139.
6. MINTZ, I. M., M. E. ADAMS, & B. P. BEAN. Neuron 9: 85–95.

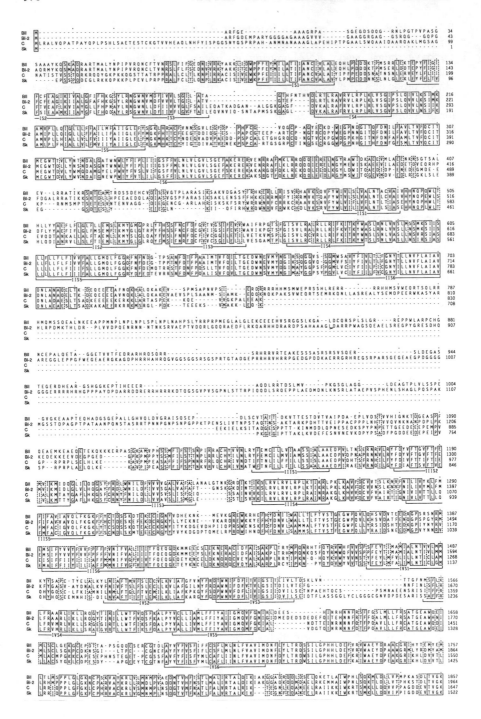

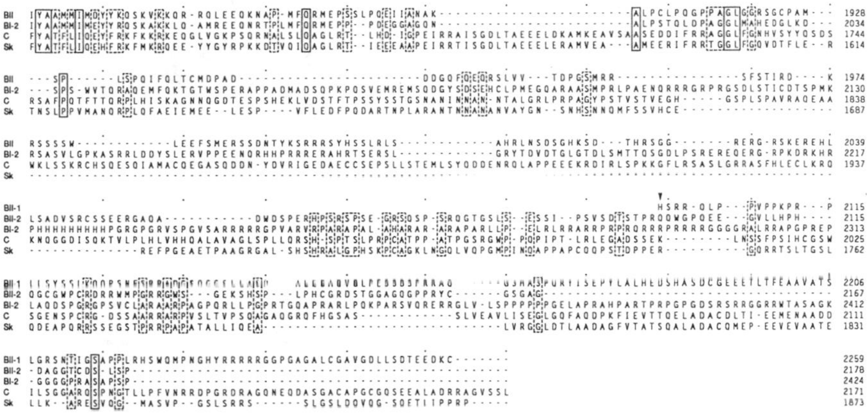

FIGURE 1. (*Left and above*) Amino acid sequence (in one-letter code) of the rabbit brain calcium channel BII, deduced from the cDNA sequence, and alignment of the amino acid sequence of different calcium channels. For the cloning procedures see Niidome *et al.*[1] An inserted sequence of 816 nucleotide residues, which contains in-frame translational termination codon, is found in two clones (λCB236 and λCB244), whereas it is missing in another clone (λCB264). This insertion-deletion, which probably results from alternative RNA splicing, gives rise to two distinct mRNAs encoding two isoforms of the BII calcium channel (BII-1 and BII-2). The four sequences (or five after divergence of the BII-1 and BII-2 sequences) compared (from top to bottom) are as follows: BII[1]; BI-2[4]; rabbit cardiac DHP-sensitive calcium channel[3] (C); rabbit skeletal muscle DHP-sensitive calcium channel[2] (Sk). Amino acid residues are numbered from the initiating methionine, and numbers of the amino acid residues on each line are given on the right-hand side. The arrowhead indicates where the BII-1 and BII-2 proteins diverge in sequence; residues 2,040–2,110 of BII-1, which immediately precede the position indicated by the arrowhead, are not displayed. Sets of four (or five) identical residues at one position are enclosed with solid lines, and sets of four (or five) identical or conservative residues at one position with broken lines. The putative transmembrane segments S1–S6 in each of the repeats I–IV are shown.

FIGURE 2. Autoradiograms of blot hybridization analysis of RNA from different rabbit tissues and different regions of rabbit brain with a BII cDNA probe (A) or a BI cDNA probe (B). (A) 10 μg of poly(A)⁺ RNA (a) or 15 μg of total RNA (b) was applied to each lane. The probe was the 0.60 kb SalI(1,298)/BamHI(1,988) fragment derived from λCBP201[1] (a) or the 2.6 kb EcoRI(3,689)/EcoRI(6,293) fragment from λCBA240[1] (b) and labeled by random primer method and the nick translation method, respectively. Similar results were obtained in the experiment in which the 2.6 kb EcoRI(3,689)/EcoRI(6,293) fragment from λCBA240 was used as a probe (data not shown). Autoradiography was at – 70°C for 3 days (a) or 7 days (b) with an intensifying screen. (B) 10 μg of poly(A)⁺ RNA (a) or total RNA (b) was applied to each lane. The probe was BbvI(3,843)/BbvI(5,088) fragment derived from λCBP107[4] and labeled by nick translation method. Autoradiography was at – 70°C for 24 h (a) or 7 days (b) with an intensifying screen. An RNA ladder (Bethesda Research Laboratories) was used for size markers (in kilobases).

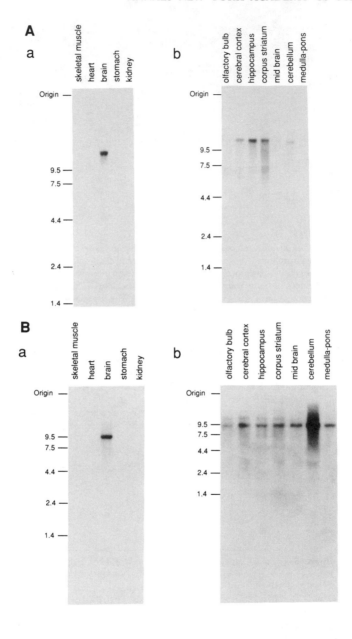

Synaptocanalins (N-type Ca Channel-associated Proteins) Form a Complex with SNAP-25 and Synaptotagmin

TERUO ABE, HIDEO SAISU,
AND HIROSHI P. M. HORIKAWA

Department of Neurochemistry
Brain Research Institute
Niigata University
Niigata 951, Japan

Two monoclonal antibodies (MAbs), SPM-1 and SPM-2, immunoprecipitate some of ω-conotoxin-sensitive (N-type) Ca channels in the brain.[1] SPM-1 and SPM-2 recognize nervous system–specific proteins of 36 kD and 28 kD, respectively. These proteins are not subunits of N-type Ca channels but are associated with them. We identified several isoforms of the 36 kD protein (named synaptocanalins) (SCs) and cloned cDNA for one isoform.[2] Very similar proteins (HPC-1/syntaxins) have been identified from different lines of research.[3,4] Here we show that the 28 kD band contains two proteins and that one is identical to SNAP-25, a protein in presynaptic terminals.[5] We also present evidence for the association of SCs with a synaptic vesicle protein, synaptotagmin (ST).

On SDS-PAGE in the presence of urea, the 28 kD bands in the SDS eluates from SPM-1- and SPM-2-Sepharose were separated into two components (28a and 28b). Both components reacted with SPM-2. The 28a component was cleaved with CNBr and fractionated by HPLC. The amino acid sequences of major fragments are given in TABLE 1. The exact sequences of all these fragments were contained in the sequence of SNAP-25. Moreover, the 28a component reacted with antisera raised against two partial amino acid sequences of SNAP-25. These results indicate that the 28a compo-

TABLE 1. Amino Acid Sequences of Fragments Obtained by CNBr Cleavage of the 28a

Fragment	Sequence
1	A-I-S-G-G-F-I-R-R-V-T-N
2	L-Q-L-V-E-E-S-K-D-A
3	D-E-N-L-E-Q-V-S-G-I

The amino acid sequences of CNBr fragments purified by HPLC are shown. The exact sequences of all these fragments are contained in the sequence of SNAP-25.[5]

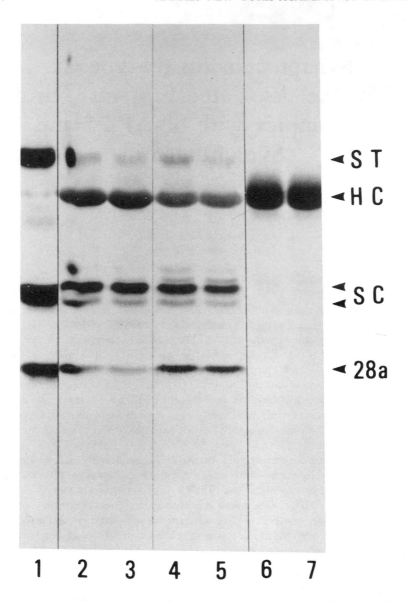

FIGURE 1. Coimmunoprecipitation of synaptocanalins, the 28a component (SNAP-25) and synaptotagmin. A CHAPS extract of the bovine brain lysed P2 fraction was incubated with SPM-1-Sepharose (lanes 2 and 3) or SPM-2-Sepharose (lanes 4 and 5) or normal mouse IgG-Sepharose (lanes 6 and 7) for 2 h at 4°C as described previously,[1] and the Sepharose beads were washed with 20 mM Tris-HCl buffer–0.2% CHAPS containing 0.15 M NaCl (lanes 2, 4, and 6) or 0.5 M NaCl (lanes 3, 5, and 7). The precipitated materials were then analyzed by immunoblotting with a mixture of MAbs (SPM-1, SPM-2, and anti-synaptotagmin (mAb1D12)). Lane 1, CHAPS extract. ST, synaptotagmin; SC, synaptocanalins; HC, heavy chains of IgGs released from the beads.

nent was identical to SNAP-25. The 28b component differed from the 28a component (SNAP-25), as it did not react with either of the antisera.

ST was not immunoprecipitated by SPM-1 or SPM-2, when a digitonin extract of bovine brain membranes was used. However, it was precipitated when CHAPS was used instead of digitonin (FIG. 1), in agreement with previous reports.[4,6] Conversely, small amounts of SCs and the 28a component were precipitated by a MAb against ST (mAb1D12, kindly supplied by M. Takahashi, Mitsubishi Kasei Institute of Life Sciences).

Together with our previous findings,[1,2] the above results indicate the association of SCs with SNAP-25 and ST. Thus, SCs may play roles in the regulation of N-type Ca channels and neurotransmitter release at the presynaptic terminals.

REFERENCES

1. SAISU, H., K. IBARAKI, T. YAMAGUCHI, Y. SEKINE & T. ABE. 1991. Biochem. Biophys. Res. Commun. 181: 59–66.
2. MORITA, T., H. MORI, K. SAKIMURA, M. MISHINA, Y. SEKINE, A. TSUGITA, S. ODANI, H. P. M. HORIKAWA, H. SAISU & T. ABE. 1992. Biomed. Res. 13: 357–364.
3. INOUE, A., K. OBATA & K. AKAGAWA. 1992. J. Biol. Chem. 267: 10613–10619.
4. BENNETT, M. K., N. CALAKOS & R. H. SCHELLER. 1992. Science 257: 255–259.
5. OYLER, G. A., G. H. HIGGINS, R. A. HART, E. BATTENBERG, M. BILLINGLY, F. E. BLOOM & C. WILLSON. 1989. J. Cell Biol. 109: 3039–3052.
6. YOSHIDA, A., C. OHO, A. OMORI, R. KUWAHARA, T. ITO & M. TAKAHASHI. 1992. J. Biol. Chem. In press.

Immunohistochemical Localization of the Proteins Associated with Brain N-Type Calcium Channels[a]

H. ISHIKAWA, O. SHIMADA, T. MURAKAMI, H. SAISU[b] AND T. ABE[b]

Department of Anatomy
Gunma University School of Medicine
Maebashi 371, Japan

[b]Department of Neurochemistry
Brain Research Institute
Niigata University
Niigata 951, Japan

N-type, ω-conotoxin GVIA-sensitive, calcium channels are believed to be directly involved in the release of neurotransmitters at many synapses in the peripheral and central nervous systems. Two membrane proteins (36 kD and 28 kD) associated with brain N-type calcium channels were identified using two monoclonal antibodies.[1] They may be involved in the regulation of the channel. Most recently, the 36 kD protein has been demonstrated to contain several isoforms, which are collectively named as "synpatocanalins."[2]

We have examined the localization of the 36 kD and 28 kD proteins in the rat brain by immunofluorescence, confocal laser, and immunoelectron microscopy using two monoclonal antibodies, SPM-1 and SPM-2. The pattern of immunostaining for these two proteins was almost identical, suggesting that they were colocalized. These proteins was widely distributed throughout the brain, though there were marked differences in staining intensity in various parts of the brain. In general, the cortex and nuclei in all parts of the brain were preferentially stained, whereas the medulla showed no weak immunostaining. The perikarya of neurons were usually immunonegative.

In the cerebellum, the whole molecular layer was intensely stained to show a lattice-like pattern of fluorescent dots as observed with a confocal laser microscope (FIG. 1). No immunostaining was detected in the perikarya and dendrites of Purkinje cells. In immunoelectron microscopy, most of, but not all, presynaptic regions were positively stained, more intensely on the presynaptic membrane, and the postsynaptic

[a] This study was supported in part by Grants-in-Aid for Scientific Research from the Ministry of Education, Science and Culture, Japan, and the Brain Science Foundation.

376

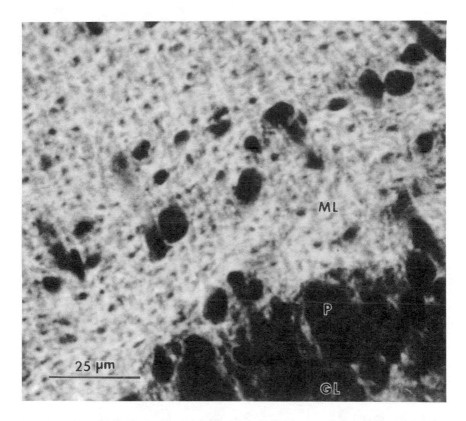

FIGURE 1. Confocal image of the cerebellar cortex immunostained for the 36 kD protein. The molecular layer (ML) shows a lattice-like pattern of fluorescence, while the perikarya of Purkinje cells (P) are not stained. In the granular layer (GL), scattered aggregates of fluorescent dots are observed.

regions in the perikarya and dendrites including dendritic spines of Purkinje cells were immunonegative (FIG. 2). The granular layer was less intense in immunoreactivity with scattered aggregates of immunopositive synapses. Interestingly, granular cells appeared to be immunopositive in their perikarya. The cerebellar medulla was only weakly stained.

In the cerebrum, the cortex was intensely stained to show a dense network with neuronal perikarya and vertically running fibers unstained. An intense immunostaining was also observed in the limbic system, especially in the hippocampus, where the molecular layer of the hippocampus was most intensely stained. The olfactory bulb and the median eminence also showed an intense immunostaining. In the brain stem most of the nuclei were stained as typified by the solitary nucleus and the nuclei of N. facialis, N. trigeminus, and N. vestibulocochlearis.

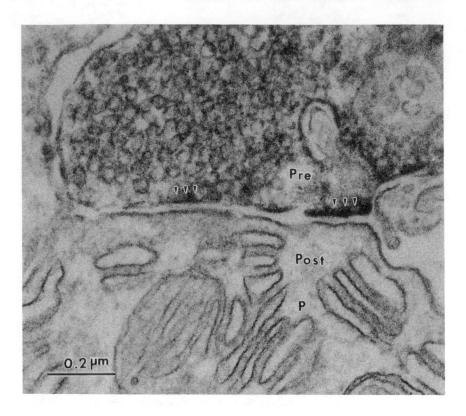

FIGURE 2. Immunoelectron micrograph of the molecular layer of the cerebellum. Presynaptic regions (Pre) are immunostained, more intensely on the presynaptic membrane (arrowheads). Note that the dendrites of the Purkinje cell (P) including the postsynaptic region (Post) are immunonegative.

These observations suggest that in the rat brain the 36 kD synaptocanalin and 28 kD protein coexist as important presynaptic membrane components associated with N-type channels.

REFERENCES

1. SAISU, H., K. IBARAKI, T. YAMAGUCHI, Y. SEKINE & T. ABE. 1991. Monoclonal antibodies immunoprecipitating ω-conotoxin-sensitive calcium channel molecules recognize two novel proteins localized in the nervous system. Biochem. Biophys. Res. Commun. **181:** 59–66.
2. MORITA, T., H. HORI, K. SAKIMURA, M. MISHINA, Y. SEKINE, A. TSUGITA, S. ODANI, H. P. M. HORIKAWA, H. SAISU & T. ABE. 1992. Synaptocanalin I, a protein associated with brain N-type calcium channels. Biomed. Res. **13:** 357–364.

Diversity of Calcium Channel Subtypes in Presynaptic Terminals of the Chick Ciliary Ganglion

HIROMU YAWO

Department of Physiology
Kyoto University Faculty of Medicine
Kyoto, 606-01 Japan

Voltage-gated Ca^{2+} channels in the presynaptic nerve terminal have been shown to be crucial molecules in the process of transmitter release. In contrast to the neuronal somata, the diversity of presynaptic Ca^{2+} channels were studied in limited preparations.

We have presented direct evidence that two subpopulations of high-threshold Ca^{2+} channels coexit in the caliciform presynaptic terminal of chick ciliary ganglion.[1] The major type was sensitive to ω-conotoxin GVIA (ω-CgTx) but insensitive to dihydropyridines (DHPs). This subtype was little affected by the change in holding potential. The other Ca^{2+} channel subtype was resistant to both ω-CgTx and DHPs and was easily inactivated at depolarized holding potentials. It has been shown that in mammalian neurons, a spider venom toxin, ω-agatoxin IVA (ω-Aga-IVA) specifically blocks a subcomponent of high-threshold Ca^{2+} currents (P-type) that is resistant to both ω-CgTx and DHP.[2,3] In this study I have tested whether the ω-CgTx-resistant Ca^{2+} current in the chick ciliary presynaptic terminal is sensitive to ω-Aga-IVA or not.

Large presynaptic terminals in the chick ciliary ganglion were identified and prepared as previously described.[1] Whole-cell Ba^{2+} current was recorded by using a patch pipette containing CsCl in a Na^+-free extracellular solution containing tetraethylammonium and tetrodotoxin. The Ba^{2+} currents were activated by step depolarizations to 0 mV from a holding potential of -80 mV (FIG. 1,A). ω-Aga-IVA (100 nM, a gift from Dr. M. E. Adams, Department of Entomology, University of California Riverside, USA) partly blocked the Ba^{2+} current (FIG. 1,A and B). The remaining current was further blocked by 10 μM ω-CgTx (FIG. 1,A and B). However, about 3% of total Ba^{2+} current was observed in the presence of both ω-Aga-IVA and ω-CgTx. The remaining current was abolished by 0.5 mM Cd^{2+} (FIG. 1,B).

When ω-CgTx was first applied, a large part of the Ba^{2+} current was irreversibly blocked (FIG. 1,C). The remaining current was, however, insensitive to 100 nM ω-Aga-IVA (FIG. 1,C). The ω-CgTx-resistant current was completely abolished by 0.5 mM Ca^{2+} (FIG. 1,C). The ω-CgTx-resistant component of Ba^{2+} current was also unaffected by ω-Aga-IVA in two other experiments.

It concluded from the above results that in the chick ciliary presynaptic terminal, the DHP- and ω-CgTx-resistant subtype of Ca^{2+} channels was not sensitive to ω-Aga-

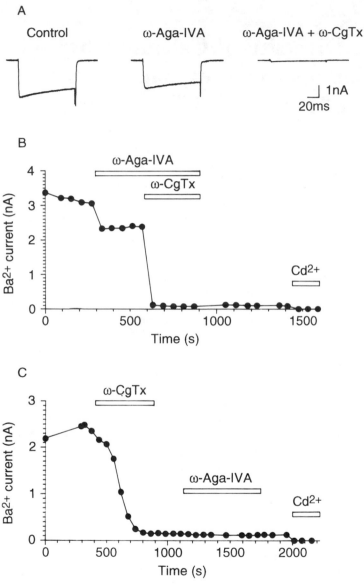

FIGURE 1. Effects of ω-Aga-IVA on Ca²⁺ channel currents in the chick ciliary presynaptic terminals. (A) Ba²⁺ currents were elicited by 100 msec pulses to 0 mV from a holding potential of −80 mV. The leak current was digitally subtracted. Left: control. Middle: in the presence of 100 nM ω-Aga-IVA. Right: in the presence of both ω-Aga-IVA and 10 μM ω-CgTx. (B) Time course of inhibition of the Ba²⁺ current by 100 nM ω-Aga-IVA, with subsequent addition of 10 μM ω-CgTx, same cell as A. The Ba²⁺ current was completely blocked by the final addition of 0.5 mM Cd²⁺. (C) Change of the Ba²⁺ current by 10 μM ω-CgTx, with subsequent application of 100 nM ω-Aga-IVA. The ω-CgTx-resistant component, which was insensitive to ω-Aga-IVA, was completely abolished by 0.5 mM Cd²⁺.

IVA. On the other hand, ω-Aga-IVA blocked a subcomponent of ω-CgTx-sensitive Ca^{2+} conductance. It is unclear where the discrepancies between the mammalian neurons and the chick ciliary presynaptic terminal come from.

REFERENCES

1. Yawo, H. & A. Momiyama. 1993. Re-evaluation of calcium currents in pre- and postsynaptic neurons of the chick ciliary ganglion. J. Physiol. (Lond) **460:** 153–172.
2. Mintz, I. M., V. J. Venema, K. M. Swiderek. T. D. Lee, B. P. Bean & M. E. Adams. 1992. P-type calcium channels blocked by the spider toxin ω-Aga-IVA Nature (Lond) **355:** 827–829.
3. Mintz, I. M., M. E. Adams & B. P. Bean. 1992. P-type calcium channels in rat central and peripheral neurons. Neuron **9:** 85–95.

Synaptotagmin Associates with Presynaptic Calcium Channels and Is a Lambert-Eaton Myasthenic Syndrome Antigen[a]

OUSSAMA EL FAR, NICOLE MARTIN-MOUTOT,
CHRISTIAN LEVEQUE, PASCALE DAVID,
BÉATRICE MARQUEZE, BETHAN LANG,[b]
JOHN NEWSOM-DAVIS,[b] TOSHIMITSU HOSHINO,[c]
MASAMI TAKAHASHI,[c]
AND MICHAEL J. SEAGAR[d]

INSERM U374
Faculté de Médecine Secteur Nord
Bd. Pierre Dramard
13916 Marseille Cedex 20

[b] Institute of Molecular Medicine
John Radcliffe Hospital
Headington, Oxford OX3 9DU

[c] Mitsubishi Kasei Institute of Life Sciences
Machida-shi, Tokyo 194

INTRODUCTION

Lambert-Eaton myasthenic syndrome (LEMS) is an autoimmune disorder of neuromuscular transmission characterized by muscle weakness and autonomic dysfunction.[1] LEMS autoantibodies affect presynaptic mechanisms by downregulating the voltage-gated calcium channels that trigger transmitter release.[1,2] N-type calcium channels are oligomeric proteins[3] that play a major role in excitation-secretion coupling. They can be labeled with ^{125}I-ω-conotoxin-GVIA (^{125}I-ωCgTx) and are specifically immunoprecipitated by LEMS antibodies.[4] We describe identification of a LEMS antigen associated with N-type calcium channels. Our results have novel implications for the mechanism of synaptic vesicle docking at release sites.

[a] This research was supported by grants from the "Association Française contre les Myopathies," the "Assistance Publique de Marseille," and Bayer Pharma.
[d] To whom all correspondence should be addressed.

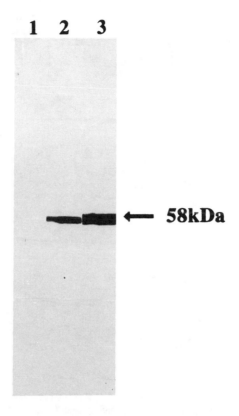

FIGURE 1. LEMS IgG and mAb 1D12 bind to closely related proteins. 1D12 antigen was immunoaffinity purified from rat brain synaptic membranes on a mAb 1D12-Sepharose 4B column. Western blots were probed with normal human IgG (lane 1), LEMS IgG (lane 2), and mAb 1D12 (lane 3) and immunoreactive proteins were detected using anti-IgG peroxidase and Amersham ECL reagents.

RESULTS AND DISCUSSION

Solubilized ^{125}I-ωCgTx prelabeled calcium channels were partially purified by a two-step affinity procedure.[5] Proteins were separated by SDS-PAGE, transferred to a nitrocellulose membrane and probed with a range of human plasma or IgG. IgG from LEMS patients that immunoprecipitate solubilized ^{125}I-ωCgTx receptors at high titre recognize a 58 kD protein in immunoblots.[5] This antigen does not appear to be a channel subunit. Western blotting experiments with crude neuronal homogenates suggested the 58 kD polypeptide was a major constituent of synaptic membranes. Sucrose density gradient centrifugation demonstrated that most of the 58 kD protein sedimented at low velocity, and only a small fraction was associated with the more rapidly sedimenting calcium channel.[5] We therefore concluded that this LEMS antigen is a major synaptic protein that can functionally interact with the channel.

The monoclonal antibody 1D12 has similar properties to these LEMS IgG. It

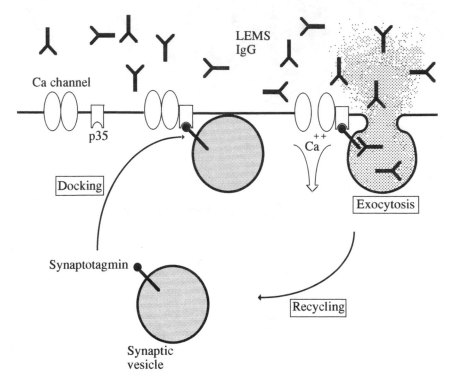

FIGURE 2. The association between synaptotagmin and calcium channels may play a role in docking synaptic vesicles at exocytotic sites in the presynaptic plasma membrane. Other proteins could be involved in this interaction including p35 (syntaxin). During exocytosis the intravesicular N-terminal domain of synaptotagmin would become accessible to circulating LEMS IgG.

immunoprecipitates calcium channels and recognizes a 58 kD band in immunoblots of rat brain synaptic membranes and partially purified ωCgTx receptor.[5,6] 1D12 antigen was purified on a mAb1D12-Sepharose 4B immunoaffinity column. LEMS IgG was shown to bind to this protein (FIG. 1) whereas healthy human IgG did not. The protein was identified, by immunoscreening a rat brain λgt11 library with mAb1D12,[5] as synaptotagmin (p65) a synaptic vesicle membrane protein thought to be involved in exocytosis.[7]

Our results indicate that synaptotagmin remains associated with N-type calcium channels during an approximately 300-fold purification. A monoclonal antibody (1D12) initially selected for its ability to immunoprecipitate ^{125}I-ωCgTx-labeled channels in fact binds to synaptotagmin. Antibodies from patients with LEMS, an autoimmune disease in which neurotransmitter release is defective, also bind to synaptotagmin. These findings are consistent with the hypothesis illustrated in FIGURE 2 and suggest physical interaction between synaptotagmin and the calcium channel. Other proteins may be involved, including a 35 kD polypeptide (syntaxin, p35), which is a synaptotagmin and N-type channel–associated plasma membrane protein.[8,9] Association of synaptic vesicle proteins with calcium channels would presumably site the calcium sensor protein(s) that trigger(s) exocytosis in a favorable zone rapidly accessible to calcium transients. Following calcium-triggered exocytosis the intravesicular N-terminal domain of synaptotagmin would be exposed at the cell surface and LEMS antibodies could then bind to a synaptotagmin-calcium channel complex.

REFERENCES

1. VINCENT, A., B. LANG & J. NEWSOM-DAVIS. 1989. Trends Neurosci. 12: 496–502.
2. KIM, Y. & E. NEHER. 1988. Science 239: 405–408.
3. McENERY, M. W., A. M. SNOWMAN, A. H. SHARP, M. E. ADAMS & S. H. SNYDER. 1991. Proc. Natl. Acad. Sci. USA 88: 11095–11099.
4. SHER, E., C. GOTTI, N. CANAL, C. SCOPETTA, A. EVOLI & F. CLEMENTI. 1989. Lancet 2: 640–643.
5. LEVEQUE, C., T. HOSHITO, P. DAVID, Y. SHOJI-KASSAI, K. LEYS, A. OMORI, B. LANG, O. EL FAR, K. SATO, N. MARTIN-MOUTOT, J. NEWSOM-DAVIS, M. TAKAHASHI & M. J. SEAGAR. 1992. Proc. Natl. Acad. Sci. USA 89. 3625 3629.
6. TAKAHASHI, M., Y. ARIMATSU, S. FUJITA, Y. FUJIMOTO, S. KONDO, T. HAMA & E. MIYAMOTO. 1991. Brain Res. 551: 279–292.
7. PERIN, M. S., V. A. FRIED, G. A. MIGNERY, R. JAHN & T. C. SUDHOF. 1990. Nature 345: 260–263.
8. BENNET, M. K., N. CALAKOS & R. SCHELLER. 1992. Science 257: 255–259.
9. YOSHIDA, A., O. CHIKARA, A. OMORI, R. KUWAHARA, T. ITO & M. TAKAHASHI. 1992. J. Biol. Chem. 267: 24925–24928.

Immunological Characterization of Proteins Associated with the Purified Omega-conotoxin GVIA Receptor

M. W. McENERY,[a,b] A. M. SNOWMAN,[c]
M. J. SEAGAR,[d] T. D. COPELAND,[e]
AND M. TAKAHASHI[f]

[b]Case Western Reserve School of Medicine
Department of Physiology and Biophysics
E521 10900 Euclid Avenue
Cleveland, Ohio 44106

[c]The Johns Hopkins School of Medicine
Department of Neuroscience
WBSB 803 725 N. Wolfe Street
Baltimore, Maryland 21205

[d]Neurobiologie des canaux ioniques
INSERUM U374
Faculté de Médicine Nord
Marseille, France

[e]National Cancer Institute
Frederick R & D Center
ABL-Basic Research Program
Frederick, Maryland 21701-1201

[f]Mitsubishi-Kasei Institute for Life Science
11, Minamiooya, Machida-shi
Tokyo, Japan

The omega-conotoxin GVIA receptor (ωCTXR) from rat forebrain yields five proteins upon purification.[1,2] The subunit composition of the CTX-sensitive N-type voltage-dependent calcium channel (VDCC) is similar to the L-type VDCC of skeletal muscle[3,4] and brain.[5] The apparent molecular weights of the associated proteins and

[a] To whom all correspondence should be addressed.

possible identities are: 240 kD protein (alpha$_1$); 140 kD protein (alpha$_2$); 110 kD protein; 70 kD protein (beta$_2$); and 58/60 kD protein(s) (beta$_1$). There is also evidence of an unidentified 100 kD protein associated with the neuronal DHP receptor.[5] In addition to proteins possibly conserved in all VDCC, there is recent evidence that certain proteins, which localize to synaptic vesicle and plasma membranes, may be associated tightly with ^{125}I-labeled CTX binding sites. Anti-synaptotagmin[6] and anti-syntaxin (HPC-1/synaptocanalin I) antibodies[7-10] are capable of immunoprecipitating solubilized ^{125}I-labeled CTX receptor, indicating an association with a crude detergent-solubilized ωCTXR preparation. With the goal of defining the component proteins tightly associated with ωCTXR, we have used specific antisera to determine the extent of co-purification of the ωCTXR with the 110 kD protein, synaptotagmin, and a novel 58 kD protein.

MATERIALS AND METHODS

Monoclonal antibodies were produced in SJL mice, immunized with partially purified digitonin-solubilized chick brain membrane protein,[11] and hybridomas were generated and screened as described.[12] A monoclonal hybridoma cell line (mAb 9A7) was established by two successive clonings by limiting dilution and the antibody was purified from the serum-free culture media (SFM-101, Nissui, Tokyo) of the hybridoma cells by ammonium sulfate precipitation. The isotype of the immunoglobulin was G1. Anti–58 kD antibodies from rabbit serum were affinity purified against 58 kD protein transferred to PVDF, eluted with 8 ml 0.2 M glycine, pH 2.5, and immediately neutralized by the addition of 2 ml 100 mM Tris, pH 7.5, and 12% BSA.[13] Purification of the ωCTXR was according to published methods,[1,2] which include negative chromatography followed by hydroxyapatite chromatography (HA). The peak fractions of the HA column were pooled and applied sequentially to succinylated wheat germ agglutin (sWGA)–coupled Sepharose and wheat germ agglutinin (WGA)–coupled Sepharose and eluted with 0.5 M N-acetylglucosamine in the presence of asolectin. The radioligand binding assay for ^{125}I-labeled CTX was according to published methods.[2,14]

RESULTS AND CONCLUSIONS

Monoclonal antibody mAb 9A7 immunoprecipitates approximately 20% of ^{125}I-labeled CTX binding from sucrose gradient–purified detergent extracts and identifies a 100–110 kD protein predominant in brain (FIG. 1). However, it is also present in non-neuronal tissues, although detected at lower amounts (data not shown). Western analysis of the 100–110 kD protein recognized by mAb 9A7 and a 58 kD protein recognized by affinity purified anti–58 kD antibodies indicates that both proteins copurify with the ωCTXR throughout purification (FIG. 2). Synaptotagmin also associates with the ωCTXR throughout purification (FIG. 2,d); however the fractionation of synaptotagmin does not parallel the recovery of ^{125}I-labeled CTX binding sites. Unlike the 110 kD and 58 kD proteins, the association of synaptotagmin with the ωCTXR is highly dependent upon ionic strength (data not shown).

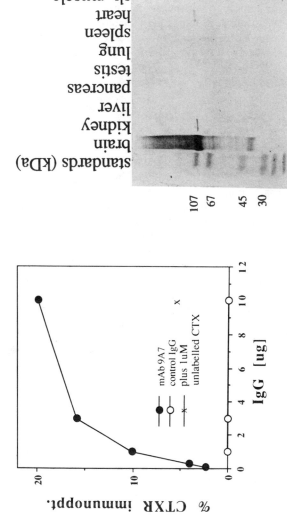

FIGURE 1. Immunoprecipitation of rat brain ωCTX receptor by mAb 9A7, which identifies a 110 kD antigen present predominantly in brain. (*Left*) Rat brain P2 membranes (5 mg/ml) were solubilized in 1% CHAPS, 10% (vol/vol) glycerol, 0.16 M sucrose, 5 mM EDTA, 1 mM EGTA, and 50 mM HEPES-Tris, pH 7.3. The solubilized proteins were separated by ultracentrifugation through a 5–20% sucrose linear gradient. The fraction with the highest specific activity of [^{125}I]CTX binding was mixed with 0.076 nM [^{125}I]CTX in the presence or absence of unlabeled 1 μM CTX and various amounts of either 9A7 or a control mouse IgG (MAC-L1) for 2 h at 4°C. The antigen-antibody complex was adsorbed onto anti-mouse IgG/protein A Sepharose CL-6B (Pharmacia) complex and 50 mM HEPES-Tris (pH 7.3), and the radioactivity in the immunoprecipitate was measured by gamma counting. The total amount of the solubilized receptor was determined by trapping onto polyethylenimine (0.3%)–treated Whatman GF/C filters. (*Right*): Membranes from various tissues were resolved on 4–17% gradient gels, transferred to polyvinylidene difluoride (PVDF) filters, probed with mAb 9A7, and detected with peroxidase-conjugated goat-anti-mouse IgG using 4-chloro-2-naphthol as substrate. Upon visualization with enhanced chemiluminescence (ECL) (Amersham), the 110 kD mAb 9A7 antigen was also detected in non-neuronal tissues. Immunohistochemical staining with mAb 947 indicates a plasma membrane location in brain, liver, adrenal medulla and cortex, skeletal muscle, pituitary, and intestine (data not shown).

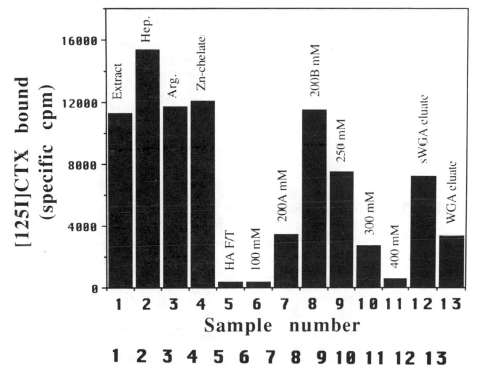

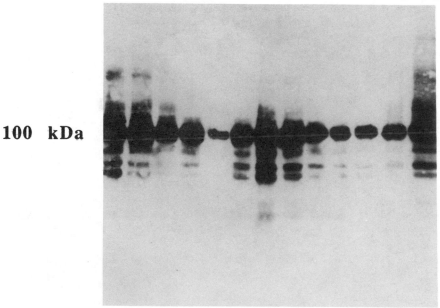

FIGURE 2. (a) top; (b) bottom. See page 391 for legend.

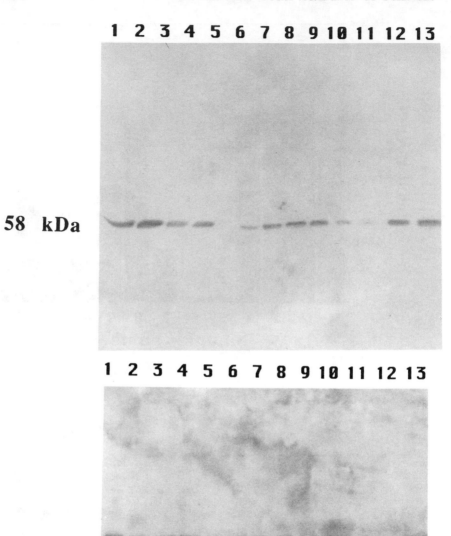

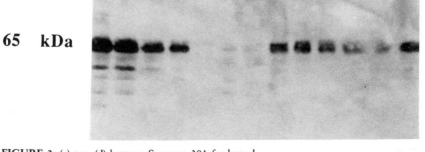

FIGURE 2. (c) top; (d) bottom. See page 391 for legend.

These results and previous studies indicate that synaptotagmin,[6] the 110 kD, and 58 kD components of the ωCTXR remain associated with the active ωCTXR by both immunoprecipitation (110 kD) and Western analysis of the purified ωCTXR. The tight association among the active ωCTX binding site, the 110 kD protein, and the 58 kD proteins suggests a functional or structural complex that may pre-exist in native neuronal plasma membranes and is preserved in the detergent-solubilized and highly purified ωCTXR preparation. The retention of ^{125}I-labeled CTX binding after the dissociation of the 110 kD protein from the ωCTXR complex[2] indicates the 110 kD protein is not required for an active ωCTXR and furthermore may have a cellular role other than an intrinsic subunit of the CTX-senstivie VDCC.

REFERENCES

1. McENERY, M. W., A. M. SNOWMAN, A. H. SHARP, M. E. ADAMS & S. H. SNYDER. 1991. Proc. Natl. Acad. Sci. USA **88:** 11095–11099.
2. McENERY, M. W. 1993. Methods Pharmacol. **7:** 3–39.
3. CURTIS, B. M. & W. A. CATTERALL. 1984. Biochemistry **23:** 2113–2118.
4. TAKAHASHI, M., M. J. SEAGAR, J. F. JONES, B. F. REBER & W. A. CATTERALL. 1987. Proc. Natl. Acad. Sci. USA **84:** 5478–5482.
5. AHLIJANIAN, M. K., R. E. WESTENBROEK & W. A. CATTERALL. 1990. Neuron **4:** 819–832.
6. LEVEQUE, C., T. HOSHINO, P. DAVID, Y. SHON-KASAI, K. LEYS, A. OMORI, B. LANG, O. EL FAR, K. SATO, N. MARTIN-MOUTOT, J. NEWSOM-DAVIS, M. TAKAHASHI & M. J. SEAGAR. 1992. Proc. Natl. Acad. Sci. USA **89:** 3625.
7. BENNETT, M. K., N. CALAKOS & R. H. SCHELLER. 1992. Science **257:** 255–259.
8. YOSHIDA, A., C. OHO, A. OMORI, R. KUWAHARA, T. ITO & M. TAKAHASHI. 1992. J. Biol. Chem. **267:** 24925–24928.
9. SAISU, H., K. IBARAKI, T. YAMAGUCHI, Y. SEKINE & T. ABE. 1991. Biochem. Biophys. Res. Commun. **181:** 59–66.
10. MORITA, T., H. MORI, K. SAKIMURA, M. MISHINA, Y. SEKINE, A. TSUGITA, S. ODANI, H. P. M. HORIKAWA, H. SAISU & T. ABE. 1992. Biomed. Res. **13:** 357–364.
11. TAKAHASHI, M. & Y. FUJIMOTO. 1989. Biochem. Biophys. Res. Commun. **163:** 1182–1188.
12. TAKAHASHI, M., Y. ARIMATSU, S. FUJITA, Y. FUJIMOTO, S. KONDO, T. HAMA & E. MIYAMOTO. 1991. Brain Res. **551:** 279–292.
13. HARLOW, E. & D. LANE. 1988. Antibodies: A Laboratory Manual. pp. 175–195. Cold Spring Harbor Laboratory. Cold Spring, N.Y.
14. WAGNER, J. A., A. M. SNOWMAN, A. BISWAS, B. M. OLIVERA & S. H. SNYDER. 1988. J. Neurosci. **8:** 3354–3359.

FIGURE 2. Co-fractionation of 100–110 kD and 58 kD throughout purification of ωCTXR. Representative fractions obtained throughout the purification of the rat brain CTXR[1] were resolved on a 4–17% gradient gel and transferred overnight to PVDF membrane. The total amount of specific [125I]CTX bound/sample is shown in (a). The PVDF filters were probed with mAb 9A7 (1/10,000 dilution), (c) affinity purified anti-58 kD antiserum (1/50 dilution) and (d) anti-65 kD/synaptotagmin (mAb 1D12) antibodies (1/500 dilution). Filters incubated with mAb 9A7 and anti-synaptotagmin were developed via ECL; anti-58 kD reactivity was detected with alkaline phosphatase coupled to goat anti-rabbit IgG.

Irregular Activity in the Giant Neurons from *Shaker* Mutants Suggests that the *Shaker* Locus May Encode Non-A-Type K$^+$ Channel Subunits in *Drosophila*

M. SAITO, M.-L. ZHAO, AND C.-F. WU

Department of Biology
University of Iowa
Iowa City, Iowa 52242

A gene that codes for a K$^+$ channel protein was first identified in *Drosophila* as a result of previous studies of behavior mutants of the *Shaker* (*Sh*) locus. Recent molecular studies show that the *Sh* gene can produce multiple transcripts by alternative splicing.[1-3] These transcripts have been expressed in *Xenopus* oocytes to form distinct voltage-activated K$^+$ channels. Some transcripts produce typical A-type currents (fast inactivating, I$_A$) and others produce delayed rectifier type currents (slow inactivating, I$_K$).[4,5]

In muscle, *Sh* mutations affect a particular type of K$^+$ current, the A-type current. However, molecular heterogeneity may be present in the A-type channels. For example, I$_A$ in the majority of neurons is independent of the effect of *Sh* mutations.[6,7] Nevertheless immunohistochemical staining and *in situ* hybridization suggest that *Sh* messages are present in a large portion of the nervous system in *Drosophila*.[8-10] In the present study, we examine the possibility that the *Sh* gene products can participate in forming not only the A-type but also the delayed rectifier type of K$^+$ channels in the nervous system.

Whole-cell voltage-clamp and current-clamp experiments are performed on the cultured giant *Drosophila* neurons derived from cell-division arrested embryonic neuroblasts.[11-13] The most extensively studied allele, *Sh*^{KS133} was used in this study. Because the site of mutation is localized in the conserved pore-forming H5 region in all *Sh* splicing variants,[5] the mutation should affect all neurons expressing *Sh* gene products. Voltage-activated K$^+$ currents were isolated, and the degree of inactivation was compared between wild type and *Sh*^{KS133} (FIG. 1). Interestingly, we observed that *Sh* neurons tended to show a more pronounced inactivation. A degree of inactivation less than 70% was not found in *Sh*^{KS133} neurons, but was not uncommon in wild-type neurons (16%).

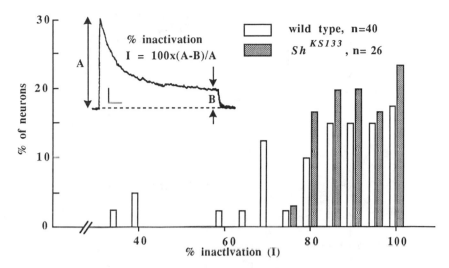

FIGURE 1. Sh^{KS133} neurons show more pronounced inactivation. Outward K⁺ current, induced by membrane depolarization to -10 mV, was isolated as differences between the currents attainable following preconditioning pulses of -100 and -20 mV.[11,12] Percent of inactivation was calculated (see inset for an example) as the ratio between the peak (A) and the amount of inactivation (A-B). Note that a degree of inactivation less than 70% was not found in Sh^{KS133} neurons, but not uncommon in wild-type neurons. *Open bars*: wild-type neurons. *Shaded bars*: Sh^{KS133}. Na⁺ and Ca²⁺ influxes were blocked by 100 nM TTX and 100 µM Cd²⁺, respectively. Scale bars: 80 pA, 50 msec.

Since voltage-activated K⁺ currents in different neurons represent mixtures of I_A and I_K, the ratio between these components is a key factor in determining the degree of inactivation of the total K⁺ current. A reduced I_K component in comparison to I_A could account for the Sh-dependent effects; all Sh neurons studied inactivated more than 50% (FIG. 1). It is consistent with the idea that some Sh gene products are utilized to form delayed rectifier type of K⁺ channels in embryonic neurons observed in this study (in contrast to previous reports in adult and prepupal neurons.[6,14]

Complex neuronal activity reflects contributions from a diversity of K⁺ channels. Under voltage clamp, it is difficult to distinguish and isolate Sh-dependent currents from the rest of K⁺ currents because of overlap in the biophysical properties. However, current-clamp experiments may be more effective in revealing defects in Sh-dependent currents, since a slight imbalance in the inward and outward currents may result in striking changes in action potential shapes. Voltage-responses in Sh^{K1333} neurons (16%) showed abnormalities such as plateau potentials and irregular oscillations, which were not seen in wild-type neurons (FIG. 2). Interestingly, similar irregular voltage-responses could be induced in wild-type neurons by applying blockers (TEA) of the delayed rectifier K⁺ channels, but not by 4-AP, which preferentially blocks I_A.[12]

Both voltage- and current-clamp studies reveal that a subpopulation (16%) of Sh^{KS133} neurons showed clear deficiency. Our results suggest that in the embryonic giant neurons, Sh gene products can be utilized to form non-A-type K⁺ channels. Since the same mutation disrupts I_A in a similar percentage of neurons in the pupal stage,[6]

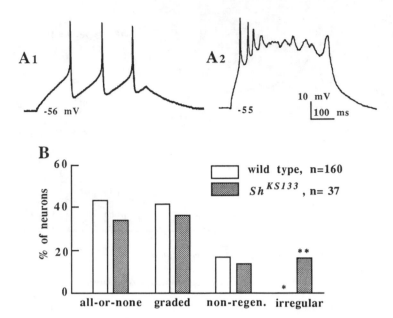

FIGURE 2. Irregular electrical activity in Sb^{KS133} neurons. A_1 and A_2 show sample records from wild-type and Sb^{KS133} neurons, respectively. (A_1) Normal all-or-none action potentials observed in a wild-type neuron. (A_2) Irregular action potentials in a Sb^{KS133} neuron. Voltage-response patterns are classified as 1 = all-or-none; 2 = graded; and 3 = non-regenerative. (For details, see refs. 11 and 12). Irregular responses were seen only in Sb^{SK133} neurons (16% or 6/37, **), but not in wild-type neurons (*). *Open bars:* wild-type neurons, *shaded bars:* Sb^{KS133}.

it is possible that *Sb* splicing products can be regulated differently at the embryonic and pupal stages.

REFERENCES

1. BAUMANN, A., A. GRUPE, A. ACKERMAN & O. PONGS. 1988. EMBO J. **7**: 2457–2464.
2. KAMB, A., J. TSENG-CRANK & M. TANOYUE. 1988. Neuron **1**: 421–430.
3. TIMPE, L. C. T. SCHWARTZ, B. TEMPEL, D. PAPAZIAN, Y. N. JAN & L. JAN. 1988. Nature **331**: 143–145.
4. IVERSON, L. E. & B. RUDY. 1990. J. Neurosci. **10**: 2903–2916.
5. STOCKER, M., W. STUHMER, R. WITTKA, X. WANG, R. MULLER, A. FERRUS & O. PONGS. 1990. Proc. Natl. Acad. Sci. USA **87**: 8903–8907.
6. BAKER, K. & L. SALKOFF. 1990. Neuron **2**: 129–140.
7. SOLC, C. K., W. N. ZAGOTTA & R. W. ALDRICH. 1987. Science **236**: 1094–1098.
8. PONGS, O., N. KECSKEMETHY, R. MULLER, I. KRAH-JENTGENS, A. BAUMANN, H. H. KILTZ, I. CANAL, S. LLAMAZARES & A. FERRUS. 1988. EMBO J. **7**: 1087–1096.
9. SCHWARZ, T. L., D. M. PAPAZIAN, R. C. CARRETTO, Y. N. JAN & L. Y. JAN. 1990. Neuron **2**: 119–127.
10. TSENG-CRANK, J., J. A. POLLOCK, I. HAYASHI & M. A. TANOYUE. 1991. J. Neurogenetics **7**: 229–239.

11. SAITO, M. & C. F. WU. 1991. J. Neurosci. 11: 2135–2150.
12. SAITO, M. & C. F. WU. 1993. In Comparative Molecular Neurobiology. Y. Pichon, Ed.: 365–388. Birkhauser Verlag.
13. WU, C. F., K. SAKAI, M. SAITO & Y. HOTTA. 1990. J. Neurobiol. 21: 499–507.
14. TANOYUE, M., A. FERRUS & S. C. FUJITA. 1981. Proc. Natl. Acad. Sci. USA 78: 6548–65552.
15. LICHTINGHAGEN, R., M. STOCKER, R. WITTKA, G. BOHEIM, W. STUHMER, A. FERRUS & O. PONGS. 1990. EMBO J. 9: 4399–4407.

Single-Channel Recordings of an ATP-Sensitive K Channel in Rat Cultured Cortical Neurons

T. OHNO-SHOSAKU AND C. YAMAMOTO

Department of Physiology
Faculty of Medicine
Kanazawa University
Kanazawa 920, Japan

The presence of ATP-sensitive K^+ (K_{ATP}) channels in the central nervous system has been demonstrated by electrophysiological studies and radioligand binding studies with specific K_{ATP} channel inhibitors.[1] There are several reports on properties of single K_{ATP} channels in the nervous tissue and most of them have indicated that K_{ATP} channels in central neurons are quite different in properties from the "classical" K_{ATP} channels originally found in pancreatic β cells and cardiac muscle cells.[2] In this study, we demonstrate that rat cortical neurons have a K_{ATP} channel quite similar to that in pancreatic β cells.

Primary cultures were prepared from the cortex of 1- to 5-day-old neonatal rats, and the inside-out, outside-out, and whole-cell configurations of the patch-clamp technique were applied to the cultured cells 4–12 days after seeding. FIGURE 1 shows an example of single-channel recordings from inside-out patches containing ATP-sensitive channels and current-voltage relationships of these channels under two different ionic conditions. As shown in FIGURE 1(A), the channel with a single-channel conductance of 65 pS (in 145 mM/145 mM K^+) was completely inhibited by application of 0.2–1 mM ATP to the cytoplasmic surface of the inside-out patch. The current-voltage relationships of the ATP-sensitive channel indicated that the channel is a K^+-selective channel, namely, a K_{ATP} channel. The K_{ATP} channel was inhibited by a K_{ATP} channel inhibitor, tolbutamide (0.5 mM), and activated by a K_{ATP} channel activator, diazoxide (0.5 mM), when these drugs were applied to the extracellular surface of outside-out patches (FIG. 2,A). These properties are quite similar to those of the K_{ATP} channel in pancreatic β cells. In whole-cell recordings, effects of these drugs on the membrane potential and the conductance were examined. Extracellular application of 0.5 mM diazoxide induced a pronounced hyperpolarization (FIG. 2,B) concomitant with an increase in membrane conductance, whereas 0.5 mM tolbutamide evoked a depolarization concomitant with a decrease in conductance. The number of K_{ATP} channels per neuron was roughly estimated from the data obtained in the whole-cell and the single-channel recordings, on the assumptions that all the overall tolbutamide- and diazoxide-sensitive K^+ conductance originated from the K_{ATP} channel characterized

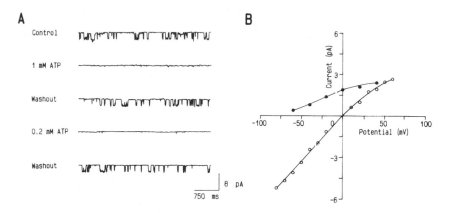

FIGURE 1. Blocking effects of cytoplasmic ATP on K⁺ channels and the current-voltage relationships obtained with inside-out patches. (A) The pipette was filled with a high-K⁺ solution containing (mM) 145 KCl, 1 MgCl₂, 2 CaCl₂, and 10 HEPES/NaOH (pH 7.3), and the bath solution was a high-K⁺/low-Ca²⁺ solution containing 125 KCl, 1 MgCl₂, 1 CaCl₂, 20 KOH, 10 HEPES, and 5 EGTA (pH 7.3) with or without 0.2–1 mM ATP. The membrane potential was clamped at −50 mV. (B) The pipette contained the high-K⁺ solution (*open circles*) or a normal solution containing 140 NaCl, 5 KCl, 1 MgCl₂, 2, CaCl₂, 10 HEPES/NaOH, 10 glucose, and 1 μM tetrodotoxin (pH 7.3). The slope conductances were 65 pS (*open circles*) and 26 pS (*closed circles*). The lines were fitted by eye.

here, and that the open probability of the channel is 100% in the presence of diazoxide and 0% in the presence of tolbutamide. The estimated number of K_{ATP} channels ranged between 20 and 150 per neuron. If neurites or presynaptic terminals also contain K_{ATP} channels, a real number of K_{ATP} channels per neuron will be higher than the estimated number.

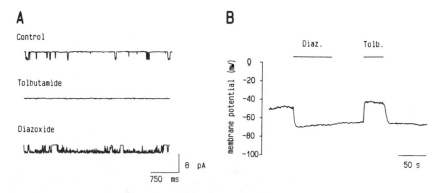

FIGURE 2. Effects of tolbutamide and diazoxide on a single K_{ATP} channel in an outside-out patch (A) and on the membrane potential of a cortical neuron (B). The pipette solution was the high-K⁺/low-Ca²⁺ solution, and the bath contained the high-K⁺ solution with or without 0.5 mM tolbutamide or 0.5 mM diazoxide. (A) The membrane potential was clamped at −50 mV. (B) The membrane potential was measured at the current clamp mode with the whole-cell configuration.

REFERENCES

1. MOURRE, C., Y. BEN ARI, H. BERNARDI, M. FOSSET & M. LAZDUNSKI. 1989. Brain Res. **486**: 159–164.
2. ASHFORD, M. L. J., N. C. STURGESS, N. J. TROUT, N. J. GARDNER & C. N. HALES. 1988. Pfluegers Arch. **412**: 297–304.

Mechanical Modulation of M-Current in Cultured Bullfrog Sympathetic Ganglion Cells

S. HARA AND K. KUBA[a]

Department of Physiology
Saga Medical School
5-1-1, Nabeshima
Saga 849, Japan

Under the whole-cell patch clamp condition, local flow of a solution (identical to the bathing solution) from a micropipette to a cell (FIG. 1), but not other mechanical stimuli, produced a non-inactivating outward (in 34 cells out of 141) or inward (in 70 cells) current ($I_{f(out)}$ or $I_{f(in)}$, respectively) depending on the cell (FIG. 2,Aa and Ba).

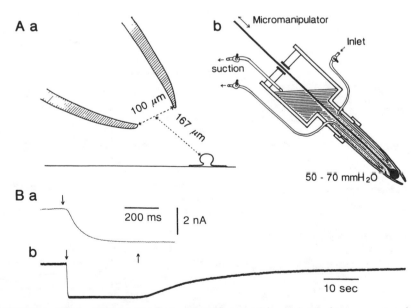

FIGURE 1. The local solution flow device (**A**) and the rate of changes in membrane current produced by local flow of a high K^+ (20 mM) solution (**B**). The tip of the micropipette (Aa) was opened or closed by pulling or pushing a rod with an oil-operated micromanipulator (Ab).[2]

[a] To whom correspondence should be addressed.

Both $I_{f(out)}$ and $I_{f(in)}$ appeared at voltages more positive than -60 mV (FIG. 2,Ab and Bb), accompanied by increases and decreases, respectively, in membrane conductance and current relaxation (FIG. 2,Ac and Bc) to a voltage jump between -30 mV and -55 mV without a change in its time constant (whose value was similar to that of a voltage-dependent non-inactivating K^+ current known as I_M[1]), reversed at a membrane potential close to the equilibrium potential for K^+, and blocked by Ba^{2+} (4–8 mM), and muscarine (10 µM: the latter produced either an "apparent inward" or outward current).[2] Thus $I_{f(out)}$ and $I_{f(in)}$ are generated by the activation and inhibition, respectively, of I_M by the local flow of solution. Although current responses were not seen, the mechanical effects took place even at -70 mV, as $I_{f(out)}$ or $I_{f(in)}$ appeared on shifting membrane potential to -30 mV immediately after the local flow.[2] A transient outward current activated by a voltage jump from -85 mV (or -75 mV) to -30 mV was little affected by local flow of a solution that produced $I_{f(out)}$ or $I_{f(in)}$. Since negative pressure to, or the lateral, forward, or backward movement of, a recording patch pipette affected little the membrane current of a ganglion cell, it seems that $I_{f(out)}$ and $I_{f(in)}$ were produced as a result of the flow-induced shear stress to the cell membrane.

Although the physiological functions of $I_{f(out)}$ and $I_{f(in)}$ are not known in sympathetic ganglion cells, they seem to play, in general, some roles in the regulation and modulation of the neuronal membrane excitability under a variety of mechanical stresses during development and plastic activities.

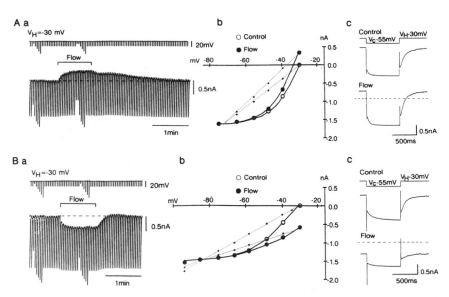

FIGURE 2. $I_{f(out)}$ (**A**) and $I_{f(in)}$ (**B**). (Aa and Ba) Membrane currents and voltages before, during, and after the local flow of Ringer. Hyperpolarizing voltage pulses (700 msec) of constant (20 mV) or various (10–65 mV) amplitudes were applied throughout the records. (Ab and Bb) Current/voltage relationships before and during a solution flow. (Ac and Bc) Current responses to a hyperpolarizing voltage pulse before and during a solution flow.

REFERENCES

1. ADAMS, P. R., D. A. BROWN & A. CONSTANTI. 1982. J. Physiol. (Lond.) **330:** 537–572.
2. HARA, S. & K. KUBA. 1992. Pflügers Arch. **422;** 305–315.

Alterations of Gating Parameters by Neutral Substitutions of Transmembrane Leu^{52} of Slow Potassium Channel

SHIGETOSHI OIKI,[a] TORU TAKUMI,[b]
YASUNOBU OKADA,[c]
AND SHIGETADA NAKANISHI[b]

[a]Department of Physiology
[b]Institute for Immunology
Kyoto University Faculty of Medicine
Kyoto 606-01, Japan

[c]Department of Cellular and Molecular Physiology
National Institute for Physiological Sciences
Okazaki 444, Japan

The slow potassium channel induced by the I_{sK} protein is unique in both structural and functional aspects.[1] The I_{sK} proteins have only 129 or 130 amino acid residues with a single transmembrane segment (FIG. 1,A) that is highly conserved among species.[2–6] Although its structure is very simple and differs completely from that of other known ion channel proteins, the slow K channel possesses all properties of the voltage-gated K$^+$ channel. The structure-function relationship of this channel has been studied using site-directed mutagenesis.[7,8] In an earlier paper,[7] we observed that a very conservative neutral substitution from leucine (Leu^{52}) to isoleucine (L52I) in the transmembrane

---

FIGURE 1. Mutational changes in the slow K channel activation. (**A**) The primary structure of the transmembrane segment of the wild I_{sK} protein and the mutants. Single-letter code is used. No net charge exists in the segment. Leu^{52} (wild-type; WT) was replaced by either Ile (L52I) or Ala (L52A) through site-directed mutagenesis. (**B**) Current records from *Xenopus* oocytes injected with the mRNA synthesized *in vitro* from a wild type cDNA or a mutant cDNA. Whole oocyte currents elicited by 14 sec-depolarizing pulses (from -20 mV to $+60$ mV) were recorded at 21°C. Currents did not reach steady-states during 14 sec pulses. The holding potential was -70 mV. Two microelectrode voltage clamp studies were performed in ND96 solution (NaCl 96 mM, KCl 2 mM, CaCl$_2$ 1.8 mM, MgCl$_2$ 1 mM, HEPES 5 mM, pH 7.4) under perfusion. Defolliculation was done just prior to the electrophysiological experiment. pClamp was used for applying the voltage command, for the sampling and the data analysis. Leakage currents were subtracted.

A

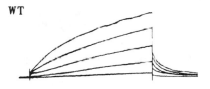

B

WT

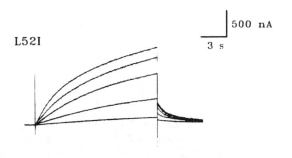

L52I

L52A

segment brought about apparently faster activation.[7] In the present study, shifts of the voltage-dependent activation curve were observed in L52I and L52A mutants in the opposite direction.

FIGURE 1(B) shows the difference of activation time course between the L52I channel and the wild type or the L52A. To see the underlying effects of the mutation quantitative kinetic analyses were performed. In FIGURE 2, two time constants (A) and the relative contribution of the fast component in the current amplitude (B) are shown as a function of voltages. Both mutations did not induce marked changes in both fast and slow time constants. The most prominent difference was a contribution of the fast component: At less depolarized potentials the fast component was negligible in the wild type and the L52A channel, whereas a considerable contribution of the fast component was found in the L52I channel even at 0 mV. FIGURE 2(C) shows the voltage-dependent activation for the sum of the two components. The activation curve was shifted to the left (more negative potentials) for the L52I mutant, whereas the right shift was observed for the L52A mutant. The effect of the mutations on the gating charge was not obvious in both mutants.

An isomeric substitution (from Leu to Ile), where the location of a methyl group is shifted only for a distance of the C-C bond, induced marked alterations in both the gating kinetics and the activation curve, whereas a relatively large volume change in the sidechain (from Leu 166.7 $\mathring{A}^3$ to Ala 88.6 $\mathring{A}^3$) induced a change only in the activation curve. There is no simple interpretation for these effects of the mutations.

FIGURE 2. Kinetic parameters and activation curves of the slow K channels. All current traces could be fitted by the double-exponential function as $I(t) = I_{const} + A_f \exp[-t/\tau_f] + A_s \exp[-t/\tau_s]$ at least during 14 sec pulses. Different symbols represent the wild type (*circles*), L52I (*triangles*), and L52A (*squares*). Data are shown as mean ± SE. Number of the experiments were 15 for the wild type, 18 for L52I, and 7 for L52A. (A) Fast (τ_f, *open symbols*) and slow (τ_s; *closed symbols*) time constants for the wild type and the mutants as a function of voltages. Voltage dependency of the τ_s was not notable. The τ_s of the L52I channel showed significantly higher value at 20 mV. There is no statistical significance for the difference of τ_s between the wild type and the L52A mutant. The τ_f became shorter with increasing the voltage in both the wild type and the mutants. There is no statistical significance for the difference between the wild type and either mutant. (B) The relative contribution of the fast component (A_f) to the sum of both fast and slow components. As the voltage became less positive the relative contribution of the fast component for the wild type and the L52A channel decreased more prominently than that of the L52I. There is no statistical difference between the wild type and the L52A. At −20 mV and 0 mV the data points for the L52I mutant are significantly different from those of the wild type. (C) Voltage-dependent activation curves. Normalized chord conductances (G) were calculated for the sum of the two components by $(A_f + A_s)/(V - E_K)$, where E_K is the equilibrium potential for potassium. Even if the current does not reach the steady-state in this time range, the equation would have the physical basis and may represent a composite function of rate constants in the partial state diagram around the initial activating process. At 0 mV the data points for both mutant channels are significantly different from those of the wild type. Activation curves were fitted by the Boltzmann distribution: $G = G_{max} / (1 + \exp[z \cdot e \cdot (V - V_{1/2})/kT])$, in which z is the effective valence, e the elementary charge, k the Boltzmann constant, T the absolute temperature and $V_{1/2}$ the half activation voltage. For the wild type G_{max} is 17 µS, z is 1.3, and $V_{1/2}$ is −14 mV. For the L52I, z is fixed as 1.3 because of the lack of points at more negative potentials. Thus, the $V_{1/2}$ value (−39 mV) is not very accurate. G_{max} is 11 µS. For the L52A, G_{max}, z and $V_{1/2}$ are 9 µS, 2.1, and, 0.2 mV, respectively.

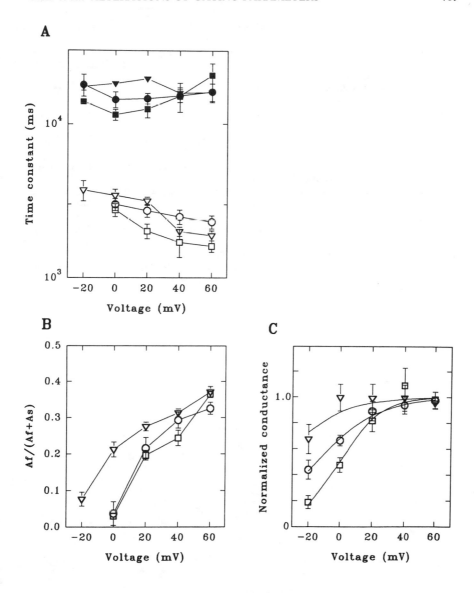

Neutral substitutions introduced in other ion channels resulted in altered gating behavior.[9-12] The mechanism has been assigned to a hydrophobic interaction, such as leucine zipper[11] (the sequence of the leucine zipper interaction is Leu > Ile > Ala) or just volumetric exclusion[12] (Leu = Ile > Ala). Neither of them can explain our results.

REFERENCES

1. TAKUMI, T., H. OHKUBO & S. NAKANISHI. 1988. Science 242: 1042–1045.
2. MURAI, T., A. KAKIZUKA, T. TAKUMI, H. OHKUBO & S. NAKANISHI. 1989. Biochem. Biophys. Res. Commun. 161: 176–181.
3. FOLANDER, K., J. S. SMITH, J. ANTANAVAGE, C. BENNETT, R. B. STEIN & R. SWANSON. 1990. Proc. Natl. Sci. USA 87: 2975–2979.
4. PRAGNELL, M., K. J. SNAY, J. S. TRIMMER, N. J. MacLUSKY, F. NAFTOLIN, L. K. KACZMAREK & M. B. BOYLE. 1990. Neuron 4: 807–812.
5. IWAI, M., M. MASU, K. TSUCHIDA, T. MORI, H. OHKUBO & S. NAKANISHI. 1990. J. Biochem. 108: 200–206.
6. HONORE, E., B., ATTALI, G. ROMEY, C. HEURTEAUX, P. RICARD, F. LESAGE, M. LAZDUNSKI & J. BARHANIN. 1991. EMBO J. 10: 2805–2811.
7. TAKUMI, T., K. MORIYOSHI, I. ARAMORI, T. ISHII, S. OIKI, Y. OKADA, H. OHKUBO & S. NAKANISHI. 1991. J. Biol. Chem. 266: 22192–22198.
8. GOLDSTEIN, S. A. N. & C. MILLER. 1991. Neuron 7: 403–408.
9. AULD, V. J., A. L. GOLDIN, D. S. KRAFTE, W. A. CATTERALL, H. A. LESTER, N. DAVIDSON & R. J. DUNN. 1990. Proc. Natl. Acad. Sci. USA 87: 323–327.
10. YOOL, A. J. & T. L. SCHWARZ. Nature 349: 700–704.
11. McCORMACK, K., M. A. TANOUYE, L. E. IVERSON, J.-W., LIN, M. RAMASWAMI, T. McCORMACK, J. T. CAMPANELLI, M. K. MATHEW & B. RUDY. 1991. Proc. Natl. Acad. Sci. USA 88:2931–2935.
12. LO, D. D., J. L. PINKHAM & C. F. STEVENS. 1991. Neuron 6: 31–40.

Ca²⁺-Activated K Channel in Vas Deferens Smooth Muscle Cells

K. MORIMOTO, F. KUKITA,[a] AND S. YAMAGISHI[a]

Shokei Junior College
Shimizumachi, Kumamoto, 860 Japan

[a]*National Institutes for Physiological Sciences*
Okazaki 444, Japan

Ca²⁺-activated K channels were observed in many smooth muscle cells[1-3] and these channels were demonstrated to be dominant K channels for the physiological functions in guinea pig vas deferens smooth muscle cells.[4] Ca²⁺-activated K channels usually

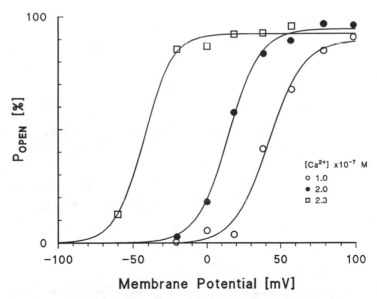

FIGURE 1. Effects of membrane potential on an open probability (P_{open}) at a different [Ca²⁺]$_i$ during single event activity. The solid line is obtained by fitting the data points with the following equation:

$$P(V) = \frac{e^{\alpha(V-\beta)}}{1 + e^{\alpha(V-\beta)}}, \quad \alpha = 2\frac{zF}{RT} \quad B = V_{\frac{1}{2}}$$

where αs are 0.085, 0.093, and 0.105 and βs (in mV) are 42.1, 14.6, and −42.6 at [Ca²⁺]$_i$ of 1 × 10⁻⁷, 2 × 10⁻⁷, and 2.3 × 10⁻⁷ M, respectively.

have sensitivities to a membrane potential and $[Ca^{2+}]_i$.[4,5] The single-channel properties of this channel is the main interest in the present paper.

The single-channel activity was analyzed with the inside-out patch clamp method. The solutions used in the experiments contained (in mM) 119 NaCl, 30 KCl, 2.3 $CaCl_2$, 1.2 $MgCl_2$, 5.8 $KHCO_3$, 5.8 HEPES, and 11.7 glucose for the external solution and 145.7 KCl, 1.4 $MgCl_2$, 5.2 HEPES, and 1 EGTA for the internal solution. The free Ca^{2+} concentration in the internal solution was calculated using the standard method. We could not obtain a pure single-channel patch, but obtained some patches containing two channels that behaved as a single channel. We analyzed only the single event activities in these patches. Single-channel recording were analyzed by pClamp (Version 5.15, Axon Instruments, Inc.).

Ca^{2+}-activated K channels in the cells were highly sensitive to the potassium concentration of both sides of the membrane and the reversal potentials in I-V relation were well within the E_K values calculated from Goldman's equation. The single-channel I-V relation did not change when $[Ca^{2+}]_i$ was changed and the single-channel conductance was approximately 90 pS.

Open probabilities (P_{open}) as a function of membrane potential at three different $[Ca^{2+}]_i$ are plotted in FIGURE 1. The experimental data were fitted by Boltzmann's distribution function and the $V_{1/2}$, which is the membrane potential where P_{open} was

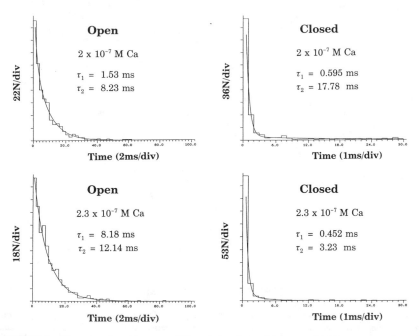

FIGURE 2. The open time histogram (*left*) and closed time histogram (*right*). τ_1 and τ_2 show the time constants of two exponential distribution of open and closed times. The open time distribution was analyzed between 0 and 100 msec with a 2 msec bin and the closed time distributions for 2 × 10^{-7} M Ca^{2+} and for 2.3 × 10^{-7} M were analyzed between 0 and 180 msec and between 0 and 25 msec, respectively, with a 1 msec bin.

half maximal, was a linear function of $\ln[Ca^{2+}]_i$. An apparent Hill coefficient in the relation of P_{open} versus $[Ca^{2+}]_i$ could be reduced from the $P_{open} - V$ relation. A slope of $V_{1/2} - \ln[Ca^{2+}]_i$ relation was around 8, showing a marked dependency of P_{open} on $[Ca^{2+}]_i$.

Both the open and the closed time histograms consist of two exponential components. In the open histogram, the time constant τ_1 of the fast component, which was a major component, increased markedly but the time constant τ_2 of the slow component increased a little, when the membrane was depolarized and $[Ca^{2+}]_i$ was increased. In the closed time histogram, the time constant τ_2 decreased when $[Ca^{2+}]_i$ was increased but τ_2 was not affected significantly by $[Ca^{2+}]_i$.

REFERENCES

1. INOUE, R., K. OKABE, K. KITAMURA & H. KURIYAMA. 1986. Pflüg. Arch. **406:** 138–143.
2. KUME, H., K. TAKAGI, T. SATAKE, H. TOKUNO & T. TOMITA. 1990. J. Physiol. (Lond.) **424:** 445–457.
3. BENHAM, C. D., T. B. BOLTON, R. J. LANG & T. TAKEWAKI. 1986. J. Physiol. (Lond.) **371:** 45–67.
4. MORIMOTO, K., F. KUKITA & S. YAMAGISHI. (In preparation).
5. BARRETT, J. N., K. L. MAGLEBY & B. S. PALLOTTA. 1982. J. Physiol. (Lond.) **331:** 211–230.

Action and Binding of Calcium Channel Blockers on the Putative Calcium Channel of Synaptosomal Plasma Membrane from the Electric Organ of *Narke japonica*

HIROSHI TOKUMARU, CHEN-YUEH WANG,
NAOHIDE HIRASHIMA, TOMOHIRO O'HORI,
HARUHIKO YAMAMOTO,[a] AND YUTAKA KIRINO[b]

Faculty of Pharmaceutical Sciences,
Kyushu University
Maidashi 3-chome
Higashi-ku, Fukuoka 812, Japan

[a]Faculty of Science
Kanagawa University
Tsuchiya, Hiratsuka 259-12, Japan

Voltage-dependent Ca channels control Ca^{2+} influx to the presynaptic nerve terminal and the neurotransmitter is released. However, characterization and identification of Ca channel(s) need further work. Synaptosomes isolated from the electric organ of electric rays consist of cholinergic nerve terminals and are free from postsynaptic membranes, characteristics which make this preparation most suitable for pharmacological and biochemical characterizations of calcium channels of presynaptic terminal membranes.

We prepared synaptosomes from a Japanese electric ray (*Narke japonica*). Acetylcholine (ACh) released from these synaptosomes depolarized with a high concentration of potassium ions was measured using a chemiluminescence method.[1] Addition of 20 μM ω-conotoxin GVIA (ω-CgTX), known to be a blocker for N-type Ca channels, almost completely inhibited the evoked release of ACh. The concentration required for 50% inhibition (IC_{50}) was 7 μM, a value close to that (8 μM) for the intrasynaptosomal Ca^{2+} increase measured with Fura-2.[1] This means that inhibition of the release of ACh is the result of a blockade of the influx of Ca^{2+}. Assay using a radioiodinated toxin derivative revealed a single-type binding site with a dissociation constant of 8 μM, again close to the IC_{50} values above. Autoradiography with SDS-PAGE analysis

[b] To whom all correspondence should be addressed.

after covalent cross-linking of the labeled toxin, using disuccinimidyl suberate, revealed the 170 kD peptide to be an ω-CgTX receptor.

Contrary to previous reports,[2,3] dihydropyridines (DHPs) known to block L-type Ca channels also inhibited (in the micromolar range) ACh release from the synaptosomes. Addition of 10 μM S-12967 completely inhibited ACh release – the inhibition-concentration relationship is shown in FIGURE 1. IC_{50} values for S-12967 and for other DHPs are shown in TABLE 1. Assay of a DHP binding to synaptosomal plasma membrane using [³H]PN 200-110 revealed a specific binding site with $K_d = 3$ μM and $B_{max} = 147$ pmol/mg protein.

The present finding that ω-CgTX or a DHP inhibits almost completely the release of ACh indicates that these two types of Ca channel blockers act on the same and single type of Ca channel through which Ca^{2+} enters the nerve terminal. Similarity

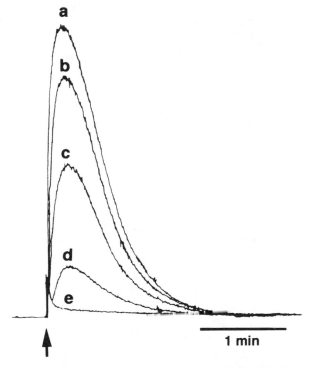

FIGURE 1. Inhibition with water-soluble DHP, (+)-S-12967 of ACh release from synaptosomes, as evoked by depolarization with a high concentration of K⁺ ion following 10 min preincubation with or without the Ca channel blocker. The concentration (in μM) of the blocker is: (a), 0; (b) 0.5; (c) 1; (d) 5; (e) 10. The released ACh was determined continuously using the chemiluminescence method. At the time indicated by the arrow, the concentration of KCl had increased from 3 mM to 40 mM by injection of 10 μl of 1 M KCl into 250 μl of synaptosomal suspension (200 μg protein/ml) present in physiological medium consisting of 280 mM NaCl, 3 mM KCl, 1.8 mM MgCl₂, 3.4 mM CaCl₂, 5.5 mM glucose, 50 mM urea, 400 mM sucrose, 40 mM HEPES-Na (pH 7.4), and chemiluminescence agents (20 U/ml choline oxidase, 6 μg/ml peroxidase, 50 μM luminol). The sharp peak at the beginning is an artifact related to injection of the stimulant.

TABLE 1. Characteristics of Ca Channel Blockers for Inhibition of Evoked ACh Release from Synaptosomes and for Their Binding to Synaptosomal Plasma Membranes[a]

Ca Channel Blocker	IC_{50}^{b} (μM)	K_d (μM)	B_{max} (pmol/mg protein)
ω-CgTX	7	8	200
DHPs			
PN 200-110		2.6	147
Nicardipine[b]	8		
(+)-S-12967[c]	2		
(−)-S-12968[c]	2		

[a] Binding assay of Ca channel blockers was done using [^{125}I]-labeled ω-CgTX or [^{3}H]PN 200-110. The binding of the labeled blocker to synaptosomal plasma membranes was measured using a rapid filtration method and a glass-fiber filter. The difference between bindings observed in the absence or presence of excess amounts of the non-labeled compound ((+)-S-12967 for [^{3}H]PN 200-110) was taken as specific binding. Dissociation constant (K_d) and maximum binding capacity (B_{max}) was obtained from Scatchard plots.

[b] From Yamanouchi Pharmaceutical Co., Tokyo, Japan.

[c] From Institut de Recherches Internationales Servier, Courbevoie, France.

in the density of receptors of the two types of Ca channel blockers may support this notion. The Ca channel probably contains a subunit of the 170 kD peptide with ω-CgTX binding site(s).

REFERENCES

1. O'HORI, T., C.-Y. WANG, H. TOKUMARU, L.-C. CHEN, K. HATANAKA, N. HIRASHIMA & Y. KIRINO. 1993. Neuroscience. 54: 1043–1050.
2. FARIÑAS, I., C. SOLSONA & J. MARSAL. 1992. Neuroscience 47: 641–648.
3. YEAGER, R. E., D. YOSHIKAMI, J. RIVIER, L. CRUZ & G. P. MILJANICH. 1987. J. Neurosci. 7: 2390–2396.

Developmental Changes and Modulation through G-Proteins of the Hyperpolarization-Activated Inward Current in Embryonic Chick Heart

HIROYASU SATOH AND NICHOLAS SPERELAKIS[b]

Department of Pharmacology
Nara Medical University
Kashihara, Nara 634, Japan

[b]*Department of Physiology and Biophysics*
University of Cincinnati
College of Medicine
Cincinnati, Ohio 45267

Hyperpolarization-activated inward current (I_f) is one of the inward currents contributing to membrane depolarization to threshold during diastole. Young embryonic chick ventricular myocytes also exhibit spontaneous activity, which generates slow-rising action potentials. In embryonic chick ventricular myocytes, whole-cell voltage-clamp experiments were performed to examine modulation of the I_f. Long-duration (3 sec) hyperpolarizing pulses were applied from holding potential of -30 mV to steps of -40 to -120 mV. Intracellular Ca^{2+} concentration was chosen to give a pCa of 7 to facilitate the I_f amplitude, according to an equation. I_f was marked in 3-day-old cells, diminished at 10 days, and almost completely gone at 17 days (Fig. 1,A–C). The I_f current density (at -120 mV) was -6.8 ± 1.2 pA/pF ($n = 23$) in 3-day-old cells, -3.1 ± 1.1 pA/pF ($n = 17$) in 10-day cells, and -2.2 ± 0.7 pA/pF ($n = 12$) in 17-day cells (Fig. 1,D). The reduction of I_f paralleled the decrease in spontaneous activity. The voltage of the half-activation (measured from the tail currents) was -92.0 ± 2.4 mV ($n = 3$) in 3-day cells, and was -87.2 ± 2.1 mV ($n = 3$) in 10-day cells. In 3-day-old cells, the threshold potential (V_{th}) was -50 to -60 mV, and the reversal potential was -13.4 ± 1.3 mV ($n = 3$). The amplitude of I_f was enhanced by $12.1 \pm 1.8\%$ (at -120 mV) at 30°C compared with 20°C. The time course of activation was also temperature dependent, and was fitted by a single exponential: τ was 1.3 ± 0.4 sec at 20°C and was 0.7 ± 0.4 sec at 30°C. Cs^+ (3 mM) completely blocked I_f and had a negative chronotropic effect. The rate of spontaneous action potentials (with less than -60 mV of the maximum diastolic potential) was

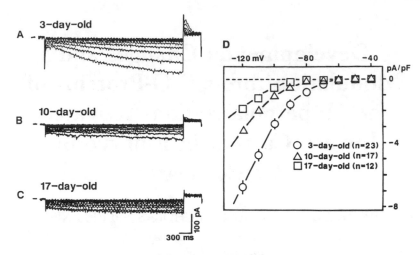

FIGURE 1. Developmental change in the hyperpolarization-activated inward current (I_f) in young embryonic chick ventricular cells. Test pulses were applied between -40 to -120 mV, in 10 mV increments, from a holding potential of -30 mV. Then, the pulses were constantly stepped to $+20$ mV to measure the tail current. (**A**) Presence of large I_f in a 3-day-old cell. (**B**) Smaller I_f in a 10-day-old cell. (**C**) Absence of I_f in a 17-day-old cell. (**D**) Current-voltage relationships for I_f current density in the three developmental stages. Symbols used are: 3-day (*open circles*), 10-day (*triangles*), and 17-day-old cells (*squares*). The values are plotted as mean $\pm$ SEM. The capacitances were 12.7 ± 2.0 pF ($n = 17$) in 3-day cells, 9.9 ± 2.1 pF ($n = 14$) in 10-day cells, and 8.5 ± 2.6 pF ($n = 14$) in 17-day-old cells. External solution included 3 mM $BaCl_2$, 10 μM TTX, and 3 mM $CdCl_2$. The short line at the left of the current records in **A–C** represents the zero current level.

decreased by $60.9 \pm 3.4\%$. Therefore, the I_f current may not be a major contributor to the pacemaker potential.

Isoprenaline (ISO; 1 μM) caused a positive chronotropic effect ($17.1 \pm 2.9\%$, $n = 5$) and increased I_f by $65.2 \pm 5.6\%$ (at -120 mV; $n = 7$). This agonist also re-started automaticity in Cs^+-induced quiescent preparations. Carbachol (CCh; 0.1 μM) had a negative chronotropic effect ($26.3 \pm 3.4\%$, $n = 5$), and decreased I_f by $41.2 \pm 1.3\%$ (at -120 mV; $n = 7$). τ at -120 mV was reduced to 1.0 ± 0.2 sec ($n = 5$) by ISO, and increased to 1.5 ± 0.4 sec ($n = 3$) by CCh. CCh (0.1 μM) also reversed the enhancement of I_f produced by ISO ($n = 3$). Intracellular application of 100 μM GTP-γS (non-hydrolyzable GTP analogue) decreased basal I_f by $35.2 \pm 5.0\%$ ($n = 17$), but potentiated the stimulant effect of 1 μM ISO (by $37.8 \pm 4.7\%$, $n = 9$) and the inhibitory effect of 0.1 μM CCh ($21.2 \pm 4.3\%$, $n = 9$) (FIG. 2, A and B). Effects of ISO and CCh on I_f were potentiated further by GTP-γS. It is likely that direct G-protein pathways connect both muscarinic and β-adrenergic receptors to I_f channels. In young (3- to 5-day-old) embryonic chick hearts, cAMP level is very high, whereas cGMP level is very low. The present results suggest that young embryonic cells possess high sensitivity to muscarinic receptors, even though the 3-day-old heart is not yet innervated by parasympathetic nerve. G-proteins inhibitory to I_f were dominant over stimulatory G-proteins in young embryonic chick heart. The known dominance of the vagal effect may also be reflected by the dominance of the inhibitory

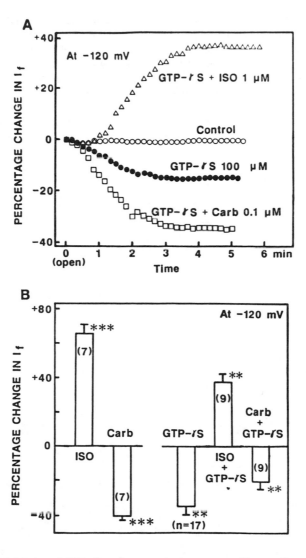

FIGURE 2. Modulation of GTP-γS on I_f current in the presence of isoprenaline and carbachol. (A) Percentage changes in I_f in a 3-day-old cell as a function of time. Test pulses were applied to -120 mV from the holding potential of -30 mV. Control (*open circles*), GTP-γS 100 μM (*filled circles*), GTP-γS + 1 μM isoprenaline (ISO) (*triangles*), and GTP-γS + 0.1 μM carbachol (Carb) (*squares*). GTP-γS (100 μM) was added into the pipette solution. External solution included 3 mM BaCl₂, 10 μM TTX, and 3 mM CdCl₂. (B) Summarized percentage changes in I_f (at -120 mV). Left panel shows effects by ISO (1 μM) or Carb (0.1 μM) alone. On the right panel, the changes by GTP-γS (100 μM), GTP-γS + ISO (1 μM), and GTP-γS + Carb (0.1 μM) are shown. Vertical bars represent the mean values and the SEM values are superimposed. The numbers in parentheses indicate the number of experiments. **$p<0.01$; ***$p<0.001$, with respect to control value.

G-proteins activated by GTP-γS. Direct G-protein coupling between autonomic receptors and I_f channels may partly account for the ability of the autonomic nervous system to produce its effects within a single heartbeat, and an indirect coupling via second messengers can account for the more persistent effects.

In conclusion, the hyperpolarization-activated I_f current (Cs^+ sensitive) may contribute to the electrogenesis of the pacemaker potential of embryonic chick heart cells and decreases during development. β-adrenergic agonists stimulate I_f, whereas muscarinic cholinergic agonists inhibit I_f and reverse β-adrenoceptor stimulation. Thus, G-proteins directly and indirectly couple autonomic receptors to I_f channels in embryonic chick ventricular cells.

Blockers and Activators for Stretch-Activated Ion Channels of Chick Skeletal Muscle[a]

MASAHIRO SOKABE, NOBORU HASEGAWA,
AND KIMIKO YAMAMORI

Department of Physiology
Nagoya University School of Medicine
65 Tsurumai
Nagoya 466, Japan

Stretch-activated (SA) ion channels are ubiquitously distributed virtually in all organisms and supposed to have important roles in fundamental cell functions such as cell volume regulation.[1] However, because of the absence of useful blockers (except for the inorganic cation gadolinium) and activators the mechanism of activation or physiological role of the channel is unclear. In the present study, a number of drugs have been subjected to screening by using the conventional patch-clamp technique. We have found that aminoglycoside antibiotics, such as neomycin, strongly blocked the cation-permeable SA channels of chick skeletal muscle cells, and that a certain class of amphipaths, such as chlorpromazine, could activate the SA channels in the absence of mechanical stimuli.

There are two types of SA channels in cultured chick skeletal muscles.[2] One (sSA) is a voltage-independent nonselective cation channel with a conductance of 60 pS at 150 mM KCl in the pipette. The other type (lSA) with a larger conductance (190 pS) is a voltage-dependent nonselective cation channel (FIG. 1,a). We found that both types of SA channels were blocked by externally applied aminoglycoside antibiotics in a dose- and voltage-dependent manner (FIG. 1,b). Dose dependence of the blockade indicated that the drug interacts with the channel in a 1:1 fashion (not shown). From the analysis of the voltage-dependent blockade, the zero voltage dissociation constants of several aminoglycosides in sSA and lSA channels were determined as follows (in μM); neomycin (2.4, 2.4), streptomycin (20.7, 25.4), ribostamycin (32.4, 55.9), dibekacin (47.5, 17.5), and kanamycin (54.7, 22.1). Although these drugs block other ion channels like Ca or K channels,[3] their dissociation constants for the SA channels are much lower than those for other channels.

A certain class of amphipaths is known to accelerate the opening of SA channels from *E. coli* spheroplasts.[4] Those drugs are thought to penetrate into the lipid bilayer, generating strain in it. We also have found that several amphipaths (chlorpromazine

[a] Supported by grants from the Ministry of Education, Science and Culture of Japan (0344073, 03454558, 04263291).

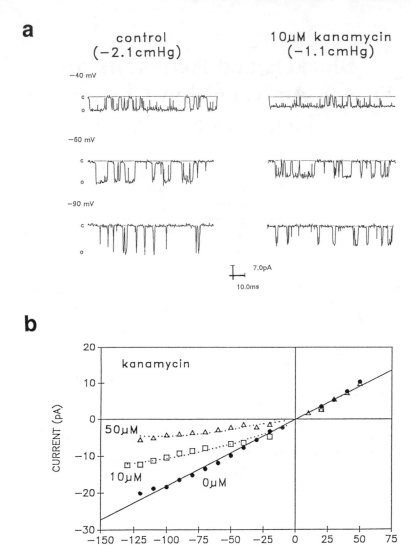

FIGURE 1. Voltage-dependent blockade of ISA channel by kanamycin. (a) Single channel currents of voltage-dependent ISA channel at different voltages in the absence (*left*, control) and in the presence of 10 μM kanamycin (*right*) in the pipette. (b) Single channel currents are plotted against bath voltage (membrane potential) at various kanamycin concentration in the pipette, where the pipette potential is defined as zero voltage. All the data were collected from excised inside-out patches of cultured chick skeletal muscle cells at room temperature (20–25°C). Bath solution (in mM): (150 NaCl; 5 KCl; 2 CaCl₂; 1 MgCl₂; 10 HEPES, pH 7.4); pipette solution: (150 KCl; 5 NaCl; 1 MgCl₂; kanamycin; 10 HEPES, pH 7.4).

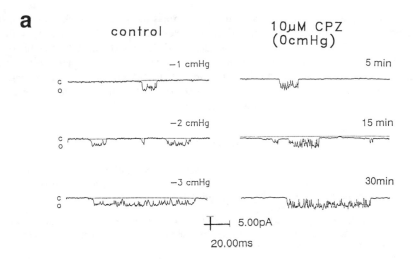

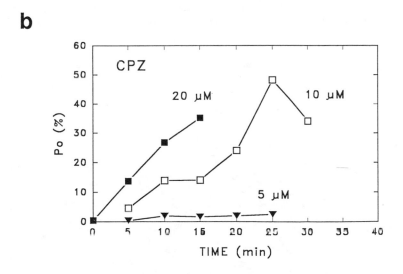

FIGURE 2. Time-dependent effect of chlorpromazine (CPZ) on the open probability of single sSA channels. (a) Activation of single sSA channel by negative pressure in the pipette (*left*, control) and by 10 μM CPZ in the pipette without suction (*right*). (b) Plot of single channel open probability (P_O, %) against time after gigaseal formation at various CPZ concentrations. All the data were obtained from cell attached patches of cultured chick skeletal muscle cells. Essentially the same results could be obtained from excised patches. All other conditions are the same as in FIGURE 1, except that pipette solutions contained CPZ instead of kanamycin.

and trinitrophenol) could activate the SA channels of cultured chick muscles in the same manner as mechanical stimuli (FIG. 2,a). For example, chlorpromazine (CPZ) in the pipette dose- and time-dependently activated the SA channels (FIG. 2,b), which might arise from the progressive stress caused by time-dependent penetration of the drug into the lipid bilayer. As the activation of the SA channels in our preparation has been suggested to relate to cytoskeletal structures,[5] both cytoskeletons and lipids may contribute to the SA channel activation.

REFERENCES

1. MORRIS, C. E. 1990. Mechanosensitive ion channels. J. Membr. Biol 113: 93–107.
2. SOKABE, M. & F. SACHS. 1992. Towards molecular mechanism of activation in mechanosensitive ion channels. *In* Comparative Aspects of Mechanoreceptor Systems. F. Ito, Ed.: 55–77. Springer Verlag. Heidelberg.
3. NOMURA, K., K. NARUSE, K. WATANABE & M. SOKABE. 1990. Aminoglycoside blockade of Ca^{2+} activated K^+ channel from rat brain synaptosomal membranes incorporated into planar bilayers. J. Membr. Biol. 115: 241–251.
4. MARTINAC, B., J. ADLER & C. KUNG. 1990. Mechanosensitive ion channels of *E. coli* activated by amphipaths. Nature (Lond.) 348: 261–263.
5. SOKABE, M., F. SACHS & Z.-G. JING. 1991. Quantitative videomicroscopy of patch clamped membranes: stress, strain, capacitance and stretch channel activation. Biophys. J. 59: 722–728.

Spatial Diversity of Chloride Transporters in Hippocampal Neurons

MITSUYOSHI HARA, MASAFUMI INOUE,
TOHRU YASUKURA, SUMIO OHNISHI,
AND CHIYOKO INAGAKI

Department of Pharmacology
Kansai Medical University
Fumizono-cho 1
Moriguchi, Osaka 570, Japan

The opening of $GABA_A$ receptor–linked Cl^- channels in hippocampal pyramidal neurons hyperpolarizes the perikaryonic membrane and depolarizes the dendrites.[1,2] The differences in the local intracellular Cl^- concentrations ($[Cl^-]_i$) among the cellular regions are considered to be responsible for such diverse effects, but have not been confirmed by conventional techniques. In this study, the regional $[Cl^-]_i$ and the distribution of Cl^- transporters in cultured rat hippocampal neurons were investigated using a Cl^--sensitive fluorescence dye, N-(6-methoxyquinolyl)acetoethyl ester (MQAE).

Hippocampal neurons from 17-day-old Wistar rat embryos were cultured and loaded with MQAE as described elsewhere.[3,4] The microscopic MQAE fluorescence intensity was continuously recorded with a multichannel window photometer, which enabled the simultaneous observation of $[Cl^-]_i$ in central and peripheral regions of the same perikaryon of a pyramidal cell-like neuron. The $[Cl^-]_i$ in neural processes was separately measured at a fiber bundle consisting of processes from clustering neurons.

After the transmembrane Cl^- gradient was minimized with ionophores, the MQAE fluorescence intensity was decreased with stepwise increases in the medium Cl^- concentration (FIG. 1). The stationary level of $[Cl^-]_i$ in each region was estimated from the initial fluorescence level in the corresponding calibration curve of fluorescence intensity against the Cl^- concentration. TABLE 1 shows the estimated values and their changes by inhibitors of Cl^- transport systems working in neurons[4]: ethacrynic acid for outwardly directed Cl^- pump,[5] furosemide and bumetanide for inwardly directed $Na^+/K^+/2Cl^-$ cotransporters.

The stationary $[Cl^-]_i$ was lower in the central region of the perikarya than in the periphery. The $[Cl^-]_i$ in the neural process was much higher than that in the perikarya.

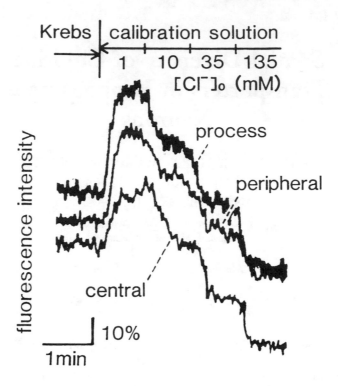

FIGURE 1. Intracellular calibration of MQAE fluorescence intensity. The regional fluorescence intensities were continuously recorded in the central and peripheral regions of a perikaryon and in a bundle of neural processes of MQAE-loaded hippocampal neurons perfused with Krebs' solution. To control the $[Cl^-]_i$, the transmembrane Cl^- gradients were minimized by the calibration solution containing 135 mM KNO_3, 10 μM valinomycin, 5 μM nigericin, 10 μM tributyltin, and definite concentration of Cl^- ($[KCl]$ + $[KNO_3]$ = 135 mM). The stationary level of $[Cl^-]_i$ in each region was estimated by the initial fluorescence intensity using the calibration curve against log $[Cl^-]$.

The increase in $[Cl^-]_i$ by the inhibition of Cl^- pump was greater in the central region than in the periphery, but not detected in the neuronal processes. After the inhibition of $Na^+/K^+/2Cl^-$ cotransporter, the $[Cl^-]_i$ was lowered prominently in the peripheral region, moderately in the central region, and weakly in the peripheral region.

These findings suggest that the heterogeneous regional $[Cl^-]_i$ in hippocampal neurons is maintained by the spatially diverse distribution and function of Cl^- transporters; the ethacrynic acid–sensitive Cl^- pumps are distributed mainly on the perikaryonic membrane and work more effectively in the central region than in the periphery, and the furosemide- and bumetanide-sensitive $Na^+/K^+/2Cl^-$ cotransporters are distributed in the entire cell but work more efficiently in the peripheral region.

TABLE 1. The Effects of Cl$^-$ Transport Inhibitors on the Regional Intracellular Cl$^-$ Concentration ([Cl$^-$]$_i$)^a

	Regional [Cl$^-$]$_i$ (mM)		
Drugs	Center of perikarya	Periphery of perikarya	Neural process
None (stationary)	11.4 ± 2.7	17.4 ± 2.8	35.0 ± 7.8
Ethacrynic acid (0.3 mM)	33.7 ± 7.6*	30.4 ± 4.9*	35.4 ± 2.3
Furosemide (1 mM)	4.4 ± 1.9*	0.4 ± 0.3*	17.3 ± 2.5*
Bumetanide (50 µM)	1.8 ± 0.8*	0.3 ± 0.2*	21.1 ± 2.1*

^{a} The stationary values of regional [Cl$^-$]$_i$ were estimated by respective calibration experiments. [Cl$^-$]$_i$ values were estimated at 5 min after the perfusion with or without drugs. The number of determinations: 4–8.

* $p < 0.05$ compared with each stationary value.

REFERENCES

1. ANDERSON, P., R. DINGLEDINE, L. GJERSTAD & I. A. LANGMON. 1980. J. Physiol. 305: 279–296.
2. Newberry, N. & R. A. NICOLL. 1985. J. Physiol. 360: 161–185.
3. HARA, M., M. INOUE, T. YASUKURA, S. OHNISHI, Y. MIKAMI & C. INAGAKI. 1992. Neurosci. Lett. 143: 135–138.
4. INOUE, M., M. HARA, X-T. ZENG, T. HIROSE, Y. YASUKURA, T. URIU, K. OMORI, A. MINATO & C. INAGAKI. 1991. Neurosci. Lett. 134: 75–78.
5. SHIROYA, T., R. FUKUNAGA, K. AKASHI, N. SHIMADA, Y. TAKAGI, T. NISHINO, M. HARA & C. INAGAKI. 1989. J. Biol. Chem. 264: 17416–17421.

Identification of an Anion Channel Protein from Transverse Tubules of Rabbit Skeletal Muscle

JUN HIDAKA, TORU IDE, TAKAHISA TAGUCHI,
AND MICHIKI KASAI

Department of Biophysical Engineering
Faculty of Engineering Science
Osaka University
Toyonaka, Osaka 560, Japan

The transverse tubule (TT) of skeletal muscle is an interface that transduces the membrane depolarization to Ca^{2+} release signal from sarcoplasmic reticulum. Hence, the stability of membrane potential permits TT to function normally. Since some kinds of Cl^- channels are considered to have a role to stabilize the potential, it is important to characterize the channel in TT membrane to understand the mechanism of excitation-contraction coupling.

The TT vesicles were prepared from rabbit skeletal muscle by the homogenization with a blender and six of the centrifugation steps including the step with sucrose density gradient. To characterize Cl^- channels in TT vesicles, these vesicles were incorporated into artificial planar lipid bilayers and the electric currents through the channels were analyzed as reported previously.[1]

The Cl^- channel of TT membrane was found to have the following biophysical properties. (*1*) The single channel conductance is 40 pS in choline Cl solution (*cis* side; 300 mM choline Cl, *trans* side; 100 mM choline Cl). (*2*) The open probability does not depend on the membrane potential in the range between -80 and $+60$ mV, and it is not affected by Ca^{2+}. (*3*) The channel is blocked from both sides of the membrane by 50 μM stilbene derivatives (DIDS, SITS), which are known as the inhibitors of the voltage-dependent Cl^- channel of *Torpedo* electric organs. (*4*) 9-anthracenecarboxylic acid (9AC), an inhibitor of the voltage-dependent Cl^- channel from muscle, inhibits the channel activity from *cis* side at 300 μM

To elucidate the biochemical structure of the Cl^- channel in TT, we used the polyclonal antibody against the 180 K protein which was identified as a Cl^- channel protein in the electric organ of *Narke japonica*.[1] The western blot analysis with this antibody (FIG. 1) revealed that the antibody recognized the 90 K and 60 K proteins, or only a 140 K protein under the condition of electrophoresis with or without 2-mercaptoethanol, respectively.

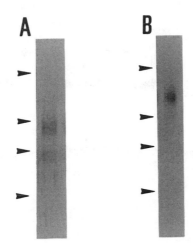

FIGURE 1. Western blot analysis of TT proteins using a polyclonal antibody against the 180 K protein from electric organ of *Narke japonica*. The proteins of TT membrane prepared from rabbit skeletal muscle were separated by sodium dodecylsulfate–7.5% polyacrylamide gel electrophoresis (SDS-PAGE). These proteins were then electrically transferred onto a nitrocellulose sheet using a semi-dry transfer system. After incubation with the anti-Cl⁻ channel (anti-180 K protein) antibody for 3 h, the sheet was reincubated with horseradish peroxidase–conjugated anti-rabbit IgG antibody. Proteins with the antibodies on the sheet was visualized by incubation with dimethylaminoazobenzene and H_2O_2. In **A** and **B**, proteins were separated by SDS-PAGE with and without 2-mercaptoethanol, respectively. Four arrowheads indicate the molecular weight of 200 K (*top*), 94 K, 67 K, and 45 K (*bottom*).

To confirm that the antibody recognizes the Cl⁻ channel in TT, we tested whether the antibody were able to modulate the channel activity. As shown in FIGURE 2, the antibody added in the solution on each side completely blocked the fluctuation of single Cl⁻ current through the channel incorporated in the planar lipid bilayer. This phenomenon was observed within 5 min after the application of the antibody and continued for more than 40 min

These results indicate that the antibody recognizes the Cl⁻ channel protein in TT membrane as well as the channel in the electric organ of *Narke japonica* and the channel in TT is composed of at least a 140 K protein, which contains two subunits (90 K and 60 K) crosslinked with a disulfide bond. To determine the complete subunit of channel, purification of the functional protein is indispensable and is now in progress.

Blatz *et al.* have characterized three Cl⁻ channels on a surface membrane of cultured rat skeletal muscle.[2] Although the unitary conductance of the Cl⁻ channel derived from TT membrane (40 pS) is close to that of one of these channels, the channel reported in this study is considered to be different from these channels because there is no voltage dependency and there is inhibition of 9AC from the *cis* side. The channel of TT, therefore, should be classified as a "background" type of Cl⁻ channel.[3]

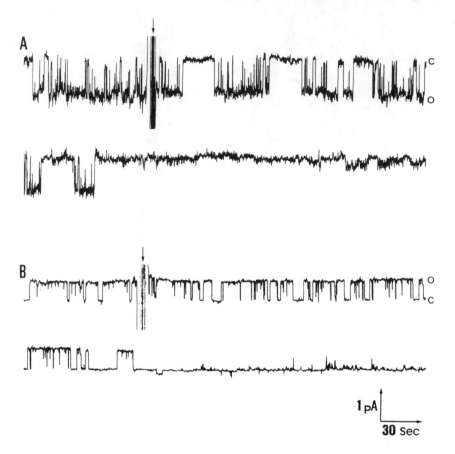

FIGURE 2. Effects of the antibodies on the Cl⁻ channel functions. The TT vesicle suspension (10 μg protein/ml) containing Cl⁻ channels was applied to the *cis* side solution of the artificial planar lipid bilayer membrane formed by the painting method. The *cis* and *trans* solutions were composed of 300 mM choline Cl, 15 mM HEPES-Tris (pH 7.1), and 1 mM $CaCl_2$, and 100 mM choline Cl and 15 mM HEPES-Tris (pH 7.1), respectively. After the stable Cl⁻ current was observed, the antibody was applied at the time indicated by the arrow. In **A**, the antibody of 12.5 μg/ml (final) was added to the *cis* side. Current fluctuation was measured at the holding potential of −25 mV. In **B**, the antibody of 30 μg/ml (final) was added to the *trans* side. Current fluctuation was measured at −10 mV. The letters of "O" and "C" in the figure show the open state and the closed state (baseline) of the channel, respectively.

REFERENCES

1. TAGUCHI, T., T. KAWASAKI & M. KASAI. 1992. Immunological identification of a Cl⁻ channel protein in electric organs of *Narke japonica*. Biochem. Biophys. Res. Comm. **188:** 1228–1234.
2. BLATZ, A. L. & K. L. MAGLEBY. 1985. Single chloride-selective channels at resting membrane potentials in cultured rat skeletal msucle. Biophys. J. **47:** 119–123.
3. FRANCIOLINI, F. & A. PETRIS. 1990. Chloride channels in biological membranes. Biochim. Biophys. Acta **1031:** 247–259.

Spider Toxin-Binding Protein: Functional and Immunohistochemical Study

KUNIKO SHIMAZAKI, MASAHIRO SOKABE,[a]
AND NOBUFUMI KAWAI

Department of Physiology
Jichi Medical School
Minamikawachi-machi
Tochigi-ken, 329-04, Japan

[a]Department of Physiology
School of Medicine
Nagoya University
Showa-ku
Nagoya 466, Japan

Joro spider toxin (JSTX) is a specific blocker of non-NMDA glutamate receptors.[1,2] The distribution of JSTX-binding sites revealed by biotinylated JSTX is similar to the distribution of non-NMDA receptors.[3] The binding is not inhibited by 1 mM glutamate, 5 mM spermine, or 1 M NaCl. Consequently, (1) JSTX does not interfere with the agonist binding, (2) the blocking activity of JSTX is not due to binding of the polyamine moiety in the toxin molecule, and (3) JSTX does not bind through the cationic charges in its structure.

Recently, we affinity-purified JSTX-binding protein from bovine cerebellum, with the final purified fraction containing a major protein band with an apparent molecular weight of 130 kD on SDS-PAGE.[4] Functional reconstitution of the purified protein was achieved in planar bilayer membranes (FIG. 1). Planar bilayers were formed from asolectin (15 mg/ml in *n*-decane) using the brush method.[5] Following incorporation of the purified protein, addition of 100 μM glutamate produced an increase in conductance associated with channel transitions. The activity and conductance of the glutamate-induced channels did not appear to be affected by changing the membrane voltage from +70 mV to −70 mV, consistent with voltage-independent channels. When JSTX (40 μM) was added on one side (*cis*) of the membrane, channel openings were greatly reduced in frequency without change in the conductance. This indicates that JSTX acts as a slow blocker of the glutamate-activated channels of the incorporated

427

The planar lipid bilayer system

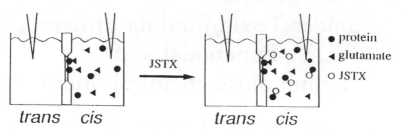

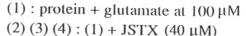

(1) : protein + glutamate at 100 μM
(2) (3) (4) : (1) + JSTX (40 μM)

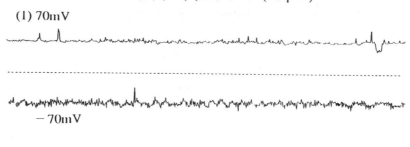

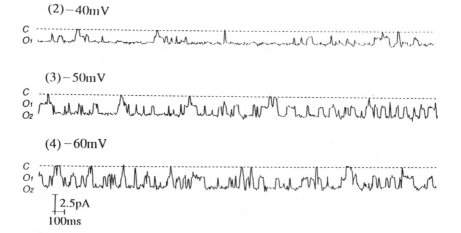

FIGURE 1. Schematic diagram of the planar bilayer system. The side to which proteins, glutamate, and JSTX were added is defined as the *cis* side. The voltage is referred to the *trans* side, opposite to the *cis* side. Single channel currents of glutamate-sensitive channels reconstituted into the artificial lipid bilayer. Glutamate (100 μM) induced channel activity, which was similar at +70 mV and −70 mV (*1*). Addition of JSTX (40 μM) to the *cis* side blocked channels in a voltage-dependent manner, and blockade became more frequent with increasing the voltage (*2–4*).

JSTX

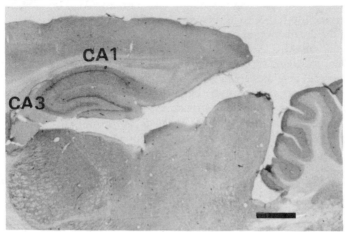

Ab#56→JSTX

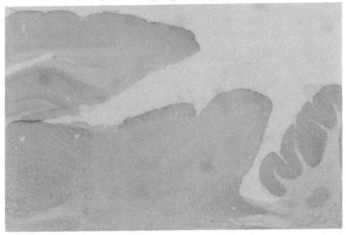

FIGURE 2. Immunohistochemical study using a monoclonal antibody to the affinity-purified protein. Sagittal section of a part of the rat brain. Specific binding of biotinylated JSTX to the cerebellum and hippocampus (*top*) was inhibited by treatment with the antibody (*bottom*). Bar = 1 mm.

protein. This inhibition appeared to be voltage dependent, since more frequent channel blockade was observed upon increasing the membrane voltage on the *cis* side. The results suggest that JSTX binds to a site in the ion-conducting pathway at which a voltage drop occurs.

A monoclonal antibody was prepared against the affinity-purified protein. The antibody was obtained as the product of a hybridoma cell clone isolated after immunization of BALB/c mice. Using the monoclonal antibody (Ab #56), which specifically recognizes the 130 K protein in Western blot, we carried out an immunohistochemical study of the distribution of binding sites in the rat brain. The immunoreactive sites revealed by the ABC method were Purkinje cell bodies and their dendrites in the cerebellum, pyramidal cells in the CA1-CA3 layers, and the dentate gyrus in the hippocampus. These sites are in good agreement with the distribution of JSTX (and AMPA)-binding sites. The specific binding of biotinylated JSTX to the cerebellum and hippocampus was inhibited by treatment with the antibody (FIG. 2).

Our results indicate that the monoclonal antibody (Ab #56) to the 130 kD protein reacts with JSTX-binding sites. The histological investigation of the distribution in the brain and the reconstitution of channel activity shows that the protein is a constituent of the glutamate receptor.

REFERENCES

1. ABE, T., A. KAWAI & A. MIWA. 1983. Effects of a spider toxin on the glutaminergic synapse of lobster muscle. J. Physiol. 339: 243–252.
2. KAWAI, N. 1991. Spider toxin and pertussis toxin differentiate post- and presynaptic glutamate receptors. Neurosci. Res. 12: 3–12.
3. SHIMAZAKI, K., Y. HIRATA, T. NAKAJIMA & N. KAWAI. 1990. A histochemical study of glutamate receptor in rat brain using biotinyl spider toxin. Neurosci. Lett. 114: 1–4.
4. SHIMAZAKI, K., H. P. C. ROBINSON, T. NAKAJIMA, N. KAWAI & T. TAKENAWA. 1992. Purification of AMPA type glutamate receptor by a spider toxin. Mol. Brain Res. 13: 331–337.
5. NOMURA, K. & M. SOKABE. 1991. Anion channels from rat brain synaptosomal membrane incorporated into planar bilayers. J. Membr. Biol. 124: 53–62.

Ionic Regulation and Signal Transduction System Involved in the Induction of Gametogenesis in Malaria Parasites

F. KAWAMOTO[a]

Department of Medical Zoology
Nagoya University School of Medicine
Nagoya 466, Japan

When malarial gametocytes are taken up into the mosquito midgut as intra-erythrocytic parasites, gametogenesis is induced within 15 min after digestion of blood-meal. These sexual parasites escape from the infected erythrocytes, during which the male (micro-)gametocyte passes through three mitotic divisions and assembles eight flagella, each snapping off with the attached nucleus to form a microgamete (sperm), in the process of "exflagellation," and fertilization ensues. In contrast, the female (macro-)gametocyte does not undergo any dramatic, visible process such as exflagellation during escape from erythrocytes, but transforms into a single macrogamete (egg). In a rodent malaria parasite, *Plasmodium berghei*, microgametogenesis can be subdivided into five stages (FIG. 1).[1,2] This process can be reproduced *in vitro* at an alkaline pH (8.0 ± 0.3) and at a temperature 5–15° C lower than that of the vertebrate hosts.

Induction of exflagellation in *P. berghei* is composed of two distinct mechanisms.[1,2] The first is low-temperature–dependent DNA synthesis (FIG. 1, stage B), and the second is pH-dependent control of subsequent development (FIG. 1, stages C–E). DNA replication may be triggered by one of the two external "positive" signals, temperature fall alone, since it is inhibited at 37° C and is independent of external ionic and pH conditions. On the other hand, the induction of exflagellation absolutely requires Na^+ and HCO_3^- or monovalent cation and Cl^- in the medium.[2] Determination of intracellular ionic concentrations by a new method of electron microscopy X-ray microanalysis of cryopreserved, unfixed, unsectioned, and unstained samples revealed that in the $NaHCO_3$ medium, external Na^+ (and probably HCO_3^-) enters the gameto-cytes by exchange with internal Cl^- (and probably H^+), whereas in Cl^--containing media external unspecified cation and Cl^- influx by exchange probably with H^+ and HCO_3^- (FIG. 2). It is therefore suggested that two separate anion exchangers, i.e., Na^+-dependent HCO_3^-(influx)/Cl^-(efflux) and non-specific monovalent cation-dependent Cl^-(influx)/HCO_3^-(efflux) exchangers, are involved in the induction mecha-

[a] This study was supported by Grant-in-Aid of the Ministry of Education, Science and Culture, Japan (No. 01570212 and 04670227).

431

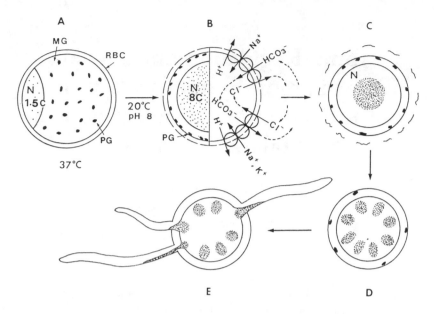

FIGURE 1. A schematic drawing of developmental events during formation of male gametes in *Plasmodium berghei* (Modified from Kawamoto *et al.*[2]). Mature, male gametocytes circulate in the host blood as non-activated forms (stage A), but are quickly activated after transfer from the vertebrate host into the bloodmeal of the mosquito. Their nuclei become enlarged and rapid DNA synthesis, from 1.5C to 8C level, occurs within 10 min (stage B). Then, the newly synthesized DNA becomes condensed (stage C), divides into 7–8 masses to form gamete nuclei (stage D), each of which incorporates into each flagellum, and exflagellation begins (stage E). MG, male gametocyte; PG, pigment granules; N, nucleus; RBC, red blood cell.

nism (FIG. 1,B). Furthermore, the presence of both class of anions in the medium enhanced exflagellation activity and increased Na^+ uptake more than that in the NaCl or $NaHCO_3$ medium alone (FIG. 2,E). To explain this enhancement, we have proposed a "recycling" model (FIG. 1,B) of cytoplasmic pH (pH_i) regulation by two contra-active anion exchangers, in which HCO_3^- and Cl^- are exchanged between the cells and the media, resulting in the acceleration of monovalent cation/H^+ exchange. These anion exchangers seem to be novel types since they are amiloride-sensitive (FIG. 2,F), and male gametocytes treated with amiloride cease development at stage B. Thus, cultivation at alkaline pH, another "positive" signal, may activate these exchangers in the presence of suitable ions, and a rise in pH_i by exchange of internal H^+ with external Na^+ or monovalent cation could allow the development to proceed beyond stage B.

In eukaryotic cells, it is well known that external "positive" signals are transduced into the cells by second messengers such as Ca^{2+}, cAMP, and/or cGMP, where these messengers activate various protein kinases and then initiate DNA synthesis. In malarial gametogenesis, its induction was found to be regulated by second messengers of Ca^{2+} and cGMP.[3] Treatment with TMB-8 (a Ca^{2+} release inhibitor) and W-7/W-66 (calmodulin inhibitors) blocked exflagellation by inhibiting DNA synthesis from 1.5C to

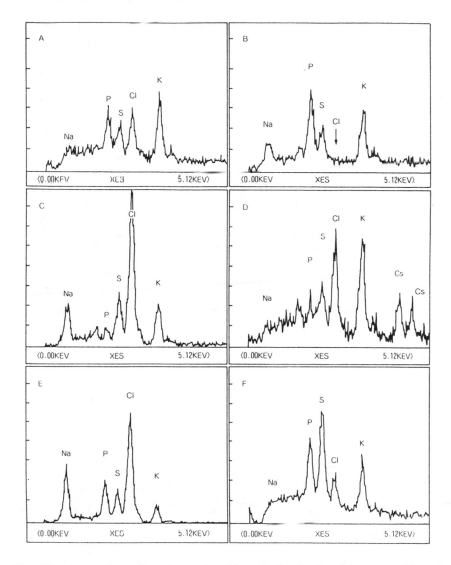

FIGURE 2. X-ray microanalyses of non-activated (control, **A**) and activated gametocytes (Modified from Kawamoto *et al.*[4]). Gametocytes are cultivated in isotonic $NaHCO_3$ (**B**), NaCl (**C**), CsCl (**D**), $NaHCO_3$ + NaCl (1:1, **E**) and treated with 0.5 mM amiloride in $NaHCO_3$ + NaCl (**F**). Note that the Cl^- content has decreased to the background noise level in **B** (arrow).

8C level.[4] In contrast, inhibitors of cGMP-dependent protein kinases such as H-8, H-87, H-89, and staurosporine also ceased the development, but DNA synthesis in male gametocytes occurred as in the controls. Electron microscopy study[4] revealed that male gametocytes treated with TMB-8 and W-7 failed to enlarge nuclei and to form axonemes in the cytoplasm. In female gametocytes, treatment with both Ca^{2+}

antagonists resulted in a dramatic morphological change in the endoplasmic reticulum (ER), which is thought to be a Ca^{2+} store. The ER network condensed near nuclei and was laminated by the abnormal attachment of ribosomes between two ER membranes. On the other hand, male gametocytes treated with protein kinase inhibitors or amiloride had enlarged nuclei and axonemes, but failed to develop further. The ER network in female gametocytes treated with these inhibitors was similar to that in the controls.

These results taken together may suggest that the rapid DNA synthesis and axoneme formation in male gametocytes may be regulated by Ca^{2+}/calmodulin, and that further development, leading to exflagellation, may be induced by cGMP-dependent pathways and/or an increase in pH_i. Low temperature–dependency of DNA synthesis in male gametocytes may imply regulation by heat-shock proteins (HSPs) during its switching on mechanism for transfer from vertebrate to invertebrate hosts. Since in malaria parasites HSPs are expressed only in the vertebrate stages, it is quite possible that mature gametocytes in the vertebrate blood may be down-regulated by HSPs and "arrested" as non-activated forms (G1 state ?).

REFERENCES

1. KAWAMOTO, F., R. ALEJO-BLANCO, S. L. FLECK & R. E. SINDEN. 1991. Exp. Parasitol. 72: 33–42.
2. KAWAMOTO, F., N. KIDO, T. HANAICHI, M. B. A. DJAMGOZ & R. E. SINDEN. 1992. Parasitol. Res. 78: 277–284.
3. KAWAMOTO, F., R. ALEIJO-BLANCO, S. L. FLECK, Y. KAWAMOTO & R. E. SINDEN. 1990. Mol. Biochem. Parasitol. 42: 101–108.
4. KAWAMOTO, F., H. FUJIOKA, R. MURAKAMI, SYAFRUDDIN, M. HAGIWARA, T. ISHIKAWA & H. HIDAKA. 1993. Eur. J. Cell Biol. 60: 101–107.

Desensitization of Nicotine Receptor through CGRP-Enhanced Non-Contractile Ca²⁺ in Neuromuscular Synapse

HIROSHI TSUNEKI, KATSUYA DEZAKI,
IKUKO KIMURA,[a] AND MASAYASU KIMURA

Department of Chemical Pharmacology
Faculty of Pharmaceutical Sciences
Toyama Medical and Pharmaceutical University
2630 Sugitani
Toyama 930-01, Japan

Two important roles for Ca^{2+} in skeletal muscle are excitatory and inhibitory ones. Ca^{2+} released from the sarcoplasmic reticulum brings about muscle contraction. Ca^{2+} influx through the sarcolemmal membrane, however, represses nicotinic acetylcholine receptor (nAChR) biosynthesis,[1] and promotes nAChR desensitization.[2] We have found that non-contractile Ca^{2+} (not accompanied by twitch tension) is mobilized via nAChR by nerve stimulation with 0.3 μM neostigmine in mouse diaphragm muscles.[3-5] The mechanism and the functional role of non-contractile Ca^{2+} were investigated using Ca^{2+} aequorin luminescence.

RESULTS

The duration of non-contractile Ca^{2+} transients was prolonged by calcitonin gene–related peptide (CGRP, 0.3–10 nM) and AA373 (300 μM), a protein kinase A activator (FIG. 1).[6] $CGRP_{8-37}$ (10–20 μM), a CGRP antagonist, and H-89 (0.1–1 μM), a protein kinase A inhibitor, decreased the non-contractile Ca^{2+} transients. 12-O-tetradecanoyl-phorbol 13-acetate (TPA, 0.3–1 μM), a protein kinase C activator, decreased contractile (accompanied by twitch tension) but not non-contractile Ca^{2+} transients. Phospholipase A_2 and cholera toxin increased only contractile components. Neither calmodulin nor phospholipase C affected both type of Ca^{2+} transients. The peak amplitude of non-contractile transients was increased by 4-aminopyridine (100 μM) and decreased by hexamethonium (10–100 μM).

---

[a] To whom all correspondence should be addressed.

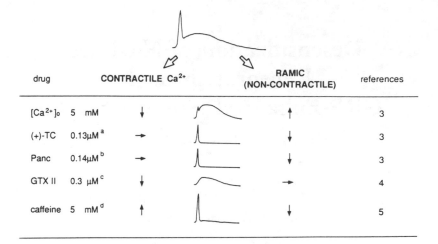

FIGURE 1. Characterization of the mobilization of contractile Ca^{2+} and RAMIC. ↑:increase, ↓:decrease, →:no effect
[a] tubocurarine: nicotinic competitive antagonist, [b] pancuronium: nicotinic competitive antagonist, [c] geographutoxin II: muscle Na^+ channel blocker, [d] Ca^{2+} releaser from SR.

To elicit contractile Ca^{2+} transients during generation of non-contractile transients, the phrenic nerve was stimulated by double pulse with delay (msec: 150, 300, 600, and 1,000). Contractile Ca^{2+} transients were suppressed at shorter delays as non-contractile Ca^{2+} transients increased, and were restored at longer delays as non-contractile transients decreased. The extent of the suppression was dependent on the peak amplitude of non-contractile Ca^{2+} transients regulated by external $[Ca^{2+}]_o$, demonstrating that non-contractile Ca^{2+} suppresses the contractile Ca^{2+} mobilization.

DISCUSSION

Non-contractile Ca^{2+}, RAMIC (receptor-activity modulating intracellular Ca^{2+}), is mobilized depending on the accumulated amounts of acetylcholine in the synaptic cleft. The RAMIC mobilization was caused by a physiological function of CGRP through the activation of protein kinase A (FIG. 2). RAMIC may be released locally in submembrane from a Ca^{2+} pool triggered by Ca^{2+} influx through nAChR.

The decrease in contractile Ca^{2+} transients by protein kinase C activator TPA is correlated with nAChR desensitization promoted by protein kinase C–induced phosphorylation.[7] RAMIC may be utilized for activating protein kinase C and affect contractile Ca^{2+} transients evoked subsequently. RAMIC may stimulate 43 kD protein like annexin beneath the postsynaptic membrane via protein kinase C for its inhibition.

The increase in contractile components by phospholipase A_2 is at least in part due to fatty acid–induced Ca^{2+} release from the sarcoplasmic reticulum.[8] RAMIC may activate phospholipase A_2, thereby enhancing contractile Ca^{2+} transients.

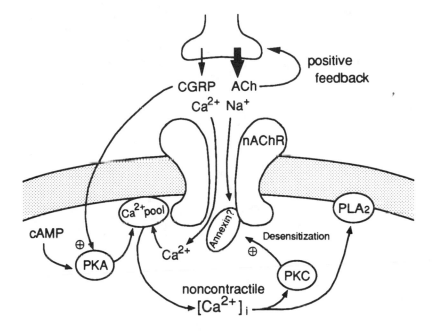

FIGURE 2. Putative mechanisms of nAChR desensitization in neuromuscular synapse.

In conclusion, endogenous CGRP is released from motor nerve terminal via positive feedback mechanism under a desensitizable condition, and promotes the RAMIC mobilization by stimulating protein kinase A. RAMIC then accelerates nAChR desensitization presumably through protein kinase C–induced phosphorylation. Its desensitization may be also modulated by phospholipase A_2.

ACKNOWLEDGMENTS

We thank Dr. Hiroyoshi Hidaka (Nagoya University School of Medicine), Dr. Ken-ichiro Mitsui (Toyama Medical and Pharmaceutical University), and Asahi-Kasei Industrial for providing samples.

REFERENCES

1. KLARSFELD, A., R. LAUFER, B. FONTAINE, A. DEVILLERS-THIERY, C. DUBREUIL & J.-P. CHANGEUX. 1989. Neuron **2**: 1229–1236.
2. PARSONS, R. L., D. E. COCHRANE & R. M. SCHNITZLER. 1973. Life Sci. **13**: 459–465.
3. KIMURA, I., T. KONDOH & M. KIMURA. 1989. Brain Res. **507**: 309–311.
4. KIMURA, M., I. KIMURA, T. KONDOH & H. TSUNEKI. 1991. J. Pharmacol. Exp. Ther. **256**: 18–23.

5. KIMURA, I., T. KONDOH, H. TSUNEKI & M. KIMURA. 1991. Neurosci. Lett. **127**: 28–30.
6. KIMURA, I., H. TSUNEKI & M. KIMURA. 1992. *In* Calcium Inhibition: A New Mode for Ca^{2+} Regulation. K. Kohama, Ed.: 43–67. Japan Science Society Press. Tokyo. CRC Press. Boca Raton, FL.
7. EUSEBI, F., F. GRASSI, C. NERVI, C. CAPORALE, S. ADAMO, B. M. ZANI & M. MOLINARO. 1987. Proc. R. Soc. Lond. B **230**: 355–365.
8. CHEAH, K. S. & A. M. CHEAH. 1981. Biochim. Biophys. Acta **634**: 70–84.

In Vitro Phosphorylation of Rat Brain Nicotinic Acetylcholine Receptor by cAMP-Dependent Protein Kinase

HITOSHI NAKAYAMA, HIROTSUGU OKUDA,
AND TOCHIKATSU NAKASHIMA

Department of Pharmacology
Nara Medical University
Kashihara, 634 Japan

Nicotinic acetylcholine receptors (nAChRs) from electric organ and muscle consist of four distinct subunits assembled into $\alpha_2\beta\gamma(\epsilon)\delta$ pentamer. To date, participation of cAMP-dependent protein kinase (PKA) in phosphorylation of nAChRs in electric organ and muscle is precisely characterized *in vitro* and *in vivo*. One of its physiological roles is to regulate the rate of desensitization. In contrast, cAMP has been reported to enhance the ACh-induced response in chick ciliary ganglion neurons.[1] Treatment of chick ciliary ganglion neuron in culture with cAMP analogs leads to an increase in the phosphorylation of the α3 agonist binding subunits,[2] which contrasts with rapid phosphorylation of the non-agonist binding subunits in electric organ and muscle. In this case, however, whether PKA directly phosphorylates nAChRs or activates other protein kinases, which in turn phosphorylate nAChRs, has not been investigated. Here, we show the *in vitro* phosphorylation of the α4 agonist binding subunits of rat brain nAChR by PKA.

nAChR was highly purified from rat brains as previously described.[3] Purified preparations contained α4 and β2 subunits. Purified nAChR was phosphorylated by partially purified PKA, immunoprecipitated with monoclonal antibody 299 (mAb 299) against α4 subunits of nAChR, and subjected to sodium dodecylsulfate-polyacrylamide gel electrophoresis (SDS-PAGE) followed by autoradiography. The reaction mixture contained 50 mM Tris-HCl buffer, pH 7.4, 10 mM $MgCl_2$, 0.2% Triton X-100, 4 μM cAMP, 18 pmol units of PKA, 2 μM [γ-^{32}P]ATP (5×10^4 cpm/pmol), and 98 fmol of ACh binding sites of purified nAChR in a final volume of 50 μl. mAb 299 was kindly supplied by Dr Jon M. Lindstrom. Incorporation of ^{32}P into the immunoprecipitate increased with time. Most of the ^{32}P in the immunoprecipitate was found to be incorporated into the 80 kD band corresponding to the α4 subunits on the autoradiogram. When cAMP was removed from the reaction mixture, the 80 kD band disappeared on the autoradiogram (Fig. 1, lane 3). H8, an inhibitor of PKA, inhibited completely the phosphorylation (lane 5). These results show the specific phosphoryla-

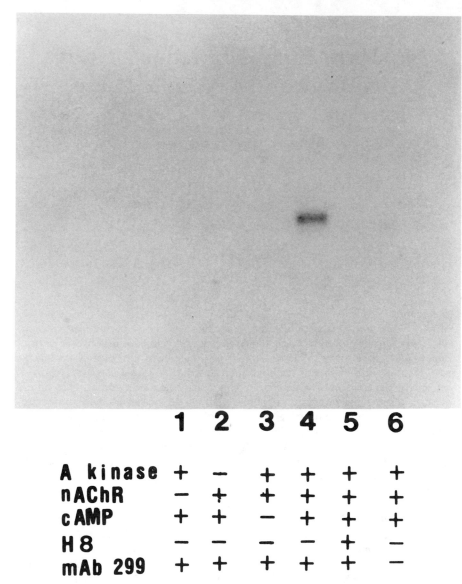

	1	2	3	4	5	6
A kinase	+	−	+	+	+	+
nAChR	−	+	+	+	+	+
cAMP	+	+	−	+	+	+
H 8	−	−	−	−	+	−
mAb 299	+	+	+	+	+	−

FIGURE 1. SDS-PAGE of phosphorylated nAChR. Purified nAChR was phosphorylated *in vitro* by partially purified PKA for 30 min at 25°C under the indicated conditions. H8 was used at 80 μM. After phosphorylation, ^{32}P-labeled nAChR was first incubated with control IgG–Protein G–agarose complex, centrifuged and washed twice by suspending in the washing buffer. The combined supernatant was incubated with mAb 299–Protein G–agarose, and then washed five times. In the case of lane 6, control IgG–Protein G–agarose instead of mAb 299–Protein G–agarose was used. The precipitated immunocomplex was incubated with SDS-sample buffer and then centrifuged for 1 min at 10,000 × g. The obtained supernatant was subjected to SDS-PAGE followed by autoradiography.

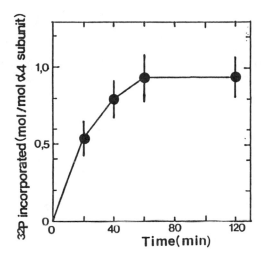

FIGURE 2. Time course of phosphorylation of the α4 subunits by PKA. Purified nAChR was incubated with the catalytic subunits of PKA at 30°C for the indicated time periods. After SDS-PAGE, regions of dry gels corresponding to α4 subunits were excised and the radioactivities were determined by Cerenkov counting. Each point represents mean ± SD of four experiments.

tion of the α4 subunits by PKA. FIGURE 2 shows the time course of phosphorylation of the α4 subunits. The incorporation of ^{32}P into α4 subunits was found to plateau at 2 h. Under the conditions used, a maximal stoichiometry of the phosphorylation of the α4 subunits was 0.95 mol of phosphate/mol of ACh binding subunits. After SDS-PAGE, ^{32}P-labeled α4 subunits were excised from the dry gels and subjected to limited digestion with *S. aureas* V8 protease followed by SDS-PAGE autoradiography. The resultant phosphopeptide maps revealed three distinct phosphopeptide bands, one major and two minor bands. When the phosphorylated α4 subunits were cleaved with cyanogen bromide, a single major and two minor bands were also detected on SDS-PAGE. Phosphoamino acid analysis showed that the α4 subunits were phosphorylated exclusively on serine residues. Based on these results, we conclude that PKA directly phosphorylates the α4 agonist binding subunits of nAChR purified from rat brains.

In chick ganglion neuron in culture, the cAMP-dependent process has been postulated to enhance the conversion of nonfunctional nAChRs to functional one.[1] In contrast, treatment of bovine adrenal chromaffin cells with cAMP analogs has shown to result in no increased nicotinic response.[4] The present study opens the possibility that PKA participates in the regulation of brain nAChRs.

REFERENCES

1. MARGIOTTA, J. F., D. K. BERG & V. E. DIONNE. 1987. Proc. Natl. Acad. Sci. USA **84**: 841–846.

2. VIJAYARAGHAVAN, S., H. A. SCHMID, S. W. HALVORSEN & D. K. BERG. 1990. J. Neurosci. **10:** 3255–3262.
3. NAKAYAMA, H., M. SHIRASE, T. NAKASHIMA, Y. KUROGOCHI & J. M. LINDSTROM. 1991. Mol. Brain Res. **7:** 221–226.
4. DUBIN, A. E., M. M. RATHOUZ, K. S. MAPP & D. K. BERG. 1992. Brain Res. **586:** 344–347.

Calcium-Independent Regulation of Transmitter Release at the Frog Neuromuscular Junction

HIROMASA KIJIMA, NORIKO TANABE,[a] JIRO SATO,
AND SHOKO KIJIMA

Department of Physics
Faculty of Science
Nagoya University
Nagoya 464-01

[a]*Daiichi Hoiku Junior College*
Dazaifu 818-01, Japan

In the non-neuronal secretory cells, such as mast cells and platelets, it has been known that secretion can be regulated by Ca^{2+}-independent mechanisms or by the direct modification of secretory apparatus, as well as by the internal Ca^{2+} concentration ($[Ca^{2+}]_i$).[1] The regulation of secretion in neuronal synapses has only recently been elucidated.

We loaded a Ca^{2+}-chelator, bis (*O*-aminophenoxy)ethane-*N*,*N*,*N'*,*N'*-tetraacetic acid (BAPTA), into the presynaptic nerve terminal of frog neuromuscular junction (NMJ) by incubating a nerve–sartorius muscle preparation of *Rana nigromaculata* with acetoxy methyl ester of BAPTA. Then, the facilitation measured as an increase in the amplitude of end-plate potentials (EPPs) during a train of 10 impulses were greatly decreased (FIG. 1, A).[2] The facilitation recovered when external Ca^{2+} was introduced into the nerve terminal by making use of an ionophore, X537A.[2,3] This suggested strongly that the $[Ca^{2+}]_i$ at the release site was well buffered by BAPTA and the increase of $[Ca^{2+}]_i$ upon stimulation decreased rapidly to the basal level, leaving little residual $[Ca^{2+}]_i$. Using this BAPTA-loaded condition, we investigated whether four components of nerve stimulation-induced increases in transmitter release, known in this synapse to occur due to an increase of $[Ca^{2+}]_i$ at the release site. The fast component of facilitation was completely lost upon BAPTA-loading, but three slower components (the slow facilitation, augmentation, and potentiation) remained unaffected (FIG. 1,B).[2,4,5] Under external Ca^{2+}-free condition, no stimulation-induced phasic release occurred, but both augmentation and potentiation of miniature end-plate potential (MEPP) frequency were clearly observed after intensive stimulation, and were unaffected by loaded BAPTA (FIG. 1,C).[5] These results summarized in TABLE 1 strongly suggested "the residual Ca^{2+} hypothesis" on the fast component of facilitation and that the three slower components occur independently of internal Ca^{2+}. An amino residue–

A

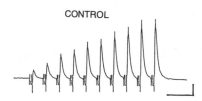

CONTROL   +BAPTA

B

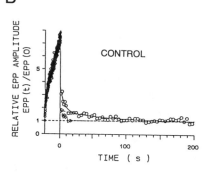

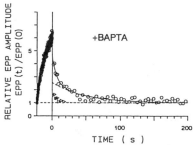

C

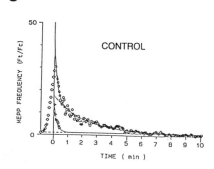

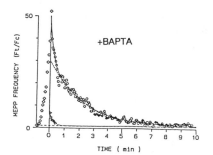

TABLE 1. Ca^{2+}-Dependence of Four Components of Stimulation-Induced Increase of Transmitter Release at the Frog Neuromuscular Junction

Components	Decay Time Constant (sec)	Effect of BAPTA-Loading	Internal Ca^{2+} Dependence
Fast facilitation	~ 0.05	Complete loss	Dependent
Slow facilitation	0.3–0.5	No Change	Independent
Augmentation	5–10	No Change	Independent
Potentiation	50–100	No Change	Independent

modifying reagent, TNBS, a dye, erythrosin B, and hypertonicity were also suggested to increase the release independently of $[Ca^{2+}]_i$.[2,3] The EPP amplitudes were little decreased by BAPTA-loading in most cases (13/19), suggesting that the peak of Ca^{2+} concentration around the Ca^{2+} channels at the active zone is little affected by BAPTA due to rapid supply of Ca^{2+} and that it can trigger the release. Computer simulation of Ca^{2+} distribution supported this inference.[6]

It has recently been reported that Ca^{2+}-insensitive protein kinase C is abundant in vertebrate presynaptic terminal and it is activated by arachidonic acid, as well as phorbol esters.[7] In order to elucidate the molecular mechanism of Ca^{2+}-independent regulation of transmitter release, we investigated the effect of arachidonic acid on the transmitter release at frog NMJ. The phorbol ester is known to augment the transmitter release at this synapse.[8] Preliminary results showed that tens μM of arachidonic acid added in the external solution caused about twofold increase of transmitter release, but reduced greatly the augmentation and moderately the potentiation (Sato and Kijima, in preparation). The protein kinase C and/or arachidonic acid might be responsible for Ca^{2+}-independent regulation of transmitter release in this synapse.

After our work, many reports have appeared, presenting evidence of the Ca^{2+} independent regulatory mechanisms in both invertebrate and vertebrate synapses. Among them are increase of release by serotonin in the crayfish NMJ,[10] regulation

---

FIGURE 1. Stimulation-induced increases of transmitter release in BAPTA-loaded neuromuscular junctions of the frog. (**A**) Facilitation of EPP amplitudes in a train of 10 stimulations at 100 Hz.[4] EPPs were surface recorded and 15 traces were averaged in both normal (control) and BAPTA-loaded NMJs. Calibration: 20 msec × −0.2 mV. (**B**) Changes in EPP amplitudes during and after tetanic 450 stimulations at 20 Hz. After tetanic stimulation, the test stimulations were delivered once every 1.5 sec 6 times and then once every 4 sec 60 times. Only augmentation (▷·······▷) and potentiation (−−−−) were resolved and EPP amplitudes were expressed by product of both components (O———O). EPPs were surface recorded and 9–13 records were averaged. Ordinate is the relative value of EPP, setting the average of EPP amplitude before stimulation as unity. There were differences in the growth profiles of EPP during stimulation, but no significant differences in augmentation and potentiation between BAPTA-loaded and normal NMJs. (**C**) Augmentation (△······△) and potentiation (−−−−) of MEPP frequency under external Ca^{2+}-free condition after 5,000 stimulations at 100 Hz.[5] Normal plots during (◇) and after (O, $t>0$) nerve stimulations. A moving bin display was used, setting the bin size 9 sec and the step size 3 sec. MEPP frequency was expressed as the product of augmentation and potentiation and relative value is shown, setting the MEPP frequency before stimulation to unity. Both augmentation and potentiation were similar to BAPTA-loaded and control NMJs.

by serotonin and FMRFamide in the synapse of *Aplysia* responsible for short-term memory,[9] and on the synapses of cultured hippocampal neurons.[11] Ca^{2+}-independent regulation mechanisms of transmitter release may be very important for synaptic plasticities relating to memory and higher brain functions.

REFERENCES

1. PENNER, R. & E. NEHER. 1988. J. Exp. Biol. **139**: 329–345.
2. KIJIMA, H. & N. TANABE. 1988. J. Physiol. (Lond.) **403**: 135–149.
3. TANABE, N. & H. KIJIMA. 1988. Neurosci. Lett. **92**: 52–57.
4. TANABE, N. & H. KIJIMA. 1989. Neurosci. Lett. **99**: 147–152.
5. TANABE, N. & H. KIJIMA. 1992. J. Physiol. (Lond.) **455**: 271–289.
6. KIJIMA, S. & H. KIJIMA. 1990. Biophysics (Kyoto) **30**: S328.
7. TANAKA, C. & N. SAITO 1992. Neurosci. Res. Suppl. **17**: S9.
8. SHAPIRA, R., S. D. SILBERBERG, S. GINSBORG & R. RAHAMIMOFF. 1987. Nature **325**: 58–60.
9. DALE, N. & E. R. KANDEL. 1990. J. Physiol. (Lond.) **421**: 203–222.
10. DELANEY, K. R., D. W. TANK & R. S. ZUCKER. 1991. J. Neurosci. **11**: 2631–2643.
11. MALGAROLI, A. & R. W. TSIEN. 1992. Nature **357**: 134–139.

Development of Inhibitory Synaptic Currents in Rat Spinal Neurons

AKIKO MOMIYAMA AND TOMOYUKI TAKAHASHI

Department of Physiology
Kyoto University Faculty of Medicine
Kyoto 606, Japan

In rat spinal neurons, the time course of glycinergic inhibitory postsynaptic currents (IPSCs) becomes faster during the first two postnatal weeks. This is due to a switch in the glycine receptor α subunit.[1] We have examined whether the quantal size of IPSCs may also change developmentally. Using whole-cell recording techniques in thin slices,[2] we recorded strychnine-sensitive spontaneous miniature IPSCs (mIPSCs) from dorsal horn neurons (cell capacitance 18–57 pF) in the presence of tetrodotoxin, CNQX, and bicuculline. External Ca^{2+} was replaced by Mn^{2+} to exclude the possibility of events caused by presynaptic Ca^{2+} spikes. Also, mIPSCs with rise time (10–90 %) slower than 1 msec were excluded from analysis. The mean conductance of mIPSCs recorded from one-day-old rat spinal neurons was 555 ± 164 pS (standard deviation, at -80 mV; with equimolar Cl^- in external and internal solutions, 8 cells). In 15-day-old rats, the conductance was significantly larger (1,695 $\pm$ 561 pS, 7 cells, $p < 0.01$, t-test). The coefficient of variation was 0.65 ± 0.18 in one-day-old animals and 0.61 ± 0.18 in 15-day-old animals. In contrast with the amplitude, no significant difference was observed for the mean rise time (10–90 %) of mIPSCs between one-day-old (0.75 ± 0.09 msec) and 15 day old (0.70 ± 0.12 msec) animals. It is suggested that the quantal size of glycinergic IPSCs increases in the first two postnatal weeks.

To determine whether or not the functional density of glycine receptors changes during ontogenesis, glycine (100 μM) was bath-applied in spinal neurons. The mean amplitude of the glycine-induced current was 1.85 ± 0.49 nA (-80 mV) in one-day-old animals (5 cells each with capacitance of 15–30 pF), comparable with that seen in 15-day-old animals (2.17 ± 0.99 nA, 5 cells with capacitance of 14–22 pF). Thus, the developmental increase in the mean amplitude of mIPSCs was not correlated with somatic glycine receptor density. The single channel conductance of glycine receptors in spinal neurons in two-week-old rats is not larger than that in 20-day embryonic rat.[1] Therefore, the developmental increase in quantal size may be produced either by an accumulation of subsynaptic glycine receptors or by an increase in vesicular transmitter content.

REFERENCES

1. TAKAHASHI, T., A. MOMIYAMA, K. HIRAI, F. HISHINUMA & H. AKAGI. 1992. Functional correlation of fetal and adult forms of glycine receptors with developmental changes in inhibitory synaptic receptor channels. Neuron 9: 1155–1161.
2. EDWARDS, F. A., A. KONNERTH, B. SAKMANN & T. TAKAHASHI. 1989. A thin slice preparation for patch clamp recordings from neurones of the mammalian central nervous system. Pflueg. Arch. 414: 600–612.

Inhibitory Effect of Methylxanthines on Glycine-Induced Cl Current in Dissociated Rat Hippocampal Neurons

KAZUYOSHI KAWA, HISAYUKI UNEYAMA,
AND NORIO AKAIKE[a]

Department of Neurophysiology
Tohoku University School of Medicine
Sendai 980, Japan

The glycine receptor (GlyR) linked to Cl^- channel is a major candidate for the inhibitory receptors as well as the GABA receptor in the central nervous system (CNS) of the mammal. The activation of the glycine receptor inhibits the neuronal firing by increasing the postsynaptic membrane Cl^- conductance.[1] The fact that a convulsant alkaloid, strychnine, potently antagonizes glycine on the GlyR makes clear the importance of the GlyR in the CNS. In the present experiments, a pyramidal neuron was acutely dissociated from area CA1 of the rat hippocampus and monitored the glycine-gated Cl^- current of the neuron under the voltage-clamped condition. We report here that the GlyR-operated Cl^- current is blocked by the well-known psychostimulant drugs, methylxanthines such as caffeine and theophylline.

METHODS

Pyramidal neurons were enzymatically and mechanically dissociated from area CA1 of the hippocampus of 2-week-old Wistar rats, according to the previous technique.[2] The ionic composition of normal external solution was (in mM): 150 NaCl, 5 KCl, 1 $MgCl_2$, 2 $CaCl_2$, 10 N-2-hydroxyethylpiperazine-N'-2-ethanesulfonic acid (HEPES), and 10 glucose. Patch-pipette solution contained (in mM): 30 NaCl, 70 KCl, 70 K-gluconate, 0.25 $CaCl_2$, 1 $MgCl_2$, 5 ATP-Mg, 5 EGTA, and 10 HEPES. Electrical measurements were performed mainly in a whole-cell mode with the conventional

[a] Address all correspondence to: N. Akaike, Ph.D., Department of Neurophysiology, Tohoku University School of Medicine, Sendai 980, Japan.

449

whole-cell configuration.[3] Drugs were applied using a rapid application method termed the "Y-tube" technique. All experiments were performed at room temperature (20–22°C).

RESULTS AND DISCUSSIONS

Dissociated CA1 pyramidal neurons were perfused with normal external and internal solutions containing 161 and 82.5 mM Cl^-, respectively. Under this condition, the external application of 0.1 mM glycine evoked Cl^- inward currents (I_{Gly}) of 592 ± 117 pA (n = 8) at a holding potential (V_H) of – 64 mV. The I_{Gly} was inhibited by applications of caffeine in a concentration-dependent manner (FIG. 1,A). Neurons were pretreated with caffeine 15 sec before simultaneous application with 0.1 mM glycine. The blocking action of caffeine on the I_{Gly} showed no significant dependence on the pretreatment time of caffeine (data not shown). We tested the effect of some compounds having methylxanthine structure on the I_{Gly} (FIG. 1,B). Consequently, caffeine and its analogues inhibited the 0.1 mM glycine-induced Cl^- current. The inhibitory potency was in the order of pentoxiphylline > theophylline > caffeine > theobromine; the half-maximum inhibitory values obtained from the concentration-inhibition curves were 0.2, 0.39, 0.45, and 1.49 mM for pentoxiphylline, theophylline, caffeine, and theobromine, respectively.

To clarify the mechanism of the blocking action of methylxanthines on glycine-induced Cl^- current, the influence of caffeine on the concentration-response relationship for I_{Gly} was investigated (FIG. 2,A). In the presence of 1 mM caffeine, the concentration-response curve of I_{Gly} was shifted to the right without affecting the maximum value. It is thus highly likely that caffeine may be an antagonist at a glycine receptor site. The voltage dependence of inhibitory action of caffeine on I_{Gly} was examined (FIG. 2,B). The current-voltage (I-V) relationships for I_{Gly} were measured in the presence and absence of 0.1 mM caffeine. In both cases, I_{Gly} reversed the direction of the current at – 15 mV. This value was very close to the Cl^- equilibrium potential (– 15.5 mV) calculated using the Nernst equation from the given intra- and extra-cellular Cl^- concentrations. Thus, caffeine had no effect on the E_{Gly} and depressed the I_{Gly} by the same degree at every V_H. Further, the inhibitory potency of caffeine did not depend on the amplitudes or direction of the current induced by glycine but on the concentrations of applied agonist (data not shown). These features make it seem less likely that caffeine acts as a channel blocker of glycine-activated Cl^-channel.

In some preparations, it is suggested that the Cl^- channels linked to GABA- or glycine receptors were negatively modulated by second messengers such as Ca^{2+} and cAMP.[4,5] The intracellular concentration of those two substances can be elevated by caffeine and other methylxanthine derivatives.[6,7] In the present preparation, however, neither the internal perfusion with a potent Ca^{2+} chelator (BAPTA) nor the pretreatment with protein kinase inhibitor (H-8) blocked the inhibitory action of caffeine significantly (data not shown). Even when the intracellular cAMP was elevated by using potent adenylate cyclase activator, forskolin (10 µM), the I_{Gly} was depressed only less than 10% (n = 5). These results suggest that the involvement of the intracellular site of caffeine action in blocking the glycine response is negligible or minor if any.

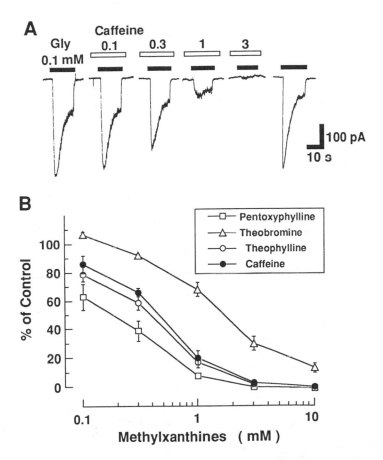

FIGURE 1. Effect of methylxanthines on glycine-induced currents. (**A**). Caffeine depressed 0.1 mM glycine-induced Cl⁻ current in a concentration-dependent manner. V_H was -64 mV. All recordings were obtained from the same neuron. (**B**) Concentration-inhibition curves of caffeine (●), theophylline (○), theobromine (△), and pentoxiphylline (□). The concentration of applied glycine was 0.1 mM. Each point and vertical bar represent mean and ±S.E.M. from 5–8 experiments.

Based on the above observations, we could conclude that methylxanthines such as caffeine might act as an inhibitor of glycine on the GlyR. Several compounds have been known to be inhibitors of the central glycine response (for instance, strychnine, avermectin B1a, and some derivatives of 4,5,6,7-tetrahydroisoxazolo[5,4-c]pyridin-3-ol).[8,9] Methylxanthines might be a new member of glycine response inhibitors in the cerebrum or at least in CA1 pyramidal cells. The blockade of glycine response in neurons may presumably result in an increase of the neuronal firing. The present results may have relevance to the behavioral observations that methylxanthines act as a kind of psychostimulant, although details are open for future study.

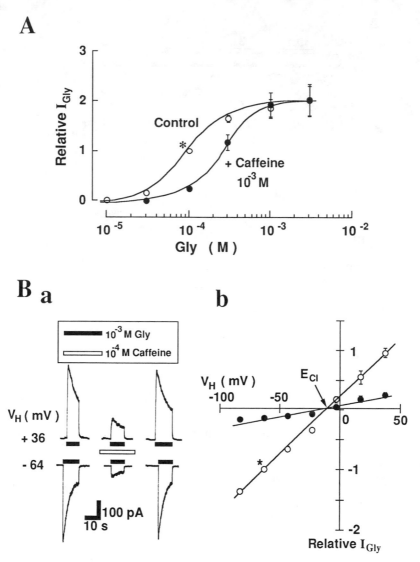

FIGURE 2. Inhibitory effect of caffeine on I_{Gly}. (**A**) Concentration-response curve for I_{Gly} in the presence (●) or absence (○) of 1 mM caffeine. Neurons were voltage-clamped at −64 mV. All responses were normalized to the peak amplitude induced by 0.1 mM glycine alone (symbol marked with asterisk). Each point and vertical bar show mean and ±S.E.M. from 6 experiments. (**B**) Effect of V_H on the blocking action of caffeine. (**a**) Current traces at V_Hs of −64 and +36 mV. (**b**) Current-voltage (I-V) relationships for I_{Gly} in the presence (●) or absence (○) of 1 mM caffeine. All responses were normalized to the peak amplitude induced by glycine at a V_H of −64 mV (*). Each point and vertical bar show mean and ±S.E.M. from 5–7 experiments. E_{Cl}, Cl⁻ equilibrium potential.

REFERENCES

1. BETZ, H. 1987. Trends Neurosci. **10**: 113–117.
2. SHIRASAKI, T., M. R. KLEE, T. NAKAYE & N. AKAIKE. 1991. Brain Res. **561**: 77–83.
3. HAMILL, O. P., A. MARTY, E. NEHER, B. SAKMANN & F. J. SIGWORTH. 1981. Pflügers Arch. **391**: 85–100.
4. INOUE, M., Y. OOMURA, T. YAKUSHIJI & N. AKAIKE. 1986. Nature **324**: 156–158.
5. DIATLOV, V. A. 1989. Neirofiziologiia **21**: 413–416.
6. BUTCHER, R. W. & E. W. SUTHERLAND. 1962. J. Biol. Chem. **237**: 1244–1250.
7. ENDO, M., M. TANAKA & Y. OGAWA. 1970. Nature **228**: 34–36.
8. PFEIFFER, D. G. F. & H. BETZ. 1982. Neurosci. Lett. **29**: 173–176.
9. BRAESTRUP, C., M. NIELSEN & P. KROGSGAARD-LARSEN. 1986. J. Neurochem **47**: 691–696.

Optical Monitoring of Glutaminergic Excitatory Postsynaptic Potentials from the Early Developing Embryonic Chick Brain Stem

YOKO MOMOSE-SATO, TETSURO SAKAI,
AKIHIKO HIROTA, KATSUSHIGE SATO,
AND KOHTARO KAMINO[a]

Department of Physiology
Tokyo Medical and Dental University School of Medicine
Bunkyo-ku, Tokyo 113, Japan

Optical techniques using voltage-sensitive dyes have made it possible to monitor electrophysiological events in living systems that are inaccessible to microelectrodes. Furthermore, optical recording methods have been developed into a powerful tool for recording electrical activity simultaneously from many sites in one living preparation. Applying these optical techniques, we have overcome traditional obstacles to functional approaches to the study of embryonic heart and central nervous system during early development. We report here some results obtained with optical techniques in the early embryonic chick brain stem.

In slice preparations of the vagus/brain stem isolated from 4- to 9-day-old chick embryos, the spatial pattern of neural responses to vagal stimulation and its development were assessed by using multiple-site optical recording of electrical activity, with a voltage-sensitive merocyanine-rhodanine dye (NK2761) and a 12 × 12-element photodiode array. Voltage-related optical (absorbance) changes evoked by vagus nerve stimulation with depolarizing (positive) square current pulses using a suction electrode were recorded simultaneously from 127 contiguous loci in the preparation (FIG. 1).

The first neural responses, viz., fast optical signals related to the action potentials were recorded in the 4-day-old embryonic brain stem preparations. The slow optical signals were detected from 7-, 8- and 9-day-old embryonic brain stem preparations. The size of the slow signal was decreased by continuous stimulation, reduced by low external calcium ion concentrations, and eliminated in the presence of manganese or cadmium ion. The slow signals were blocked by kynurenic acid. The later phase of

[a] Address correspondence to Dr. K. Kamino, Department of Physiology, Tokyo Medical and Dental University School of Medicine, 1-5-45 Yushima, Bunkyo-ku, Tokyo 113, Japan.

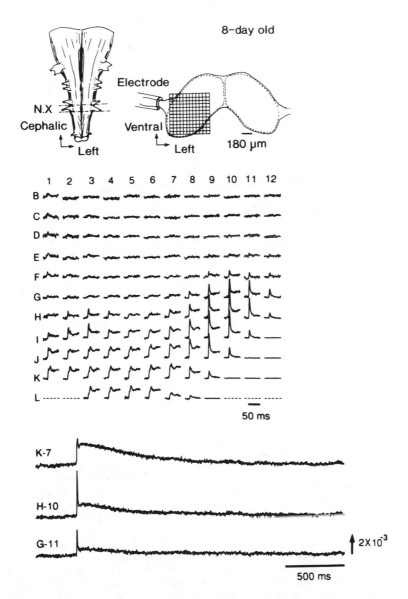

FIGURE 1. Multiple-site optical recording of neural activity evoked by vagus stimulation in an embryonic chick brain stem slice preparation. The slice preparation was made by transverse sectioning of an 8-day-old embryonic brain stem at the level of the vagus nerve (*left upper corner*). Positive square current pulses were applied to the right vagus nerve fibers by a suction electrode. The evoked optical signals were detected by the 12 × 12-element photodiode array positioned on the image of the right side area of the brain stem. The relative position of the photodiode matrix array is illustrated on the upper right corner of the optical recording. The recording was made with a 4.2 μA/7.0 msec square current stimulus. Wavelength 702 ± 13 nm. Enlargements of the optical signals obtained from three different positions (K-7, H-10, and G-11) are shown in the bottom. The recordings were made in a single sweep, at room temperature (26–28 °C).

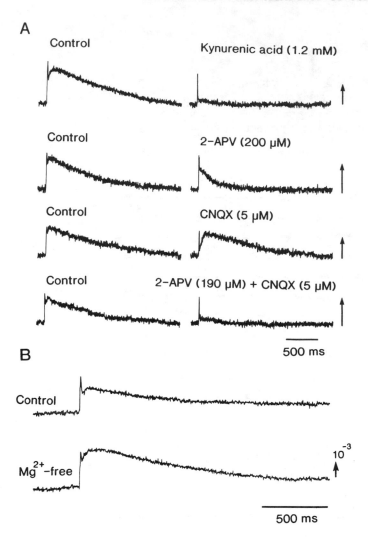

FIGURE 2. (A) The effects of kynurenic acid, 2-APV, and CNQX on the postsynaptic slow signals in 8-day-old embryonic brain stem slices. In the top traces, kynurenic acid (1.2 mM) was applied; in the second, 2-APV (200 μM) was applied; in the third, CNQX (5.0 μM) was applied; and in the last, 2-APV (190 μM) and CNQX (5.0 μM) were applied together. (B) The effect of removal of Mg^{2+} from the bathing solution in an 8-day-old embryonic brain stem slice: the normal Ringer's solution contains 0.5 mM $MgCl_2$. Recordings shown in this figure were obtained from five different preparations.

the slow signal was reduced by 2-APV (DL-2-amino-5-phosphonovaleric acid), and the initial phase was reduced by CNQX (6-cyano-7-nitroquinoxaline-2,3-dione) (FIG. 2,A). We conclude that the slow signals correspond to excitatory postsynaptic potentials, which are glutamate-mediated.[1,2]

There were differences in the regional distribution of the 2-APV and CNQX sensitivities, supporting a different spatial distribution of NMDA- and non-NMDA-receptors in the brain stem during early embryonic development. We have constructed early developmental maps of the spatial patterning of NMDA- and non-NMDA-subtype postsynaptic responses.

In addition, the later phase of the slow signal was enhanced in a Mg^{2+}-free bathing medium (FIG. 2,B), and the enhanced fraction was sensitive to 2-APV and insensitive to CNQX. There were regional differences in the enhancement of the slow signals in the Mg^{2+}-free bathing medium.

In some preparations from 6-day-old embryonic brain stems, only action potential–related fast optical signals were detected in the normal bathing solution. However, in some loci of some preparations, postsynaptic potential–related slow signals preceded by the fast signal appeared in a Mg^{2+}-free bathing medium. In other preparations from 6-day-old embryonic brain stems, very small slow signals were detected from a small restricted area in the normal Ringer's solution. When Mg^{2+} was removed from the bathing solution, the slow signals were enhanced, and new small slow signals were detected from some additional loci. Therefore, it is suggested that, at the 6-day-old embryonic stage, the NMDA-subtype receptor-mediated postsynaptic function has been generated potentially within the brain stem and, in the normal Ringer's solution, it is mostly suppressed by external Mg^{2+}.

REFERENCES

1. KOMURO, H., T. SAKAI, Y. MOMOSE-SATO, A. HIROTA & K. KAMINO. 1991. J. Physiol. (Lond.) 442: 631–648.
2. MOMOSE-SATO, Y., T. SAKAI, H. KOMURO, A. HIROTA & K. KAMINO. 1991. J. Physiol. (Lond.) 442: 649–668.

Quantal Properties of Single Glutamatergic Synaptic Boutons in Thin Slices from Rat Neostriatum

HARUO KASAI,[a] AKIHISA MORI,[b]
TETSUO TAKAHASHI,[c]
AND YASUSHI MIYASHITA[a]

[a]Department of Physiology
Faculty of Medicine
University of Tokyo
Hongo, Bunkyo-ku, Tokyo 113, Japan

[b]Pharmaceutical Research Laboratories
Kyowa Hakko Co. Ltd.
Nagaizumi-cho,
Sunto-gun, Sizuoka 411, Japan

[c]Department of Molecular Neurobiology
School of Human Science
Waseda University
Mikajima, Saitama 359, Japan

Synaptic boutons are elementary functional units of neurotransmitter release in the central nervous system.[1] Quantal properties of individual synaptic boutons, however, have been difficult to assess directly, since axons usually make multiple synaptic contact with postsynaptic neurons.[1,2] It has been suggested that a single release site could release only one vesicle at a time[2] rather than that every synaptic vesicle has a small probability for release in an independent manner.[3] The "one vesicle" hypothesis is rather surprising, considering the fact that many synaptic vesicles are closely located to the presynaptic plasma membrane at each release site.[4] Experimental testing of the one vesicle hypothesis has so far been achieved only in those synapses with multiple release sites, in which true numbers of functional release site are not known. In order to clearly prove this hypothesis, it is therefore desirable to use a synapse with only one presynaptic bouton.

Excitatory synaptic inputs from the cortex to the medium spiny neuron in the neostriatum may be suited for testing the one vesicle hypothesis. Because each cortical afferent fiber takes a relatively straight course, scarcely arborizes,[5] and forms en passant synapse on a spine head of a dendrite of the neuron.[6] This anatomical feature makes afferent fibers form very few synaptic contacts on the spiny neuron. We have therefore performed quantal analysis of the excitatory synaptic currents (EPSCs) in the medium spiny neurons using the whole-cell recording method. Fine glass pipettes were used to stimulate only one synaptic fiber. We found that the distribution of the unitary EPSCs displayed only one peak. When concentrations of extracellular Ca^{2+} were reduced, the probability of release was reduced, keeping the peak of the amplitude histogram constant. This indicates that the single peak represents the quantum and confirms that only a single presynaptic fiber was stimulated. It also suggests that the presynaptic fiber makes contact to the medium spiny neurons with one synaptic bouton, consistent with the anatomical features of cortical afferents. Furthermore, it suggests that only one vesicle was released from the single bouton. Thus, our result might support the one vesicle hypothesis. Alternatively, the apparent single peak in the amplitude histogram could be due to the relative scarcity in the postsynaptic glutamate receptors in the central nervous system.[7]

REFERENCES

1. WALMSLEY, B. 1991. Central synaptic transmission: studies at the connection between primary afferent fibers and dorsal spinocerebellar tract (DSCT) neurones in Clerke's column of the spinal cord. Prog. Neurobiol. **36**: 391–423.
2. KORN, H. & D. S. FABER. 1991. Quantal analysis and synaptic efficacy in the CNS. Trends Neurosci. **14**: 439–445.
3. KATZ, B. 1969. The Release of Neural Transmitter Substances. Liverpool University Press. Liverpool.
4. HEUSER, J. E., T. S. REESE, M. J. DENNIS, Y. JAN., L. JAN. & L. EVANS. 1979. Synaptic vesicle exocytosis captured by quick freezing and correlated with quantal transmitter release. J. Cell Biol. **81**: 275–300.
5. KEMP, J. M. & T. P. S. POWELL. 1971. The structure of the caudate nucleus of the cat: Light and electron microscopy. Phil. Trans. R. Soc. Lond. B. **226**: 383–401.
6. FOX, C. A., A. N. ANDRADE, D. E. HILLMAN & R. C. SCHWYN. 1971. The spiny neurons in the primate striatum: A Golgi and electron microscopic study. J. Himforsch. **13**: 181–201.
7. FABER, D. S., W. S. YOUNG, P. LEGENDRE & H. KORN. 1992. Intrinsic quantal variability due to stochastic properties of receptor-transmitters. Science **258**: 1494–1498.

Expression of the α1 and α2 Subunits of the AMPA-Selective Glutamate Receptor Channel in Insect Cells using a Baculovirus Vector

SUSUMU KAWAMOTO, SATOSHI HATTORI,
IZUMI OIJI, ATSUHISA UEDA, JUN FUKUSHIMA,
KENJI SAKIMURA,[a] MASAYOSHI MISHINA,[a]
AND KENJI OKUDA

Department of Bacteriology
Yokohama City University School of Medicine
3-9 Fukuura, Kanazawa-ku, Yokohama 236

[a]Department of Neuropharmacology
Brain Research Institute
Niigata University
Niigata 951, Japan

In order to produce large quantities of functional glutamate receptor channel subunits for structural and functional analyses, we[1] constructed recombinant baculoviruses carrying the α1 and α2 subunits[2] of the mouse AMPA (α-amino-3-hydroxy-5-methyl-4-isoxazole propionate)-selective glutamate receptor channel. Insect cells infected with these recombinant baculoviruses produced high levels of [³H]AMPA binding activities on their cell surfaces.

METHODS

cDNAs encoding the α1 and α2 subunits of the mouse glutamate receptor channel were inserted into the baculovirus transfer vector pJVP10Z, containing the β-galactosidase gene, to yield recombinant baculoviruses. [³H]AMPA binding activities on the cell surface were determined as described previously.[1]

RESULTS AND DISCUSSION

As shown in FIGURE 1, Sf21 insect cells infected with recombinant viruses carrying the α1 and α2 subunits of the mouse glutamate receptor channel exhibited high levels of [³H]AMPA binding activities on their cell surfaces (2 ~ 3 × 10⁵ binding sites per cell). This suggests that functional α1 and α2 subunits capable for agonist binding were successfully produced and inserted into the plasma membrane. The finding that the α2 subunit is able to bind [³H]AMPA is in accord with our previous results[3] that the mutant α2 subunit, α2-R586Q, forms highly active homomeric glutamate receptor channels, although the wild-type α2 subunit alone shows little channel activity when expressed in *Xenopus* oocytes. Effects of various agonists and antagonists on [³H]AMPA binding of the α1 subunit expressed in Sf21 cells are in good agreement with those of the α1 subunit expressed in *Xenopus* oocytes (FIG. 2). Thus the baculovirus-insect cell

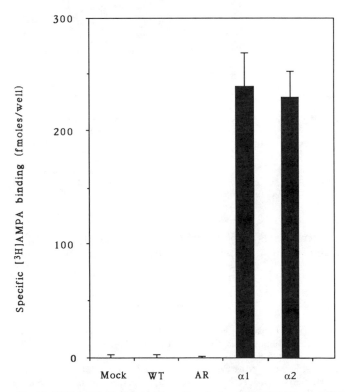

FIGURE 1. Specific [³H]AMPA binding activity of baculovirus-infected Sf21 cells. Sf21 cells were placed in 12-well dishes and infected with wild-type (WT), α1-adrenergic receptor recombinant (AR), glutamate receptor α1 subunit recombinant (α1), α2 subunit recombinant (α2) baculoviruses, or with medium only (Mock). At 5 days post-infection, a [³H]AMPA binding assay was performed. Data are presented as mean ± SEM (*n* = 3).

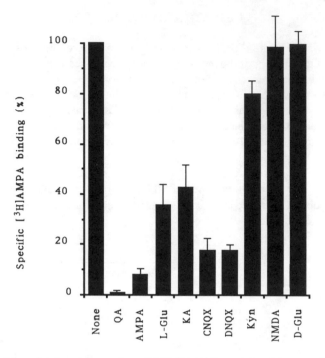

FIGURE 2. Effect of various compounds on [³H]AMPA binding activity of the recombinant gluta-mate receptor α1 subunit expressed in baculovirus-infected Sf21 cells. Sf21 cells were placed in 12-well dishes and infected with α1 subunit recombinant baculovirus. At 5 days post-infection, a [³H]AMPA binding assay was performed in the presence of each compound (1 μM). QA, quisqualate; L-Glu, L-glutamate; KA, kainate; CNQX, 6-cyano-7-nitroquinoxaline-2,3-dione; DNQX, 6,7-dinitroquin-oxaline-2,3-dione; Kyn, kynurenate; NMDA, N-methyl-D-aspartate; D-Glu, D-glutamate. Data are presented as mean ± SEM (n = 3).

expression system affords the possibility of high-efficiency expression of the glutamate receptor channel protein for biochemical, pharmacological, and electrophysiological studies.

REFERENCES

1. KAWAMOTO, S., H. ONISHI, S. HATTORI, Y. MIYAGI, Y. AMAYA, M. MISHINA & K. OKUDA. 1991. Biochem. Biophys. Res. Commun. **181**(2): 756–763.
2. SAKIMURA, K., H. BUJO, E. KUSHIYA, K. ARAKI, M. YAMAZAKI, M. YAMAZAKI, H. MEGURO, A. WARASHINA, S. NUMA & M. MISHINA. 1990. FEBS Lett. **272**(1,2): 73–80.
3. MISHINA, M., K. SAKIMURA, H. MORI, E. KUSHIYA, M. HARABAYASHI, S. UCHINO & K. NAGAHARI. 1991. Biochem. Biophys. Res. Commun. **180**(2): 813–821.

Distinct Spatio-temporal Distributions of the NMDA Receptor Channel Subunit mRNAs in the Brain

MASAHIKO WATANABE, YOSHIRO INOUE,
KENJI SAKIMURA,[a] AND MASAYOSHI MISHINA[a]

Department of Anatomy
Hokkaido University School of Medicine
Sapporo 060, Japan

[a]*Department of Neuropharmacology*
Brain Research Institute
Niigata University
Niigata 951, Japan

The ε and ζ subfamilies of the glutamate receptor channel subunit, designed as the ε1, ε2, ε3, ε4, and ζ1 subunits, constitute the *N*-methyl-D-aspartate (NMDA) receptor channel in the mouse brain.[1-4] Functional properties of the ε/ζ heteromeric NMDA receptor channels are determined by the constituting ε subunit species.[2-4] Furthermore, the four ε subunit mRNAs exhibit distinct distributions in the developing and mature brains.[2,3,5] Therefore, we suggest that the molecular diversity of the ε subunit family underlies the functional diversity of the NMDA receptor channel. In this report, we have specified the spatio-temporal changes in combinatory expressions of the NMDA receptor channel subunit mRNAs in several brain regions by *in situ* hybridization histochemistry.

Sequences of the 45-mer oligonucleotide probes and procedures for *in situ* hybridization were described previously.[5] Briefly, the probes radiolabeled by ^{35}S-dATP were hybridized to adjacent horizontal sections of the mouse brain at embryonic day 15 (E15) and postnatal days 1 (P1) and 21 (P21). After washing, sections were exposed to X-ray film for 10 days.

FIGURE 1 shows drastic changes in distribution of respective NMDA receptor channel subunit mRNAs during brain development. In general, hybridizing signals for the ζ1, ε2, and ε4 subunit mRNAs appeared in the brain at fetal stages (FIG. 1,A–C,

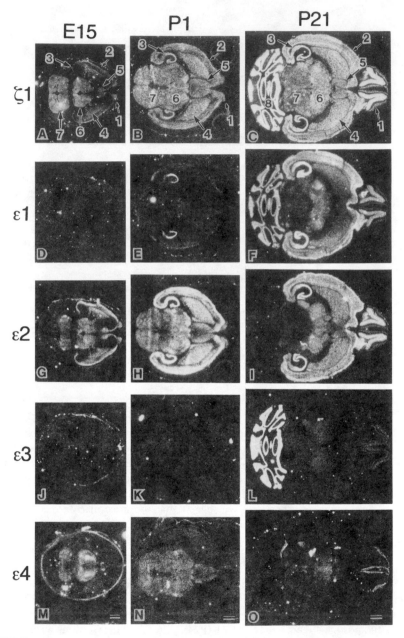

FIGURE 1. Distributions of five NMDA receptor channel subunit mRNAs in brains at E15 (A, D, G, J, M), P1 (B, E, H, K, N), and P21 (C, F, I, L, O). A–C, ζ1; D–F, ε1; G–I, ε2; J–L, ε3; M–O, ε4 subunit mRNA. 1, olfactory bulb; 2, cerebral cortex; 3, hippocampus; 4, caudate-putamen; 5, septum; 6, thalamus; 7, brainstem; 8, cerebellum. Scale bars: 1 mm.

G–I, M–O). As the brain matures, the signals for the ε2 subunit mRNA became restricted to the forebrain, and the signal levels of the ε4 subunit mRNA decreased in the brain except for the olfactory glomerular layer. On the other hand, signals for the ε1 and ε3 subunit mRNAs appeared in postnatal brains (FIG. 1,D–F, J–L). In the olfactory bulb, signals for the ζ1 and ε2 subunit mRNAs were detected in the mitral cell layer at E15, and various combinatory expressions of the five subunit mRNAs were notable in different layers of the olfactory bulb at P21. The cerebral cortex and the caudate-putamen expressed the ζ1 and ε2 subunit mRNAs at E15 and P1, and the ζ1, ε1, and ε2 subunit mRNAs at P21. The hippocampal pyramidal cells expressed the ζ1 and ε2 subunit mRNAs at E15, and the ζ1, ε1, and ε2 subunit mRNAs at P1 and P21. On the other hand, distinct signals for the ζ1, ε1, and ε2 subunit mRNAs appeared in the granule cells of the dentate gyrus at P21. In the septum, hybridizing signals for the ζ1, ε2, and ε4 subunit mRNAs were detected at E15 and P1, and weak signals for the ε1 subunit mRNA also appeared at P21. The thalamic nuclei expressed the ζ1, ε2, and ε4 subunit mRNAs at E15 and P1, and various distribution patterns of the five subunit mRNAs were observed in the thalamus at P21. The brainstem showed hybridizing signals for the ζ1, ε2, and ε4 subunit mRNAs at E15 and P1. Thereafter, weak signals for the ε1 subunit mRNA appeared in the brainstem, whereas those for the ε2 subunit mRNA disappeared. The developmental changes in combinatory expressions of the NMDA receptor channel subunit mRNAs are summarized in TABLE 1.

The present study has disclosed the region-specific and neuron type–specific changes in expressions of the four ε subunit mRNAs during brain development. In contrast, the ζ1 subunit mRNA is present ubiquitously from fetal to adult stages. These findings suggest that the distribution of the four ε subunits are highly differential in the brain *in vivo* both spatially and temporally, which would result in the functional heterogeneity of the NMDA receptor channel at various brain regions and developmental stages.

TABLE 1. Summary of Developmental Changes in Combinatory Expressions of the NMDA Receptor Channel Subunit mRNAs

Regions	E15	P1	P21
Olfactory bulb			
Periglomerular cells	–	ζ1, ε2, ε4	ζ1, ε1, ε2, ε3, ε4
Mitral cells	ζ1, ε2	ζ1, ε2	ζ1, ε1, ε2, ε3
Granule cells	–	ζ1, ε2	ζ1, ε1, ε2
Cerebral cortex	ζ1, ε2	ζ1, ε2	ζ1, ε1, ε2
Caudate-putamen	ζ1, ε2	ζ1, ε2	ζ1, ε1, ε2
Hippocampus (Pyramidal cells)	ζ1, ε2	ζ1, ε1, ε2	ζ1, ε1, ε2
Dentate gyrus (Granule cells)	–	–	ζ1, ε1, ε2
Septum	ζ1, ε2, ε4	ζ1, ε2, ε4	ζ1, ε1, ε2, ε4
Thalamus	ζ1, ε2, ε4	ζ1, ε2, ε4	ζ1, ε1, ε2, ε3, ε4
Brainstem	ζ1, ε2, ε4	ζ1, ε2, ε4	ζ1, ε1, ε4

–, not expressed.

REFERENCES

1. YAMAZAKI, M., H. MORI, K. ARAKI, K. J. MORI & M. MISHINA. 1992. FEBS Lett. 300: 39–45.
2. MEGURO, H., H. MORI, K. ARAKI, E. KUSHIYA, T. KUTSUWADA, M. YAMAZAKI, T. KUMANISHI, M. ARAKAWA, K. SAKIMURA & M. MISHINA. 1992. Nature 357: 70–74.
3. KUTSUWADA, T., N. KASHIWABUCHI, H. MORI, K. SAKIMURA, E. KUSHIYA, K. ARAKI, H. MEGURO, H. MASAKI, T. KUMANISHI, M. ARAKAWA & M. MISHINA. 1992. Nature 358: 36–41.
4. IKEDA, K., M. NAGASAWA, H. MORI, K. ARAKI, K. SAKIMURA, M. WATANABE, Y. INOUE & M. MISHINA. 1992. FEBS Lett. 313: 34–38.
5. WATANABE, M., Y. INOUE, K. SAKIMURA & M. MISHINA. 1992. NeuroReport 3: 1138–1140.

Glutamate-Induced Hyperpolarization in Mouse Cerebellar Purkinje Cells

TAKAFUMI INOUE,[a-c] HIROYOSHI MIYAKAWA,[b]
KEN-ICHI ITO,[b] KATSUHIKO MIKOSHIBA,[c]
AND HIROSHI KATO[b]

bDepartment of Physiology
Yamagata University School of Medicine
Yamagata 990-23, Japan

cDepartment of Molecular Neurobiology
The Institute of Medical Science
The University of Tokyo
4-6-1 Shirokanedai, Minato-ku, Tokyo 108, Japan

L-Glutamate (Glu) mediates a variety of neuronal responses through activation of its receptors, which are classified into several subtypes. In the vertebrate nervous system, the Glu receptors (GluRs) are generally known to cause depolarizing responses. In the cerebellum, Glu is a candidate for the neurotransmitter at the two excitatory inputs to Purkinje cells, the climbing and parallel fibers. Fast postsynaptic potentials at both synapses are mediated most likely by non-NMDA subtypes of ionotropic GluRs (iGluRs). Metabotropic GluRs (mGluRs) are also expressed in Purkinje cells, and mGluR-mediated Ca^{2+} release from intracellular stores in Purkinje cells has been reported, although their physiological role is not clear. We found a novel type of Glu response in Purkinje neurons of mouse cerebellar slices, namely glutamate-induced hyperpolarization (GH).

We applied Glu or its analogues iontophoretically to Purkinje cell dendrites in mouse cerebellar slices, and measured membrane potential through intracellular micropipettes, or, in some cases, carried out whole–cell–clamp recording. GH was observed in 30 out of 49 adult Purkinje cells (FIG. 1,A). This response was not due to activation of inhibitory interneurons, because application of tetrodotoxin (TTX), bicuculline, or strychnine did not abolish GH. In addition, GH persisted in a Ca^{2+}-free or a low-Cl^- solution, which rules out the involvement of $g_{K(Ca)}$ or $GABA_A$ mechanisms. Selective

---

a Address all correspondence to: Takafumi Inoue, M.D., Ph.D., Department of Molecular Neurobiology, The Institute of Medical Science, The University of Tokyo, 4-6-1 Shirokanedai, Minato-ku, Tokyo 108, Japan.

agonists for non-NMDA type iGluRs, AMPA and kainate, failed to induce hyperpolarization (FIGS. 1,B, and 2,B), and CNQX, which is a selective and potent antagonist
for non-NMDA type iGluRs, did not affect GH (FIG. 1,B). NMDA also failed to
induce hyperpolarization. Quisqualate (Quis, FIG. 1,B) and *trans*-1-amino-1,3-cyclopen-
tanedicarboxylic acid (tACPD), which are potent and selective agonists, respectively,
for the mGluR, failed to induce GH. L-2-amino-4-phosphonobutyric acid (L-AP4),
which is known to cause hyperpolarization in bipolar cells in the retina, was also
ineffective. Simultaneous recording of electrical activity and intracellular Ca^{2+} concentration ($[Ca^{2+}]_i$) showed that GH was not accomplished by $[Ca^{2+}]_i$ changes (FIG. 2).
Voltage clamp experiments showed that GH was due to reduction of a tonically active
conductance with a reversal potential around 0 mV (FIG. 1,C).

Two possible mechanisms are suggested for GH: (1) changes in the desensitized
steady state of iGluRs (DS) or (2) a novel Glu-mediated mechanism. If the dose-response
relationship of DS is bell-shaped[1] and background Glu concentration was maintained
by leakage from an iontophoretic tip, which might drive iGluRs on Purkinje cell
dendrites into DS, a further Glu ejection from the tip could result in reduction of
the inward current through the iGluRs. However, there are discrepancies between
our results and the bell-shaped feature of DS reported by Geoffroy *et al.* First, in our
work, CNQX was not effective against GH, but they reported that DNQX, Glu
antagonist analogous to CNQX, shifted the bell shape. Second, they showed that
Quis and AMPA also have a bell-shaped dose-response relationship in DS, but we
did not observe hyperpolarization by those drugs. Thus, it is not clear whether a shift
in DS level could in fact explain GH. There is a candidate for a novel Glu-mediated
mechanism. In depolarizing bipolar cells of the retina, Glu induces hyperpolarization
by closing a cGMP-sensitive ion channel through G-protein mediated cGMP level
control. It is reasonable to imagine that a similar process takes place in cerebellar

FIGURE 1. (A) Glutamate-induced depolarization and hyperpolarization of a Purkinje cell in a
cerebellar slice from an adult mouse. Glu was applied iontophoretically (30 nA, 50 msec; horizontal
bars under the traces) at the middle portion of the molecular layer. Initially, the response to Glu
of this cell was depolarization (1), in 18 sec, at the next Glu application, it changed to hyperpolarization
(GH, 2). The membrane potential was held below – 75 mV by constant hyperpolarizing current
(– 0.7 nA) injected through the intracellular recording electrode. Saline containing TTX (1 µM) and
bicuculline (10 µM) was superfused constantly. (B) Quis and kainate do not mimic the hyperpolarizing
effect of Glu. Glu (70 nA, 80 msec) induced hyperpolarization, while Quis (20 nA, 3 msec) or kainate
(20 nA, 20 msec) induced depolarizations (*upper traces*). Depolarization induced by Quis or kainate
was reduced by constant superfusion of saline containing CNQX (25 µM), while GH was unaffected
(*lower traces*). Drugs were ejected iontophoretically through separate barrels of the same three-barrelled
pipette. (C) Reversal potential of GH was revealed to be close to 0 mV by whole-cell mode voltage-
clamp recording. GH was observed as an outward current at negative membrane potentials, which
reversed at positive membrane potentials (*left*). The horizontal bar indicates Glu application. Current-
voltage relationships of recording from two cells (open rectangle, the same cell indicated in the left
trace; Glu, 80 nA, 150 msec, filled triangle, another Purkinje cell; Glu 70 nA, 100 msec) are indicated
(*right*). Iontophoretic pipettes were placed in the middle portions of the molecular layers, and whole-cell
clamp recordings were carried out at the somata. Both cells were in cerebellar slices from a 12-day-old
mouse, and were recorded in a solution containing TTX (1 µM) and bicuculline (10 µM).

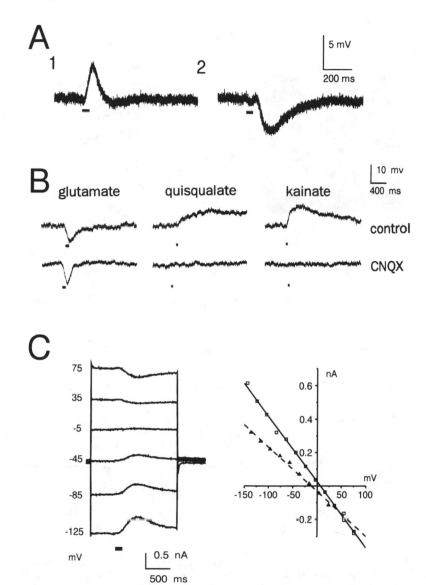

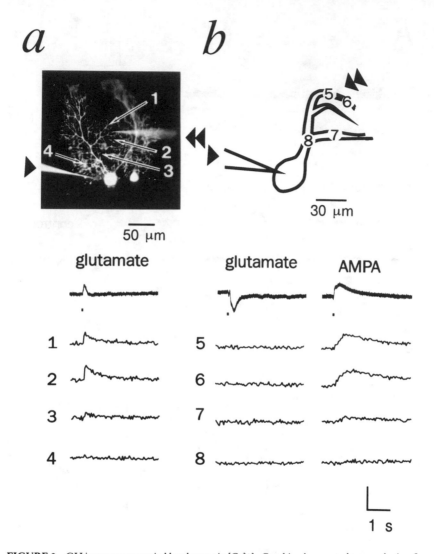

FIGURE 2. GH is not accompanied by changes in $[Ca^{2+}]_i$. Combined current-clamp and microfluorometric (fura-2) recordings were carried out from Purkinje cells superfused with saline containing TTX (0.15 μM) and bicuculline (10 μM). **(A)** When Glu-induced depolarization (*top trace*), $[Ca^{2+}]_i$ transients were measured in the dendritic region close to the iontophoretic tip (1 and 2). Only small (3) or no (4) $[Ca^{2+}]_i$ changes were detected in the dendrites distant from the iontophoretic tip. **(B)** GH was never accompanied by changes in $[Ca^{2+}]_i$. AMPA was applied from another barrel of the same multi-barrelled iontophoretic pipette, which caused depolarization (*top trace*), and $[Ca^{2+}]_i$ transients were recorded at points close to the tip of the iontophoretic pipette (5, 6, and 7). Arrow head, intracellular recording electrode; double arrow heads, iontophoretic pipette; 1–8, locations where fluorescence was measured; vertical scale, 20 mV for electrical recording and 20% change for fura-2 fluorescence.

Purkinje cells. Recently, several types of mGluRs have been identified. Some of these could have a role in the production of GH.

REFERENCE

1. GEOFFROY, M., B. LAMBOLEZ, E. AUDINAT, B. HAMON, F. CREPEL, J. ROSSIER & R. T. KADO. 1991. Mol. Pharmacol. **39:** 587–591.

Tissue-Specific Expression of Putative Domains of the Mouse Neuronal Alpha-2 and Delta Genes of the Voltage-Gated Calcium Channel in Rodent Tissues[a]

A. S. WIERZBICKI,[b-d] D. M. W. BEESON,[d]
B. LANG,[c] AND J. NEWSOM-DAVIS[c]

[c]Neurosciences Group
Institute of Molecular Medicine
John Radcliffe Hospital
Oxford OX3 9DU
United Kingdom

[d]Department of Chemical Pathology
Charing Cross and Westminster Medical School
Chelsea and Westminster Hospital
369, Fulham Road
London SW10 9NH,
United Kingdom

INTRODUCTION

The voltage-gated calcium channels (VGCCs) are a highly diverse group of ion channels found in most cells. The VGCC present in skeletal muscle consists of five subunits termed the $\alpha 1$, $\alpha 2$, β, γ, and δ.[1] The five subunits are encoded by four genes as the $\alpha 2$ and δ subunits and are derived from a common pro-protein by post-translational cleavage.[2] Alternative splicing of exons has been shown to occur in $\alpha 1$, β, and γ subunits and recently an alternative splicing mechanism has been proposed in the $\alpha 2$ gene.[3] We investigated the tissue expression profile of each of the main proposed domains of the $\alpha 2$-δ gene using clones corresponding to the full length $\alpha 2$-δ gene from the mouse neuroblastoma $\times$ rat glioma line NG108-15.

[a] A. S. Wierzbicki was supported by a Medical Research Council (UK) training fellowship.
[b] Address all correspondence to: Dr. A. S. Wierzbicki, Department of Chemical Pathology, Charing Cross and Westminster Medical School, Chelsea & Westminster Hospital, 369 Fulham Road, London SW10 9NH, UK.

METHODS

Tissue samples of brain, lung, kidney, liver, skeletal, and cardiac muscle were obtained from freshly killed mice. Total RNA was prepared from 0.5 g aliquots of homogenized tissue by the method of Chomchynski and Sacchi.[4] Twenty μg of total RNA were separated by electrophoresis and blotted by the method of Fourney et al.[5] cDNA clones corresponding to the α2-δ gene from the NG108-15 cell line were isolated from specially primed cDNA libraries by hybridization screening[6] initially with a rabbit skeletal muscle α2-δ PCR product (bp 2532–3155).[7] The approximate span of the five clones corresponding to the α2-δ gene are shown in FIGURE 1. The northern blots were hybridized at 42°C in 6 × SSC, 0.5% sodium dodecyl sulfate, 10 × Denhardt's solution, 10% dextran sulfate, 50% formamide, 100 μg/ml of sonicated boiled herring sperm DNA, and 10^6 cpm/ml of ^{32}P-dCTP labeled probe. The probes comprised the purified inserts from the cDNA clones. Blots were washed at high stringency (0.1 × SSC, 0.1% sodium dodecyl sulfate at 50°C) and autoradiographed for 5–7 days at −70°C with intensifying screens.

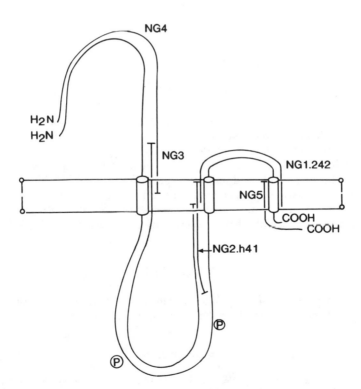

FIGURE 1. A schematic model of the alpha-2 and delta proprotein inserted into the membrane showing the segments covered by the five cDNA clones encoding the NG108-15 α2-δ gene.

TABLE 1. The Results of Northern Blots of Mouse Tissues (Brain, Lung, Kidney, Liver, Skeletal, and Cardiac Muscle) Hybridized with NG108-15 α2-δ λgt10 Library Clones Covering the Whole NG108-18 α2-δ Gene

Clone	Region Covered	Base Pairs	Tissue (+)
NG1	δ	2828–3303	brain, lung, skeltal muscle, liver
NG3	α2 intracellular	999–2768	brain
NG4	α2 N terminal	–6–1364	brain, lung, skeletal muscle, liver, heart, kidney

RESULTS

The three clones NG1.242, NG3, and NG4 were used to probe northern blots as Ng2.h41 and HG5 overlapped portions of Ng1.242, which spans almost all of the δ peptide portion of the gene. The results of screening with the probes are shown in TABLE 1. The NG108-15 α2-δ gene is transcribed as a 8.5 kb mRNA (data not shown) and hybridizing homologous bands of similar size were only found in mouse tissues. Hybridization of the δ probe NG1.242 was faint in comparison with the two α2 domain clones (NG3 and NG4).

DISCUSSION

Calcium channel diversity is generated by alternative splicing and separate genes in the case of the α1 subunit and alternative splicing have been identified in the β and γ genes.[1] A number of differing size mRNAs have been described for the α2-δ[8] but evidence for tissue variation has been limited. The α2-δ genes identified to date seem to be highly homologous over the whole protein and show little variation.[7,9] Two putatively alternatively spliced variants of the rat spinal cord α2 gene have been described differing in a deletion of 19 amino acids at positions 507–526[3,9] and with a seven amino acid insertion at lysine$_{602}$[9] or threonine$_{617}$[3] within the main intracellular domain.[3] This study confirmed previous experiments in which the NG108-15 α2-δ gene had been shown to be present in the ancestral mouse neuroblastoma line N18TG2. It suggests that variation in the intracellular domain of the 22-S gene occurs in specific tissues and the presence of a brain-specific portion of the gene.

CONCLUSION

The α2-δ gene may show similar mechanisms for generating diversity to those seen in other subunits comprising the voltage-gated calcium channel.

REFERENCES

1. TSIEN, R. W., P. T. ELLINOR & W. A. HORNE. 1992. Molecular diversity of voltage-dependent Ca^{2+} channels. Trends Pharmacol. Sci. 12:349–354.
2. JAY, S. D., A. H. SHARP, S. D. KAHL, T. S. VEDNICK, M. M. HARPOLD & K. P. CAMPBELL. 1991.

Structural characterisation of the dihydropyridine-sensitive calcium channel and associated delta peptides. J. Biol. Chem. **266**: 3287–3293.

3. KIM, H. L., H. KIM, P. LEE, R. G. KING & H. CHIN. 1992. Rat brain expresses an alternatively spliced form of the dihydropyridine-sensitive L-type calcium channel alpha 2 subunit. Proc. Natl. Acad. Sci. **89**: 3251–3255.

4. CHOMCYNSKI, P. & N. SACCHI. 1987. Single step method of RNA isolation by acid guanidinium thiocyanate phenol-chloroform extraction. Anal. Biochem. **162**: 156–159.

5. FOURNEY, R. M., J. MIGEWASHI, R. S. DAY & M. C. PATERSON. 1988. Northern blotting: efficient RNA staining and transfer. Bethesda Res. Lab. Focus **10**: 5–7.

6. WIERZBICKI, A. S., D. M. W. BEESON, B. LANG & J. NEWSOM-DAVIS. 1992. The tissue distribution of α2-δ subunit genes of voltage-gated calcium channels in rodent tissues and NG108-15 cells. Ann. N. Y. Acad. Sci. In Press.

7. ELLIS, S. B., M. E. WILLIAMS, N. R. WAYS, R. BRENNER, A. H. SHARP, A. T. LEUNG, K. P. CHAMPBELL, E. MCKENNA, W. J. KOCH, A. HUI, A. SCHWARTZ & M. M. HARPOLD. 1988. Sequence and expression of mRNAs encoding the α1 and α2 subunits of a DHP-sensitive calcium channel. Science **241**: 1661–1664.

8. BIEL, M., R. HULLIN, S. FREUNDNER, D. SINGER, N. DASCAL, V. FLOCKERZYL & F. HOFFMANN. 1991. Tissue specific expression of high voltage-activated dihydropyridine-sensitive L-type calcium channels. Eur. J. Biochem. **200**: 81–88.

9. WILLIAMS, M. E., D. H. FELDMAN, A. F. MCCUE, R. BRENNER, G. VELICELEBI, S. B. ELLIS & M. M. HARPOLD. 1992. Structure and functional expression of α1, α2 and β subunits of a novel human neuronal calcium channel subtype. Neuron **8**: 71–84.

Substance P Decreases the Non-Selective Cation Channel Conductance in Dissociated Outer Hair Cells of Guinea Pig Cochlea

SEIJI KAKEHATA,[a,b] NORIO AKAIKE,[a,c]
AND TOMONORI TAKASAKA[b]

[a]Department of Neurophysiology
[b]Department of Otorhinolaryngology
Tohoku University School of Medicine
Sendai 980, Japan

Substance P (SP) is a member of the tachykinin family and is widely distributed in the central and peripheral nervous systems. Recent immunohistochemical studies suggest the presence of SP in the guinea pig cochlea.[1] However, little is known about the physiological function of SP in the cochlea. In other preparations SP modulates K^+, Ca^{2+}, Cl^-, and non-selective cation channels.[2-5] There are several lines of evidence that SP regulates ionic channels via G-protein in either cytoplasmically dependent or independent pathways.[3,6] The aim of the present study is to elucidate the cellular mechanism of the SP-induced response in the outer hair cells (OHCs) of guinea pig cochlea using the patch clamp technique in both the conventional whole-cell and the perforated-patch configurations.

OHCs were dissociated from the organ of Corti of the guinea pig cochlea and drugs were applied using the "Y-tube" method, as previously described.[7] Under current-clamp condition, SP generated hyperpolarization in OHCs at a resting potential of -61 mV using the perforated-patch configuration. At a holding potential (V_H) of -60 mV, SP elicited outward currents in a dose-dependent manner. The average amplitude of 3×10^{-5} M SP-induced currents (I_{SPS}) was 60 ± 22 pA (mean $\pm$ S.D., $n = 37$). The half-maximum concentration was 14 μM and the Hill coefficient was 2.5. 100 μM neurokinin A induced small currents in two out of eight cells. Neurokinin B did not induce any detectable currents in concentrations up to 100 μM in four cells tested. The reversal potential of I_{SP} (E_{SP}) obtained by a ramp command potential in the conventional whole-cell configuration was -4.1 ± 0.6 mV (mean $\pm$ S.E.M.; $n = 5$). When approximately half of the intracellular Cl^- was replaced with equimolar gluconate$^-$, there was little effect on the E_{SP}. The E_{SP} was shifted negatively by about 12 mV when two-thirds of the extracellular Na^+ was substituted with $Tris^+$ (FIG. 1).

[c] Address all correspondence to: N. Akaike, Ph.D., Department of Neurophysiology, Tohoku University School of Medicine, Sendai 980, Japan.

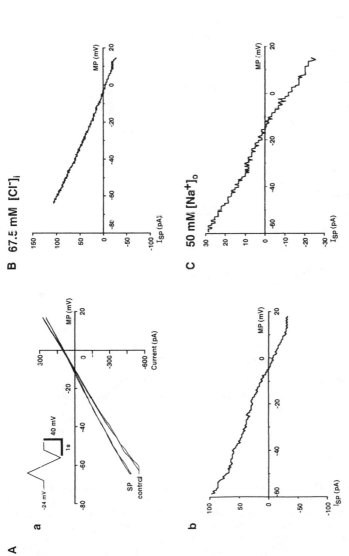

FIGURE 1. (A) The current-voltage (I-V) relationship for the I_{SP} (a) The I-V relationships in the presence and absence of SP (10^{-4} M) using the voltage ramp. Inset shows membrane potential changes produced by a ramp command potential superimposed on a V_H of -24 mV. Membrane currents were measured before (control) and during the application of SP. Both external and internal KCl were replaced with CsCl, and 10 μM La^{3+} was added. The 30 μm cell was obtained from the second turn. The resting potential and the input conductance were -10 mV and 11 nS, respectively. (b) SP-sensitive current. The I-V curve before SP application was subtracted from that during SP application in (a). Effects of internal anion (B) and external cation (C) on the E_{SP}. SP-sensitive currents obtained by the ramp voltage-clamp, as shown in (A). (B) Approximately half of the internal Cl$^-$ was replaced with gluconate$^-$. (C) Two-thirds of external Na$^+$ were replaced with an equimolar Tris.$^+$

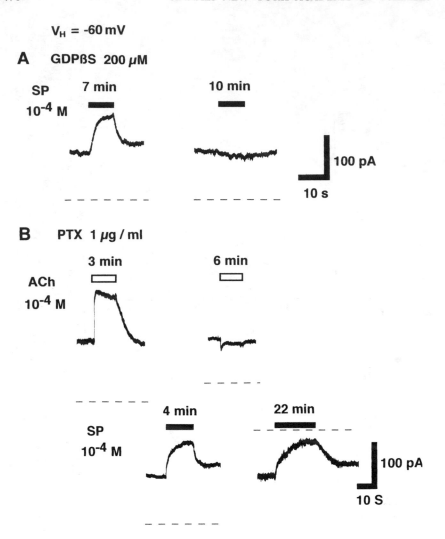

FIGURE 2. Effects of GDPβS and pertussis toxin (PTX) on I_{SP}. (A) I_{SP} was recorded using the conventional whole-cell mode with patch-pipettes containing 200 μM GDPβS at a V_H of −60 mV. I_{SP} diminished at 10 min. Horizontal solid bars indicate the application period of SP. (B) I_{SP} was recorded using the conventional whole-cell mode with patch-pipette containing 1 μg/ml PTX at a V_H of −60 mV. The I_{SP} remained relatively constant, whereas the ACh-induced outward current diminished within 6 min. Horizontal open and solid bars indicate the application period of ACh and SP, respectively. SP was applied every three minutes. Each dashed line indicates the zero current level.

these results suggest that SP decreases a non-selective cation conductance. In further studies, we, therefore, examined the ionic permeability of the I_{SP} for several monovalent cations. In this experiment, 150 mM external Na^+ was replaced with 50 mM Cs^+ or Li^+ and 100 mM $Tris^+$. Neither external Li^+ nor Cs^+ produced a significant shift in the E_{SP} compared with the one obtained using 50 mM external Na^+. To investigate Ca^{2+} permeability, 50 mM NaCl in the external solution was replaced with 25 mM $CaCl_2$. External Ca^{2+} produced a significant positive shift in the E_{SP} with the average of 8.8 ± 1.7 mV ($n = 3$). Based on the E_{SP}s obtained with extracellular ion replacement, the relative ion permeability of the channel modulated by SP was in the order of: $Ca^{2+} > Li^+ \approx Cs^+ \approx Na^+ > Tris^+$.

We next investigated the cellular mechanism of the SP-induced current. In OHCs dialyzed with 200 μM GDPβS through the patch pipette at a V_H of -60 mV, the I_{SP} diminished at approximately 10 min. When OHCs were dialyzed with 1 μg/ml pertussis toxin (PTX) through the patch pipette, the I_{SP} remained for over 20 min whereas the ACh-induced current, which is mediated via a PTX-sensitive G-protein,[7] completely diminished within 10 min (FIG. 2). These results suggest that the SP response is mediated by a PTX-insensitive G-protein. I_{SP} was little affected when OHCs were loaded with 10 mM BAPTA in the patch-pipette, suggesting that intracellular calcium is not involved in generation of the SP-induced response. Staurosporine up to 10^{-6} M did not affect the I_{SP}, suggesting that protein kinase C and protein kinase A is not involved in the SP response.

The results suggest that SP decreases a non-selective cation conductance in the OHCs of guinea pig cochlea and that this occurs via a PTX-insensitive G-protein in cytoplasmically independent pathways.

REFERENCES

1. NOWAK, R., A. DORN, H.-G. BERNSTEIN, A. RINNE, H. J. SCHOLTZ & M. ZIEGLER. 1986. Arch. Otorhinolaryngol. **243:** 36–38.
2. JONES, S. W. 1985. J. Physiol. **366:** 63–87.
3. NAKAJIMA, Y., S. NAKAJIMA & M. INOUE. 1988. Proc. Natl. Acad. Sci. USA **85:** 3643–3674.
4. MATTHEWS, G., E. NEHER & R. PENNER. 1989. J. Physiol. **418:** 105–130.
5. MURASE, K., P. D. RYU & M. RANDIC. 1989. J. Neurophysiol. **61:**854–865.
6. ABDEL-LATIF, A. A. 1986. Pharmacol. Rev. **38:** 227–272.
7. KAKEHATA, S., T. NAKAGAWA, T. TAKASAKA & N. AKAIKE. 1993. J. Physiol. **463:** 227–244.

Molecular Cloning of a Novel Type of Somatostatin Receptor and Platelet-Activating Factor Receptor cDNAs from Rat

HARUHIKO BITO, CHIE SAKANAKA,
TOMOKO TAKANO, YOSHIHISA KUDO,[a]
AND TAKAO SHIMIZU

Department of Biochemistry
Faculty of Medicine
The University of Tokyo
Bunkyo-ku, Tokyo 113

[a]Department of Neuroscience
Mitsubishi-Kasei Institute of Life Sciences
Machida, Tokyo 194, Japan

Lipid mediators constitute a distinct class of bioactive molecules that display a variety of biological activities in various organs.[1] In order to elucidate the physiological function of lipid mediators such as prostaglandins, leukotrienes, or platelet-activating factor (PAF) in the central nervous system (CNS), we investigated two candidate receptor pathways that functionally activate or are activated by these lipid mediators: somatostatin receptor (SSTR)[2] and platelet-activating factor (PAF) receptor.[3]

Based upon published nucleotide sequences for SSTR1 and SSTR2,[4,5] we isolated a new subtype of somatostatin receptor from a rat hippocampal cDNA library using PCR and cross-hybridization strategy (Bito *et al.*, unpublished data). The cloned cDNA (SS-H3) was stably expressed in Chinese Hamster Ovary (CHO) cells and the pharmacological character of the cloned receptor was examined. Both somatostatin-14 (SS14) and somatostatin-28 (SS28) displaced the binding of iodinated SS-14 at an IC_{50} of nM order, with a higher specificity for SS14. Both SS14 and SS28 strongly inhibited the cAMP accumulation activated by forskolin. Functional coupling of somatostatin receptor to arachidonic acid release through Gi protein was also shown. These data suggest a possible involvement of arachidonic acid and/or its metabolites as functional messenger(s) downstream of somatostatin receptor–mediated signal transduction in the CNS, especially in the hippocampus where the cloned receptor is highly expressed (TABLE 1).

Taking advantage of the cloning of PAF receptor as the first successful example of lipid autocoid receptor cloning,[6] we isolated the rat PAF receptor cDNA, determined its distribution in the rat CNS, and examined the PAF function in cultured rat hippo-

TABLE 1. Functional Coupling of the Cloned Hippocampal Somatostatin Receptor (SS-H3) to Various Effector Systems through G Protein

Receptor	Coupled G Protein	Coupled Effector System	Reference
SSTR1	Unknown	Unknown	8
SSTR2	Unknown	Unknown	8
Clone SS-H3	IAP-sensitive	Inhibition of cAMP synthesis arachidonate release	(unpublished data)

campal neurons. (Bito et al., unpublished data).[7] Brain PAF receptor, which was suggested to be identical to peripheral PAF receptor, was quite ubiquitously distributed in the CNS. Moreover, we discovered that PAF mobilizes Ca^{2+} mobilization in rat hippocampal neurons and observed that some PAF receptors colocalized on the same cells as NMDA receptors. PAF-mediated Ca^{2+} signaling may thus play a modulatory role in the neuronal signal transduction in the hippocampus.

From our study, we propose that different lipid mediators can be synthesized upon physiological stimuli including neurotransmitter/neuropeptide receptor activation and may act either as a second messenger within the cell or as a first messenger after release outside of the cells.

REFERENCES

1. SHIMIZU, T. & L. S. WOLFE. 1990. J. Neurochem. 5: 1–15.
2. SCHWEITZER, P., S. MADAMBA & G. R. SIGGINS. 1990. Nature 346: 464–467.
3. KORNECKI, E. & Y. H. EHRLICH. 1988. Science. 240: 1792–1794.
4. YAMADA, Y. et al. 1991. Proc. Natl. Acad. Sci. USA 89: 251–255.
5. KLUXEN, F.-W., C. BRUNS & H. LUBBERT. 1992. Proc. Natl. Acad. USA 89: 4618–4622.
6. HONDA, Z. et al. 1991. Nature 349: 342–346.
7. BITO, H. et al. 1992. Neuron 9: 285–294.
8. RENS-DOMIANO, S. et al. 1992. Mol. Pharmacol. 42: 28–34.

Endothelin Induces Phosphoinositide Metabolite-Dependent Cellular Responses in NG108-15 Hybrid Cells

M. NODA, Y. OKANO,[a] Y. NOZAWA,[a]
A. EGOROVA, AND H. HIGASHIDA

Department of Biophysics
Neuroinformation Research Institute
Kanazawa University School of Medicine
Kanazawa 920, Japan

[a]*Department of Biochemistry*
Gifu University School of Medicine
Gifu 500, Japan

A potent mammalian vasoconstrictive peptide, endothelin-1 (ET-1)[1] has a neuronal function as well as vasoconstrictive action. ET-1 causes the receptor-mediated activation of phosphoinositide-specific phospholipase C (PLC) to produce two second messengers, diacylglycerol (DAG) and inositol-1,4,5-trisphosphate (IP$_3$). DAG serves as an endogenous activator of protein kinase C (PKC), and IP$_3$ induces Ca^{2+} mobilization from intracellular stores. Also, influx of extracellular Ca^{2+} through a plasma membrane channel may contribute to an increase in intracellular Ca^{2+} concentration ([Ca^{2+}]$_i$). On the other hand, a vasoactive intestinal contractor (VIC), which differs from ET-2 in one amino acid residue and is expressed in intestinal cells, but not in endothelial cells, has differential contractile activities on mouse ileum and pig coronary artery,[2] and much less information is available about its signal pathway. In order to gain insight into the signaling mechanism of ET-1 and VIC, we have examined phosphoinositide turnover by measuring the mass contents of IP$_3$ and DAG, and also measured the [Ca^{2+}]$_i$ level in NG108-15 neuroblastoma × glioma hybrid cells. Furthermore, we observed the effects of ET-1 on membrane currents in whole-cell patch-clamped NG108-15 cell to compare with those evoked by bradykinin, which is known to activate two putative second messengers of phosphoinositide metabolites.

ET-1 and VIC caused the receptor-mediated activation of phosphoinositide-specific phospholipase C in NG108-15 cells. IP$_3$ was generated in response to 10 nM ET-1 or VIC: the mean peak value increased to 24 or 74 from 8 pmol/10^6 cells, respectively.[3] Pretreatment with pertussis toxin (PT) suppressed VIC-mediated generation of IP$_3$,

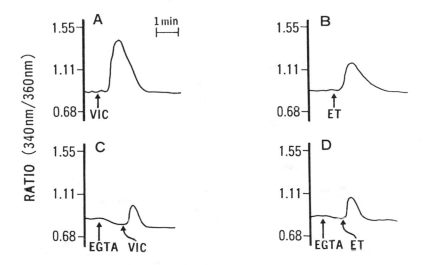

FIGURE 1. Effects of VIC and ET-1 on $[Ca^{2+}]_i$ level in NG108-15 cells. VIC (10 nM) or ET-1 (10 nM) was added to fura 2-loaded NG108-15 cells in the presence of 1.8 mM Ca^{2+} (**A, B**) or 1 mM EGTA (**C, D**). The rise in $[Ca^{2+}]_i$ induced by either ET-1 or VIC was suppressed in the presence of EGTA, suggesting that there were two pathways to produce the $[Ca^{2+}]_i$ rise; influx from outside the cell and intracellular mobilization by IP$_3$.

but not IP$_3$ production induced by ET-1, indicating the selective effect of PT on VIC response. DAG was also generated.[2] Upon stimulation with 10 nM VIC, NG108-15 cells showed the biphasic accumulation of DAG; first a transient phase and then a sustained phase. The time course of the first peaks of DAG production was consistent with those of IP$_3$ formation, indicating that the second sustained phase of agonist-induced DAG accumulation originates from phospholipid(s) other than phosphoinositide. The initial transient of DAG accumulation in 40 µM Quin 2/AM-preincubated cells to chelate intracellular Ca^{2+} was partially decreased, whereas the second phase of DAG production was completely abolished, indicating the Ca^{2+} dependency of the second DAG accumulation. $[Ca^{2+}]_i$ assay in a single cell using fura-2 revealed that both peptides increased $[Ca^{2+}]_i$ transiently (FIG. 1,A and B).[4] The rise in $[Ca^{2+}]_i$ induced by either 10 nM ET-1 or 10 nM VIC was suppressed in the presence of 1 mM EGTA (FIG. 1,C and D).[4] This suggests that there were two pathways considered to produce the increased $[Ca^{2+}]_i$ level; influx from outside the cell and mobilization from intracellular stores by IP$_3$. External application of ET-1 or VIC initially elicited the outward current associated with increased conductance and subsequently the inward current associated with decreased conductance in whole-cell voltage-clamped NG108-15 cells (FIG. 2,A). Inhibition of the voltage-dependent-M potassium current was apparently observed during the inward current induced by ET-1 (FIG. 2, A*c*). In steady-state current-voltage relationships, the outward rectification associated with M current development above −60 mV before application of ET-1 became moderate during the ET-1-induced inward current (FIG. 2,B). The ET-1 or VIC-induced current was compa-

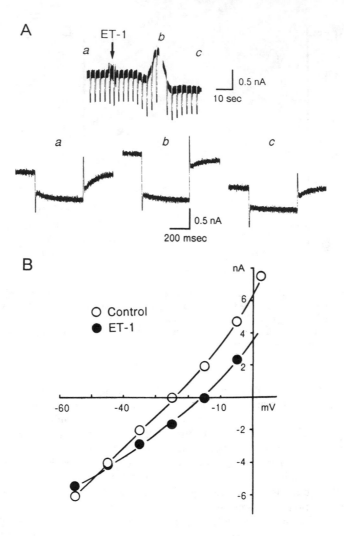

FIGURE 2. Effect of ET-1 on membrane currents in two NG108-15 cells. **(A)** An ET-1-induced current in a whole-cell voltage-clamped NG108-15 cell. The membrane potential was stepped from the holding potential of -30 mV to -50 mV for 500 msec at 3-sec intervals to measure input conductance. Upper trace is the current change induced by focal application of ET-1 (3 nl, 100 nM) on the cell at the time indicated by the arrow. The initial outward current is associated with an increase in conductance and the subsequent inward current is associated with decreased conductance. The lower traces (a–c) show expanded current records at the time indicated in upper trace. Notice the difference of holding current level in a–c. The extracellular solution was Dulbecco's modified Eagle's medium (DMEM) and the pipette solution contained 90 mM K-acetate, 20 mM KCl, 40 mM HEPES, 3 mM $MgCl_2$, and 3 mM EGTA, pH 7.4 with KOH.
(B) Current-voltage relationships obtained in another cell before adding ET-1 (O) and during the inward current induced by ET-1 (●). The current was measured as the displacement from the original holding current at the end of a series of incremental 500 msec voltage steps from a holding potential of -30 mV.

rable to that evoked by bradykinin, which is known to activate the Ca^{2+}-dependent potassium and cationic currents and inhibit voltage-dependent-M potassium current induced by IP_3 and DAG, the two putative second messengers of phosphoinositide metabolites in NG108-15 cells.[5] However, the frequency encountered in the ET-1 or VIC responses was not as much as that of the bradykinin responses in NG108-15 cells.

REFERENCES

1. YANAGISAWA, M., H. KURIHARA, S. KIMURA, Y. TOMOBE, M. KOBAYASHI, Y. MITSUI, K. GOTO & T. MASAKI. 1988. Nature 332: 411–415.
2. SAIDA, K., Y. MITSUI & N. ISHIDA. 1989. J. Biol. Chem. 264: 14613–14616.
3. FU, T., Y. OKANO, W. ZHANG, T. OZEKI, Y. MITSUI & Y. NOZAWA. 1990. Biochem. J. 272: 71–77.
4. FU, T., W. ZHANG, N. ISHIDA, K. SAIDA, Y. MITSUI, Y. OKANO & Y. NOZAWA. 1989. FEBS Lett. 257: 351–353.
5. BROWN, D. A. & H. HIGASHIDA. 1988. J. Physiol. 397: 167–184.

Intracellular Injection of Cyclic AMP Inhibits the Dopamine-Induced K⁺-Dependent Responses of *Aplysia* Ganglion Cells

MITSUHIKO MATSUMOTO, REIKO FUJITA,
SHINGO KIMURA, KAZUHIKO SASAKI,
AND MAKOTO SATO

Department of Physiology
School of Medicine
Iwate Medical University
19-1 Uchimaru
Morioka, Iwate, 020 Japan

Adenosine 3′,5′-cyclic monophosphate (cAMP) has been shown to be an intracellular messenger mediating or modulating the transmitter-induced responses of various cells in different animals. Kehoe[1] has reported that intracellularly injected cAMP blocks the K⁺-dependent response of the *Aplysia* neuron to arecoline, a muscarinic agonist. We

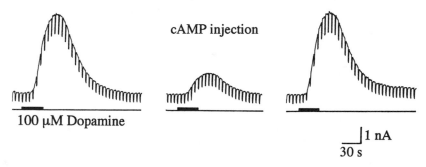

FIGURE 1. Blocking effect of an intracellular injection of cAMP on dopamine (DA)-induced K⁺ current. All records were obtained from the same cell under voltage-clamp at −60 mV. Upward is outward. DA (100 μM) was applied to the cell by perfusion. Intracellular application of cAMP was made 30 sec prior to the test trace in the middle. The injection was made by pressure pulses of 0.2–3.0 kg/cm² with a duration of 100–300 msec. The downward rectangular pulses appearing periodically from the baseline indicate the change in membrane conductance during the response.

486

are interested in testing whether the blocking effect of cAMP is specific to K⁺-dependent response of the muscarinic receptor or common to all K⁺-dependent responses.

Application of DA also induces a slow hyperpolarizing response associated with an increase in K⁺ permeability in the identified neurons of *Aplysia*. We have proven previously that this response is mediated by activation of a GTP-binding protein (G-protein) that is sensitive to pertussis toxin.[2] Intracellularly injected cAMP depressed this DA-induced K⁺-current response (FIG. 1). The cAMP-induced depression was augmented in the presence of either IBMX, a phosphodiesterase inhibitor, or okadaic acid, a phosphatase inhibitor. The DA-induced response was also depressed by intracellularly injected catalytic subunit of cAMP-dependent kinase, suggesting that the protein kinase A is involved in the depressing effect of cAMP.

Previously, we reported that the K⁺ channels regulated by DA in these cells could be activated simply by raising temperature from 22 to 32°C, without stimulating the DA receptor.[3,4] Intracellular application of GTPγS produces an opening of this K⁺ channel by activating the G-protein irreversibly. Neither warm-induced nor GTPγS-

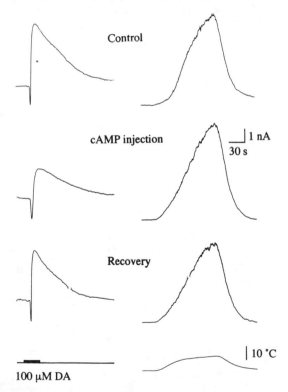

FIGURE 2. Effects of intracellularly injected cAMP on the K⁺-dependent responses to DA and raising temperature from 23 to 31°C. The change in temperature was monitored as shown in the bottom right. All recordings were obtained from the same cell voltage-clamped at −70 mV. The effect was examined 30 sec prior to each test with 100 μM dopamine or raising temperature as shown in the middle row. The recovery was observed 25 min later and shown at the bottom.

induced K^+ response was depressed by injected cAMP (FIG. 2) or the catalytic subunit of protein kinase A.

We concluded from these results that (1) the blocking effect of cAMP might be due to activation of protein kinase A and (2) the acting site of protein kinase A could be somewhere between the receptor and GTP-binding protein, but not the K^+ channels themselves.

REFERENCES

1. KEHOE, J. 1985. Synaptic block of a transmitter-induced potassium conductance in *Aplysia* neurons. J. Physiol. (Lond) **369**: 399–437.
2. SASAKI, K. & M. Sato. 1987. A single GTP-binding protein regulates K^+-channels coupled with dopamine, histamine and acetylcholine receptor. Nature **325**: 259–262.
3. HAKOZAKI, S., M. MATSUMOTO & K. SASAKI. 1989. Temperature-sensitive activation of G-protein regulating the resting membrane conductance of *Aplysia* neurons. Jpn. J. Physiol. **39**: 115–130.
4. TAMAZAWA, Y., M. MATSUMOTO, A. KUDO & K. SASAKI. 1991. Potassium ion channels operated by receptor stimulation can be activated simply by raising temperature. Jpn. J. Physiol. **41**: 117–127.

Regulation of Ca^{2+} Mobilization by β-Adrenergic Receptor in Jurkat Human Leukemia T-Cells

H. TAKEMURA, S. HATTA, K. YAMADA,[a]
AND H. OHSHIKA

Department of Pharmacology

[a]Department of Biochemistry
Sapporo Medical College
S.1, W.17, Sapporo, 060, Japan

INTRODUCTION

β-Adrenergic receptors are present in lymphocytes[1] and the agents but not β-adrenergic agonist, which raise cyclic AMP, modulate Ca^{2+} flux stimulated by concanavalin A, and alone mobilize Ca^{2+} in T-cells;[2,3] however, the role of β-adrenergic receptors in signal transduction in T-cells, which contribute to the immune system, is less known. We have examined the signal transduction system, including G protein and cytosolic Ca^{2+} ($[Ca^{2+}]_i$), by β-adrenergic receptor in human T-cell line Jurkat.

METHODS

Jurkat human leukemia T-cells were maintained under 5% CO_2–95% O_2 in RPMI 1640 medium supplemented with 10% fetal bovine serum, penicillin (50 U/ml), streptomycin (50 μg/ml), and L-glutamine (300 μg/ml). Membranes of Jurkat cells were subjected to SDS/PAGE and immunologically stained with the antibodies, RM/1, AS/7, QL, and GC/2, for α subunits of G proteins, Gs, Gi, Gq, and Go, respectively. Furthermore, enhanced chemiluminescence method was used as a detecting system. $[Ca^{2+}]_i$ was estimated from the fluorescence of fura-2–loaded Jurkat cells as described before.[4]

RESULTS AND DISCUSSION

As shown in FIGURE 1, α subunits of G proteins were detected in Jurkat cells. The antiserum with specificity for Gsα reacted with two polypeptides with apparent molecular weights of 42 and 52 kD, though their role in signal transduction in Jurkat cells is as yet unknown. The antiserum with specificity for Giα weakly reacted with

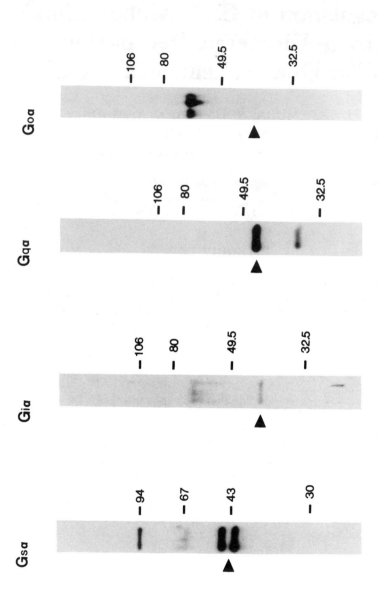

FIGURE 1. Characterization of α subunits of G proteins in Jurkat cells. Membranes of Jurkat cells were subjected to SDS/PAGE and immunologically stained with the antibodies, RM/1, AS/7, QL, and GC/2, for α subunits of G proteins, Gs, Gi, Gq, and Go, respectively. Enhanced chemiluminescence method was used as a detecting system.

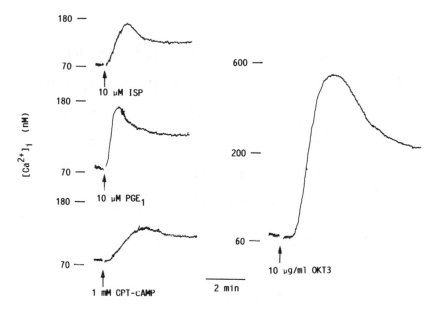

FIGURE 2. Effects of isoproterenol (ISP), prostaglandin E_1 (PGE$_1$), and 8-(4-chlorophenylthio)-cyclic AMP (CPT-cAMP) (*left*), which increase cyclic AMP accumulation, and anti-CD3 antibody OKT3 (*right*), which causes inositol phosphate production, on $[Ca^{2+}]_i$ in Jurkat cells. $[Ca^{2+}]_i$ was measured in fura-2--loaded cells as described under METHODS.

a polypeptide. Furthermore, Gqα was recognized by an antiserum with specificity for Gqα. However, no Goα was detected. FIGURE 2 shows that 10 μM of isoproterenol (ISP) caused an increase in $[Ca^{2+}]_i$, after which it declined to a sustained, elevated level. A rise in $[Ca^{2+}]_i$ induced by ISP was mimicked by 10 μM prostaglandin E_1, which increases cyclic AMP accumulation in Jurkat cells, as well as by 1 mM 8-(4-chlorophenylthio)-cyclic AMP (CPT-cyclic AMP), which is membrane-permeable cyclic AMP. On the other hand, CD3 monoclonal antibody OKT3, which increases inositol trisphosphates, caused a larger increase in $[Ca^{2+}]_i$ than that induced by ISP, PGE$_1$, or CPT-cyclic AMP.

These data suggest that the β-adrenergic receptor may couple to G protein and that cyclic AMP accumulated through the activation of G protein may regulate the immune system via Ca^{2+} in Jurkat cells.

REFERENCES

1. FUCHS, B. A., J. W. ALBRIGHT & J. F. ALBRIGHT. 1988. Cell. Immunol. **114**: 231–245.
2. LERNER, A., B. JACOBSON & R. A. MILLER. 1988. J. Immunol. **140**: 936–940.
3. KELLY, L. L., P. F. BLACKMORE, S. E. GRABER & S. J. STEWART. 1990. J. Biol. Chem. **265**: 17657–17664.
4. TAKEMURA, H., H. OHSHIKA, N. YOKOSAWA, K. OGUMA & O. THASTRUP. 1991. Biochem. Biophys. Res. Commun. **180**:1518–1526.

Propagated Calcium Modulates the Calcium-Dependent Potassium Current by the Activation of GABA$_B$ Receptor at the Axonal Branch in the Type B Photoreceptor of *Hermissenda*

M. SAKAKIBARA,[a] H. TAKAGI,[b]
T. YOSHIOKA,[c] AND D. L. ALKON[d]

[a]*Department of Neurobiological Engineering*
School of High-Technology
Tokai University
Numazu 410-03, Japan

[b]*Department of Physics*
School of Science and Engineering
Waseda University
Tokyo 169, Japan

[c]*Department of Molecular Neurobiology*
School of Human Science
Waseda University
Tokorozawa 359, Japan

[d]*Section on Neural Systems*
National Institute of Neurological Diseases and Stroke
Bethesda, Maryland 20892

Potassium channels in the type B photoreceptor of *Hermissenda* are persistently inactivated after the acquisition of associative learning and memory. A number of recent studies suggested that the monosynaptic release of GABA from pre-synaptic hair cell terminals onto a postsynaptic B cell has been considered to be the loci of the K$^+$ channels modifications.[1,2] Further, a classical inhibitory GABA-ergic synaptic response in the B cell is transformed into an excitatory response following the paired stimulus of conditioning.[3] Neither the conventional microelectrode–voltage-clamp nor the patch-clamp measurement with an axotomized type B cell is adequate to characterize the mechanism of GABA-induced K$^+$ channel modulation. We applied the slice-patch

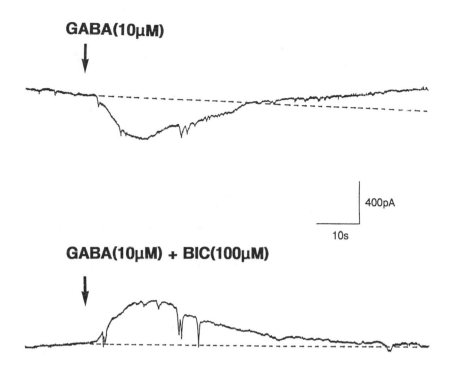

FIGURE 1. GABA induced currents at the type B photoreceptor. GABA (10 μM) induced the inward current, which blocked by 100 μM bicuculline, followed by the outward current. The reversal potentials of the inward and the outward currents were – 40 to – 60 mV and – 90 mV, respectively. The membrane was clamped at – 60 mV (absolute).

technique to the intact soma membrane of an axon bearing the type B photoreceptor to elucidate the mechanism of the postsynaptic event by the activation of GABA_B receptor. Type B cell was whole-cell clamped in Na-free ASW replaced by choline-chloride. Application of 10 μM GABA initially induced the inward current followed by the longlasting outward current at the membrane potential of – 60 mV and the initial inward current had a reversal potential of – 40 to – 60 mV, which was the intermediate equilibrium potential of Cl⁻ and K⁺. On the other hand, the cell pre-treated with GABA_A antagonist, 100 μM bicuculline, application of GABA induced solely the outward current, which reversed at the membrane potential of – 90 mV coincident with the potassium equilibrium potential. Progressive enhancement of volt-age- and calcium-dependent current I_{Ca-K} was observed in the bicuculline-containing GABA solution being untouched I_A. This enhancement was even unaffected by lowering the external calcium from 10 to 1 mM. Since GABA did not modulate the voltage-dependent calcium current across the type B soma membrane, this enhancement was thought to be due to the elevation of cytosolic calcium propagated along the

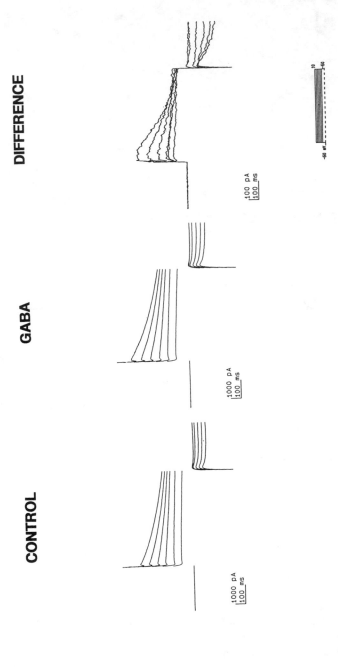

FIGURE 2. Activation of GABA$_B$ receptor enhanced the voltage-dependent potassium current (I$_{Ca-K}$) being untouched I$_A$. GABA$_A$ receptor was blocked by 100 μM bicuculline. The voltage-dependent potassium current was not modulated by the extracellular Ca^{2+} concentration.

axon from the terminal branch via GABA$_B$ receptor activation by exogenous GABA application. This calcium propagation was revealed by the Fura-2 $[Ca^{2+}]_i$ imaging. These experiments suggest that intracellular calcium elevation in the synaptic terminals affects regulation of ionic currents across the cell soma membrane 60–80 μm away. The present findings together with the calcium propagation confirm that the pairing of specific changes of the soma K^+ current is mediated by GABA-induced Ca^{2+} rise.

REFERENCES

1. MATZEL, L. D. & D. L. ALKON. 1991. GABA-induced potentiation of neuronal excitability occurs during pairings with intracellular calcium elevation. Brain Res. **554:** 77–84.
2. ALKON, D. L., J. V. SANCHEZ-ANDRES, E. ITO, K. OKA, T. YOSHIOKA & C. COLLIN. 1992. Long-term transformation of an inhibitory into excitatory GABA-ergic synaptic response. Proc. Natl. Acad. Sci. USA **89:** 11862–11866.
3. ALKON, D. L., C. COLLIN, E. ITO, T. J. NELSON, K. OKA, M. SAKAKIBARA, B. G. SCHREURS & T. YOSHIOKA. 1993. Molecular and biophysical steps in the storage of associative memory. Ann. N. Y. Acad. Sci. **707:** In press.

Inhibition of the Vesicular Release of Neurotransmitters by Stimulation of GABA$_B$ Receptor

K. TANIYAMA, M. NIWA, Y. KATAOKA,
AND K. YAMASHITA

Department of Pharmacology II
Nagasaki University School of Medicine
Sakamoto, 1-12-4
Nagasaki 852, Japan

INTRODUCTION

Since the γ-aminobutyric acid-B (GABA$_B$) receptor-mediated inhibition of noradrenaline (NA) release from the atrium was first noted in a study by Bowery *et al.*,[1] there has been much documentation on the GABA$_B$ receptor-mediated response in the central and peripheral nervous systems.[2,3] Stimulation of GABA$_B$ receptor has been shown to close the Ca^{2+} channels and to open the K^+ channels. Such GABA$_B$ receptor-mediated modulation of ion channels may correspond to the inhibition of neurotransmitter release from the nerve terminals. It is proposed that the neurotransmitters are released in the exocytotic-vesicular and non-vesicular manners: NA is released from adrenergic nerve terminals mainly in a vesicular manner, and acetylcholine (ACh) is released from cholinergic nerve terminals in vesicular and non-vesicular manners.[4] Thus, this paper was focused on the GABA$_B$ receptor-mediated modulation of the vesicular release of NA and ACh.

IDENTIFICATION OF VESICULAR AND NON-VESICULAR RELEASES OF NA AND ACH

Application of either high K^+ or ouabain to the medium evoked the release of NA from the slices of cerebellar cortex and ileal strips and the release of ACh from the ileal strips.[5,6] High K^+-evoked release was external Ca^{2+} dependent, but ouabain-evoked release was Ca^{2+} independent. There is accumulating evidence that protein kinase C (PKC) is involved in the exocytotic-vesicular release of neurotransmitters.[9-13] The phosphorylation of vesicle-associated protein by PKC as well as Ca^{2+}/calmodulin-dependent protein kinase and cyclic AMP-dependent protein kinase no doubt contributes to the exocytotic-vesicular release of neurotransmitters.[10-13] An activator of PKC, 12-*O*-tetradecanoyl-phorbol-13-acetate (TPA) potentiated the Ca^{2+}-dependent high K^+-evoked release of NA and ACh from the slices of guinea pig cerebellar cortex and ileal strips, respectively (FIG. 1).[6] The potentiating effect of TPA was not mimicked by 4α-phorbol-12,13-didecanoate, the inert derivative of phorbol ester, and was antagonized by sphingosine, a compound that inhibits the activity of PKC. On the other hand, the Ca^{2+}-independent ouabain-evoked release of NA and ACh was not potentiated by activation of PKC (FIG. 1). Thus, the depolarization of nerve terminals with high K^+ evokes the vesicular release of neurotransmitters, however ouabain evokes the non-vesicular release. Although the mechanism of ouabain action in neurotransmitter release remains unknown, ouabain is assumed to induce a non-vesicular release of

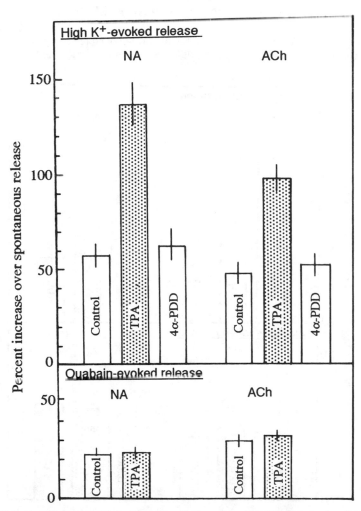

FIGURE 1. Potentiation by protein kinase-C activation of the high K^+-evoked release of NA and ACh from guinea pig cerebellar slices and ileal strips, respectively, but not of the ouabain-evoked release. TPA (10^{-7} M) or 4-α-PDD (10^{-7} M) was added to the medium 15 min before and during application of KCl (40 mM) or ouabain (0.1 mM).

neurotransmitters.[4] Ouabain depolarizes the plasma membrane of nerve terminals due
to inhibition of Na^+, K^+-ATPase, and this seems to be the primary step.[7] There is
the concept that intracellular Na^+, increased by inhibition of Na^+,K^+-ATPase with
ouabain, leads to the release of Ca^{2+} from intracellular Ca^{2+} storage sites, and this
may result in a release of neurotransmitter.[8] However, free intracellular Ca^{2+} concentrations do not seem to increase in the presence of ouabain.[7]

GABA$_B$ RECEPTOR-MEDIATED INHIBITION OF VESICULAR, BUT NOT NON-VESICULAR RELEASE

Stimulation of GABA$_B$ receptor inhibited the high K^+-evoked release of NA and ACh,
but not the ouabain-evoked release.[6] GABA inhibited the high K^+-evoked release of

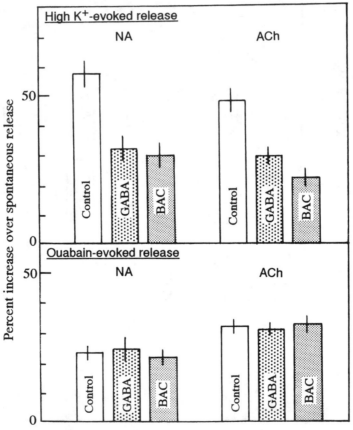

FIGURE 2. Inhibition by GABA$_B$ receptor stimulation of the high K^+-evoked release of NA and
ACh from guinea pig cerebellar slices and ileal strips, respectively, but not of the ouabain-evoked
release. GABA (10^{-5} M) or baclofen (BAC, 10^{-5} M) was added to the medium 2 min before and
during application of KCl (40 mM) or ouabain (0.1 mM).

NA from the cerebellar slices and ACh from the ileal strips (FIG. 2). The inhibitory effect of GABA was mimicked by baclofen, a $GABA_B$ agonist, but not by muscimol, a $GABA_A$ agonist, and antagonized by phaclofen, a $GABA_B$ antagonist. On the other hand, neither GABA nor baclofen inhibited the ouabain-evoked release of NA and ACh. Thus, the $GABA_B$ receptor appears to participate in inhibiting the vesicular but not the non-vesicular release of neurotransmitters. Stimulation of the $GABA_B$ receptor has been shown to decrease the Ca^{2+} current in dorsal root ganglion cells and to induce a hyperpolarization of hippocampal pyramidal cell membranes due to an increase in K^+ conductance.[2,3] The $GABA_B$ receptor expressed in oocytes by injection with mRNA from the rat cerebellum appears to be coupled to K^+ channels,[14] therefore an increase in K^+ conductance may contribute to the $GABA_B$ receptor-mediated inhibition of neurotransmitter release, especially NA from the cerebellar cortex.

REFERENCES

1. Bowery, N. G., D. R. Hill, A. L. Hudson, A. Doble, D. N. Middlemiss, J. Shaw & M. Turnbull. 1980. (−)Baclofen decreases neurotransmitter release in the mammalian CNS by an action at a novel GABA receptor. Nature 283: 92–94.

2. Bormann, J. 1988. Electrophysiology of $GABA_A$ and $GABA_B$ receptor subtypes. Trend Neurosci. 11: 112–116.

3. Bowery, N. G. 1989. $GABA_B$ receptors and their significance in mammalian pharmacology. Trend Pharmacol. Sci. 10: 401–407.

4. Cooper, J. R. & E. M. Meyer. 1984. Possible mechanisms involved in the release and modulation of release of neuroactive agents. Neurochem Int. 6: 419–433.

5. Hashimoto, S., H. Shuntoh, K. Taniyama & C. Tanaka. 1988. Role of protein kinase C in the vesicular release of acetylcholine and norepinephrine from enteric neurons of guinea pig small intestine. Jpn. J. Pharmacol. 48: 377–385.

6. Taniyama, K., M. Niwa, Y. Kataoka & K. Yamashita. 1992. Activation of protein kinase C suppresses the γ-aminobutyric acid$_B$ receptor-mediated inhibition of the vesicular release of noradrenaline and acetylcholine. J. Neurochem. 58: 1239–1245.

7. Santos, M. S., P. P. Goncalves & A. P. Carvalho. 1990. Effect of ouabain on the γ-[^{3}H]aminobutyric acid uptake and release in the absence of Ca^{2+} and K^+-depolarization. J. Pharmacol. Exp. Ther. 253: 620–627.

8. Sandoval, M. E. 1980. Sodium-dependent efflux of [^{3}H]GABA from synaptosomes probably related to mitochondrial calcium mobilization. J. Neurochem. 35: 915–921.

9. Nishizuka, Y. 1984. Turnover of inositol phospholipids and signal transduction. Science 225: 1365–1370.

10. Nestler, E. J., S. I. Walaas & P. Greengard. 1984 Neuronal phosphoproteins: Physiological and clinical implications. Science 225: 1357–1364.

11. Linstedt, A. D. & R. B. Kelly. 1987. Overcoming barriers to exocytosis. Trend. Neurosci. 10: 446–448.

12. Walker, J. H. & D. V. Agoston. 1987. The synaptic vesicle and the cytoskeleton. Biochem. J. 247: 249–258.

13. Trimble, J. H. & R. H. Scheller. 1988. Molecular biology of synaptic vesicle-associated proteins. Trend. Neurosci. 11: 241–242.

14. Taniyama, K., K. Takeda, H. Ando, T. Kuno & C. Tanaka. 1991. Expression of the $GABA_B$ receptor in Xenopus oocytes and inhibition of the response by activation of protein kinase C. FEBS Lett. 278: 222–224.

Molecular and Biophysical Steps in the Storage of Associative Memory

D. L. ALKON, C. COLLIN, E. ITO, C.-J. LEE,
T. J. NELSON, K. OKA, M. SAKAKIBARA,[a]
B. G. SCHREURS, AND T. YOSHIOKA[b]

Neural Systems
Building 9, Room 1w125
National Institute of Neurological Disorders and Stroke
National Institutes of Health
Bethesda, Maryland 20892

[a]*School of High-Technology Human Welfare*
Tokai University
Numazu 410-03, Japan

[b]*School of Human Science*
Waseda University
Tokorozawa 359, Japan

Previous studies[1,2] of molluscan and mammalian neural networks have implicated a sequence of changes in associative memory storage: (*1*) elevation of $[Ca^{2+}]_i$ and DAG (diacylglycerol); (*2*) activation of Ca^{2+}-calmodulin-dependent-type II kinase and activation of PKC (protein kinase C); (*3*) phosphorylation of a low molecular weight G-protein called cp20; (*4*) persistent reduction of postsynaptic voltage-dependent K^+

FIGURE 1. (**A**) Typical HPLC tracings of proteins from eye of a *Hermissenda* subjected to random light and rotation (Random), a naive animal (Naive), or an animal subjected to light-rotation pairing (Paired), prepared as in Table 1[3] and chromatographed on an ion-exchange (AX-300) column (0 to 0.6 M sodium acetate, pH 7.4). Each chromatogram represents proteins from a single *Hermissenda* eye. A baseline (derived from a chromatogram with no injection) has been subtracted from each digitized chromatogram. Zero is set at 0.35 for the random tracing, 0.15 for naive, and 0.0 for paired. Peaks with t_R <27 min were not analyzed. The peak with a t_R of 33.0 min was analyzed as the sum of two peaks. The peak at t_R 58.0 was variable among animals. About 80 to 85% of injected proteins eluted in the nonretained fraction between 0 and 10 min and between 25 and 28 min. (**B**) Identification of cp27 and cp20 as phosphoproteins. Seventeen naive *Hermissenda* eyes were incubated at 15°C for 6 h with 11 µCi of carrier-free [^{32}P]Pi, then homogenized, centrifuged (10,000 g, 7 sec), and chromatographed by AX-300 ion-exchange HPLC. Fractions (0.2 min) were collected

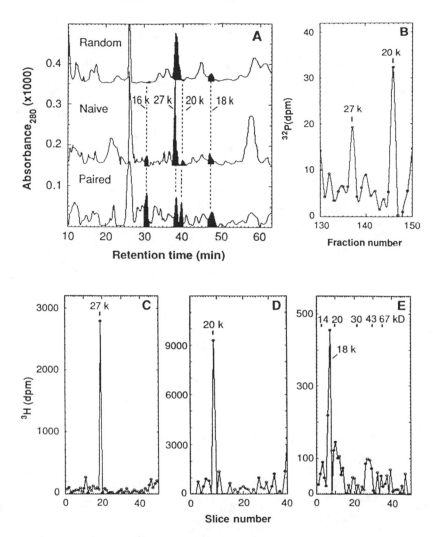

and the radioactivity determined. Distinct peaks of ^{32}P incorporation, coinciding with the absorbance peaks, were visible for cp27 and cp20 in all chromatograms ($n = 5$). Fractions 137 and 146 correspond to peaks with retention times of 38.2 to 38.4 and 40.0 to 40.2 min, respectively. This experiment was done with a different AX-300 column than the other experiments in this paper. A blank of 35 dpm was subtracted from each data point. (C–E) SDS polyacrylamide gel profiles of conditioning-associated proteins (C) cp27, (D) cp20, and (E) cp18 from *Hermissenda* eyes were combined, homogenized, and chromatographed as in (A). Fractions corresponding to each peak of interest were combined, desalted, lyophilized, dissolved in 2 μl of water, and incubated with 0.25 mCi of [^{3}H]acetic anhydride in 2 μl for 1 h at 25°C (C and E) or 1.25 2 mCi in 10 μl of dimethyl sulfoxide (DMSO) for 18 h (D). Excess acetic anhydride was removed by lyophilization and samples were subjected to SDS-PAGE (10% acrylamide). Gel slices [2 mm for (C) and (E), 2.7 mm for (D)] were solubilized and the radioactivity determined. About 1×10^{-10} g of protein are present in (D) and (E), and 5×10^{-9} g in (C). Blanks of 40 to 50 dpm have been subtracted from (C) and (E) and 1,000 dpm from (D).

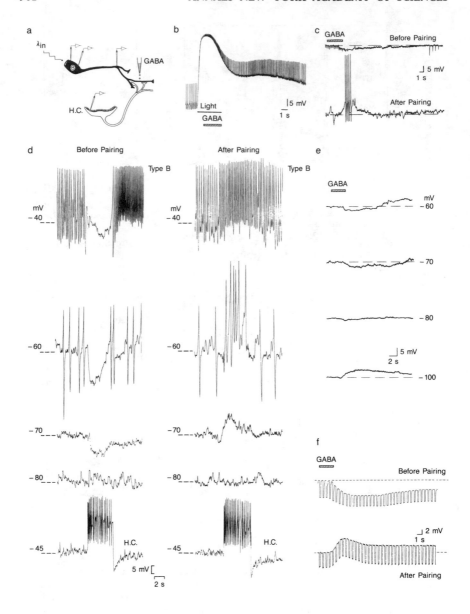

currents (I_A and I_C); (5) alteration of intraaxonal particle transport; and (6) changes in distribution of postsynaptic terminal branches. Many related studies from our laboratory indicate that the G-protein, cp20, is a critical transforming molecule during learning and memory. Cp20, for example, directly regulates voltage-dependent K^+ channels,[3] retrograde transport of intraaxonal particles, turnover of mRNA, and neuronal branching architecture. Thus, all of the profound neuronal transformations produced by learning are reproduced by cp20 elevation within neuronal targets. This transformational role of cp20 during memory storage may share common features with the transformational role of v-ras during oncogenesis. These parallel roles are suggested by the striking functional homology in the regulation of both K^+ and Ca^{2+} channels demonstrated by cp20 and v-ras protein.[4] In search of neurotransmitter pathways that lead to activation of PKC, calcium/calmodulin type II kinase, and then cp20, in both invertebrate and vertebrate networks, we have now identified GABA as an important neurotransmitter between neuronal pathways reponding to temporally associated sensory stimuli. Recent evidence, for example, indicates that in the visual-vestibular network of *Hermissenda*, statocyst hair cells release GABA onto type B visual cell terminal branches. At this synapse we found that a classical inhibitory GABAergic synaptic response is transformed into an excitatory response following pairing of statocyst-stimulated release of GABA and light-induced depolarization of the type B cell.[5] The stimulus pairing that caused this synaptic transformation closely followed stimulus conditions used to train the intact animals. Current- and voltage-clamp experiments suggest that this synaptic transformation is due to a shift from a $GABA_A$- and $GABA_B$-mediated increase of Cl^- and K^+ conductance to a GABA-mediated decrease of conductance. Fura-2 imaging of $[Ca^{2+}]_i$ suggests that pairing-specific prolongation of $[Ca^{2+}]_i$ elevation contributes to this synaptic transformation as well as previously reported reduction of K^+ channels at spatially separate membrane loci on the type B cell. Furthermore, associative learning–specific changes of these $[Ca^{2+}]_i$ signals remain for many days in the memory storage period. Still other evidence

---

FIGURE 2. Type B responses to GABA and hair cell impulses are transformed from inhibitory to excitatory after light-GABA pairings. (A) Diagram of the experimental preparation. (B) Under current-clamp conditions, after 10 min of dark adaptation, the Type B photoreceptor responds to a flash of light with a depolarizing generator potential accompanied by enhanced impulse activity One sec after the 4-sec light onset, a 3-sec puff of GABA, indicated by horizontal bar, was delivered at the terminal branches. (C) After three light-GABA applications at 90 sec intervals, the initial hyperpolarizing response to GABA (upper trace) was transformed into an excitatory response (lower trace). Dashed line indicates level (– 60 mV) of resting membrane potential. (D) Following three light-GABA pairs, the B cell's endogenous hyperpolarizing response (left) to HC stimulation with 1 sec, 1 nA current pulses (bottom traces) was transformed into depolarization (right traces). Both hyperpolarizing and depolarizing responses became negligible at about – 80 mV. The traces here are typical examples of 12 independent replications in 12 independent animals. Dashed lines indicate resting level of membrane potential. Some impulse peaks were filtered from the records. (E) Current-clamp recordings of the postsynaptic response to GABA in 0 Na^+ ASW. The early hyperpolarizaton had a reversal potential of about – 80 mV, and the later depolarizing phase was minimal at potentials more negative than – 60 mV. (F) Hyperpolarizing responses (before pairing) evoked by GABA puffs are accompanied by a decrease of input resistance of the B cell, while depolarizing responses (after pairing) are accompanied by an increase of input resistance measured by 200 msec, hyperpolarizing (– 1 nA) pulses.[5]

suggests that a similar transformation of GABAergic synaptic transmission occurs in mammalian brain. Since such GABA-synapse transformations appear to involve PKC pathways, the molecular steps implicated in a variety of earlier studies (see above) could occur within GABAergic synaptic pathways. These and other observations of biological memory networks have been used to design artificial computer-based networks with unusual capacity for storing correlated and anticorrelated events in spatial and/or temporal matrices.

REFERENCES

1. ALKON, D. L. 1989. Sci. Am. **261**: 42–50.
2. ALKON, D. L. & H. RASMUSSEN. 1988. Science **239**: 998–1005.
3. NELSON, T. J., C. COLLIN & D. L. ALKON. 1990. Science **247**: 1479–1483.
4. COLLIN, C., A. G. PAPAGEORGE, D. R. LOWY & D. L. ALKON. 1990. Science **250**: 1743–1745.
5. ALKON, D. L., J. V. SANCHEZ-ANDRES, E. ITO, K. OKA, T. YOSHIOKA & C. COLLIN. 1992. Proc. Natl. Acad. Sci. USA **89**: 11862–11866.

Characterization of Metabotropic Glutamate Receptors in Cultured Purkinje Cells

M. YUZAKI, K. MIKOSHIBA,[a] AND Y. KAGAWA

Department of Biochemistry
Jichi Medical School
Minamikawachi-machi
Tochigi 329-04, Japan

[a]*Department of Neurobiology*
Institute of Medical Science
University of Tokyo
Shirokanedai, Minato-ku
Tokyo 108, Japan

Metabotropic glutamate receptor (mGluR) is highly expressed in cerebellar Purkinje cells. Several lines of evidence suggest that mGluRs play an important role in synaptic plasticity, as in long-term depression in the cerebellum. Little is known, however, about the functions and pharmacology of mGluRs in single neurons. We used Ca^{2+} imaging with fura-2 in cultured Purkinje cells, identified immunocytochemically, to record the direct effect of drugs in stable conditions. This preparation also permitted the investigation of change of functional mGluR during maturation.

Purkinje cells, cultured in serum-free medium for up to 5–6 weeks, followed a developmental pattern similar to that *in vivo* to a considerable degree: they showed spontaneous synaptic activity,[1] expressed mGluR and inositol trisphosphate receptor (IP_3R) in the same way, and developed rich dendritic arborization. In Ca^{2+}-free medium, quisqualate increased intracellular Ca^{2+} concentration, but AMPA, kainate, and NMDA did not (Fig. 1,A). This response was thus considered as quisqualate-induced mGluR activation causing Ca^{2+} mobilization. The mGluR responses in Purkinje cells were insensitive to CNQX (Fig. 1,B), D,L-2-amino-3-phosphonopropionic acid (AP3) (Fig. 1,C), and pertussis toxin (IAP) (Fig. 1,D). These results were quite different from earlier studies measuring phosphoinositide turnover in brain slices and those measuring Ca^{2+}-activated Cl^- current in *Xenopus* oocytes injected with mGluR mRNA,[2] but in good agreement with a recent study with CHO cells transfected with mGluR cDNA.[3] The mGluR agonists showed the rank order of potency: quisqualate > glutamate > ibotenate = trans-ACPD (Fig. 1,G). Asparate, one of the candidates for a neurotransmitter between climbing fiber and Purkinje cells, caused no mGluR responses. The dose-response relationship showed an all-or-none tendency (Fig. 1,F). This may be explained by Ca^{2+}-sensitization of IP_3R.[4] The mGluR responses changed

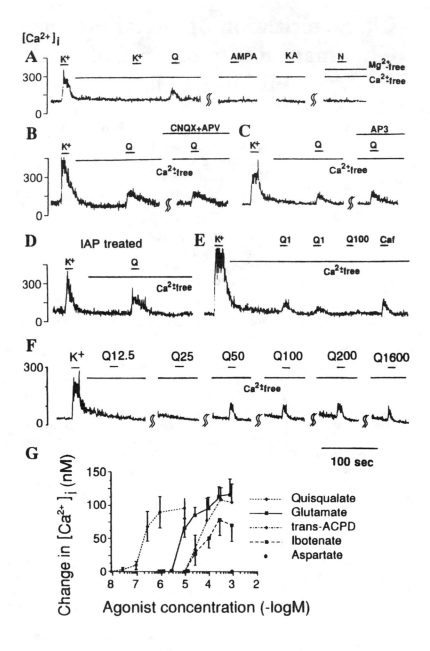

markedly during development. The percentage of cells showing mGluR response decreased after Day 4 of culture in Purkinje cells, but not in non-Purkinje cells (FIG. 2,A). The amplitude of mGluR response in responding Purkinje cells decreased during development in culture, but that of ionotropic response (iGluR) did not (FIG. 2,B). The change was thus specific to mGluR in Purkinje cells. This apparent desensitization of mGluR was not blocked by inhibition of protein kinase C (PKC) or ADP-ribosyltransferase.[5] The mGluR responses were mainly localized to the center of the somata of Purkinje cells (FIG. 2,C), whereas both receptor proteins were expressed immunocytochemically throughout the cell. These results suggest that the function of mGluR is spatially and developmentally controlled by a posttranslational mechanism involving a mechanism other than phosphorylation by PKC or ADP-ribosylation.

REFERENCES

1. YUZAKI, M., Y. KUDO, K. AKITA, A. MIYAWAKI & K. MIKOSHIBA. 1990. MK-801 blocked the functional NMDA receptors in identified cerebellar neurons. Neurosci. Lett. **119**: 19–22.
2. SUGIYAMA, H., I. ITO & M. WATANABE. 1989. Glutamate receptor subtypes may be classified into two major categories: a study on *Xenopus* oocytes injected with rat brain mRNA. Neuron **3**: 129–132.
3. ARAMORI, I. & S. NAKANISHI. 1992. Signal transduction and pharmacological characteristics of a metabotropic glutamate receptor, mGluR1, in transfected CHO cells. Neuron **8**: 757–765.
4. MIYAZAKI, S., M. YUZAKI, K. NAKADA, H. SHIRAKAWA, S. NAKANISHI, S. NAKADE & K. MIKOSHIBA. 1992. Block of Ca^{2+} wave and Ca^{2+} oscillation by antibody to the inositol 1,4,5-trisphosphate receptor in fertilized hamster eggs. Science **257**: 251–255.
5. YUZAKI, M. & K. MIKOSHIBA. 1992. Pharmacological and immunocytochemical characterization of the metabotropic glutamate receptors in cultured Purkinje cells. J. Neurosci. **12**: 4253–4263.

FIGURE 1. Pharmacology of mGluR response in single Purkinje cells. Temporal changes in the fluorescence image from fura-2 loaded cerebellar neurons were recorded on video tape. After immunocytochemical identification of Purkinje cells, $[Ca^{2+}]_i$ at the soma of each cell was calculated from video tapes with a computer. (A) Quisqualate (1 µM), but not AMPA (100 µM), kainate (20 µM), or NMDA (100 µM), induced increase in $[Ca^{2+}]_i$ in Ca^{2+}-free solution. The insensitivity to high K^+ (50 mM)-induced depolarization confirmed that the Ca^{2+} concentration in the solution was low after washing the cells with Ca^{2+}-free medium for 90 sec. (B) A combination of CNQX (50 µM) and APV (200 µM) did not block quisqualate (1 µM)-induced Ca^{2+} mobilization. (C) D,L-AP3 (1 mM) was ineffective in blocking Ca^{2+} mobilization by quisqualate (1 µM). The trace is representative for 74 of 80 cells tested on different culture days. Six cells showed partial irreversible blockage by 2 mM L-AP3. (D) Pre-treatment of Purkinje cells with 10 µg/ml IAP for 20–22 h did not affect the sensitivity of the mGluR responses. The trace is a response to 1 µM quisqualate and is representative of the responses to various doses of other mGluR agonists including trans-ACPD, ibotenate, glutamate. (E) After the cell became refractory to 100 µM quisqualate by repeated application of 1 µM quisqualate, caffeine (Caf) (10 mM) induced Ca^{2+} mobilization. (F) Typical all-or-none responses to increasing doses of quisqualate of 12.5, 25, 50, 100, 200 nM, and 1.6 µM. (G) Summary of averaged dose-response relationship. Neurons were exposed to increasing concentrations of the indicated agonists in Ca^{2+}-free medium. The average amplitudes of Ca^{2+} mobilization are plotted against the log concentrations of the agonists. Values for quisqualate, glutamate, trans-ACPD, and ibotenate are means ± SEM for 10, 14, 5, 6 Purkinje cells, respectively.

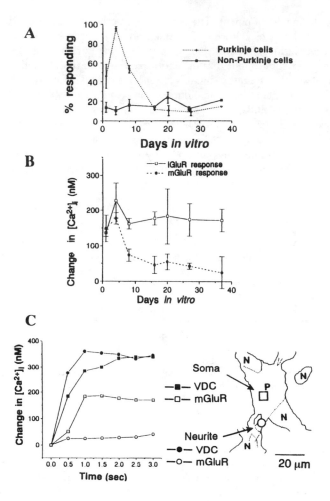

FIGURE 2. Developmental and spatial pattern of expression of the functional mGluR. (**A**) Percentages of Purkinje (dotted line) and non-Purkinje (solid line) cells exhibiting Ca^{2+} mobilization in response to 5 μM quisqualate in Ca^{2+}-free medium are plotted against the days in culture. (**B**) Amplitudes of Ca^{2+} mobilization induced by 5 μM quisqualate (mGluR response) in responding Purkinje cells. The amplitudes of $[Ca^{2+}]_i$ increases in response to 20 μM kainate in medium containing Ca^{2+} (iGluR response) in these cells are also shown for comparison. Data in **A** and **B** are means ± SEM from at least 3 independent cultures except for those on Days 16 and 37, which are for single cultures. Number of cells tested at each day ranges from 17 to 84 for Purkinje cells and from 44 to 237 for non-Purkinje cells. (**C**) Spatial change in $[Ca^{2+}]_i$ after exposure to 1 μM quisqualate in Ca^{2+}-free medium (mGluR response) (○, □) and to KCl (50 mM) in Ca^{2+}-containing medium (Voltage Dependent Channel-VDC response) (●, ■). The outlines of the cells were traced after immunological staining. Circles (○, ●) are from the proximal dendrite and squares (□, ■) from the center of the soma. This cell was examined on Day 4 of culture, and are representative of five Purkinje cells examined. We did not analyze the response in neurites from more developed Purkinje cells at later days in culture because of technical limitations with our method.

Diversity of Glutamate Receptor Subtypes Responsible For Increase in $[Ca^{2+}]_i$ in Hippocampal Cells

Y. KUDO, M. TAKITA, K. NAKAMURA, K. SUGAYA,
M. NAKAZAWA, AND A. OGURA

Mitsubishi Kasei Institute of Life Sciences
Machida
Tokyo 194, Japan

Since the discovery of Ca^{2+} permeability through N-methyl-D-aspartate–activated glutamate receptor (NMDA receptor), a scenario giving a principal role in the synaptic plasticity including long-term potentiation (LTP) to the NMDA receptor has gained popularity.[1] But recent reports including ours have shown that some of non-NMDA subtypes of ionotropic glutamate receptors have considerably high Ca^{2+}-permeability.[2-5]

Metabotropic glutamate receptors are also capable of increasing $[Ca^{2+}]_i$.[6] Moreover, recent molecular biological studies have revealed the polymorphism of both ionotropic and metabotropic glutamate receptors.[7-10] Basic information is, therefore, required about the topographical patterns of Ca^{2+}-mobilizable receptors distribution in the hippocampus.

We applied Ca^{2+} imaging techniques established by us to various preparations of hippocampal cells. Those include dissociated cells, fresh and organotypic cultured hippocampal slices. The microscopic fluorescence images of those fura-2–loaded preparations were detected and analyzed by a highly sensitive video camera and an image processor.

We found that approximately one third of the examined cells responded to both NMDA and non-NMDA receptor agonists with an increase in $[Ca^{2+}]_i$. *trans*-ACPD, a metabotropic receptor agonist induced two types of $[Ca^{2+}]_i$ elevation, a fast transient and a slow, longlasting one; the former was found in neurons while the latter was seen predominantly in astrocytes. We found peculiar patterns of distribution in those receptor subtypes in the fresh and organotypically cultured hippocampal slices. The NMDA- and AMPA-responsive neurons were mainly distributed in CA1 region, while the kainate-responsive neurons were found in the CA3 region. Kainate evoked the $[Ca^{2+}]_i$ increase in CA3 glial cells as well. *trans*-ACPD caused an increase in $[Ca^{2+}]_i$ in the glial cells, which migrated out of the organotypic culture.

Thus, not only NMDA receptors but also non-NMDA receptors should be taken into account for the increase in $[Ca^{2+}]_i$ during the tetanic stimulation leading to LTP. Contribution of glial receptors should not be overlooked either.

REFERENCES

1. COLLINGRIDGE, G. L. & T. V. P. BLISS. 1987. NMDA receptors—their role in long term potentiation. Trends Neurosci. 10: 288–293.
2. OGURA, A., K. AKITA & Y. KUDO. 1990. Non-NMDA receptor mediates cytoplasmic Ca^{2+} elevation in cultured hippocampal neurones. Neurosci. Res. 9: 103–113.
3. OGURA, A., M. NAKAZAWA & Y. KUDO. 1992. Further evidence for calcium permeability of non-NMDA receptor channels in hippocampal neurons. Neurosci. Res. 12: 606–616.
4. IINO, M., S. OZAWA & K. TSUZUKI. 1990. Permeation of calcium through excitatory amino acid receptor channels in cultured hippocampal neurones. J. Physiol. (Lond.) 424: 151–165.
5. KUDO, Y., E. ITO & A. OGURA. 1991. Heterogeneous distribution of glutamate receptor subtypes in hippocampus as revealed by calcium fluorometry. Adv. Exptl. Med. Biol. 287: 431–440.
6. SUGIYAMA, H., I. ITO & M. WATANABE. 1989. Glutamate receptor subtypes may be classified into two major categories: A study on Xenopus oocytes injected with rat brain mRNA. Neuron 3: 129–132.
7. HOLLMAN, M., A. O'SHEA-GREENFIELD, S. W. ROGERS & S. HEINEMANN. 1989. Cloning by functional expression of a member of the glutamate receptor family. Nature 342: 643–648.
8. KEINANEN, K., W. WISDEN, B. SOMMER, P. WERNER, A. HERB, T. A. VERDOORN, B. SAKMANN & P. H. SEEBURG. 1990. A family of AMPA-sensitive glutamate receptors. Science 249: 556–560.
9. SAKIMURA, K., H. BUJO, E. KUSHIYA, A. ARAKI, M. YAMAZAKI, H. MEGURO, A. WARASHINA, S. NUMA & M. MISHINA. 1990. Functional expression from cloned cDNAs of glutamate receptor species responsible to kainate and quisqualate. FEBS Lett. 272: 73–80.
10. MASU, M., Y. TANABE, K. TSUCHIDA, R. SHIGEMOTO & S. NAKANISHI. 1991. Sequence and expression of a metabotropic glutamate receptor. Nature 349: 760–765.

Metabotropic Glutamate Response in Acutely Dissociated Hippocampal CA3 Neurons of the Rat

NOBUTOSHI HARATA AND NORIO AKAIKE[a]

Department of Neurophysiology
Tohoku University School of Medicine
1-1 Seiryo-cho, Aoba-ky
Sendai 980, Japan

Glutamate and its analogues are the predominant excitatory neurotransmitters in the central nervous system (CNS). Metabotropic glutamate receptors (mGluR) are one category of the glutamate receptors, insensitive to ionotropic glutamate receptor antagonists, such as 6-cyano-7-nitroquinoxaline-2,3-dione (CNQX) and D-2-amino-5-phosphonovaleric acid (D-AP5). They have been shown to be critically involved in various functions of the CNS, including postsynaptic excitation[1,2] and presynaptic inhibition.[3] In addition, induction of long-term potentiation (LTP) in the hippocampus and induction of long-term depression (LTD) in the cerebellum are also attributed to mGluR.[3] In order to understand these multifaceted actions of mGluR, clarification of cellular responses to direct mGluR stimulation is inevitable.

Recently developed "perforated" patch-clamp[4] maintains the intracellular environment intact during electrophysiological recording and thus greatly facilitates the analysis of receptor-mediated responses with a patch-clamp technique. The aim of the present paper is to characterize mGluR in acutely dissociated CA3 hippocampal pyramidal neurons of the rat with the use of perforated patch-clamp technique.

Neurons were obtained from 7- to 10-day-old Wistar rats by mechanical and enzymatic dissociation. The composition of the external solution was (in mM): NaCl 150, KCl 5, $CaCl_2$ 2, $MgCl_2$ 1, glucose 10, and N-2-hydroxyethylpiperazine-N'-2-ethanesulfonic acid (HEPES) 10. The composition of the internal solution was (in mM): KCl 150 and HEPES 10. The pH of the external and internal solutions were adjusted to 7.4 and 7.2, respectively, with tris[hydroxymethyl]aminomethane (Tris-base). For perforated patch-clamp recording, nystatin was added to the internal solution at 150 μg/ml. The drugs were applied by a rapid application system, termed the

[a] Please address all correspondence to: Norio Akaike, Ph.D., Professor and Chairman, Department of Neurophysiology, Tohoku University School of Medicine, 1-1 Seiryo-cho, Aoba-ku, Sendai, 980, Japan.

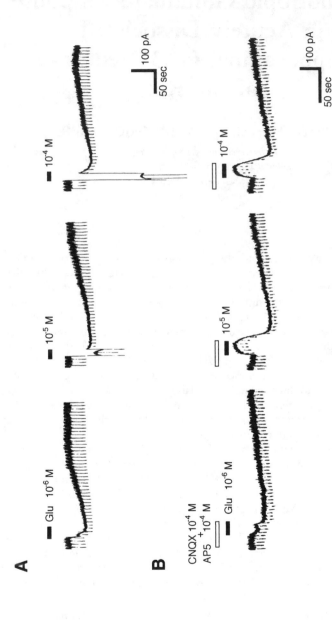

FIGURE 1. Two types of mGlu responses in CA3 pyramidal neurons. (**A**) Glutamate (Glu) application induced a rapid inward current accompanied by conductance increase, and a slow inward current accompanied by conductance decrease after washout. (**B**) Metabotropic glutamate response in another neuron was an outward current followed by a slow inward current. Ionotropic glutamate responses were blocked by 10^{-4} M CNQX and 10^{-4} M D-AP5. Holding potential was -44 mV.

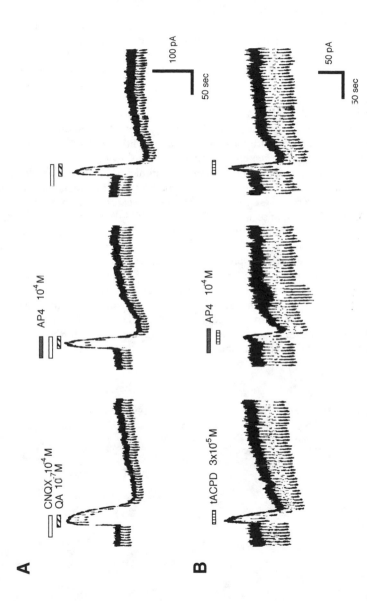

FIGURE 2. Different sensitivity of outward and inward mGlu responses to DL-AP4. (A) Outward and inward mGlu responses induced by QA were insensitive to a putative mGlu receptor antagonist, DL-AP4. The neuron was pretreated with 10^{-4} M DL-AP4 for 20 to 30 sec. (B) The inward mGlu response elicited by tACPD was also insensitive, while the outward response was reversibly blocked by DL-AP4. Holding potential was −44 mV.

"Y-tube" method. The exchange of external solution surrounding a neuron can be completed within 10 to 20 msec.

Application of glutamate (Glu) concentration-dependently induced a rapid inward current with increased membrane conductance, and a slow inward current with decreased conductance under a voltage-clamp at a holding potential of -44 mV (FIG. 1,A). The rapid inward current was observed at a concentration higher than 10^{-6} M and was blocked in the presence of 10^{-4} M CNQX and 10^{-4} M D-AP5, compatible with ionotropic Glu response. In contrast, the slow inward current was insensitive to CNQX and D-AP5, indicating the metabotropic glutamate (mGlu) response. This type of mGlu response was observed in more than 50% of the neurons tested. Interestingly, the threshold of slow inward mGlu response was lower than that of ionotropic Glu response. In other neurons, Glu induced another type of mGlu response, consisting of an outward current followed by a slow inward current (FIG. 1,B). The threshold of the outward mGlu response was higher than that of the slow inward mGlu response. Both the outward and inward mGlu responses were mimicked by quisqualic acid (QA) and ($\pm$)-1-aminocyclopentane-*trans*-1,3-dicarboxylic acid (tACPD) (data not shown).

In order to characterize these mGlu responses, a receptor antagonist study was performed. The mGlu response induced by 10^{-7} M QA was not affected in the presence of 10^{-4} M DL-2-amino-4-phosphonobutyric acid (DL-AP4) (FIG. 2,A). The inward current elicited by 3×10^{-5} M tACPD was also insensitive to DL-AP4 but the outward current was markedly suppressed by DL-AP4 (FIG. 2,B).

These results show that there are two types of mGlu responses in hippocampal CA3 pyramidal neurons: slow inward current and outward current. Moreover, the outward currents induced by QA and tACPD seem to be mediated by distinct receptors. The difference in thresholds of the currents, and differential sensitivity to antagonists add heterogeneity to the features of mGlu responses and would aid in further clarifying the properties of mGluR in the mammalian CNS.

REFERENCES

1. CHARPAK, S., B. H. GAHWILER, K. Q. Do & T. KNOPFEL. 1990. Nature 347: 765–767.
2. CHARPAK, S. & B. H. GAHWILER. 1991. Proc. R. Soc. Lond. B 243: 221–226.
3. ANWYL, R. 1991. Trends Pharmacol. Sci. 12: 324–326.
4. HORN, R. & A. MARTY. 1988. J. Gen. Physiol. 92: 145–159.

Mode of Interactions between Metabotropic Glutamate Receptors and G Proteins in *Xenopus* Oocyte

HIROYUKI SUGIYAMA, KOJI NAKAMURA,
AND TOSHIHIDE NUKADA[a]

Department of Biology
Faculty of Science
Kyusyu University 33
Fukuoka 812, Japan

[a]Department of Biochemistry
Institute of Brain Research
Faculty of Medicine
University of Tokyo
Tokyo 113, Japan

Metabotropic glutamate receptor-1 (mGluR1), when expressed in *Xenopus* oocytes, activates phospholipase C (PLC) through endogenous G proteins and causes chloride current responses that are mediated by inositol phospholipid metabolism and intracellular calcium mobilization.[1,2] In this study, we implanted various G protein subunits into *Xenopus* oocyte together with mGluR1, by injecting mRNAs obtained by *in vitro* transcription from respective cDNAs. We examined the effects of G protein subunits on the mGluR1-evoked chloride current responses. Four types of bovine G protein α subunits (Gsα, Go1α, Gi1α, and GL2α), and Gβ1 and Gγ2 subunits were examined (abbreviated as αs, αo1, αi1, αL2, β1, and γ2, respectively).

Expression of αL2,[3] which is isolated from a bovine liver cDNA library and is highly homologous to the mouse G14 α subunit,[4] potentiated the responses significantly, whereas αs and αo1 suppressed them. Expression of αi1 did have a significant effect (FIGS. 1 and 2).

Effects of β1 and γ2 subunits were then examined. Expression of β1γ2 subunits enhanced the potentiation effects of αL2, whereas the suppression by αs was completely reversed by β1γ2 subunits (FIG. 2). In the case of αo1, however, expression of β1γ2 not only reversed the suppression but also caused even the potentiation (FIG. 2).

These results were readily explained by the following model: α subunits, either endogenous or exogenous, compete with each other for βγ subunits, either exogenous or endogenous, to form functional heterotrimers. In the absence of exogenous β1γ2,

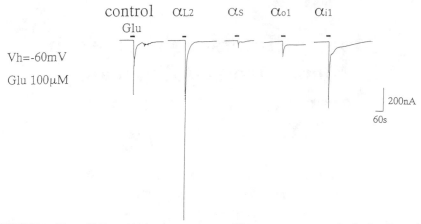

FIGURE 1. The mGluR1-evoked current responses of *Xenopus* oocytes expressing bovine G protein α subunits. The control means the oocytes injected with mGluR1 transcript alone (50 pg/oocyte). The others are the oocytes injected with transcripts of mGluR1 and bovine α subunits indicated (mGluR1 50 pg, α subunit 5 ng/oocyte). The current responses to glutamate (Glu, 100 μM) were measured 3–4 days after the injection under voltage-clamp (– 60 mV) conditions.

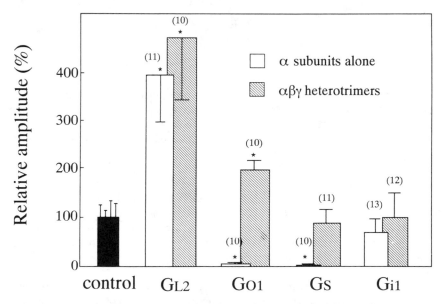

FIGURE 2. The effects of bovine G protein subunits on the mGluR1-evoked current responses of the oocytes. The peak amplitudes of the current responses, as exemplified in FIG. 1, were expressed relative to the control as mean ± SEM (*n*). The control (*solid column*) means the oocytes where only mGluR1 was implanted. In the experimental oocytes, the α subunits of the indicated G proteins were implanted without (*open columns*) or with (*shaded columns*) β1γ2 subunits, together with mGluR1. **p* < 0.01 by *t*-test.

exogenous α can potentiate (or suppress) mGluR1 responses if it can activate PLC more (or less) efficiently than α subunits of endogenous PLC-coupled G proteins, whereas in the presence of supplemented βγ subunits, exogenous α subunits can potentiate them if it can activate PLC, irrespective of its relative efficiency compared to endogenous PLC-coupled G proteins.

REFERENCES

1. SUGIYAMA, H., I. ITO & C. HIRONO. 1987. Nature 325: 531–533.
2. MASU, M., Y. TANABE, K. TSUCHIDA, R. SHIGEMOTO & S. NAKANISHI. 1991. Nature 349: 760–765.
3. NAKAMURA, F., K. OGATA, K. SHIOZAKI, K. KAMEYAMA, K. OHARA, T. HAGA & T. NUKADA. 1991. J. Biol. Chem. 266: 12676–12681.
4. STRATHMANN, M., T. M. WILKIE & M. I. SIMON. 1989. Proc. Natl. Acad. Sci. USA 86: 7407–7409.

The t-ACPD-Induced Current Response in Rat Cerebellum

HIROSHI TAKAGI, JEAN DE BARRY,[a]
YOSHIHISA KUDO,[b] AND TOHRU YOSHIOKA

Department of Molecular Neurobiology
School of Human Sciences
Waseda University
Tokorozawa 359, Japan

[a]*Centre de Neurochemie du CNRS*
5 rue B. Pascal
F67084 Strasbourg, Cedex, France

[b]*Mitsubishi Kasei Institute of Life Sciences*
Machida 194, Japan

G protein coupled metabotropic glutamate receptor (mGluR) is thought to play an important role in long-term depression (LTD) in cerebellum,[1,2] because LTD was demonstrated when the desensitization of the AMPA receptor is induced by co-activation of metabotropic receptors and non-NMDA receptors.[3,4] It was well known that quisqualate (QA) and t-ACPD were the agonists of mGluR. Recently Aramori and Nakanishi[5] have cloned a mGluR sensitive to QA but very weakly sensitive to t-ACPD in CHO cells, when RNA was transfected with cDNA isolated from rat cerebellum. Furthermore, recent studies have suggested that t-ACPD receptor is localized presynaptically in hippocampal CA1 neurons.[6]

In this report, we examined the postsynaptic and presynaptic distribution of mGluR in cerebellum at different developmental stages by patch-clamp recording and by $[Ca^{2+}]_i$ imaging, using t-ACPD and QA as agonists.

When QA was applied to cerebellar slice of postnatal day (PND) 21, the current response was found in Purkinje cell. In the presence of 100 μM CNQX, response continued. This response was not eliminated by removal of external Ca^{2+}. When L-AP3 was applied extracellularly, however, the QA-induced response was greatly reduced.

The response of the same cells to t-ACPD (up to 1 mM) was observed. But it was not affected by CNQX. This t-ACPD induced current almost disappeared, when external Ca^{2+} was removed. This is surprising since one would expect that responses to both QA and t-ACPD in the presence of CNQX are due to the activation of the same mGluR. However, QA- and t-ACPD-induced responses were both sensitive to 200 μM L-AP3. These results suggest that QA and t-ACPD may stimulate different subtypes of the receptor, though both sites are L-AP3 sensitive, and that the mGluR in Purkinje cells is far less sensitive to t-ACPD than to QA.

In order to verify the presence of a QA-activated metabotropic receptor in Purkinje cells by observing agonist-induced increases in $[Ca^{2+}]_i$, we injected 200 µM rhod-2 triammonium salt, which has a lower affinity than fura-2 with Ca^{2+} ($K_d \cong 1$ µM), into the Purkinje cell soma. We measured the effect of QA and t-ACPD on the $[Ca^{2+}]_i$ of dye-injected cells in slices isolated from PND 14 rat. QA (10 µM) application in nominally Ca^{2+} free conditions resulted in an obvious $[Ca^{2+}]_i$ elevation at the dendrite region. On the other hand, t-ACPD (up to 1 mM) showed little effect in the same region. These results appear to agree with the electrophysiological results.

When t-ACPD was applied to a slice isolated from a PND 7 animal, at a stage when synaptic input to Purkinje cells is poorly developed, QA induced a single phase of large inward current associated with a slight increase in spontaneous miniature currents (FIG. 1,A). On the other hand, t-ACPD showed no significant effect on the same preparation (FIG. 1,B). In slices obtained from PND 9 rat, QA induced current responses showed three phases (FIG. 1,C), although the amplitude of the outward current was still small. At this stage, QA resulted in an increase in frequency as well as in the amplitude of spontaneous miniature current. On PND 9 and later, t-ACPD strongly enhanced the frequency and amplitude of miniature currents (FIG. 1,D). These developmental changes in the effects of t-ACPD on the amplitude and frequency of spontaneous miniature current were observed with a holding potential of − 90 mV.

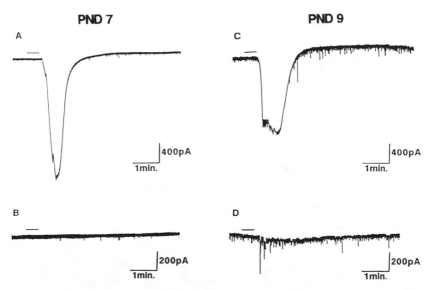

FIGURE 1. Effect of QA and t-ACPD on Purkinje cells at PND 7 and PND 9. In young animals QA induced a large and slowly activating current in Purkinje cells, but caused neither fast inactivating inward current nor outward current (**A** and **C**). On the other hand, t-ACPD had no effect on PND 7 cells (**B**) but produced small bursting inward currents at PND 9 (**D**)

TABLE 1. Effect of t-ACPD on mEPSCs and mIPSCs

	mEPSCs		mIPSCs		
	Amplitude (pA)	Frequency (Hz)	Amplitude (pA)	Frequency (Hz)	n
Control	26.7 ± 4.8	0.54 ± 0.26	34.7 ± 5.3	0.53 ± 0.25	5
t-ACPD	21.2 ± 3.6	0.75 ± 0.18***	12.6 ± 2.9***	0.20 ± 0.06***	5

Data are expressed as mean value ± standard deviation. Numbers of experiment are 5 ($n = 5$).
Difference between control and t-ACPD was tested with the paired t-test (***$p<0.001$).

To examine the presynaptic action of t-ACPD in more detail, we separated mEPSCs (detected as inward current) and mIPSCs (detected as outward current) in PND 14 preparations. As shown in TABLE 1, after administration of t-ACPD, the frequencies of mEPSCs and mIPSCs were significantly augmented and diminished, respectively (paired t-test, $p < 0.001$, $n = 5$). This suggests that t-ACPD acts differently on the presynaptic regions of excitatory and inhibitory synapses. Furthermore, the mean value of the mEPSCs amplitude was not affected by the application of t-ACPD, while that of the mIPSCs was significantly reduced, suggesting that t-ACPD also affects the postsynaptic inhibitory synaptic site. The detailed mechanisms of the separate effects of t-ACPD on the inhibitory and excitatory synapses will appear elsewhere soon.

REFERENCES

1. ITO, M. & L. KARACHOT. 1990. Neuro Rep. 1: 129–132.
2. MARR, D. 1969. J. Physiol. 202: 437–470.
3. KANO, M. & M. KATO. 1987. Nature 325: 276–279.
4. LINDEN, D. J., M. H. DICKINSON, M. SMEYNE & J. A. CONNOR. 1991. Neuron 7: 81–89.
5. ARAMORI, I. & S. NAKANISHI. 1992. Neuron 8: 757–765.
6. BASKYS, A. & R. C. MALENKA. 1991. Eur. J. Pharmacol. 193: 131–132.

Cerebellar Long-Term Depression Enabled By Nitric Oxide, A Diffusible Intercellular Messenger

KATSUEI SHIBUKI[a]

Laboratory for Neural Networks
RIKEN
Wako, Saitama 351-01, Japan

Purkinje cells, the output neurons from the cerebellar cortex, receive parallel and climbing fiber inputs (FIG. 1). Coactivation of these two evokes long-term depression (LTD) in the parallel fiber/Purkinje cell synapses.[1] This LTD is the possible cellular mechanism for cerebellar motor learning.[1] Since the two inputs terminate at different loci on Purkinje cell dendrites, heterosynaptic interaction connecting between synapses is required for inducing LTD (FIG. 1,A).

One possibility is that climbing fiber input evokes marked depolarization in Purkinje cell dendrites (FIG. 1,C), which induces Ca^{2+} spikes at the postsynaptic sites of the parallel fiber synapses to evoke LTD.[2] Another possibility is the diffusible intercellular messenger, nitric oxide (NO) is produced by climbing fiber input so that the climbing fiber information is conveyed to the parallel fiber synapses by diffusion of NO.

In fact, NO release is observed following climbing fiber input,[3] suppression of NO signaling blocks LTD,[3] and exogenous NO/cGMP can replace climbing fiber input in inducing LTD.[3,4] However, NO is not required for LTD in cultured Purkinje cells[5] or in slices treated with fluorocitrate, a gliotoxic metabolic inhibitor (unpublished data). In Purkinje cells injected with Ca^{2+} chelators, parallel fiber stimulation plus cGMP application, which normally evokes LTD, induces long-term potentiation (LTP).[4] Therefore NO/cGMP signaling cannot replace Ca^{2+} signaling. These results can be explained by assuming two important messengers in LTD: NO enables both of LTD and LTP by acting on Bergmann glia cells, which respond to NO with a marked cGMP increase,[6] while Ca^{2+} signaling in Purkinje cell dendrites determines the direction of plastic changes (FIG. 2).

[a] Present address: Department of Neurophysiology, Brain Research Institute, Niigata University, Asahi-machi, Niigata-shi 951, Japan.

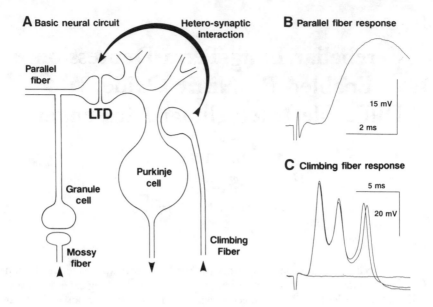

FIGURE 1. (A) Basic neural circuit for inducing LTD in cerebellar cortex. (B and C) Parallel and climbing fiber responses of Purkinje cell dendrites.

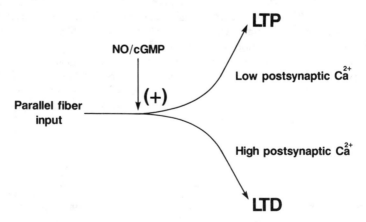

FIGURE 2. Two important messengers controlling cerebellar synaptic plasticity.

REFERENCES

1. Ito, M. 1989. Ann. Rev. Neurosci. **12:** 85–102.
2. Sakurai, M. 1990. Proc. Natl. Acad. Sci. USA **87:** 3383–3385.
3. Shibuki, K. & D. Okada. 1991. Nature **349:** 326–328.
4. Shibuki, K. & D. Okada. 1992. NeuroReport **3:** 231–234.
5. Linden, D. J. & J. A. Connor. 1992. Eur. J. Neurosci. **4:** 10–15.
6. De Vente, J., J. G. J. M. Bol & H. S. Berkelmans. 1990. Eur. J. Neurosci. **2:** 845–862.

Molecular Evolution of Ion Channels in Central Nervous System: Possible Primitive mRNA Segments Conserved in Glutamate and Acetylcholine Receptor Channel Genes

KOJI OHNISHI

Department of Biology
Faculty of Science
Ikarashi-2
Niigata 950-21, Japan

The poly-tRNA hypothesis[1] has recently proposed that *Bacillus subtilis* (BSU) *trrnD* operon is a relic of an early protein-synthesizing RNA molecule (*trrnD* trRNA) consisting of three rRNAs and a tRNA cluster (including 16 tRNAs). A primitive mRNA (*trrnD* mRNA) was hypothesized to be complementary to a 48-base RNA consisting of 16 triplet anticodons in the order of the tRNA anti-codons in *trrnD* operon. The poly-tRNA structure and rRNAs in *trrnD* trRNA would have functioned in making a 16-amino acid peptide (*trrnD* peptide) in association with *trrnD* mRNA.

Amino acid and base sequence segments potentially homologous to the *trrnD* peptide and *trrnD* mRNA were searched for in PIR and GenBank databases, using FORTRAN programs that output n amino acid (base)-segments sharing m or more amino acids (bases) with an inputted sequence. Similarity levels were evaluated by computing $P_{nuc}(m,n) = \Sigma_{i=m}^{n} [n!/i!(n-i)!](1/4)^i(3/4)^{n-i}$, which gives probability by

---

FIGURE 1. Alignment of *trrnD* mRNA with tRNAs, 5S rRNAs, and U1 snRNA, and with gene segments encoding glycyl- and glutaminyl-tRNA synthetases (GlyRS, GlnRS), *E. coli* ribonuclease P protein (*rnpA*), ion-receptor channel proteins, and albumin/α-fetoprotein. Base complementaries between triplet anticodons in *trrnD* operon transcript and the hypothetical *trrnD* mRNA are indicated by colons(:). Bases in spacer regions (of *rnnD*) and in intron (3'-side of exon 13 in albumin gene) are given in lowercase letters. Di- and longer oligo-nucleotides and deduced amino acids shared by GlyRS and one or more of other sequences are boxed. Other amino acid matches are italicized. Sequence data are obtained from GenBank Database. Alignment is yet tentative in tRNAs (bases 36–51) and corresponding region of 5S rRNAs (3'-term). Abbreviations: EC, *E. coli*; BSU, *Bacillus subtilis*; BST, *B. stearothermophilus*; RNase P, ribonuclease P; NMDA R1, rat NMDA receptor 1: GluR, glutamate receptor; AchR acetylcholine receptor.

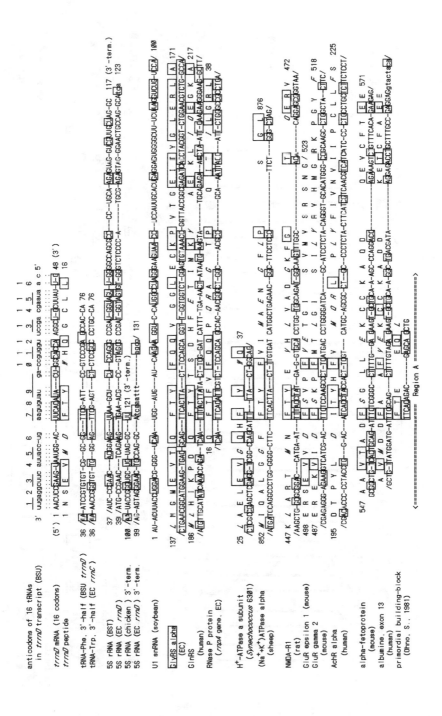

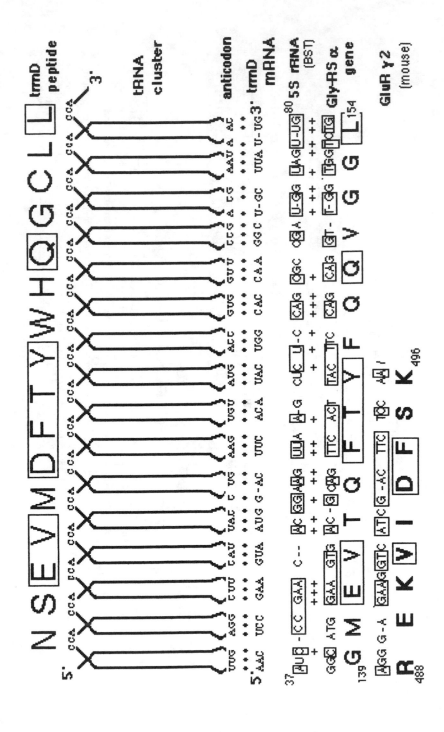

chance for the occurrence of *m* or more base-matches in *n*-base alignment. The *E. coli* (EC) glycyl-tRNA synthetase (GlyRS) α subunit (amino acids 141–154) was found to be highly homologous to *trrnD* peptide.[1] This and other segments (proteins, genes, and RNAs) thus found were aligned with each other. Extensive visual comparisons were also made in refining alignments.

In the resulting alignment (FIG. 1), the *trrnD* mRNA gives base-match levels [and P_{nuc}] of 63% [$P_{nuc}(46,29) = .56 \times 10^{-7}$] to GlyRs α, 60% [$.52 \times 10^{-5}$] to *B. stearothermophilus* (BST) 5S rRNA (bases 37–), 59 and 48% [$.50 \times 10^{-3}, .92 \times 10^{-3}$] to H+-ATPase a subunit and glutamate receptor (GluR) γ 2 subunit, 48% to EC ribonuclease P protein (*rnpa*), and 48% [$.86 \times 10^{-3}$] to albumine exon 8, but shows lower base-match levels, 41% [.020] and 36%, to the 3' halves of EC tRNA^{Trp} and BSU tRNA^{Phe}. On the other hand, the BST 5S rRNA (bases 37–117) shares 55, 60, 42, and 41% bases [$.13 \times 10^{-7}, .31 \times 10^{-7}, .0026, .0025$] with GlyRs α, *rnpa*, NMDAR1 channel receptor, and albumin, respectively. GlyRs α shares 49% bases [.0012] with tRNA^{Trp}, 53% [$P_{nuc}(52,99) = .42 \times 10^{-8}$] with U1 snRNA, 65% [$P_{nuc}(43,66) = .73 \times 10^{-11}$] with *rnpA*, 44, 44 and 39% with (Na+, K+)-ATPase α, NMDR1, and GluR γ, and 47% [$.52 \times 10^{-5}$] with acetylcholine receptor (AchR) α, 58% [$.91 \times 10^{-4}$] with 3'-terminal region (99–131) of 5S rRNA in EC *rrnD*. (Na+,K+)-ATPase shows 45 and 55% matches [$.25 \times 10^{-3}, .42 \times 10^{-8}$] with GluR γ 2 and AchR α.

These results strongly suggest that the peptide-encoding gene segments (and therefrom transcribed mRNA segments) corresponding to the region A in FIGURE 1 are true homologues not only of *trrnD* MRNA but also of the segments from 5S rRNAs, U1 snRNA, and tRNAs. The *trrnD*-like mRNA segments must have more likely derived from direct homologue of primitive 5S rRNA, than from primitive (or ancestral) tRNA. U1 snRNA is a close homologue of these mRNA regions and 5S rRNA, and therefore seems to have emerged after the origin of 5S rRNA. Since 5S rRNA seems to have emerged by duplication of ancestral tRNA,[1,2] and ancestral tRNA (proto-tRNA) having (5') "CCA" (3') in anticodon region must have emerged by duplication of semi-tRNA,[1-3] first mRNA would have derived from an ancestral 5S rRNA, or from an RNA closely related to 5S rRNA.

Early proteins would have been made by cooperative interaction of ancestral 5S rRNA-like RNA (ancestral *trrnD* mRNA) with a *trrnD*-like poly-tRNA structure in *trrnD* trRNA. The interaction occurred in a way more or less like that shown in FIGURE 2, where the triplet anticodon region (probably "CCA" in the beginning) in the *k*th tRNA (*k* = 1,2, . . .,16) consistently interacted with the *k*th triplet region

---

FIGURE 2. Poly-tRNA model of early protein synthesis by interaction of the tRNA cluster of *trrnD*-trRNA and a 5S rRNA-like RNA. Bases and amino acids identical to those in *trrnD*-mRNA and *trrnD*-peptide are boxed. Asterisks (*) indicate Watson-Crick type base-pairings between anticodons on tRNAs and triplet codons on *trrnD*-mRNA. Base matches between 5S rRNA and glycyl-tRNA synthetase (GlyRS) α gene are indicated by "+". This alignment of *trrnD* mRNA, 5S rRNA, and two protein gene segments is exactly the same as that shown in FIG. 1. GluR, glutamate receptor; BST, *B. stearothermophilus*.

of ancestral 5S rRNA-like (*trrnD* mRNA-like) RNA, towards generating more base-pairing between the former triplet and the latter triplet (for every *k*) by natural selection.

Accordingly, *trrnD* mRNA-like segments in ion channel receptor genes including not only AchR and GluR genes but also H$^+$- and (Na$^+$,K$^+$)-ATPase subunit genes are relics of ancient proteins before the establishment of triplet anticodons, when *trrnD* peptide-like peptide was made by interaction of poly-tRNA structure in *trrnD*-trRNA and primitive 5S rRNA-like RNA via ancestral anticodons and codons. Ohno's "18-base primordial building-block" found in α-fetoprotein/albumin genes[4] is now demonstrated to be a short region within this *trrnD*-like most ancient gene, as shown in Figure 1. Codon table must have been so made that *trrnD*-mRNA could code for *trrnD*-peptide.

ACKNOWLEDGMENT

The author acknowledges Dr. H. Yanagawa for useful discussions.

REFERENCES

1. Ohnishi, K. 1993. *In* Endocytobiology V, H. Ishikawa *et al.*, Eds. Tubingen University Press. Tubingen. in press.
2. Ohnishi, K. 1991. *In* Symmetries in Science IV. B. Gruber *et al.*, Eds.: 147–176. Plenum Press. New York.
3. Ohnishi, K. 1992. Endocytobiosis Cell Res. **8**: 109–120.
4. Ohno, S. 1981. Proc. Natl. Acad. Sci. USA **78**:7657–7661.

Induction of TRE and CRE Binding Activities in Cultured Granule Cells Stimulated via Glutamate Receptors

MASAAKI TSUDA

Department of Microbiology
Faculty of Pharmaceutical Sciences
Tsushima-naka
Okayama 700, Japan

In order to understand the genetic responses of neuronal cells to synaptic activation, it is important to know the relationship betweeen synaptic transmission and gene expression in neuronal cells. In concert with several types of seizures or electrical stimulation of neuronal path fibers in brain, immediate early genes (IEGs), which encode transcriptional factors, are rapidly induced in brain neuronal cells. We are thus now focusing on the c-*fos* induction, which can be evoked via stimulation of cultured mouse cerebellar granule cells with glutamate receptor agonists.

We prepared primary cultures of cerebellar granule cells from 1-week-old mouse, when cerebellar granule cells receive glutamatergic inputs from mossy fibers during postnatal development of mouse cerebellum. Using nuclear mini-extracts prepared from cultured cerebellar granule cells, we carried out a gel-shift assay to examine the changes in DNA-binding activities of transcriptional factors and found that exogenous NMDA (*N*-methyl-D-aspartate) or kainate (100 μM) increased both TRE (TPA-responsive element)- and CRE (cyclic-AMP–responsive element)-binding activity specifically though NMDA or non-NMDA receptors.[1] The increase of TRE-binding activities appeared to be caused by the c-*fos* induction followed by *de novo* synthesis of c-Fos proteins. Dose dependencies of the increase in TRE-binding activity and the Ca^{2+} uptake into the cells showed a good coincidence when the cells were stimulated with NMDA or kainate, indicating that influxes of extracellular Ca^{2+} into the cells could trigger the induction of TRE-binding activities. Since treatment of the cells with calphostin C, an inhibitor for protein kinase C (PKC), inhibited the increases of TRE-binding activities induced by NMDA or kainate, an activation of PKC could be involved in the c-*fos* induction evoked via glutamate receptors.

Competition and proteolytic bandshift experiments revealed that the increases in TRE- and CRE-binding activities were both mediated by the same DNA-binding complexes whose binding affinity was higher to CRE than to TRE. The super-gel shift assay including anti–c-Fos antiserum revealed that the DNA-binding complexes formed not only on TRE but also on CRE involved c-Fos or Fos-related proteins. In addition, the proteins that can form heterodimers with c-Fos appeared to be different between cerebellar granule cells and hippocampal neuronal cells.

We have developed a method for preparing the nuclear mini-extracts from small punches of 1–2 mm-thick brain slices (approximately 1–3 mg wet weight), which were taken by a microcapillary (1 mm diameter). Using this method, we found that intraperitoneal administration of NMDA or kainate caused the increases in CRE-binding activities as well as TRE-binding activities in the CA1, CA3, and dentate gyrus of hippocampus. The increase of TRE-binding activities induced by NMDA was inhibited by the intraperitoneal administration of NMDA-receptor antagonist MK801. This TRE-binding activity showed a strong binding to the TRE of nerve growth factor (NGF) gene but only a weak binding to the TRE of proenkephalin gene. We are now characterizing these kinds of DNA-binding activities expressed in mouse brain and investigating the relationship with the expression of NGF gene or other genes whose expression could be affected by an activation of glutamate receptors.

REFERENCE

1. SAKURAI, H., R. KURUSU, K. SANO, T. TSUCHIYA & M. TSUDA. 1992. Stimulation of cultured cerebellar granule cells via glutamate receptors induces TRE- and CRE-binding activities mediated by common DNA-binding complexes. J. Neurochem. 59(6): 2067–2075.

Molecular Cloning of Novel Small GTP-Binding Proteins in Rat Liver

H. NAGAHARA, M. UEDA,[a] AND H. OBATA

Department of Medicine
Institute of Gastroenterology

[a]Institute of Neurology
Tokyo Women's Medical College
Tokyo 162, Japan

A family of small GTP-binding proteins with molecular weight 20–30 kD has been identified, and it is revealed that their biological functions are coupling to vesicular transport, superoxide production, formation of cellular matrix, and transformation. The purpose of our study is to identify how many small G proteins are expressed in the liver tissue and what role these proteins play in liver regeneration, hepatocyte growth, and hepatocarcinogenesis.

MATERIALS AND METHODS

A cDNA library was constructed from rat liver poly(A) RNA by random primer annealing and reverse transcription and used as a template for polymerase chain reaction (PCR). PCR primers were synthesized corresponding to GVGKSCLL and DTAGQEE, which amino acid residue were conserved in small G proteins (ras and rab family). After 35 cycle PCR, DNA products were purified, ligated to plasmid vectors, and cloned. Sequencing of cloned DNA by dideoxy chain termination reaction confirmed that its DNA belongs to a small G protein family. Cloned DNA fragments were used as probes for screening of cDNA library to obtain full length cDNA clone.

RESULTS AND DISCUSSION

Amplified 150-bp fragments were cloned using pTZ19R vector. By the sequencing analysis of 48 clones, it was revealed that 13 clones had 150-bp length DNA fragments and contained both GVGKSCLL and DTAGQEE amino acid sequence. Comparing these 50 amino acid sequences to those of previously reported small G proteins, it was suggested that three clones were identical to rab2, two were homologous to human G25K, one was homologous to rab11, one was homologous to rac1 or 2, one was homologous to rhoA, and the remaining five clones were a novel rab family. From these results, it was shown that this cloning strategy may be useful for identifying novel small G proteins.

Using the clones homologous to human G25K, rat liver cDNA library were screened, and four cDNA clones were obtained. The nucleotide sequences of one cDNA clone revealed that it was encoding 191 amino acids and identical to the human G25K, which have been identified yeast cell cycle gene CDC42. These results suggest that G25K protein is expressed in rat liver tissue and involved in hepatocyte regeneration.

We reported here a set of small G proteins other than G25K also expressed in rat liver tissue, and within a few years it will be revealed how these small G proteins play a role in signal transduction, intracellular vesicular transport, hepatocyte growth, and carcinogenesis.

Immunohistochemical Analysis of Signal Transduction System in Developing Rat Purkinje Cell by Using Antibodies for Signaling Molecules

ATSUO MIYAZAWA, MEGUMI TAKAHASHI,[a]
TETSURO HORIKOSHI,[b] AND TOHRU YOSHIOKA

Department of Molecular Neurobiology
Waseda University
Tokorozawa 359

[a]*School of Medicine*
Yokohama City University
Yokohama

[b]*Department of Physiology*
Tsurumi University
Yokohama, Japan

In order to know the localization and interaction between molecules involved in signal transduction in neurons, we studied cellular distribution of phosphoinositide (PI) turnover–related molecules, such as phosphatidylinositol 4,5-bisphosphate (PIP_2), inositol-1,4,5-trisphosphate receptor (IP_3R), phosphatidyl serine (PS), and γ-type protein kinase C (γPKC) in developing rat cerebellum by immunohistochemical method. In the Purkinje cell, PIP_2 immunostaining was strong around postnatal day (PND) 7 then weakened until around PND 15.[1] The change in the reduction of PIP_2 was confirmed by ^{32}P-labeling experiment. The content of PIP_2, phosphatidylinositol 4-monophosphate (PIP), and phosphatidic acid (PA) in the cerebellum were found to be reduced remarkably around PND 15 (FIG. 1) and gradually recovered. These data suggest that PI turnover system in Purkinje cells might be a functional disorder around PND 15. Although γPKC-immunostaining density was found to be constantly independent of Purkinje cell development, PS-immunostaining was reduced between PND 14 and 21. Since PS-immunostaining patterns inversely reflect the PKC activity in the cell, we can estimate that there is a poor relationship between PI turnover and

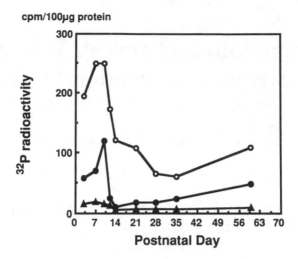

cpm/100µg protein

FIGURE 1. PIP_2, PIP, and PA content in developing rat cerebellum. O: PA, ●: PIP, ▲: PIP_2. These contents were analyzed quantitatively by [^{32}P]ATP incorporation. Around PND 15, PIP and PA content was found to be reduced remarkably. PIP_2 content was also changed in the same fashion with PIP and PA.

activation of PKC during the development of rat cerebellum.[2] To find out the cause of transient reduction of PI turnover on PND 15 in the Purkinje cell, expression of c-myc was studied, because we have already found that c-myc expression could be closely coupled with the deficit of PI turnover in some cancer cell lines.[3] In this experiment, we found that the developing rat cerebellum showed the transient expression of c-myc mRNA around PND 7–10 by northern blotting (FIG. 2,a). By western blotting, c-MYC protein showed peak level around PND 10–14 in cytosol, while the protein in nucleus was maintained in high level around PND 14–21 (FIG. 2,b). These data suggest that expression of c-myc mRNA and c-MYC protein in cytosol reached maximum levels prior to reduction of PI turnover but in parallel with the activation of PKC. Therefore it is strongly suggested that there is a close correlation between PI turnover and c-myc expression during postnatal development in rat cerebellar Purkinje cells.

REFERENCES

1. ITO, E., A. MIYAZAWA, H. TAKAGI, T. YOSHIOKA, T. HORIKOSHI, K. YANAGISAWA, T. NAKAMURA, Y. KUDO, M. UMEDA, K. INOUE & K. MIKOSHIBA. 1991. Neurosci. Res. **11:** 179–188.
2. MIYAZAWA A., H. INOUE, T. YOSHIOKA, T. HORIKOSHI, K. YANAGISAWA, M. UMEDA & K. INOUE. 1992. J. Neurochem. **59:** 1547–1554.
3. KUBOTA, Y., T. SHUIN, M. YAO, H. INOUE & T. YOSHIOKA. 1987. FEBS Lett. **212:** 159–162.

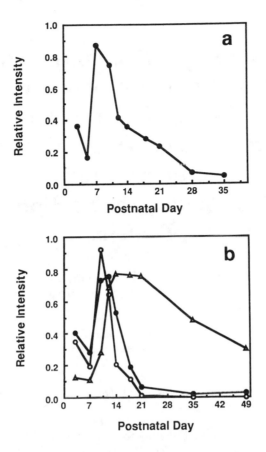

FIGURE 2. (a) c-myc mRNA expression in developing rat cerebellum. Transient c-myc mRNA expression was shown around PND 7 by northern blotting. (b) c-MYC protein expression in developing rat cerebellum. O: total cell extract, ●: cytosol fraction, △: nuclear fraction. c-MYC protein was detected by anti–c-MYC antibody. The total cell and cytosol c-MYC protein showed peak level around PND 10–14, while in the nucleus the protein showed constantly high levels around PND 14–21.

Anti-Idiotypic Antibody Identifies a Consensus Recognition Site for Phosphatidylserine Common to Protein Kinase C and Other Cellular Phosphatidylserine-Binding Proteins

KOJI IGARASHI, MASATO UMEDA, REZA FAROOQ,
SHIGERU TOKITA, YOSHINORI ASAOKA,[a]
YASUTOMI NISHIZUKA,[a] AND KEIZO INOUE

The Department of Health Chemistry
Faculty of Pharmaceutical Sciences
The University of Tokyo
7-3-1 Hongo, Bunkyo-ku
Tokyo 113, Japan

[a]*Department of Biochemistry*
Kobe University School of Medicine
Kobe 650, Japan

[b]*Biosignal Research Center*
Kobe University
Kobe 657, Japan

Phosphatidylserine (PS) in membranes contributes to many regulatory processes of biological responses. The well known function of PS is its ability to promote blood coagulation[1] and to regulate enzymatic activity of protein kinase C.[2] Protein kinase C is a family of Ca^{2+}-phospholipid–dependent kinases that bind to the plasma membrane in response to receptor-mediated generation of diacylglycerol and Ca^{2+}. Membrane-association of protein kinase C is mediated by the interaction with multiple acidic phospholipids in the presence of Ca^{2+}, while the enzymatic activity displays the strict structural requirement for 1,2-diacyl-*sn*-glycero-3-phospho-L-serine (PS).[3] Although these observations suggest the existence of specific binding site for PS, the identification of the site has been very difficult since the enzyme interacts with multiple phospholipid molecules during activation.

In order to elucidate the molecular mechanisms underlying the specific PS-protein interaction, we have undertaken structural and idiotypic analyses of an anti-PS mAb, PS4A7.[4] The mAb PS4A7 binds to 1,2-diacyl-sn-glycero-3-phospho-L-serine (PS) but not to 1,2-diacyl-*sn*-glycero-3-phospho-D-serine or 1,2-diacyl-*sn*-glycero-phospho-L-homoserine, showing a similar phospholipid specificity with that required for the activation of protein kinase C.[3] We have established a series of anti-idiotypic mono-clonal antibodies against the combining site of PS4A7 according to a method previously described.[5] Three subspecies of protein kinase C were purified to homogeneity and the binding of the anti-idiotypic antibodies to the plate-coated protein kinase C was examined by ELISA. Among 34 anti-idiotypic mAb established, one anti-idiotypic antibody, named Id8F7, showed an extensive cross-reaction with protein kinase C. Id8F7 bound to the three subspecies of protein kinase C almost equally and effectively inhibited the enzymatic activities. The binding of Id8F7 to protein kinase C was specifically inhibited by PS, but not by other phospholipids such as phosphatidyletha-nolamine, phosphatidylinositol, and phosphatidylcholine (FIG. 1). The inhibitory activ-ity of PS was strictly dependent on the structure of PS, since the synthetic PS analogs such as 1,2-diacyl-*sn*-glycero-3-phospho-D-serine and 1,2-diacyl-*sn*-glycero-phospho-L-homoserine showed no significant inhibitory effect on the binding. In contrast, the binding was significantly enhanced by the presence of 1,2-dioleoyl-glycerol and also by sphingosine. These findings clearly indicate that the anti-idiotypic antibody recognizes a

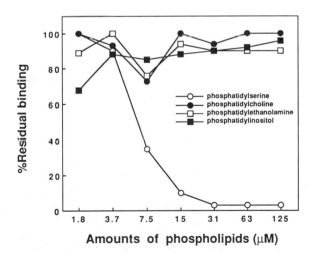

FIGURE 1. Inhibition of the binding of anti-idiotypic mAb Id8F7 to protein kinase C by phosphati-dylserine. Purified protein kinase C α (1 μg/ml) was coated onto the microtiter plates and was preincubated with various concentrations of phospholipid vesicles for 1 h at room temperature. Then the anti-idiotypic mAb (2 μg/ml) was added to each well and incubated for 1.5 h at room temperature. The anti-idiotypic mAb bound was detected with biotinylated anti-mouse immunoglobulin and peroxi-dase-conjugated streptavidin.

consensus structure between the PS-specific mAb and protein kinase C, which may be responsible for the specific interaction with PS.

In order to define the molecular structure recognized by Id8F7, we first determined the amino acid sequences of the heavy- and light-chain variable regions of PS4A7 and tried to identify the epitope on PS4A7. We found that the anti-idiotypic antibody bound strongly to a 12 amino acid–residue synthetic peptide derived from the third complementarity determining region (CDR) of the heavy chain of PS4A7 (AREGDYD-GAMDY, amino acid residue 93–101, referred to as CDR3-H), but not to the synthetic peptides derived from other CDRs. In this assay, the synthetic peptides derived from the CDRs of the heavy- and light-chains of PS4A7 were coupled to bovine serum albumin (BSA) via a functional sulfhydryl group at the carboxyl terminus and the binding of 8F7 to the plate-coated peptide-BSA complexes were examined by ELISA. Furthermore, the CDR3-H peptide was shown to bind to PS with similar specificity to that of PS4A7, although its affinity was markedly lower (FIG. 2). No significant binding was observed with the synthetic peptides derived from other CDRs. The binding was effectively inhibited by PS, and slightly by phosphoserine, but not by serine.

Our recent analyses have shown that the anti-idiotypic antibody Id8F7 effectively inhibited the activities of both blood coagulation factor V and VIII, and bound strongly to the purified coagulation factor V but not to prothrombin nor coagulation factor X. These findings indicate that the CDR3-H peptide, which is recognized by Id8F7, may represent a novel PS-recognizing peptide motif that may be common among the cellular PS-binding proteins.

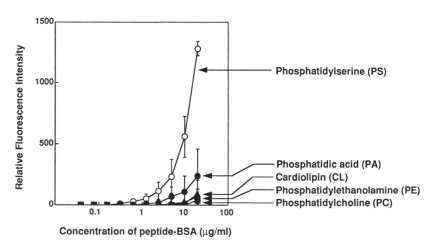

FIGURE 2. Specific binding of the CDR3-H peptide to phosphatidylserine. The CDR3-H peptide (amino acid residue 93–105 of PS4A7 H-chain, AREGDYDGAMDY) was coupled with biotinylated BSA via an additional cysteine at the carboxyl terminus. The microtiter wells were coated with various phospholipids (10 μM) and incubated with the peptide-BSA complex. The complex bound was detected with alkaline phosphatase–conjugated streptavidin. Each point represents the mean value of three different experiments.

REFERENCES

1. BEVERS, E. M., J. ROSING & R. F. A. ZWALL. 1987. *In* Platelets in Biology and Pathology III. MacIntyre and Gordon, Eds. 3: 27–159. Elsevier Science Publishers. New York.
2. NISHIZUKA, Y. 1988. Nature 334: 661–665.
3. LEE, M-H. & R. M. BELL. 1989. J. Biol. Chem. 264: 14797–14805.
4. UMEDA, M., K. IGARASHI, K. S. NAM & K. INOUE. 1989. J. Immunol. 143: 2273–2279.
5. UMEDA, M., I. DIEGO, E. D. BALL & D. M. MARCUS. 1986. J. Immunol. 136: 2562–2567.

Cloning and Mapping of a Ryanodine Receptor Homolog Gene of *Caenorhabditis elegans*[a]

YASUJI SAKUBE, HIDEKI ANDO,
AND HIROAKI KAGAWA[a]

Department of Biology
Faculty of Science
Okayama University
Okayama, 700 Japan

INTRODUCTION

Excitation-contraction coupling is one of the interesting mechanisms to be elucidated from the point of view of calcium signal transduction in muscle contraction.[1] Calcium channel genes were cloned and sequenced as the receptors of dihydropyridine[2] and ryanodine receptor in mammals.[3,4] Takeshima *et al.*[3] confirmed that the ryanodine receptor has a molecular mass of 500,000 daltons with a large portion at the N-terminal, which was observed as a foot structure by electron microscopy, and the transmembrane domain at the C-terminal. The molecular architecture of the receptor was close to that of rabbit, mink, and human. The functional loss of the molecule causes malignant hyperthermia of swine.[5] The ryanodine binding site and the channel-forming transmembrane structure were clarified by several other observations.

The soil nematode *Caenorhabditis elegans* is a model animal for behavior genetics,[6] especially for their convenience of the complete cell lineage, complete neurocellular network, and a good correlation of the genome map to the physical map of the chromosome.[7] We reported that a single-charge change of the muscle protein paramyosin, encoded by *unc-15* of *C. elegans*, disrupts charge interaction between paramyosin molecules and thick filament assembly, leading to muscle paralysis.[8] This result encouraged us to examine how neuromuscular systems might be affected by a single mutation in the genome of the animal. It is worthwhile to solve the question of how the excitation-contraction coupling is controlled by different molecules and reactions. By using the genetically handy worm, we could understand gene expression, gene product, mutant, and suppressor mutant of the molecules that functioned in the excitation-contraction coupling.

In this report, we cloned and sequenced the ryanodine receptor homolog gene of the worm. cDNA clone was obtained from cDNA library by using cDNA fragment

[a] This work was supported by the Grant-in-Aid for Priority Area 04270212 to H.K.
[b] To whom correspondence should be addressed.

of rabbit skeletal ryanodine receptor as a probe. Deduced amino acid sequence shows that the ryanodine receptor of the nematode had a conserved amino acid sequence similar to mammals not only in the transmembrane domain but in the foot structure of the molecule.

MATERIALS AND METHODS

DNA Recombinant Techniques

DNA recombinant technique was the standard method.[9] cDNA clone of rabbit skeletal ryanodine receptor was kindly provided by Takeshima *et al.*[3] Southern hybridization of genomic and cloned fragments and screening cDNA library were performed by the method of ECL (enhanced chemiluminescence, Amersham). Shotgun sequence was followed by MRC protocol[10] using a DNA sequencer of Applied Biosystems. Sequenced data were processed by DNASIS (Hitachi). RNA analysis was done by the conventional method. Genome mapping was kindly performed by Coulson.[11]

The Worm Handling

C. elegans strain N2 was grown by the established procedures and used for preparation of material.[6,7]

RESULTS AND DISCUSSION

Southern Blot Analysis with Rabbit cDNA Clones

To clone the gene, genomic Southern blot analysis was performed by using three cDNA clones of rabbit RYR, pRR229, pRR203, and pRR616.[3] Three unique fragments, EcoRI; 8.4, 6.1, and 5.4 kb, were clearly stained with pRR616, which corresponds to transmembrane domain of the rabbit ryanodine receptor. We used the cDNA clone of pRR616 for cloning genomic fragments. More than three genomic clones were obtained from genomic libraries, which were constructed with sized fragments and plasmid vectors (data not shown). Sequence results of deduced amino acids show that there was no sequence homology with the ryanodine receptor. One clone was exactly the gene of collagen. We stopped to clone genomic fragments. The correct genomic fragment of *ryr-1* gene (EcoRI-2.1 kb) was stained with rabbit cDNA probe but was weak. Cross-reactivity with the rabbit clone might be caused by some repeated sequence or the sequence encoding coiled-coil structure in transmembrane region.

cDNA Cloning and Sequencing

cDNA library of the nematode was screened with a pRR616 clone of rabbit cDNA of RYR as a probe. Four clones were obtained and had the same origin as the transcript.

The longest clone, pCERR-1, having a 1.9 kb insert, was sequenced and encoded the one open reading frame, which had 43% amino acid homology to the C-terminal part of rabbit ryanodine receptor, although the insert had reversed orientation in the vector. Correct molecular mass of protein was produced in *E. coli* after being processed by exon fusion of a cDNA fragment.[8] Fusion protein with β-galactosidase could be used for antibody preparation followed by cytohistochemistry.

Conserved Transmembrane Domain of the Ryanodine Receptor

The amino acid residues of putative transmembrane segments of CERYR had 45.5%, 41.6%, 39.4%, 65%, 75%, and 75% identical to the rabbit RYR protein in M5-M9 and M10 domains, respectively. It was noted that the *C. elegans* ryanodine receptor (CERYR) was similar to the cardiac type of RYR in mammals (TABLE 1). Although the C-terminal end of CERYR molecule had one additional amino acid residue, high homology of this region might be derived from the functional importance for channel formation. Kim *et al.*[12] have demonstrated with a biochemical approach that the CERYR protein has a functional homology to those in mammals. Our result was consistent with this observation.

Genome Mapping and the Number of the Ryanodine Receptor Gene

The gene *ryr-1* position of CERYR was determined by screening a yeast artificial chromosome (YAC) filter covering almost all of the genome[11] with a cDNA fragment as probe (Alan Coulson). Two clones of Y44A7 and Y37G2 on the central of the chromosome V were found at the position of *ryr-1* gene. The position was close to the position of the cDNA clone cm16c2, which had signals in the YAC clones, Y44A7, Y37G2, and Y57E12 on the chromosome V (Ed Maryon *et al.* and Alan Coulson, personal communications). We concluded that the both cDNA clones came from the single gene, *ryr-1*.

TABLE 1. Comparison of Putative TM Segments of CERYR with Rabbit RYR Proteins

TM Segment	CE/Cardiac(%)	CE/Skeletal(%)
M5	45.5	54.5
M6	41.6	33.3
M7	39.4	36.9
M8	65.0	60.0
M9	75.0	70.0
M10	75.0	75.0
Foot region[a]	42.3	39.8
Total[b]	43.0	40.8

TM segments M5 ~ M10 were deduced from the cDNA clone c511/pCERR-1.
[a,b] Data were obtained by comparing sequenced 3,000 amino acid residues of C-terminal and that of rabbit RYR. About 30% of foot region were revealed from the genomic sequences.

The difference of cross-reactivity of both cDNA clones came from the position of genome location. That means Y57E12 locates upstream of *ryr-1* gene. The genome sequence result confirmed that cm16c2 encoded the region of 3,250–3,700 and pCERR-1 encoded the C-terminal end (FIG. 1).

Southern analysis shows that only one band, EcoRI-2.1 kb (HindIII-5.1 kb) cross-reacted with a cDNA clone, pCERR-1. A fragment (>10 kb) was cross-reacted with cDNA clone pCERR-1 on Northern analysis. These results taken together indicate that *C. elegans* had only one *ryr-1* gene. Unfortunately we could not find any mutant worm at that map region. Recent progress of microinjection techniques in *C. elegans*,[13,14] can express a gene in the animal although inserted DNA was extrachromosomal. Processed DNA could be used for monitoring how the *ryr-1* gene affected development of the animal.

Genome Organization of the ryr-1 Gene

We finally located the *ryr-1* gene on one cosmid clone, M04C11, and are sequencing the genome fragments from this cosmid. Current data are shown in FIGURE. 1. Coding efficiency (exon/total length) was about 63% (9.5 kb/15 kb). Twenty-nine exons, corresponding to amino acid residues of 1,878–4,969, were separated into many small sized introns, 46–100 bp, and a few longer introns were observed in other genes of the nematode. The 3'-noncoding region of the gene was only 330 bp. This means the transcript might be translated without complex regulations. The open reading frame of *odc-1* gene (ornithine decarboxylase, by P. Coffino) was found at 1,730 bp down stream of the *ryr-1* gene. It was noted that CERYR had about 42% identical to the mammal RYR even in the foot structure. This result indicates that RYR could be common in animals from invertebrate to mammals. We are still working on cloning another calcium channel, the dihydropyridine receptor gene, with a similar procedure.

ACKNOWLEDGMENTS

We thank Dr. H. Takeshima and Prof. S. Numa for generously providing us a cDNA clones. This work was dedicated to Prof. S. Numa who died in the spring of 1992. cDNA library was kindly provided by Dr. Bob Barstead and Prof. Bob Waterston. Dr. Ed Maryon is acknowledged for communication about their work on ryanodine receptor before publication. Drs. Alan Coulson and John Sulston are also acknowledged for their mapping of the gene.

Note added in proof: One small exon was found in the 3' end of large intron beside the M' domain in FIG. 1 by a recent experiment with RT-PCR (reverse transcriptase-polymerase chain reaction) method.

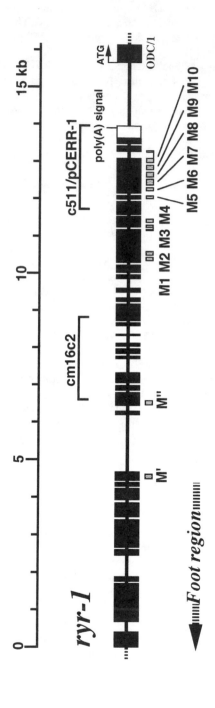

FIGURE 1. The genomic structure of *ryr-1* gene of *C. elegans*. The ryanodine receptor gene, *ryr-1* of *C. elegans* was located on the central left of the chromosome V. The cosmid clone M04C11 was processed and was partially sequenced. cDNA clones cm16c2 and c511 (pCERR-1) were cloned by cDNA project of the nematode and in this report, respectively. Exon and intron were determined with comparison to the sequenced cDNA of mammals and the genome sequence of *ryr-1*. Putative transmembrane domains were shown by the numbering as was described. More than 5 kb of cDNA sequence of the 5' end was missing. It was noted that the *ryr-1* gene was close to the cardiac type of the RYR in mammals.

REFERENCES

1. EBASHI, S. & M. ENDO. 1986. Progr. Biophys. Mol. Biol. **18**: 123–183.
2. TANABE, T., H. TAKESHIMA, A. MIKAMI, V. FLOCKERZI, H. TAKAHASHI, K. KANGAWA, M. KOJIMA, H. MATSUO, T. HIROSE & S. NUMA. 1987. Nature **336**: 134–139.
3. TAKESHIMA, H., S. NISHIMURA, T. MATSUMOTO, H. ISHIDA, K. KANGAWA, N. MINAMINO, H. MATSUO, M. UEDA, M. HANAOKA, T. HIROSE & S. NUMA. 1989. Nature **339**: 439–445.
4. ZORZATO, F., J. FUJII, K. OTSU, M. PHILLIPS, N. M. GREEN, F. N. LAI, G. MEISSENER & D. H. MACLENNAN. 1990. J. Biol. Chem. **265**:2244–2256.
5. MACLENNAN, D. H. & M. S. PHILLIPS. 1992. Science **256**: 789–794.
6. BRENNER, S. 1974. Genetics **77**: 71–94.
7. WOOD, W. B. 1988. The nematode *Caenorhabditis elegans*. Cold Spring Harbor Laboratory. Cold Spring Harbor, N.Y.
8. GENGYO-ANDO, K. & H. KAGAWA. 1991. J. Mol. Biol. **219**: 429–441.
9. MANIATIS, T., E. F. FRITSCH & J. SAMBROOK. 1982. Molecular Cloning. Cold Spring Laboratory Press. Cold Spring Harbor, N.Y.
10. BANKIER, A. T. & B. G. BARRELL. 1983. Techniques in the Life Sciences. Elsevier. Ireland.
11. COULSON, A., R. H. WATERSTON, J. E. KIFF, J. E. SULSTON & Y. KOHARA. 1988. Nature **335**: 184–186.
12. KIM, Y.-K., H. H. VALDIVIA, E. B. MARYON, P. ANDERSON & R. CORONADO. 1992. J. Physiol. **63**: 1379–1384.
13. FIRE, A., D. ALBERTSON, S. W. HARRISON & D. G. MOERMAN. 1991. Development **113**:503–514.
14. MELLO, C. C., J. M. KRAMER, D. STINCHCOMB & V. AMBROS. 1991. EMBO J. **10**: 3959–3970.

Effects of 9-methyl-7-bromoeudistomin D (MBED), a Powerful Ca^{2+} Releaser, on Smooth Muscles of the Guinea Pig

Y. IMAIZUMI, S. HENMI, Y. UYAMA,
M. WATANABE, AND Y. OHIZUMI[a]

Department of Chemical Pharmacology
Faculty of Pharmaceutical Sciences
Nagoya City University
Nagoya 467, Japan

[a]*Department of Natural Products Pharmacology*
Pharmaceutical Institute
Tohoku University
Sendai 980, Japan

MBED is a derivative of eudistomin D, a natural marine product that releases Ca^{2+} from sarcoplasmic reticulum (SR) in skinned skeletal muscle fiber in a manner similar to caffeine but 100 times more effectively (ED_{50} = 10 µM).[1,2] Effects of MBED on smooth muscle were first examined in the present study, using skinned strips of the ileal longitudinal layer and mesenteric artery, and single smooth muscle cells enzymatically isolated from ileal longitudinal muscle layers and whole urinary bladder of the guinea pig.

When 1–300 µM MBED was applied to skinned smooth muscle strips after storage sites were filled with Ca^{2+}, no contractile response was observed (FIG. 1). Subsequent addition of 30 mM caffeine elicited a large Ca^{2+} release as assessed by the contractile response, which was not affected significantly by the pretreatment with 1–100 µM MBED. A contraction induced by a solution of pCa 6.3 was not affected siginificantly by addition of 100 µM MBED. Application of 30 mM caffeine to intact strips elicited a contraction, whereas that of 1–300 µM MBED did not. It has been reported that cyclic AMP phosphodiesterase activity is not affected by 100 µM MBED.[2]

When single smooth muscle cells were depolarized from a holding potential of −60 to 0 mV under whole-cell clamp, an initial inward Ca^{2+} current and a subsequent outward current were recorded.[3] An early large transient component of the outward current is mainly Ca^{2+}-dependent K^+ current (I_{K-Ca}) through K^+ channels that have

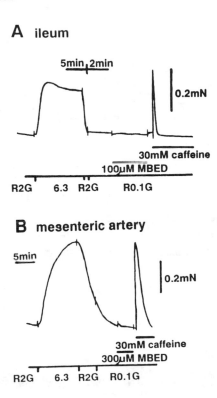

A ileum

30mM caffeine
100μM MBED

R2G   6.3 R2G   R0.1G

B mesenteric artery

30mM caffeine
300μM MBED

R2G   6.3 R2G R0.1G

FIGURE 1. Effects of 30 mM caffeine and 100 or 300 μM MBED on muscle tension of an ileal longitudinal strip (**A**) and a mesenteric arterial strip (**B**) skinned by β-escin. Bathing solutions were shown below the line at the bottom. R2G and R.0.1G indicate relaxing solutions containing 2 and 0.1 mM EGTA, respectively. The solution "6.3" was prepared adjusting pCa to 6.3 with 5 mM EGTA and corresponding Ca^{2+}, and was used for Ca^{2+}-loading of intracellular storage sites. Muscle tension was developed in the "6.3" solution. Application of 100 or 300 μM MBED did not change muscle tension in R0.1G. Subsequent addition of 30 mM caffeine in R0.1G induced a large transient contraction, indicating Ca^{2+} release from storage sites.

a large conductance (BK channels). The activation of I_{K-Ca} is mediated by Ca^{2+} release from storage sites via Ca^{2+}-induced Ca^{2+}-release, which is triggered by Ca^{2+}-influx through voltage-dependent Ca^{2+} channels, and therefore is markedly reduced by ryanodine or cyclopiazonic acid.[3] Application of 30 μM MBED almost abolished I_{K-Ca} (FIG. 2) in a manner similar to 10 mM caffeine. Transient enhancement of I_{K-Ca} just after the application of 30 μM MBED was occasionally observed but not in FIGURE 2. The reduction of I_{K-Ca} occurred in an all-or-none manner after the application of MBED in a concentration range of 10 and 30 μM. The effect was completely reversible. Spontaneous transient outward currents (STOCs),[4] which may be elicited by spontaneous Ca^{2+} release from local SR, were transiently enhanced and thereafter suppressed by application of 30 μM MBED in all cells examined (FIG. 2, C; $n = 3$). Activities

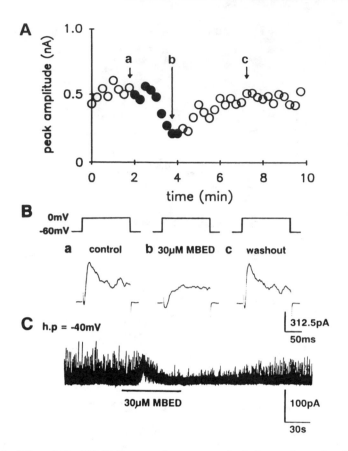

FIGURE 2. Effects of 30 µM MBED on membrane currents in single smooth muscle cells isolated from urinary bladder of the guinea pig under whole-cell voltage clamp. (A) Cells were depolarized from – 60 to 0 mV for 150 msec once every 15 sec. Open and closed symbols indicate amplitude of peak outward currents in the absence and presence of 30 µM MBED, respectively, and are plotted against time. Traces shown in "Ba, b and c" were recorded at the time indicated correspondingly in A. Note that the initial transient component of the outward current was markedly reduced by application of MBED and completely recovered after washout of MBED. (C) Spontaneous transient outward currents (STOCs) were recorded at holding potential of – 40 mV. STOCs were transiently enhanced by application of 30 µM MBED and suppressed thereafter.

of BK channels recorded under outside-out patch clamp were not affected by 30 µM MBED. Voltage-dependent Ca^{2+} and K^+ currents were not affected by 30 µM MBED.

These results suggest that MBED effectively releases Ca^{2+} from intracellular storage sites available for the activation of BK channels but not from those for contractile system in smooth muscles cells.

REFERENCES

1. KOBAYASHI, J., M. ISHIBASHI, U. NAGAI & Y. OHIZUMI. 1989. Experientia **45**: 782–783.
2. SEINO, A., M. KOBAYASHI, J. KOBAYASHI, Y. FANG, M. ISHIBASHI, H. NAKAMURA, K. MOMOSE & Y. OHIZUMI. 1991. J. Pharmacol. Exp. Ther. **256**: 861–867.
3. SUZUKI, M., K. MURAKI, Y. IMAIZUMI & M. WATANABE. 1992. Br. J. Pharmacol. **107**: 134–140.
4. BENHAM, C. D. & T. BOLTON. 1986. J. Physiol. **381**: 385–406.

Abnormal Development of Cone Cells in Transgenic Mice Ablated of Rod Photoreceptor Cells

JIRO USUKURA, WILSON KHOO,[a]
THORU ABE,[b] TOSHIMICHI SHINOHARA,[b]
AND MARTIN BREITMAN[a]

Department of Anatomy
Nagoya University
School of Medicine
65 Tsurumai, Showa-ku
Nagoya, 466 Japan

[a]*Mount Sinai Hospital*
Toronto, Canada

[b]*NEI*
National Institutes of Health
Bethesda, Maryland 20895

The retina is an integral component of the central nervous system. It arises during development from the optic stalk, an embryonic outgrowth of the brain, and subsequently invaginates to form the optic cup. The neural retina derived from the inner layer of the cup is comprised of photoreceptor cells, other interconnecting neurons, and functionally integrated glia. The photoreceptor cells, which convert light energy to electrical signals, are of two morphological types, rods and cones; these differ in relative number, anatomical location, timing of appearance in the retina, and in the complement of specialized phototransducing protein they produce.[1] However, both rods and cones show a polarized pattern of cyto-differentiation and at their apical surface develop characteristic inner and outer segments. It is suggested that retinal cell precursors become committed to the development of specific phenotypes rather early in embryogenesis.[2-6] Commitment as well as subsequent cellular differentiation and maturation are thought to be programmed by a hierarchy of molecular events. However, the precise nature of these events and the particular cellular interactions involved remain virtually unknown. We have taken a transgenic approach to investigate cellular interdependencies during development by examining how retinal cell differentiation and maturation proceed when developing rod cells are ablated. For this purpose, transgenic mice were generated carrying the cytotoxic Diphtheria toxin A fragment (*dt-a*) gene driven by human opsin promoter sequences (− 1046 to + 36) according to the methods in previous works.[7-9]

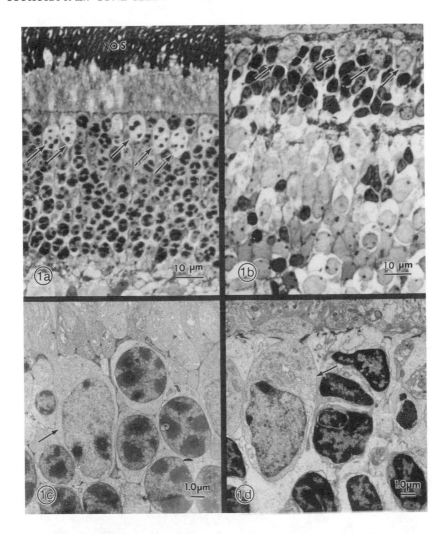

FIGURE 1. (a and b): Light micrographs of normal and transgenic mice retina cut through the central region. Arrows indicate cone cells. Cone cells are characterized by lucent large nuclei found close to outer limiting membrane. (a) postnatal day 15 control mouse retina. OS: outer segment (b) postnatal day 11 transgenic mouse retina. Number of cone cells per micron in control and transgenic retinas are 1.0/10 μm and 0.87/10 μm respectively. However, these numbers are variable depending on the sections. (c and d): Electron micrographs taken from the same specimen block used in above figures. (c) postnatal day 15 control mouse retina, (d) postnatal day 11 transgenic mouse retina. Arrow indicates cone cell. In general, rod cells have round electron-dense nuclei smaller than cones. In transgenic mouse retina, rod shows much denser nucleus with irregular shape because of damage by expression of toxin gene. (Original figure reduced to 70%.)

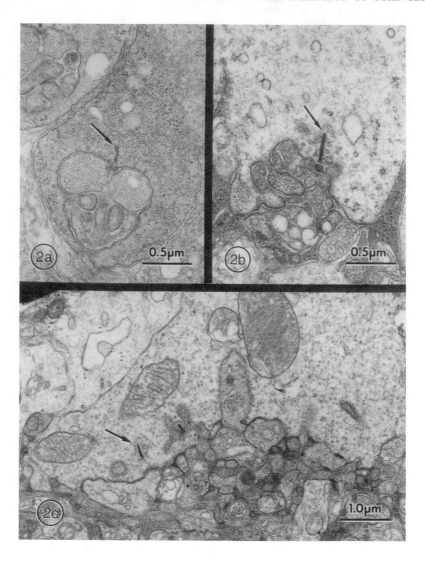

FIGURE 2. Electron micrographs of synaptic terminals of photoreceptor cells. Arrows indicate synaptic ribbons, a characteristic structure found in the photoreceptor synaptic terminal. Photographs are not printed at the same magnification, please note the scale bar. (a) Typical rod synaptic terminal in 3 months postnatal normal mouse retina. (b) Synaptic terminal of the transgenic photoreceptor cell in postnatal day 15. Number of synaptic vesicles are reduced remarkably with the dilation of the terminal. (c) Synaptic terminal of the transgenic photoreceptor cell in postnatal day 11. The large synaptic terminals with healthy cytoplasmic structure are found frequently in the central retina. This is presumed to be cone synaptic pedicle, judging from the size and location. Even in postnatal day 11, this synaptic terminal retains many synaptic vesicles and large synaptic ribbons, and maintains good synaptic organization with secondary neurons. (Original figure reduced to 90%.)

Mice expressing this construct showed progressive degeneration of rod photoreceptor cells commencing at birth, with obvious depletion of such cells by postnatal day 7. Although all rod photoreceptors were lost between 15 days and 3 months, a small number of photoreceptors remained, particularly in central region of retina, that on the basis of morphology and staining characteristics could be readily identified as cone cells: these were similar in number to cone cells of the normal retina, but lacked outer segments. Namely, ablation of rod photoreceptor cells in the transgenic retina was accompanied by failure of developing cone cells to elaborate outer segments, although all other aspects of cone cell cyto-differentiation appeared normal (FIG. 1). Ultrastructural analysis of rod cell maturation in opsin-*dt-a* mice revealed that *dt-a* expression had a more pronounced effect on outer segment formation than photoreceptor synaptogenesis. Whereas formation of outer segment, which normally begins between Day 5 and Day 7, was completely inhibited by action of the transgene, photoreceptor synaptogenesis appeared to proceed normally until Day 9, after which time synaptic termini degenerated (FIG. 2). These findings presumably relate to both the relative tolerance of outer segment formation and synaptogenesis to reduced protein synthetic capacity as well as the relative time frame required for threshold level of toxin to accumulate.

Our experimental results suggest that the 1.0 kb opsin promoter segment contains photoreceptor cell type specificity and that cone cells require maturation of rod cells to complete the late stages of their terminal differentiation.

REFERENCES

1. CHABRE, M. & P. DETERRE. 1989. Molecular mechanism of visual transduction. Eur. J. Biochem. 179: 255–266.
2. ADLER, R. 1986. The differentiation of retinal photoreceptors and neurons in vitro. *In* Progress in Retinal Research. N. Osborne & J.Chader, Eds.: 1–26. Pergamon Press. Oxford.
3. ADLER, R. 1992. Cellular and molecular mechanisms regulating retinal cell differentiation. *In* The Visual System. R. Lent, Ed.: 3–17. Boston Inc. Boston.
4. GRUN, G. 1982. The development of the vertebrate retina: A comparative study. Adv. Anat. Embryol. Cell Biol. 78: 7–85.
5. TURNER, D. L. & C. L. CEPKO. 1987. A common progenitor for neurons and glia persist in rat retina late in development. Nature 328: 131–136.
6. TURNER, D. L., E. Y. SNYDER & C. L. CEPKO. 1990. Lineage dependent determination of cell types in the embryonic mouse retina. Neuron 4: 833–845.
7. LEM, J., M. L. APPLEBURY, J. D. FALF, J. G. FLANNERY & M. I. SIMON. 1991. Tissue-specific and development regulation of rod opsin chimeric genes in transgenic mice. Neuron 6: 201–210.
8. NATHANS, J. & D. S. HOGNESS. 1984. Isolation and nucleotide sequence of the gene encoding human rhodopsin. Proc. Natl. Acad. Sci. USA 81: 4851–4855.
9. ZACK, D. J., J. BENNETT, Y. WANG, C. DAVENPORT, B. KLAUNBERG, J. GEARHART & J. NATHANS. 1991. Unusual topography of bovine rhodopsin promoter lacZ fusion gene expression in transgenic mouse retina. Neuron 6: 187–199.

Immunohistochemical Localization of Metabotropic and Ionotropic Glutamate Receptors in the Mouse Brain

Y. RYO, A. MIYAWAKI, T. FURUICHI,
AND K. MIKOSHIBA

Department of Molecular Neurobiology
The Institute of Medical Science
The University of Tokyo
4-6-1 Shirokanedai, Minato-ku
Tokyo 108, Japan

INTRODUCTION

L-Glutamate acts as a major excitatory neurotransmitter in the mammalian central nervous system. Glutamatergic excitatory neurotransmission is thought to play crucial roles in brain function (the regulation of neurotransmitter release, memory and learning, development, and synaptogenesis) as well as in dysfunctions (epilepsy, stroke, and neurodegenerative disorders).[1] These functionally diverse responses to L-glutamate are mediated by two major classes of glutamate receptor family: ionotropic receptors (NMDA types and non-NMDA types (GluR)) and metabotropic receptors (mGluR). We prepared the polyclonal antibodies against mGluR1α and GluR1 and examined immunohistochemically localization of both receptors in the mouse brain during postnatal development and in cerebellar ataxic mutants.

MATERIALS AND METHODS

Anti-glutamate receptor antibodies were prepared by using the synthetic peptides corresponding to the carboxyl termini 15 amino acid residues of rat mGluR1α[2] and the carboxyl termini 14 residues of mouse GluR1,[3] and were purified to IgG. Western blot analysis was performed as described[4] using 5% discontinuous polyacrylamide gel. For immunohistochemistry, brains from ICR mice and mutant mice, fixed by perfusion with Bouin's solution, were used. Paraffin sections (10 μm) were sequentially incubated with either of anti-glutamate receptors antibodies for 3 h, with goat biotinylated anti-rabbit IgG (Vector) for 2 h, with avidin D-conjugated HRP (Vector) for 1 h, and with DAB solution. The anti-inositol 1,4,5-trisphosphate receptor (IP_3R) monoclonal antibodies[5,6] were also used for immunostaining.

RESULTS AND DISCUSSION

Western Blot Analysis

Western blot analysis showed both receptor proteins are glycosylated predominantly in an asparagine-linked manner and are enriched in cerebellar postsynaptic membrane.

Immunohistochemistry

Localization of mGluR1α and GluR1 in the adult mouse brain (FIG. 1). In adult brain, most prominent expression of mGluR1α was observed in glomerulus of olfactory bulb, thalamus, and molecular layer of cerebellum, and intermediate expression was observed in hippocampus. Whereas GluR1 expression was prominent in glomerulus of olfactory bulb, lateral septal nucleus, induseus griseum, hippocampus, dentate gyrus, and molecular layer of cerebellum. In cerebellum, mGluR1α was intensively expressed in Purkinje neurons, while GluR1 was in Bergmann glial cells and probably in Purkinje cell dendrites.

Within the first three postnatal weeks, mGluR1α and GluR1 expression were drastically changed in time and space namely in hippocampus and cerebellum. These spatiotemporal expression patterns appear to be correlated with the postnatal ontogenesis and the establishment of glutamatergic neurotransmission system in hippocampus and cerebellum: cell migration, dendritic and axonal growth, spine formation, and synaptogenesis.

mGluR1α and GluR1 are fairly expressed in *weaver* mutant cerebellum even suffering from degeneration of granule cells. However, the intrinsic expression levels of

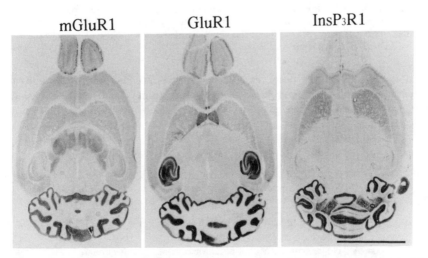

FIGURE 1. Distribution of mGluR1α and GluR1 throughout adult mouse brain (6 month old). Horizontal brain sections immunoreacted with M1 (*left*), A1 (*middle*), and anti-IP₃R1 (*right*) antibodies. Bar = 5 mm. (Original figure reduced to 65%.)

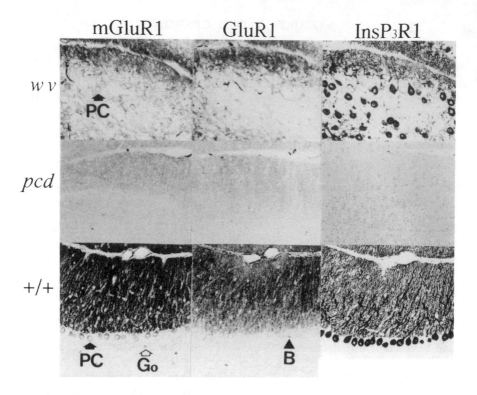

FIGURE 2. mGluR1α and GluR1 expression in cerebellar molecular layers of cerebellar ataxic mutant mice. Sagittal sections of *weaver* (*wv*, 3 month old) (*upper*), *Purkinje-cell-degeneration* (*pcd*, 5 month old) (*middle*), and wild-type mouse (*lower*) immunostained with M1 (*right*), A1 (*middle*), and anti-Ip₃R1 (*left*) antibodies. PC, Purkinje cell body; B, Bergmann glial cell body; Go, Golgi cell body. Bar = 0.1 mm. (Original figure reduced to 80%.)

GluR1 as well as mGluR1α are remarkably reduced in cerebellums of Purkinje cell–deficient mutant mice, *Purkinje-cell degeneration*, suggesting that GluR1 expression in Bergmann glial cells may depend upon the sustained interaction with adjacent Purkinje neurons.

REFERENCES

1. McDonald, J. W. & M. V. Johnston. 1990. Brain Res. Rev. **15:** 41–70.
2. Masu, M., Y. Tanabe, K. Tsuchida, R. Shigemoto & S. Nakanishi. 1991. Nature **349:** 760–765.
3. Sakimura, K., H. Bujo, E. Kushiya, K. Araki, M. Yamazaki, M. Yamazaki, H. Meguro, A. Warashina, S. Numa & M. Mishina. 1990. FEBS Lett. **272**(1,2): 73–80.
4. Miyawaki, A., T. Furuichi, N. Maeda & K. Mikoshiba. 1990. Neuron **5:** 11–18
5. Maeda, N., M. Niinobe, K. Nakahira & K. Mikoshiba. 1988. J. Neurochem. **51**(6): 1724–1730.
6. Maeda, N., M. Niinobe, Y. Inoue & K. Mikoshiba. 1989. Dev. Biol. **133:** 67–76.

GTP-Binding Proteins Coupling to Glutamate Receptors on Bovine Retinal Membranes

S. KIKKAWA, M. NAKAGAWA, T. IWASA,
AND M. TSUDA

Department of Life Science
Himeji Institute of Technology
Harima Science Garden City
Ako-gun, Hyogo 678-12, Japan

INTRODUCTION

In the visual system of vertebrates, light reduces the release of glutamate from photoreceptor cells, resulting in depolarization of ON-bipolar cells by opening of cation channels on the plasma membrane of the cells.[1] It has been reported that cGMP is the second messenger mediating the action of glutamate on ON-bipolar cells and that the receptor sends signals to an effector via a GTP-binding protein (G-protein).[2-4] However, the property of the G-protein that couples to the glutamate receptor still remains unknown.

We investigated the effects of glutamate ligands on ADP-ribosylation of G-proteins catalyzed by bacterial toxins with membranes from bovine retina. Since transducin, a G-protein sensitive to both pertussis toxin and cholera toxin, is located in rod outer segments (ROS), the membranes were prepared through sucrose-flotation method to deplete ROS. The membranes were incubated with pertussis toxin or cholera toxin in the presence of [^{32}P]NAD. Then the reaction mixtures were subjected to SDS-PAGE, followed by autoradiographic detection of labeled proteins. Pertussis toxin selectively ADP-ribosylated 39–40 kD protein(s) on ROS-depleted membranes (FIG. 1). Cholera toxin catalyzed ADP-ribosylation of 45 and 52 kD polypeptides, probably α-subunits of G_s, but not the 39–40 kD subunit. Cholera toxin clearly ADP-ribosylated 39 kD transducin α-subunit in ROS, thus contamination of transducin in the membrane preparation seems considerably low, and it is plausible that the α-subunit(s) that is/ are ADP-ribosylated by pertussis toxin is/are different subtype(s), possibly $α_o$ and/ or $α_i$. 2-Amino-4-phosphonobutyric acid (APB), a glutamate analogue selective for the receptor on the ON-bipolar cell,[5] reduced in part the ADP-ribosylation by pertussis toxin at the concentration of 1 mM in the presence of Mg^{2+} and guanosine 5'-(βγ-imido)triphosphate (Gpp(NH)p). The results imply that the APB receptor couples to and activates the pertussis toxin substrate.

557

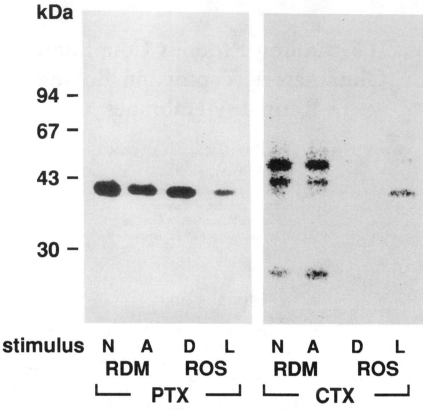

FIGURE 1. ADP-ribosylation of ROS-depleted bovine retinal membranes and ROS by pertussis or cholera toxin. Membrane preparations (RDM: ROS-depleted membranes, ROS: rod outer segment disk membranes; 1 mg protein/ml) were incubated at 30°C for 30 min with either pertussis toxin (PTX; 5 μg/ml) or cholera toxin (CTX; 200 μg/ml), which was activated prior to incubation,[6] in a buffer containing 2 μM [^{32}P]NAD and 1 mM NADP under conditions shown below. The reaction was terminated by addition of equal volume of sample buffer and subjected to SDS-PAGE on 10% gel followed by autoradiography. Conditions; N: no additives, A: 1 mM APB, D: dark, L: light.

We next examined whether GTPγS-binding to the membranes, an activation step of G-proteins, is affected by glutamate ligands. The membranes incubated in the presence or absence of glutamate ligands in a buffer containing 1 μM [^{35}S]GTPγS. The radioactivity bound to the membranes was measured after bound/free separation by filtration. GTPγS bound to the membranes in a time-dependent manner and binding reached apparent saturation. APB at 1 mM concentration accelerated the time course of GTPγS-binding. APB (1 mM) enhanced GTP-γS-binding to the membranes about 30% at 15 min (Fig. 2). Furthermore, glutamate analogues, which are effective on cellular responses of the ON-bipolar cell, ibotenic acid and 2-phosphono-5-valeric acid, increased GTPγS-binding at 1 mM (Fig. 2). While native L-glutamic acid did not have any effect on GTPγS-binding at 1 mM, it enhanced GTPγS-binding to the

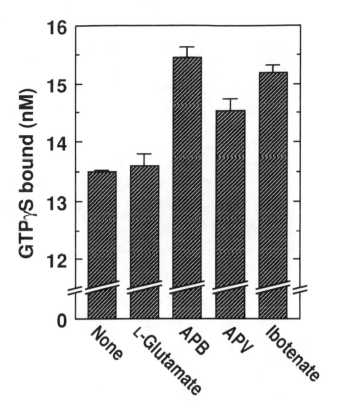

FIGURE 2. Effects of various glutamate ligands on GTPγS-binding to ROS-depleted membranes. Membrane preparation (0.5 mg total protein/ml) was incubated at 30°C in the presence or absence of various glutamate ligands (1 mM) in a buffer containing 1 μM [³⁵S]GTPγS, 10 μM GDP, and 100 μM adenosine 5'-(βγ-imido)triphosphate (App(NH)p). After 12 min, the mixtures were diluted with 10 volumes of an ice-cold Tris buffer (20 mM; pH 8.0) containing 25 mM MgCl₂ and 100 mM NaCl. The diluted mixtures were immediately filtered through glass filters (GF/B, Whatman), and the filters, after being washed three times with the same buffer, were counted for ³⁵S.

membranes at a higher concentration (10 mM) (data not shown). It has been reported that APB is more potent ligand than native L-glutamate for the ON-bipolar cell.[5] Our results suggest that the glutamate receptor, which is linked to G-protein, is present on bovine retinal membranes, probably on the ON-bipolar cell, and that the G-protein coupled with the receptor is pertussis toxin–sensitive but not cholera toxin–sensitive.

ACKNOWLEDGMENT

We thank Professor A. Kaneko of Keio University School of Medicine for helpful discussion.

REFERENCES

1. MURAKAMI, M., T. OHTSUKA & H. SHIMAZAKI. 1975. Vision Res. 15: 456–458.
2. Nawy, S. & C. E. JAHR. 1990. Nature 346: 269–271.
3. SHIELLS, R. A. & G. FALK. 1990. Proc. R. Soc. Lond. B 242: 91–94.
4. SHIELLS, R. A. & G. FALK. 1992. Proc. R. Soc. Lond. B 247: 17–20.
5. SLAUGHTER, M. M. & R. F. MILLER. 1981. Science 211: 182–185.
6. IIRI, T., Y. OHOKA, M. UI & T. KATADA. 1991. Eur. J. Biochem. 202: 635–641.

Index of Contributors